Principles *of*

ENGINEERING PHYSICS

VOLUME I

Principles *of*
ENGINEERING PHYSICS

VOLUME I

[For BTech students of West Bengal University of Technology and also for BSc (Honours/General) of Indian Universities, NET, SET, and GATE examiners]

Prof. (Dr) S P Kuila MSc, PhD (IIT, Kharagpur)
Professor and Head, Department of Applied Physical Sciences
Bengal Institute of Technology and Management, Santiniketan, WB
Former Professor of Physics, Abhedananda Mahavidyalaya
Sainthia, Birbhum, WB
Author of *Perspective of Quantum Mechanics* and
Vector Analysis, Its Applications, Principles of Engineering Physics (Vol II) etc.

New Central Book Agency (P) Ltd
LONDON
HYDERABAD ERNAKULAM BHUBANESWAR
NEW DELHI KOLKATA PUNE GUWAHATI

NCBA

REGD OFFICE

8/1 Chintamoni Das Lane, Kolkata 700 009, India
email: ncbapvtltd@eth.net

OVERSEAS

NCBA (UK) Ltd, 149 Park Avenue North, Northampton, NN3 2HY, UK
email: ncbauk@yahoo.com

EXPORT

NCBA Exports Pvt. Ltd, 212 Shahpur Jat, New Delhi 110 049
email: ncbaexports@ncbapvtltd.com

BRANCHES

212 Shahpur Jat
New Delhi 110 049
email: ncbadel@ncbapvtltd.com

Shop Nos. 3 & 4, Vinayak Towers
681/B Budhwar Peth, Appa Balwant Chowk
Pune 411 002
email: ncbapune@ncbapvtltd.com

Shop No. 15, Bharati Towers, Ground Floor
Forest Park, **Bhubaneswar** 751 001
email: ncbabub@ncbapvtltd.com

House No. 3-1-315, 1st Floor
Nimboliadda, Kachiguda, **Hyderabad** 500 027
email: ncbahydb@ncbapvtltd.com

GSS Shopping Complex, 1st Floor
Opposite College Post Office, Convent Road
Ernakulam 682 035
email: ncbaernk@ncbapvtltd.com

Radhamayee Complex (Second Floor)
Opp. SBI, Panbazar, Jaswanta Road
Guwahati 781 001
email: ncbaguwahati@ncbapvtltd.com

PRINCIPLES OF ENGINEERING PHYSICS [Volume I] • Prof. (Dr) S P Kuila

First Published: January 2009
Thoroughly Revised and Enlarged Second Edition (New Syllabus): January 2012
Reprinted: April 2013
With Additions: March 2014

PUBLISHER AND TYPESETTER
New Central Book Agency (P) Ltd
8/1 Chintamoni Das Lane, Kolkata 700 009

PRINTER
New Central Book Agency (P) Ltd
Web-Offset Division, Dhulagarh, Sankrail, Howrah

COVER DESIGNER
Soumen Paul

TECHNICAL EDITOR
Dipan Roy

PROJECT TEAM
Abhijit Sen, Koushik Samanta, and Prabhat Jas

ISBN 978 81 7381 527 0

Price: ₹ 350.00

Dedicated to Spiritual Teacher
Swami Satyanandadev

(b) Sketch the Fermi distribution for $T = 0$ and $T > 0$ and explain.

(c) What is the occupation probability at $E = E_F$?

(d) Calculate how the degeneracy function $g(E)$ depends on E for a Fermionic gas.

(e) Express the Fermi level in a metal in terms of free electron density.

12. (a) Give two examples of bosons.

(b) Define Einstein's A, B coefficients and deduce their mutual relation.

(c) Show that the ratio of spontaneous and stimulated emission is proportional to the cube of the frequency.

(d) Explain the construction and action of the optical resonator in ruby laser.

(e) If the energy levels of K, L, M states of Pt are 78 keV, 12 keV and 3 keV, respectively, then calculate the wavelengths of K_α and K_β lines emitted from platinum.
($h = 6.626 \times 10^{-34}$ J-s; $c = 3 \times 10^8$ m/s)

June 2009
Semester 2
Group A: Multiple Choice Type Questions

1. Choose the correct alternatives for any *ten* of the following:

(i) Newton's ring experiment is based on (a) division of amplitude (b) division of wave-front (c) none of these.

(ii) de Broglie wavelength of a particle of mass m and kinetic energy E is (a) $\lambda = \frac{h}{2mE}$ (b) $\frac{h}{\sqrt{2mE}}$ (c) $\frac{\sqrt{2mE}}{h}$.

(iii) Mass of a photon of frequencies ν is given by (a) $\frac{h\nu}{c}$ (b) $\frac{h\nu}{c^2}$ (c) $\frac{h\nu^2}{c}$.

(iv) The eigenvalue of the eigenfunction e^{ix} for the operator $\frac{d^2}{dx^2}$ is (a) 1 (b) 0 (c) -1.

(v) Miller indices of plane which cut intercepts of 2, 3 and 4 units along the three axes are (a) $(2, 3, 2)$ (b) $(2, 3, 4)$ (c) $(6, 4, 3)$.

(vi) In a plane transmission grating, light (a) diffracts to produce the resultant pattern (b) diffracts and interferes to produce the resultant pattern (c) interferes to produce the resultant pattern.

(vii) The atomic radius of a face centred cubic crystal of lattice constant a is (a) $\frac{a}{2}$ (b) $\frac{\sqrt{3}a}{4}$ (c) $\frac{\sqrt{2}a}{4}$.

(viii) An X-ray tube is subjected to a potential difference of 50 kV with the corresponding current of 8 mA through it. The number of electrons striking per second on the target material is (a) 5×10^{16} (v) 6×10^{11} (c) none of these.

(ix) In He-Ne laser, the laser light emits due to the transition from (a) $3s \to 2p$ (b) $3s \to 3p$ (c) $2s \to 2p$.

(x) For an optical fibre, if n_0, n_1 and n_2 are the refractive indices of air, core cladding region, respectively, then (a) $n_0 > n_2 > n_1$ (b) $n_1 > n_2 > n_0$ (c) $n_2 > n_1 > n_0$.

(xi) If the speed of an electron increases, the specific charge (a) increases (b) decreases (c) remains constant.

(xii) One milligram of matter converted into energy will give (a) 90 joules (b) 9×10^{10} joules (c) none of these.

(xiii) Relative velocity of two particles moving with velocity (c) of light in opposite direction is (a) c, (b) $2c$, (c) 0.

Group B: Short Answer Type Questions
Answer any *three* of the following.

2. (a) What is the difference between temporal coherence and spatial coherence?

(b) If the amplitudes of two coherent light waves are in the ratio 1 : 4, find the ratio of maximum and minimum intensity in the interference pattern.

3. (a) What is Compton effect? Calculate the Compton wavelength for an electron.

(b) Why does the unmodified line appear in Compton scattering?

4. (a) Deduce the formulae for interplanar spacing of a simple cubic crystal.

(b) Why X-ray diffraction is used for crystal structure analysis?

5. Find the possible arrangements of two particles in three cells for

(a) Bose-Einstein statistics

(b) Fermi-Dirac statistics.

6. (a) Describe briefly the working principle of laser action.

(b) Mention at least three main advantages of optical fibre over wire or cable?

7. (a) What are the basic postulates of special theory of relativity?

(b) Show that for $V \ll c$, Lorentz transformation reduces to the Galilean transformation.

8. (a) Discuss the important characteristics of nuclear forces.

(b) Why nuclear fission reaction must be done first before nuclear fusion?

Group C: Long Answer Type Questions
Answer any *three* of the following questions.

9. (a) What is double refracting crystal?

(b) Discuss Nicol prism as polarizer and analyzer.

(c) Determine the Brewster's angle for glass of refractive index 1.5 immersed in water of refractive index 1.33.

(d) Prove that the intensities of secondary maxima formed for Fraunhofer diffraction at a single slit are of decreasing order.

(e) In a plane transmission grating the angle of diffraction for 2nd order maxima for wavelength 5×10^{-5} cm is 30°. Calculate the number of lines in one centimetre of the grating surface.

10. (a) State and explain de Broglie hypothesis.

(b) Prove that the product of phase velocity and group velocity for a de Broglie wave is equal to the square of the velocity of light.

(c) Compute the smallest possible uncertainty in the position of an electron moving with velocity 3×10^7 m/s. The rest mass of electron is 9.1×10^{-31} kg.

(d) Derive the Wein's displacement law from Planck's radiation law.

11. (a) Write down the postulates of Fermi-Dirac statistics.

(b) Plot electron distribution function governed by Fermi-Dirac statistics in metal at $T = 0\,\text{K}$ and $T > 0\,\text{K}$. Explain their physical significance.

(c) Why Compton effect cannot be observed with visible light but can be observe due to X-rays?

12. (a) Discuss the operation of a Ruby laser with the help of energy level diagram.

(b) Find the numerical aperture and angle of acceptance of a given optical fibre. [Given, the refractive indices of core and cladding are 1.562 and 1.497, respectively.]

(c) In a He-Ne laser transition from $3s$ to $2p$ level gives a laser beam of wavelength 632.8 nm. If the $2p$ level has energy equal to 15.2×10^{-19} J, calculate the required pumping energy (assuming no loss of energy).

(d) What is the role of optical resonator in laser production?

13. (a) Write down Schrödinger equation for one-dimensional motion of a free particle in a one-dimensional potential box. Find its eigenfunction and eigenenergy.

(b) Prove that the first excited energy state of a free particle in a cubical box has three-fold degeneracy.

(c) Copper has FCC structure and the atomic radius is 0.1278 nm. Find its density and the interplanar spacing for (321) planes. The atomic weight of copper is 63.5.

14. (a) What are the differences between inertial and non-inertial frame of references?

(b) Deduce and expression of time dilation on the basis of Lorentz transformation equation.

(c) Find the mass and speed of 2 MeV electron.

(d) If the total energy of a particle is thrice of its rest energy, find the velocity of the particle.

15. (a) What is nuclear binding energy?

(b) Draw a binding energy curve. What informations do yŏu set from such a curve?

(c) Find the energy released for the following nuclear fusion reaction:

$4\,{}_1H^1 \rightarrow {}_2He^4 + 2\,{}_{+1}e^0 +$ energy.

Given that

mass of ${}_1H^1$ atom $= 1.007825$ amu

mass of ${}_2H^4$ atom $= 4.0026$ amu

mass of ${}_{+1}e^0$ (positron) $= 0.00055$ amu.

(d) Differentiate conductor, insulator and semiconductor on the basis of energy band diagram.

2009

Semester 1

Group A: Multiple Choice Type Questions

1. Choose the correct alternatives of the following:

(i) The centre of mass of a body is a point at which (a) only translational motion occurs, (b) both translational and rotational motions occur, (c) none of these.

(ii) If a system has f degrees of freedom the number of Hamilton's equation for the system is (a) 2, (b) f, (c) $2f$.

(iii) The value of $\bar{\nabla} \times \bar{r}$ is equal to (a) 1, (b) 0, (c) −1.

(iv) The electric flux through the surface vector $\vec{S}$ equal to $6\hat{j}$ in a region of electric field $3\hat{i} + \hat{j}$ is (a) $10\,\mathrm{Nm^2C^{-1}}$, (b) $6\,\mathrm{Nm^2C^{-1}}$, (c) none of these.

(v) In free space Poisson's equation reduces to (a) $\nabla^2 V = 0$, (b) $\nabla^2 V = \rho/\epsilon_0$, (c) $\nabla^2 V = -\rho/\epsilon_0$.

(vi) The divergence of a magnetic flux density $(\bar{B})$ is (a) 0, (b) 1, (c) −1.

(vii) The differential form of Faraday's laws of electromagnetic induction is (a) $\bar{\nabla} \times \bar{E} = \partial\bar{B}/\partial t$, (b) $\bar{\nabla} \cdot \bar{E} = 2\bar{B}$, (c) $\bar{\nabla} \times \bar{E} = -\partial\bar{B}/\partial t$.

(viii) When a spring with spring constant K is cut into three equal parts, the force constant of each of the parts would be (a) $K/3$, (b) $3K$, (c) K.

(ix) If α is the force constant of an oscillating body of mass m, the Q-factor is (a) $Q = C\sqrt{am}$, (b) $Q = C\sqrt{m/a}$, (c) $Q = C\sqrt{a/m}$, where $C =$ relaxation time.

(x) Superposition of two SHMs of equal time period and equal amplitude with phase difference $\phi = \pi/2$ forms (a) circle, (b) ellipse, (c) parabola.

Group B: Short Answer Type Questions

Answer any *three* of the following.

2. (a) Show that, $y = ae^{i(\omega t - kx)}$. Do the solution of wave equation.

(b) The potential energy of a particle with mass 10 g is given by, $V(x) = 32x^2 + 0.2$, where x is in metre and V is in joule. Write down the equation of motion and solve it.

3. (a) Prove that $\bar{E} = \cos(y - t)\hat{k}$ and $\bar{B} = \cos(y - t)\hat{i}$ constitute possible electromagnetic wave.

(b) Define displacement current.

4. (a) Calculate the magnetic field intensity just outside and inside of a hollow cylinder of radius 4 cm carrying 50 A current.

(b) Differentiate between electrostatic field and magnetic field.

5. Prove that $\bar{A} \times \bar{B} \times \bar{C} = \bar{B} \times (\bar{A} \times \bar{C}) - \bar{C} \times (\bar{A} \times \bar{B})$.

6. (a) What is cyclic coordinate? Explain with an example.

(b) Derive the Lagrangian for a particle falling freely under the influence of gravity.

Group C: Long Answer Type Questions

Answer any *three* of the following.

7. (a) Define Hamiltonian function and explain its general significance.

(b) If Lagrangian of a system is given as $L = \frac{1}{2}\dot{x}^2 + \dot{x} - \frac{x^2}{2}$, find the Hamiltonian and equation of motion.

(c) At what angle must a body be incident on a perfectly hard plane, so that the angle between the directions before and after may be a right angle?
[Coefficient of restitution, $e = \frac{1}{3}$.]

8. (a) Distinguish between scalar and vector fields with example.

(b) If the potential of a field is given by $V(x, y, z) = (4x^2 + 2y^2 + z^2)^{1/2}$, find the field intensity at the point $(1, 1, 1)$.

(c) Prove that curl grad $\phi = 0$.

(d) Show that the potential function $x^2 - y^2 + z$ satisfies Laplace's equation.

(e) Find the potential of a uniformly charged sphere of radius R having a constant charge density ρ at a distance r from the centre of the sphere, where $r > R$.

9. (a) State and prove Ampere's cricuital law.

(b) Define magnetic vector potential.
If the vector potential $\bar{A} = (10x^2 + y^2 - z^2)\hat{j}$ at any position, then find the magnetic field at the point $(1, 1, 2)$.

(c) Write down the condition of steady state current. Show that the Ampere's law implies that the current is in the steady state.

(d) Calculate the magnetic field intensity just outside a hollow cylinder of radius 4 cm.

10. (a) Write down Maxwell's equations for free space.

(b) Show that for free space the electromagnetic wave equation for $\bar{E}$ is $\nabla^2 \bar{E} = \mu_0 \epsilon_0 \frac{\partial^2 \bar{E}}{\partial t^2}$, where symbols have their usual significance. Prove that electromagnetic wave moves with the velocity of light in free space.

(c) Prove that $\bar{E} = \sin(y - t)\hat{k}$ and $\bar{B} = \sin(y - t)\hat{\imath}$ constitute a possible electromagnetic wave.

(d) Establish the integral form of Faraday's law of electromagnetic induction.

11. (a) Establish the differentiation equation of damped harmonic motion.

(b) Solve the equation for light damping and prove that the amplitude of vibration decreases exponentially with time.

(c) A cubical block of side L cm and density d is floating in water of density $\rho(\rho > d)$. The block is slightly depressed and released. Show that it will execute simple harmonic motion and hence determine the frequency of oscillation.

2010

Semester 1

Group A: Multiple Choice Type Questions

1. Choose the correct alternatives for any *ten* of the following.

(i) If a particle is executing simple harmonic motion with frequency ν, then its potential energy (a) remains constant over time, (b) is oscillating with a frequency ν, (c) is oscillating with a frequency $\nu/2$, (d) is oscillating with a frequency 2ν.

(ii) The quality factor Q for an LCR circuit is (a) $\frac{\omega R}{L}$, (b) $\frac{\omega L}{R}$, (c) $\frac{\omega}{LR}$, (d) $\frac{R}{\omega L}$.

(iii) An external force $F = F_0 e^{i\omega x}$ is applied to a slightly damped oscillator of natural frequency ω_0, then in steady state it will oscillate with a frequency (a) ω_0, (b) ω, (c) $\omega_0 - \omega$, (d) $\sqrt{\omega_0^2 - \omega^2}$.

(iv) The intensity of principal maximum in the Fraunhofer diffraction spectrum produced by a grating with N number of lines is proportional to (a) $\frac{1}{N}$, (b) N, (c) N^2, (d) $\frac{1}{N^2}$.

(v) In the propagation of light wave the angle between the plane of polarization and plane of vibration is (a) $0°$, (b) $\frac{\pi}{2}$, (c) $\frac{\pi}{4}$, (d) π.

(vi) In case of simple cubic crystal the effective number of atoms per unit cell is (a) 2 (b) 1 (c) 3 (d) none of these.

(vii) In order to produce X-ray by striking electron beam on a metal target the kinetic energy of the electron should be (a) less than 1 eV, (b) of the order of eV, (c) of the order of keV, (d) of the order of GeV.

(viii) If the energy of a particle is much higher than its rest energy, then the energy is (a) independent of the momentum, (b) proportional to the momentum, (c) proportional to square of the momentum, (d) proportional to the square root of the momentum.

(ix) If a wave packet is described by $\phi(x) = A \exp\left(-\frac{x^2}{2\sigma^2}\right)$, then the momentum uncertainty is proportional to (a) $h\sigma$, (b) $\frac{h}{\sigma}$, (c) $h\sigma^2$, (d) $\frac{h}{\sigma^2}$.

(x) Given that the temperature of a younger star is higher than that of an older one (a) a blue star is younger than a red star (b) a blue star is older than a red star, (c) both of them are of same age, (d) colour of the star cannot be correlated to the age of the star.

(xi) In Ruby laser the active medium is (a) solid, (b) liquid, (c) gas, (d) a mixture solid and gas.

(xii) In a Nicol-prism of O-ray is totally internally reflected and the E-ray is transmitted. This statement is (a) true, (b) false, (c) partly true, (d) partly false.

(xiii) Two sources are said to be coherent when the waves produced by them are of (a) same wavelength, (b) same wavelength and same phase, (c) same wavelength and constant phase difference, (d) same amplitude and constant phase difference.

(xiv) In an arrangement for viewing Newton's ring, if the lens which rests on a glass plate is moved upwards by one wavelength (of the viewing light), which of the following will be observed? (a) The central spot becomes bright. (b) No change of fringe pattern is observed. (c) The rings shift towards the centre. (d) The rings move out from the centre.

(xv) Holography is based on (a) the interference of reference and object waves which are coherent, (b) superposition of object and the reference waves which are of slightly different wavelengths, (c) recording of superposed images of two different wavelengths, (d) recording the phase information of the resultant of the reference and the object waves.

(xvi) For a laser action to occur, the medium used must have at least (a) 4 energy levels, (b) 2 energy levels, (c) 3 energy levels, (d) one energy level.

Group B: Short Answer Type Questions

Answer any *three* of the following.

2. A vibrator of mass 10 gm is acted on by a restoring force of 5 dynes/cm and a damping force 2 dynes-sec/cm. Find whether the motion is over-damped or oscillatory. If at $t = 0$ the vibrator was at position $x = 0$, when a velocity 1 cm/sec is imparted to it, then calculate the maximum deviation along positive x-axis.

3. (a) In a Newton's ring experiment, the diameter of a dark ring is 0.32 cm, when the wavelength of monochromatic light be 6000 Å, What would be the diameter of that ring when the wavelength of light changes to 5000 Å?

(b) Write down the expression for the intensity of light due to Fraunhofer diffraction in a transmission grating and hence find the condition for secondary minima in the interference pattern.

4. (a) Two polarizers are placed at crossed position—angle between the polarizing planes are 90°—a third polarizer with angle θ with the first one is placed between them. An unpolarized light of intensity I is incident on the first one and passes through all three polarisers. Find the intensity of the light that comes out.

(b) What is population inversion in the context of LASER? Does it violate Maxwell-Boltzmann distribution law?

5. (a) An electron is observed moving at 50% of the speed of light, $v = 1.5 \times 10^8$ m/s. What is the relativistic mass of the electron? What is the kinetic energy of the electron?

(b) Write down the conservation laws in Compton scattering. Show that in Compton scattering while the photon can be scattered at any angle between 0° to 180°, the recoil electron can only be emitted at angles between 0° and 90°.

6. (a) The linear absorption coefficient of an element of X-rays having $\lambda = 0.3$ Å is $135\,\text{m}^{-1}$. Find the half value thickness for the material.

(b) Cs metal (atomic weight 130) has a cubic unit cell of lattice constant 0.6 nm. If the density of Cs is 2 gm/cc, determine whether the unit cell is simple, body-centred or face-centred.

Group C: Long Answer Type Questions

Answer any *three* of the following.

7. (a) Starting from the equation of motion and after solving it show that, for a forced oscillator in the steady state, the displacement amplitude at low frequencies ($\omega \to 0$), the velocity amplitude at velocity resonance ($\omega = \omega_0$) are independent of the frequency of the driving force.

(b) Explain the terms logarithmic decrement and quality factor of a damped oscillatory system. How are they related?

8. (a) Derive the intensity distribution of diffraction of Fraunhofer class of light due to a single slit. Sketch the intensity distribution.

(b) A single slit forms diffraction pattern of Fraunhofer class with white light. The second maximum in the pattern for red light of wavelength 7000 Å coincides with the third maximum of an unknown wavelength. Calculate the unknown wavelength.

(c) In Young's double slit experiment the distance between two slits is 0.5 mm. The wavelength of light is 5000 Å and the separation between the sources and the screen is 50 cm. Calculate the fringe width in this case.

9. (a) How can you get an elliptically polarized light from a linearly polarized light by using an optical device?

A plane polarized light is incident on a piece of quartz cut parallel to the axis. Find the least thickness for which the ordinary and the extraordinary rays comine to form plane polarized light given that $\mu_0 = 1.5442$ and $\mu_e = 1.5533, \lambda = 5 \times 10^{-5}$ cm.

(b) Define Einstein A, B coefficients of absorption and emission. Find the relations among them.

(c) Explain how holographic images are reconstructed from the holograms.

10. (a) Describe the function of an optical resonator? What is the use of such a device in the context of LASER generation?

(b) What is coordination number? Determine the coordination number for *sc* and *bcc* structures. Draw necessary sketch. 'Atomic packing factor' increases with coordination number. Justify the statement.

(c) Find the short-wave limit of continuous X-ray spectrum. Does it shift by 0.5 nm when the voltage applied to X-ray tube is doubled?

11. (a) State and explain de Broglie's hypothesis. Show that the relativistic de Broglie wavelength is given by

$$\lambda_{\text{relativistic}} = \frac{h}{\sqrt{E_k(E_k + 2m_0c^2)}}$$

(The notations used have their usual significance).

(b) Describe an experiment which verified de Broglie's hypothesis.

(c) Write down the expression of the number of possible modes of cavity waves of frequency ν to $\nu + d\nu$. Using Planck's hypothesis about the energy quantization of the cavity oscillators, find out the average energy of an oscillators at a temperature T and the energy density within the frequency range ν to $d\nu$. Explain why the energy density decreases at very high values of the frequency at a finite temperature.

June 2011

Semester 2

Group A: Multiple Choice Questions

1. Choose the correct alternatives for any *ten* of the following:

(i) If the damping force on an one-dimensional harmonic oscillator of natural frequency ω_0 is $2bmv$, where m is the mass and v is the instantaneous velocity of the oscillator, then the frequency of oscillation (when $b \ll \omega_0$) is (a) ω_0 (b) b (c) $\omega_0\left(1 - \frac{b^2}{2\omega_0^2}\right)$ (d) $\omega_0\left(1 + \frac{b^2}{2\omega_0^2}\right)$.

(ii) Two mutually perpendicular oscillations with same frequency, amplitude but phase difference δ will produce closed curve with nonzero area enclosed (a) for all values of δ except $\delta = 0$ (b) only for $\delta = \frac{\pi}{2}$ (c) for all values of δ except $\delta = 0$ and $\delta = \pi$ (d) for all values of $\delta > \frac{\pi}{2}$.

(iii) Example of weakly damped harmonic oscillator is (a) Dead-beat galvanometer (b) Tangent galvanometer (c) Ballistic galvanometer (d) Discharge of a charged capacitor through a resistance.

(iv) Missing orders are found in case of double slit diffraction patterns due to (a) unequal value of two slit widths (b) superposition of diffraction minima and interference maxima (c) superposition of diffraction maxima and interference minima (d) oblique incidence of light.

(v) How does reducing the slit separation affect the appearance of the fringes in the double slit interference? (a) The fringe width increases and the fringes are brighter. (b) The fringe width increases but the brightness remains unchanged. (c) The fringe width decreases and the fringes are brighter. (d) The fringe width is unchanged but the fringes are less bright.

(vi) Resolving power of microscope objective placed in air with light of wavelength λ_1 is the same when immersed in oil of refractive index μ and wavelength λ_2. Assuming the semi-vertical angle to be the same in these two cases, then (a) $\lambda_1 = \mu\lambda_2$ (b) $\lambda_2 = \mu\lambda_1$ (c) $\lambda_1\lambda_2 = \mu(\lambda_1 + \lambda_2)$ (d) $\lambda_1 + \lambda_2 = \mu\lambda_1\lambda_2$.

(vii) For a doubly refracting crystal, the refractive indices for the ordinary and extraordinary rays are denoted by μ_0 and μ_e. Which of the following relations is valid along the optical axis of the crystal? (a) $\mu_e = \mu_0$ (b) $\mu_e < \mu_0$ (c) $\mu_e > \mu_0$ (d) $\mu_0 \le \mu_e$.

(viii) Which of the following schemes does not produce lasing action? (a) Two level scheme (b) Three level scheme (c) Four level scheme (d) Five level scheme.

(ix) In holography, the 3D images are formed obeying the principle of (a) interference (b) diffraction (c) polarization (d) dispersion.

(x) Number of oscillation modes for the electromagnetic standing waves of frequency γ for the cavity radiation, is proportional to (a) $\gamma^{1/2}$ (b) γ (c) γ^4 (d) γ^2, where γ is the frequency of the wave.

(xi) How fast must a particle travel so that its mass becomes twice its rest mass? (a) $0.5c$ (b) $2c$ (c) $\frac{\sqrt{3}c}{2}$ (d) c.

(xii) Origin of continuous X-rays is due to the process of (a) ionization (b) inner orbital transition (c) bremsstrahlung (d) none of these.

(xiii) The effective number of atoms per unit cell of an FCC lattice is (a) 4 (b) 6 (c) 14 (d) 1.

(xiv) The interplanar distance d_{111} for a (111) plane of simple cubic crystal is (a) $\frac{a}{\sqrt{3}}$ (b) $\frac{a}{\sqrt{2}}$ (c) a (d) $2a$.

(xv) A proton accelerated through a potential difference of V has a certain de Broglie wavelength. In order to have the same de Broglie wavelength, an α-particle must be accelerated through a potential difference of (a) $8V$ (b) $4V$ (c) $\frac{V}{4}$ (d) $\frac{V}{8}$.

Group B: Short Answer Type Questions

Answer any *three* of the following:

2. A damped harmonic oscillator, with natural frequency ω_0 and damping force $2mbV$ (V = velocity, m = mass, b = a constant) is driven by a periodic force $Fe^{i\omega t}$, then the steady state displacement is given by

$$x = \frac{F}{m}\frac{e^{i\omega t}}{\sqrt{(\omega_0^2 - \omega^2) + i2b\omega}}.$$

Use this fact to show that at the velocity resonance the phase of the periodic force and the velocity is the same.

3. A mica sheet of thickness t and refractive index μ is introduced in the path of one of the interfering beams in Young's double slit experiment. Find out the linear displacement of the nth bright fringes in terms of μ and t.

4. (a) What is polaroid ? Give instances of its practical application.

(b) A quartz plate with thickness of 0.1436 mm is used as phase retardation plate. For what wavelengths in the visible region (450–800 nm) will it act as a quarter wave plate (μ_0 = 1.5443, μ_e = 1.55333).

5. Prove that the maximum recoil energy of a free electron of rest mass m_0 when struck by a photon of wavelength λ is given by $E_{\max} = \frac{2m_0c^2\lambda_c^2}{\lambda^2 + 2\lambda_c\lambda}$. Here λ_c is the Compton wavelength of the electron.

6. (a) The spacing between principal planes of NaCI crystal is 2.82 Å. It is assumed that the first order Bragg reflection occurs at an angle 10°. Calculate the wavelength of X-rays.

(b) Find the atomic packing fraction of a BCC structure.

Group C: Long Answer Type Questions

Answer any *three* of the following.

7. (a) A cubical block of side L cm and density d is floating in water of density $\rho (\rho > d)$. The block is slightly depressed and released. Show that it will execute simple harmonic motion. Determine the frequency and time period of oscillation.

(b) Find the displacement as a function of time of a particle of mass m which is subjected to over-damped harmonic motion with natural frequency ω_0 and damping force γV (V being the instantaneous velocity), given that the displacement is zero initially and the initial velocity is V_0. Sketch the displacement as a function of time.

8. (a) Newton's ring arrangement is used with a source emitting two wavelengths $\lambda_1 = 600$ nm and $\lambda_2 = 590$ nm. It was found that the nth dark ring due to λ_1 coincides with the $(n+1)$th dark ring due to λ_2. If radius of curvature of the lens is 0.9 m, then find out the value of n and the diameter of the nth dark ring due to λ_2.

(b) A plane transmission grating having 1500 lines/inch is being used under normal incidence of light.

(i) What is the longest wavelength of light for which a spectrum can be seen?

(ii) What is the highest order spectrum that can be seen for the light of 589.3 nm wavelength?

(iii) The spectral line of 589.3 nm in the second order spectrum overlaps with another spectral line in the next order. Find wavelength of the other spectral line.

(iv) Find the expression of the dispersive power for a given order.

(v) If 90% of the width of the grating is covered, how the width of the spectral lines are changed?

9. (a) Define Einstein's A, B coefficient for spontaneous and stimulated emission. Derive the relations among them.

(b) At what angle should light be incident on a glass plate of refractive index $\mu = 1.5697$ to get a plane polarized light by reflection.

(c) An analyzing nicol examines two adjacent plane polarized beams A and B whose planes of polarization are mutually perpendicular. In one position of the analyzer the beam B shows zero intensity. From this position, a rotation of 30° shows the two beams as of equal intensity. Deduce the ratio of two intensities I_A/I_B of the two beams.

(d) Why is it necessary that the object beam and reference beam in holography are highly coherent?

10. (a) Discuss the action of optical resonator in a Baser device. What is the relation of the length of the resonator with the frequency of the laser beam?

(b) A ruby laser emits light of 693.95 nm wavelength. If 1 mole of Cr^{3+} ions are involved in population inversion process in a pulse, calculate the pulse energy in eV.

(c) Calculate average energy of a cavity oscillator with frequency γ. How does this average energy vary with frequency in the high frequency limit?

(d) Derive Wien's displacement law from Planck's law of Black body radiation.

11. (a) What voltage must be applied to an X-ray, tube for it to emit X-ray with a minimum wavelength 30 pm. Deduce the formula that you use $\left[\frac{hc}{e} = 1.24 \times 10^{-4}\,\text{Vm}\right]$.

(b) What are characteristic X-rays? How are they produced?

(c) Describe the relation between the lattice constants and the angles between them in any one Bravais lattice other than cubic crystal system.

(d) Find the Miller indices of all the principal planes of a cubic crystal system.

(e) Aluminium is an *fcc* crystal with lattice constant $a = 0.405$ nm. Calculate the number of unit cells present in an aluminium foil of 0.005 cm thickness and two sides of 25 cm length.

12. (a) What is de Broglie hypothesis of matter wave? Discuss how Davisson and Germer proved this by an experiment.

(b) If a particle of charge e and rest mass m_0 is accelerated by a potential V, then what is the de Broglie wavelength of the particle viewing the relativistic energy momentum relation.

(c) The atoms in a solid possess a certain minimum zero point energy even at 0 K while no such restriction holds for the molecules in an ideal gas. Use uncertainty principle to explain this statement.

(d) An electron has de Broglie wavelength 2 pm. Find the phase and group velocities of its de Broglie wave, given that the rest energy of an electron is 511 eV.

2011
Engineering Physics

Time Allotted: 3 Hours *Full Marks*: 70

The figures in the margin indicate full marks.

Candidates are required to give their answers in their own words as far as practicable.

Group A
(Multiple Choice Type Questions)

1. Choose the correct alternatives for any *ten* of the following: $10 \times 1 = 10$

(a) A ray of light is incident on the surface of a glass plate at an angle of incidence equal to the Brewster's angle ϕ. If μ represents the refractive index of glass w.r.t. air, then the angle between reflected and refracted rays is

(i) $90° + \phi$
(ii) $\sin^{-1}(\mu \cos \phi)$
(iii) $90°$
(iv) $90° - \sin^{-1}(\mu \sin \phi)$.

(b) The first minimum due to a Fraunhofer diffraction using light of wavelength 500 nm and a slit width of 0.5 mm will be formed at an angle

(i) 0.084°
(ii) 0.057°
(iii) 0.001°
(iv) 3.42°.

(c) Two waves having amplitudes 5 : 1 produce interference. The ratio of the maximum to minimum intensity in the interference pattern is

(i) 25 : 1
(ii) 6 : 4
(iii) 3 : 1
(iv) 9 : 4.

(d) A grating has 5000 lines/cm. The maximum order of diffraction visible with light of wavelength 600 nm is

(i) 4
(ii) 3
(iii) 2
(iv) 1.

(e) In a ruby laser population inversion is achieved by

(i) chemical reactions
(ii) inelastic collision between atoms
(iii) optical pumping
(iv) applying strong electric field.

(f) The coordination number in a *fcc* lattice is

(i) 4
(ii) 6
(iii) 8
(iv) 12.

(g) Velocity of a particle when its mass become twice its rest mass is

(i) 0.5 c
(ii) 0.72 c
(iii) 0.866 c
(iv) c.

(h) The resolving power of a grating having N number of lines exposed to nth order is
  (i) N/n
  (iii) $n+N$
  (ii) n/N
  (iv) nN.

(i) The wavelength λ at which a blackbody emits maximum about of radiation is proportional to
  (i) T
  (iii) T^4
  (ii) $1/T$
  (iv) e^T.

(j) The maximum velocity of an electron emitted by light of wavelength λ incident on the surface of a metal of work function ϕ is
  (i) $\sqrt{\dfrac{2(hc+\lambda\phi)}{m\lambda}}$
  (iii) $\sqrt{\dfrac{2(h\lambda-\phi)}{m}}$
  (ii) $\sqrt{\dfrac{2(hc-\lambda\phi)}{m\lambda}}$
  (iv) $\dfrac{2(hc+\lambda\phi)}{m\lambda}$.

(k) Electron of masses m, when accelerated through a potential difference V has a de Broglie wavelength λ. The de Broglie wavelength associated with a proton of mass M accelerated through the same potential difference will be
  (i) $\lambda\dfrac{m}{M}$
  (iii) $\lambda\sqrt{\dfrac{m}{M}}$
  (ii) $\lambda\dfrac{M}{m}$
  (iv) $\lambda\sqrt{\dfrac{M}{m}}$.

(l) The commutation relation $[x^2, p_x]$, where x and p_x are respectively the quantum mechanical position and momentum operators along x-axis, is equal to
  (i) $2i\hbar x$
  (iii) $2i\hbar$
  (ii) $i\hbar$
  (iv) $2i\hbar x^2$.

(m) For a particle in a potential the two lowest states have energies ϵ_1 and ϵ_2 and the corresponding eigenfunctions are ψ_1 and ψ_2 respectively. At one instant the wave function is represented by $\psi = \dfrac{\sqrt{3}}{2}\psi_1 + \dfrac{1}{2}\psi_2$. The expectation value of the total energy corresponding to the wave function ψ is
  (i) $\frac{\sqrt{3}}{2}\epsilon_1 + \frac{1}{2}\epsilon_2$
  (iii) $\frac{3}{4}\epsilon_1 + \frac{1}{2}\epsilon_2$
  (ii) $\frac{3}{2}\epsilon_1 + \frac{1}{2}\epsilon_2$
  (iv) $\epsilon_1 + \epsilon_2$.

(n) In the X-ray spectrum emitted from a tungsten anode, if I_α and I_β be the intensities and λ_α and λ_β are the wavelengths of the K_α and K_β lines respectively, then
  (i) $I_\alpha > I_\beta$ and $\lambda_\alpha > \lambda_\beta$
  (iii) $I_\alpha < I_\beta$ and $\lambda_\alpha < \lambda_\beta$
  (ii) $I_\alpha > I_\beta$ and $\lambda_\alpha < \lambda_\beta$
  (iv) $I_\alpha < I_\beta$ and $\lambda_\alpha > \lambda_\beta$.

(o) The number of ways in which four identical bosons can be distributed in three different energy states is
  (i) 15
  (iii) 144
  (ii) 6
  (iv) 24.

Group B
(Short Answer Type Questions)

Answer any *three* of the following: $3 \times 5 = 15$

2. (a) Write the basic difference between Fresnel and Fraunhofer type of diffraction.

   (b) In a two-slit Fraunhofer diffraction pattern, with equal slit width $b = 8.8 \times 10^{-3}$ cm and the slit separation $d = 7.0 \times 10^{-2}$ cm, find the number of interference minima occurring between the two diffraction minima occurring on either side of the central maximum. The wavelength of light used $\lambda = 6328$Å.

3. Assuming that the oscillators can have only discrete energy values $\epsilon = nh\nu$, where $n = 0, 1, 2, 3, \ldots$ (all positive integer values) and follow the Boltzmann distribution function show that the average energy of each oscillator at temperature T is $\bar{\epsilon} = \dfrac{h\nu}{e^{h\nu/kT} - 1}$.

4. (a) An electron is inside an atom of radius 5×10^{-11} m. Calculate from uncertainty principle the minimum kinetic energy (KE) of the electron.

   (b) Using the relation $\epsilon^2 = p^2c^2 + m_0^2c^4$; $\epsilon = mc^2$ for a relativistic particle, show that group velocity is equal to particle velocity. Also find the phase velocity of the particle.

5. (a) Find the miller indices of a crystal plane, which intercepts the a-axis at $3a$, b-axis at $4b$ and is parallel to c-axis; a, b and c being primitive vectors of the lattice.

   (b) Copper has *fcc* structure and the atomic radius is 0.128 nm. Calculate the density of copper taking the atomic weight of copper as 63.5.

   (c) Calculate the glancing angles on the set of parallel planes (110) of cubic crystal with lattice constant $\alpha = 2.814$Å corresponding to second order diffraction maximum for X-rays of wavelength 0.710Å.

6. What are the essential requirements for the construction of an optical fibre? Calculate the numerical aperture of an optical fibre in which the refractive index of core and cladding are 1.58 and 1.5 respectively. Find the maximum acceptance angle if the fibre is placed in a medium of refractive index 1.33.

Group C
(Long Answer Type Questions)

Answer any *three* of the following: $3 \times 15 = 45$

7. (a) State and explain Malus' law.

   (b) What are half wave and quarter wave plates?

   (c) How can you convert plane polarised light to circularly polarised light?

   (d) A beam of unpolarised light of intensity I is passed first through a tourmaline crystal A and then through another tourmaline crystal B oriented such that its principal plane is parallel to that of A. If A is now rotated by 45° in a plane parallel to the direction of incident ray, find the intensity of the emergent light.

   (e) The critical angle for a particular pair of medium is $\sin^{-1}$ (3/5). Find the polarising angle for the pair of mediums.

8. (a) Explain the terms spatial coherence and temporal coherence.

(b) Find an expression for the intensity distribution when two sinusoidal coherent waves with amplitudes A_1 and A_2 and phase difference ϕ superpose to produce interference.

(c) Find an expression for the fringe width in the interference pattern for Young's double slit experiment.

(d) A thin sheet of glass of thickness 6 micron (refractive index 1.5) is introduced in the path of one of the interfering beams in a double slit experiment. The shift pattern shifts by 5 fringes. Find the wavelength of light used in the experiment.

9. Consider a particle of mass m that can move along X-axis in a region between $x = 0$ and $x = L$ in a potential given as:

$$V(x) = \begin{cases} 0 & \text{for } 0 < x < L \\ \infty & \text{for } x < 0 \text{ and } x > L \end{cases}$$

(a) Write the time-independent Schrödinger equation for both regions and derive expressions for the energy eigenvalues and corresponding normalised wave functions.

(b) Draw the wave functions corresponding to lowest two states.

(c) Find the expectation value of the position operator in the ground state.

(d) An electron is confined in the infinite potential box with length $L = 1.0$Å. Calculate energy of the ground state.

2011

Physics I — Sem I

Time Allotted: 3 Hours *Full Marks*: 70

The figures in the margin indicate full marks.

Candidates are required to give their answers in their own words as far as practicable.

Group A
(Multiple Choice Type Questions)

1. Choose the correct alternatives for any *ten* of the following: $10 \times 1 = 10$

(a) Resultant of two perpendicular simple harmonic motions of equal frequency and equal amplitude but a phase difference of π radian is

(i) a straight line
(ii) an ellipse
(iii) a circle
(iv) a spiral collapsing inward.

(b) If the equation of motion of an oscillator is given by $\ddot{x} + \frac{\gamma^2}{4}x + \gamma\dot{x} = 0$, then the motion is

(i) simple harmonic without damping
(ii) a critically damped simple harmonic
(iii) an over-damped simple harmonic
(iv) an under-damped simple harmonic.

(c) An oscillator with natural frequency ω_0 experiences a damping force and also a periodic force $A\cos(\Omega t)$. Amplitude of displacement velocity of the resulting steady state forced oscillation is studied with slow variation of Ω, when $\Omega = \omega_0$

(i) the displacement amplitude shows a maximum
(ii) the velocity amplitude shows a maximum

(iii) amplitude of both displacement and velocity show maxima

(iv) amplitude of displacement shows minimum while the amplitude of velocity shows maximum.

(d) Diffraction occurs when the width of the narrow slit D is

(i) comparable to the wavelength of the light used

(ii) equal to the wavelength of the light used

(iii) much greater than the wavelength of the light used

(iv) much smaller than the wavelength of the light used.

(e) In Newton's ring experiment the diameters of the dark rings are proportional to

(i) odd natural numbers

(ii) square root of the natural numbers

(iii) square root of the odd natural numbers

(iv) square of the natural numbers.

(f) A diffraction pattern is obtained with a plane transmission grating using a beam of red light. What happens after red light is replaced by blue light?

(i) The lines will shift towards lower angles

(ii) The lines will shift towards higher angles

(iii) The lines will disappear

(iv) No visible changes will be noticed.

(g) A rotating calcite crystal (a doubly refracting crystal) is placed over an ink dot. On seeing through the crystal, one finds

(i) two stationary dots

(ii) two dots moving along a straight line

(iii) one dot rotating about the other

(iv) both dots rotating about a common axis.

(h) In He-Ne laser, Ne atom obtains energy

(i) on collision with He atom

(ii) from chemical reaction

(iii) from optical pumping

(iv) from electrical pumping.

(i) Which of the following statements is not true about reconstruction of the holographic image?

(i) The hologram must be illuminated with a partially incoherent light

(ii) Both real image and virtual image are formed

(iii) The frequency of the illumination should be equal to that of the reference wave

(iv) The reconstructed image is a virtual image.

(j) A proton, electron and a helium nucleus move with equal velocity. Rank their de Broglie wavelengths from longest to shortest.

(i) Helium nucleus, proton, electron

(ii) Proton, electron helium nucleus

(iii) Helium nucleus, electron, proton

(iv) Electron, proton, helium nucleus.

(k) The shift in wavelength in Compton effect is equal to Compton wavelength when the scattering angle is

(i) zero
(ii) $\pi/2$
(iii) π
(iv) $\pi/4$.

(l) If the kinetic energy (KE) of a particle is equal to its rest energy then

(i) its mass is equal to its rest mass
(ii) its mass is double of its rest mass
(iii) its mass is half of its rest mass
(iv) its mass is $\sqrt{2}$ times its rest mass.

(m) Number of oscillation modes for electromagnetic standing waves of frequency between ν to $\nu + d\nu$ in case of cavity radiation is proportional to

(i) ν^3
(ii) ν^2
(iii) ν
(iv) $\frac{h\nu}{e^{\frac{h\nu}{kT}}-1}$.

(n) The spacing between (340) planes in a face centred cubic crystal of lattice constant 10 nm is

(i) 1 nm
(ii) 2 nm
(iii) 4 nm
(iv) 7 nm.

(o) A X-ray tube, while operated with an applied potential difference of 80000 V, shows a current of 8 mA. The number of electrons striking per second on the target material is

(i) 6.4×10^3
(ii) 1×10^7
(iii) 5×10^{11}
(iv) 5×10^{16}.

Group B
(Short Answer Type Questions)

Answer any *three* of the following: $3 \times 5 = 15$

2. (a) Derive the expression for the total energy of a simple harmonic oscillator and show that it is constant and proportional to the square of the amplitude.

(b) If an oscillator's motion is described by $x = e^{-\beta t} \sin \Omega t$, where t is the time, then calculate its logarithmic decrement.

3. How are coherent sources produced in Young's double slit experiment. Show that the law of conservation of energy is not violated in case of interference.

4. (a) Find the ratio of Einstein's coefficients of spontaneous and stimulated emission.

(b) Describe how the optical resonator in a laser setup is used to ensure that the frequency spread of the output light is very small.

5. A particle of mass m is constrained to move in one dimension and trapped in an infinite potential well of width a. Estimate its ground state energy from uncertainty principle. Sketch a graph of this energy with the width a.

6. Draw a diagram showing the behaviour of X-ray intensity with wavelength. Distinguish between continuous and characteristic X-ray spectra. Derive short wavelength limit of continuous spectra. How are the characteristic spectra generated?

Group C
(Long Answer Type Questions)

Answer any *three* of the following: $3 \times 15 = 45$

7. (a) Find the frequency of oscillation of an under-damped harmonic oscillator of mass m, in terms of its natural frequency ω_0 and damping constant γ (where $-\gamma v$ is the damping force, v being the velocity).

(b) If the same oscillator is driven by an external force, $F = F_0 \cos \Omega t$, then find the phase difference between the displacement and the driving force in the steady state.

(c) A series LCR circuit is driven by ac source of frequency Ω. Write down the differential equation the charge across the capacitor should satisfy. Identity the parameters of forced damped oscillator in terms of LCR. Find the frequency at which the current in this circuit in maximum. You have to deduce the necessary formula from the differential equation. (You may use your results of part (b) of this question)

8. (a) In Young's double slit experiment, the slits are 0.2 mm apart and the screen is 1.5 m away. It is observed that the distance between the central bright fringe and fourth dark fringe is 1.8 cm. Find the wavelength of light.

(b) Write the expression of intensity (explaining the symbols) distribution for Fraunhofer diffraction due to a double slit. Draw the distribution graphically.

(c) Find the conditions of diffraction minima and interference maxima from the above expression.

(d) In a double slit experiment if the slit width is 2 mm and the separation between the slits is 4 mm, then find the missing orders. Derive the formula that you use.

(e) Define resolving power of a grating. What factors does it depend upon?

9. (a) Using Brewster's law show that light incident on a transparent substance at polarising angle gives reflected and refracted rays at right angles to each other.

(b) Describe how circularly polarised light and unpolarised light may be distinguished by experiment.

(c) Unpolarised light is incident on two polarising sheets placed on top of the other. Each sheet reduces the intensity of unpolarised light by 50 %. What must be the angle between the characteristic directions of sheets if the intensity of the transmitted light is 1/3 of the intensity of the incident beam.

(d) A quarter wave plate is fabricated with smallest possible thickness for wavelength of 589.3 nm. What phase retardation will be obtained with this plate with light having wavelength of 435.7 nm. You may neglect the variations of refractive indices with wavelength for this problem.

(e) Describe how the phase information of an object wave may be recorded in a holographic plane.

10. (a) Describe the construction of a ruby laser. Also discuss how population inversion is achieved in ruby laser with clearly drawing the energy diagrams. Discuss whether this result is suitable for continuous or pulse mode operation.

(b) Show that the total energy per unit volume of a blackbody radiation is proportional to the fourth power of the absolute temperature.

(c) Write down the equation describing mass energy equivalence. From the relation show that if a particle is moving with a velocity much smaller that of light then the kinetic energy (KE) is given by $\frac{1}{2}m_0v^2$, where m_0 is the rest mass and v is the velocity of the particle.

(d) X-ray of wavelength 0.2Å is scattered by a stationary particle of Compton wavelength of 2.43 pm at an angle 45°. Calculate the wavelength of the scattered ray the kinetic energy (KE) of recoiled particle.

(e) If a beam of electron is maximally, diffracted from a crystal of inter-planar separation d, at an angle θ, then show that the accelerating voltage V for the beam of electron is given by

$$V = \frac{1}{2me}\left(\frac{h}{2d\sin\theta}\right)^2,$$

where m and $(-e)$ are the mass and charge of an electron and h is Planck's constant.

11. (a) It is found that the maximum possible kinetic energy (KE) of an electron emitted by a radioactive nucleus of radius 10^{-14} m is 4 MeV. Then show that the electron did not exist in the nucleus before the radioactive decay.

(b) Light of wavelength 5893Å falls on a material of work function, $\phi = 1.90$eV. Calculate the energy of incident photon and the wavelength of the most energetic photoelectron ($h = 6.62 \times 10^{-34}$ Js, $c = 10^8$ m/s, mass of electron = 9×10^{-31} kg, charge of electron $e = 1.6 \times 10^{-19}$ C).

(c) Discuss why the phenomenon of photoelectric effect cannot be explained from the wave theory of light.

(d) Deduce Wien's displacement law from Planck's law of blackbody radiation.

12. (a) Define primitive cell. Are all unit cells primitive?

(b) Deduce the inter-planar space of a simple cube of lattice constant a, for the planes (hkl).

(c) In a certain crystal, the crystal planes of Miller indices (110) having a separation of 1.12Å show X-ray diffraction. Calculate up to which order, Bragg's reflection can be observed.

(d) In a triclinic crystal, a lattice plane makes intercepts of length a, $2b$ and $-c/2$. Find the Miller indices of the plane.

(e) Calculate the lattice constant of a substance having *fcc* lattice, molecular weight 60.2 and density 6250 kg/m^3. Avogadro number = 6.02×10^{23}.

2011
Physics I — Sem II

Time Allotted: 3 Hours *Full Marks*: 70

The figures in the margin indicate full marks.
Candidates are required to give their answers in their own words as far as practicable.

Group A
(Multiple Choice Type Questions)

1. Choose the correct alternatives for any *ten* of the following: $10 \times 1 = 10$

(a) If the damping force on an one-dimensional harmonic oscillator of natural frequency ω_0 is $2bmv$, where m is the mass and v is the instantaneous velocity of the oscillator, then the frequency of oscillation (when $b \ll \omega_0$) is

(i) ω_0

(ii) b

(iii) $\omega_0\left(1-\frac{b^2}{2\omega_0^2}\right)$

(iv) $\omega_0\left(1+\frac{b^2}{2\omega_0^2}\right)$.

(b) Two mutually perpendicular oscillations with same frequency, amplitude but phase difference Δ will produce closed curve with non-zero area enclosed

(i) for all values of Δ except $\Delta = 0$

(ii) Only for $\Delta = \pi/2$

(iii) for all value of Δ except $\Delta = 0$ and $\Delta = \pi$

(iv) for all values of $\Delta > \pi/2$.

(c) Example of weakly damped harmonic oscillator is

(i) Dead-beat galvanometer

(ii) Tangent galvanometer

(iii) Ballistic galvanometer

(iv) Discharge of a charged capacitor through a resistance.

(d) Missing orders are found in case of double slit diffraction patterns due to

(i) unequal value of two slit widths

(ii) superposition of diffraction minima and interference maxima

(iii) oblique incidence of light.

(e) How does reducing the slit separation, affect the appearance of the fringes in the double slit interference?

(i) The fringe width increases and the fringes are brighter

(ii) The fringe width increases but the brightness remains unchanged

(iii) The fringe width decreases and the fringes are brighter

(iv) The fringe width is unchanged but the fringes are less bright.

(f) Resolving power of microscope objective placed in air with light of wavelength λ_1 is the same when immersed in oil of refractive index μ and wavelength λ_2. Assuming the semi-vertical angle to be the same in these two cases then

(i) $\lambda_1 = \mu\lambda_2$
(ii) $\lambda_2 = \mu\lambda_2$
(iii) $\lambda_1\lambda_2 = \mu(\lambda_1 + \lambda_2)$
(iv) $\lambda_1 + \lambda_2 = \mu\lambda_1\lambda_2$.

(g) For a doubly refracting crystal, the refractive indices for the ordinary and extraordinary rays are denoted by μ_0 and μ_e. Which of the following relations is valid along the optical axis of the crystal?

(i) $\mu_e = \mu_0$
(ii) $\mu_w < \mu_0$
(iii) $\mu_e > \mu_0$
(iv) $\mu_0 \leq \mu_e$.

(h) Which of the following schemes does not produce lasting action?

(i) Two level scheme
(ii) Three level scheme
(iii) Four level scheme
(iv) Five level scheme.

(i) In holography, the 3D images are formed obeying the principle of

(i) interference
(ii) diffraction
(iii) polarisation
(iv) dispersion.

(j) Number of oscillation modes for the electromagnetic standing waves of frequency γ for the cavity radiation is proportional to

(i) $\gamma^{1/2}$
(ii) γ
(iii) γ^4
(iv) γ^2

where γ is the frequency of the wave.

(k) How fast must a particle travel so that its mass becomes twice its rest mass?

(i) $0.5c$
(ii) $2c$
(iii) $\sqrt{3}c/2$
(iv) c.

(l) Origin of continuous X-rays is due to the process of

(i) ionisation
(ii) inner orbital transition
(iii) bremsstrahlung
(iv) none of these.

(m) The effective number of atoms per unit cell of an FCC lattice is

(i) 4
(ii) 6
(iii) 14
(iv) 1.

(n) The inter-planar distance d_{111} for a (111) plane of simple cubic crystal is

(i) $a/\sqrt{3}$
(ii) $a/\sqrt{2}$
(iii) a
(iv) $2a$.

(o) A proton accelerated through a potential difference of V has a certain de Broglie wavelength. In order to have the same de Broglie wavelength, an α-particle must be accelerated through a potential difference of

(i) $8V$ (iii) $V/4$

(ii) $4V$ (iv) $V/8$.

Group B
(Short Answer Type Questions)

Answer any *three* of the following: $3 \times 5 = 15$

2. If a damped harmonic oscillator, with natural frequency ω_0 and damping force $2mbV$ (V = velocity, m = mass, $b = a$ constant) is driven by a periodic force $Fe^{i\omega t}$, then the steady state displacement is given by

$$x = \frac{F}{m} \cdot \frac{e^{i\omega t}}{\sqrt{(\omega_0^2 - \omega^2) + i2b\omega}}.$$

Use this fact to show that at the velocity resonance the phase of the periodic force and the velocity is the same.

3. A mica sheet of thickness t and refractive index μ is introduced in the path of one of the interfering beams in Young's double slit experiment. Find out the linear displacement of the nth bright fringes in terms of μ and t.

4. (a) What is polaroid? Give instances of its practical application.

(b) A quartz plate with thickness of 0.1436 mm is used as phase retardation plate. For what wavelengths in the visible region (450–800 nm) will it act as a quarter wave plate ($\mu_0 = 1.5443$, $\mu_e = 1.55333$).

5. Prove that the maximum recoil energy of a free electron of rest mass m_0, when struck by a photon of wavelength λ is given by $E_{\max} = \frac{2m_0c^2\lambda_c^2}{\lambda^2 + 2\lambda_c\lambda}$. Here λ_c is the Compton wavelength of the electron.

6. (a) The spacing between principal planes of NaCl crystal is 2.82Å. It is assumed that the first order Bragg reflection occurs at an angle 10°. Calculate the wavelength of X-rays.

(b) Find the atomic packing fraction of a BCC structure.

Group C
(Long Answer Type Questions)

Answer any *three* of the following: $3 \times 15 = 45$

7. (a) A cubical block of side L cm and density d is floating in water of density $\rho (\rho > d)$. The block is slightly depressed and released. Show that it will execute simple harmonic motion. Determine the frequency and time period of oscillation.

(b) Find the displacement as a function of time of a particle of mass m, which is subjected to over-damped harmonic motion with natural frequency ω_0 and damping force γV (V being the instantaneous velocity), given that the displacement is zero initially and the initial velocity is V_0. Sketch the displacement as a function of time.

8. (a) Newton's ring arrangement is used with a source emitting two wavelengths $\lambda_1 = 600$ nm and $\lambda_2 = 590$ nm. It was found that the nth dark ring due to λ_1 coincides with the $(n+1)$th dark

ring due to λ_2. If radius of curvature of the lens is 0.9 m, then find out the value of n and the diameter of the nth dark ring due to λ_2.

(b) A plane transmission grating having 1500 lines/inch is being used under normal incidence of light.

(i) What is the longest wavelength of light for which a spectrum can be seen?

(ii) What is the highest order spectrum that can be seen for the light of 589.3 nm wavelength?

(iii) The spectral line 589.3 nm in the second order spectrum overlaps with another spectral line in the next order. Find wavelength of the other spectral line.

(iv) Find the expression of the dispersive power for a given order.

(v) If 90 % of the width of the grating is covered, how the width of the spectral lines are changed?

9. (a) Define Einstein's A, B coefficient for spontaneous and stimulated emission. Derive the relations among them.

(b) At what angle should light be incident on a glass plate of refractive index $\mu = 1.5697$ to get a plane polarised light by reflection.

(c) An analysing nicole examines two adjacent plane polarised beams A and B whose planes of polarisation are mutually perpendicular. In one position of the analyser the beam B shows zero intensity. From this position, a rotation of 30° shows the two beams as of equal intensity. Deduce the ratio of two intensities I_A/I_B of the two beams.

(d) Why is it necessary that the object beam and reference beam in holography are highly coherent?

10. (a) Discuss the action of optical resonator in a Baser device. What is the relation of the length of the resonator with the frequency of the laser beam?

(b) A ruby laser emits light of 693.95 nm wavelength. If 1 mole of Cr^{3+} ions are involved in population inversion process in a pulse, calculate the pulse energy in eV.

(c) Calculate average energy of a cavity oscillator with frequency γ. How does this average energy vary with frequency in the high frequency limit?

(d) Derive Wien's displacement law from Planck's law of blackbody radiation.

11. (a) What voltage must be applied on an X-ray tube for it to emit X-ray with a minimum wavelength 30 pm. Deduce the formula that you use $[hc/e = 1.24 \times 10^{-4}\,\text{Vm}]$.

(b) What are characteristic X-rays? How are they produced?

(c) Describe the relation between the lattice constants and the angles between them in any one Bravais lattice other than cubic crystal system.

(d) Find the Miller indices of all the principal planes of a cubic crystal system.

(e) Aluminium is an *fcc* crystal with lattice constant $a = 0.405$ nm. Calculate the number of unit cells present in an aluminium foil of 0.005 cm thickness and two sides of 24 cm length.

12. (a) What is de Broglie hypothesis of matter wave? Discuss how Davisson and Germer proved this by an experiment.

(b) If a particle of charge e and rest mass m_0 is accelerated by a potential V, then what is the de Broglie wavelength of the particle viewing the relativistic energy momentum relation.

(c) The atoms in a solid posses a certain minimum zero point energy even at 0 K while no such restriction holds for the molecules in an ideal gas. Use uncertainty principle to explain this statement.

(d) An electron has de Broglie wavelength 2 pm. Find the phase and group velocities of its de Broglie wave given that the rest energy of an electron is 511 eV.

2012
Physics I — Sem I

Time Allotted: 3 Hours *Full Marks*: 70

The figures in the margin indicate full marks.

Candidates are required to give their answers in their own words as far as practicable.

Group A
(Multiple Choice Type Questions)

1. Choose the correct alternatives for any *ten* of the following: $10 \times 1 = 10$

(a) The relaxation time (τ) of a damped harmonic oscillator with damping constant (K) is

(i) $\tau = 1/K$ (iii) $\tau = K$

(ii) $\tau = 1/2K$ (iv) $\tau = 2K$.

(b) The resonant frequency of an electrical oscillator is given by

(i) $v = 2\pi\sqrt{LC}$ (iii) $v = 2\pi/\sqrt{LC}$

(ii) $v = 1/2\pi\sqrt{LC}$ (iv) $v = 2\pi\sqrt{L/C}$.

(c) If α is the force constant of an oscillating body of mass m the Q-factor is

(i) $Q \propto \sqrt{\alpha m}$ (iii) $Q \propto \sqrt{\alpha/m}$

(ii) $Q \propto \sqrt{m/\alpha}$ (iv) $Q \propto \alpha/m$.

(d) If a thin mica sheet is placed between two interference waves, then

(i) Fringe width increases (iii) Fringe pattern gets shifted

(ii) Fringe width decreases (iv) No change in the fringe pattern.

(e) The emissive power of a blackbody kept at absolute temperature T, which is very near to the temperature of surroundings (T_0) is proportional to

(i) $(T - T_0)^4$ (iii) T^4

(ii) $(T - T_0)$ (iv) $T^{2/3}$.

(f) The miller indices of a plane having intercepts 2, 3, ∞ units along X, Y, Z axis respectively are

(i) (6, 4, 3) (iii) (2, 3, 1)

(ii) (4, 3, 6) (iv) (2, 3, 4).

(g) Compton shift $\Delta\lambda$ and Compton wavelength λ_c an equal if the angle of scattering is

(i) $\theta = 0°$ (iii) $\theta = 180°$

(ii) $\theta = 90°$ (iv) $\theta = 360°$.

(h) The relativistic energy momentum relation is

(i) $p^2 = E^2 + m_0^2c^2$ (iii) $E^2 = p^2c^2 + m_0^2c^4$

(ii) $E^2 = p^2 + m_0^2c^4$ (iv) $p^2 = E^2c^2 + m_0^2c^4$.

(i) According to Wien's displacement law

(i) $\lambda_m T = \text{constant}$ (iii) $\lambda_m T^2 = \text{constant}$

(ii) $\lambda_m / T = \text{constant}$ (iv) $\lambda_m / T^2 = \text{constant}$.

(j) Volume of a unit cell in FCC structure is

(i) $16\sqrt{2}\, r^3$ (iii) $16\sqrt{2}\, / r^3$

(ii) $4\sqrt{2}\, r^3$ (iv) $2\sqrt{2}\, r^3$.

[r = mean radius of the constituent atoms]

(k) Polarisation conclusively proves that light waves are

(i) longitudinal (iii) stationary

(ii) progressive (iv) transverse.

(l) An α-particle is 4 times heavier than proton. If a proton and α-particle are moving with the same velocity, their de Broglie wavelengths are given by

(i) $\lambda_p = \lambda_a$ (iii) $\lambda_p - \lambda_{a/2}$

(ii) $\lambda_p = 4\lambda_a$ (iv) $\lambda_p = \lambda_{a/4}$.

(m) For larger value of damping constant k the resonance curve will be

(i) Unchanged (iii) Sharper

(ii) Flatter (iv) None of these.

(n) In ruby laser, the host crystal is

(i) Al_2O_3 (iii) $CaCO_3$

(ii) MNO_2 (iv) Al_2SO_4.

(o) If we measure the energy of a particle accurately, then the uncertainty of the measurement of time becomes

(i) 0 (iii) 1

(ii) ∞ (iv) 1/2.

Group B
(Short Answer Type Questions)

Answer any *three* of the following: $3 \times 5 = 15$

2. (a) Show that for a particle executing SHM, the average kinetic energy (KE) is half of the corresponding maximum energy.

(b) Calculate the time period of the liquid column of length 1 in a U-tube, if it is depressed in one arm by x, d is the density of liquid and A is the cross-sectional area of each arm of the tube.

3. (a) What do you mean by population inversion?

(b) Draw the energy level diagram in helium and neon laser transition.

4. (a) Determine the atomic packing fraction of FCC lattice.

(b) X-rays of wavelength 1.54Å are used for the calculation of the d_{100} plane of a cubic crystal, the Bragg's angle of 1st order reflection is 10°. What is the size of the unit cell?

5. What is Einstein's A and B coefficient? Relate spontaneous and stimulated emission probabilities and hence find out the relation between field energy and A, B coefficient.

6. Define blackbody? Establish Wien's distribution law and Stefan's law from Planck's blackbody radiation law.

Group C
(Long Answer Type Questions)

Answer any *three* of the following: $3 \times 15 = 45$

7. (a) A cubical block of side L cm and density ρ is floating in water of density $\rho_0\,(\rho_0 > \rho)$. The block is slightly depressed and released. Show that it will execute simple harmonic motion. Determine the frequency and time period of oscillation.

   (b) Write down the differences between standing and progressive wave and establish the differential equation of a progressive wave.

   (c) Establish the differential equation of SHM from energy conservation principle. An oscillator executing SHM has zero displacement at time $t = 0$. If the displacements are 1 mm and 1.5 mm at instants 0.1 and 0.2 seconds. Calculate the frequency and amplitude of oscillation.

   (d) A body of mass 10 g is acted upon by a restoring force/unit displacement of 10^7 dyne/cm, a frictional force/unit velocity of 4×10^3 dyne/cm, s^{-1} and a driving force $10^5 \cos \omega t$ dyne. Find the value of maximum amplitude.

8. (a) Two independent source of light of same wavelength cannot produce interference. Justify.

   (b) Plot the intensity distribution curve of Young's double slit interference experiment and label it.

   (c) Can you measure the refractive index of a liquid by Newton's ring experiment? Explain.

   (d) In Young's experiment the width of the fringe obtained with light of wavelength 6000Å is 2.0 mm. What will be the fringe width if the entire apparatus is immersed in a liquid of refractive index 1.33.

   (e) Newton's ring experiment is performed with reflected light of wavelength 5700Å using a plano-convex lens and a plane glass plate. What would be the observation when the glass plate is moved away from the lens along the axis of the lens by 10^{-5} m?

9. (a) Distinguish between Fresnel and Fraunhofer class of diffraction.

   (b) Using the expression of single slit diffraction intensity, show that the secondary maxima are given by the equation $\tan \alpha = d$, where $\alpha = \pi e \sin \theta / \lambda$, symbols have their usual meaning.

   (c) Draw and explain the intensity distribution curve in case of Fraunhofer single slit diffraction phenomenon.

   (d) State and explain Rayleigh criterion of resolution.

   (e) An oil immersion microscope just resolves the rulings of a grating having 3900 lines/mm when light of wavelength 400 nm is employed. Find the numerical aperture of the lens.

10. (a) Calculate the distance between the adjacent parallel planes of the type [100], [110] and [111] in an FCC lattice of lattice constant a. Check the validity of the statement: *The most closely packed planes are the most widely spaced.*

(b) Establish the relation between lattice constant and density of a material of a simple cubic crystal.

(c) If an X-ray tube is subjected to a potential difference of 50 kV and the corresponding current is 8 mA, find

(i) the number of electrons striking per second the target material

(ii) speed of electron

(iii) minimum wavelength of the X-ray produced

(iv) cut-off wavelength of the X-ray produced

(v) energy of each sticking electron.

11. (a) What is holography? Why is it called a wavefront reconstruction?

(b) Explain with neat diagram, spontaneous emission, stimulated absorption and stimulated emission of radiation and deduce of expressions relating various Einstein's coefficients.

(c) Write a short note on Nicole prism and its use as polariser and analyser.

12. (a) What is the origin of modified and unmodified lines in Compton effect?

(b) Can you observe Compton effect is visible light is used instead of X-rays?

(c) Prove that the maximum recoil energy of a free electron of rest mass m_0 when struck by a photon of wavelength λ is given by $E_{\max} = \dfrac{2m_0c^2\lambda^2c}{\lambda^2+2\lambda_c\lambda}$. Here λ_c is the Compton wavelength of the electron.

(d) Show that the Compton scattering while the photon can be scattered at any angle between 0° to 180°, the recoil electron can only be emitted at angles between 0° and 90°.

2012

Physics I — Sem II

Time Allotted: 3 Hours *Full Marks*: 70

The figures in the margin indicate full marks.

Candidates are required to give their answers in their own words as far as practicable.

Group A

(Multiple Choice Type Questions)

1. Choose the correct alternatives for any *ten* of the following: $10 \times 1 = 10$

(a) The de Broglie wavelength of a moving electron subject to a potential V is

(i) $1.26 / \sqrt{V}$ (iii) $122.6 / \sqrt{V}$

(ii) $12.26 / \sqrt{V}$ (iv) $1226 / \sqrt{V}$

(consider the appropriate units).

(b) For laser action to occur, the medium used must have at least

(i) 2 energy levels (iii) 3 energy levels

(ii) 4 energy levels (iv) one energy level.

(c) The Miller indices of a plane holding the intercept $a/2, b, 3c$ is

(i) (361) (iii) (631)

(ii) (611) (iv) (321).

(d) The intensity of principal maxima in the spectrum integrating with N number of lines is proportional to

(i) $1/N$ (iii) N^2

(ii) N (iv) $1/N^2$.

(e) To get a circular Lissajous figure, the phase difference (ϕ) and the amplitude (a and b) of two superimposing, mutually perpendicular SHMs and respectively

(i) $\phi = 0, a = b$ (iii) $\phi = \pi/2, a \neq b$

(ii) $\phi = \pi/2, a = b$ (iv) $\phi = \pi/4, a \neq b$.

(f) X-ray is suitable for crystallography because

(i) X-ray is electromagnetic

(ii) X-ray is neutral

(iii) X-ray wavelength is in same order of international spacing of crystal

(iv) X-ray is hard penetrating.

(g) In Fraunhofer diffraction, the wave front is

(i) Cylindrical (iii) Plane

(ii) Spherical (iv) Circular.

(h) The refractive indices of E-ray and O-ray are respectively 1.65 and 1.45. Then, the thickness of the material required for a quarter wave plate for light of wavelength 500 nm is

(i) 250 nm (iii) 625 nm

(ii) 125 nm (iv) 740 nm.

(i) If a spring of force constant K is cut into three equal parts, then the force constant of each part is

(i) $K/3$ (iii) 1

(ii) $3K$ (iv) K.

(j) Davisson-Germer experiment is related with

(i) Interference (iii) Phosphorescence

(ii) Them-ionic emission (iv) Electron diffraction.

(k) The velocity of a particle when its mass becomes twice its rest mass is

(i) $0.5c$ (iii) $0.866c$

(ii) $0.72c$ (iv) c

(where c is the speed of light in vacuum).

(l) Which of the following pairs cannot be simultaneously measured?

(i) p_x, p_y (iii) p_x, z

(ii) y, p_y (iv) p_x, p_z.

(m) The Compton shift $\Delta\lambda$ and Compton wavelength λ_c particle are equal if the angle of scattering is

(i) 0° (iii) 180°

(ii) 90° (iv) 45°.

(n) For holography system, the exposure time is of the order of

(i) 5 seconds and depends on the colour of the objects

(ii) nearly 2 seconds

(iii) 1 second

(iv) 50 seconds.

(o) The emissive power of a blackbody kept at an absolute temperature T is proportional to

(i) T^4 (iii) T^{-1}

(ii) T^3 (iv) T.

Group B
(Short Answer Type Questions)

Answer any *three* of the following: $3 \times 5 = 15$

2. The displacement of any particle at any instant t is given by $x = 3\cos\omega t + 4\sin\omega t$. Show that the motion is simple harmonic. What is the amplitude of this oscillation? Show that its kinetic energy (KE) oscillates with angular frequency 2ω.

3. (a) What is diffraction of light?
   (b) Draw the intensity distribution pattern of diffraction grating.
   (c) Two slits in Young's double slit experiment are 1.3 mm apart from each other and the screen is placed 1.6 m away from the slits. Calculate the distance of the 4th bright fringe. The wavelength of the light used is 6000Å.

4. (a) State Bragg's law.
   (b) Cs metal (atomic weight 130) has a cubic unit cell of lattice constant 0.6 nm. If the density of Cs is 2 gm/cc, determine whether the unit cell is SC, BCC or FCC.

5. (a) Explain Brewster's law.
   (b) Show that when light is incident on a transparent substance at the polarising angle, the reflected and the refracted rays are perpendicular to each other.

6. (a) Deduce Stefan's law from Planck's radiation law.
   (b) What is population inversion? What is it needed for lasing action?

Group C
(Long Answer Type Questions)

Answer any *three* of the following: $3 \times 15 = 45$

7. (a) State Heisenberg's uncertainty principle.
   (b) Using this principle, prove that a free electron cannot exist in an atomic nucleus.
   (c) Write down the uncertainty relation for energy and time.
   (d) Find the wavelength spread of a 1 nano-second pulse from a ruby LASER with a wavelength of 630 nm.
   (e) Show that the group velocity of the wave pack representing a particle is equal to the velocity of the particle itself.
   (f) Calculate the de Broglie wavelength of a baseball mass 1 kg moving at a speed of 10 m/sec. Discuss the reason why its wave nature cannot be observed.
8. (a) State Malus law. A beam of polarised light makes angle of 60° with the axis the polaroid sheet. He much intensity of light is transmitted through the sheet?
   (b) What is a retardation plate? How can you distinguish between circularly polarised light and unpolarised light with the help of a quarter wave plate and Nicole prism.
   (c) The critical angle of glass w.r.t. air is 4. Find the refractive index of the medium and the angle of refraction of the light incident on the glass plate the polarising angle.
9. (a) What are the differences between LASER and normal visible light?
   (b) Derive the relation between Einstein's A and B coefficients.
   (c) Calculate the ratio of stimulated to the spontaneous emission at a temperature 300 K for sodium D line ($\lambda = 5890$Å).
   (d) What the help of a neat diagram, briefly describe the construction of a Ruby LASER.
   (e) Why is optical pumping not generally employed in case of a gas LASER?
10. (a) Distinguish between interference and diffraction. Is energy conserved in interference phenomenon? Explain.
    (b) In a Newton's ring arrangement, the diameter of the 5th dark ring is 0.3 cm and the diameter of 25th dark ring is 0.8 cm. If the radius of curvature of the plano-convex lens is 100 cm, find the wavelength of the light used.
    (c) What are the conditions to get sustained interference pattern?
    (d) Show that the intensity of the 1st secondary maxima formed by a single slit Fraunhofer diffraction process is nearly 4.5 % of the principal maxima.
11. (a) Show that the fractional change in the natural frequency of a damped simple harmonic oscillator is $1/8Q^2$, where Q is the quality factor of the oscillator.
    (b) The displacement of a simple harmonic oscillator is given by $x = a\sin(\omega t + \theta)$. If the oscillations started at time $t = 0$ from a position x_0 with velocity $\dot{x} = v_0$, show that $\tan\theta \frac{\omega x_0}{v_0}$ and $a = \left(x_0^2 = \frac{v_0^2}{\omega^2}\right)^{1/2}$.

(c) A particle of mass 2 kg is subjected to an elastic loan per unit displacement $0.03\,\mathrm{Nm^{-1}}$ and frictional loan per unit velocity $0.005\,\mathrm{Nm^{-1}\,s}$. If it is displacement through 2 cm and then released, find whether the resulting motion is oscillatory or not.

(d) The forced harmonic oscillators have displacement amplitude at frequencies $\omega_1 = 400\,\mathrm{sec^{-1}}$ and $\omega_2 = 600\,\mathrm{sec^{-1}}$.
Find the resonant frequency at which the displacement amplitude is maximum.

(e) Differentiate between amplitude resonance and velocity resonance for a forced harmonic oscillator.

12. (a) In the continuous X-ray spectra, λ_{min} values are obtained for higher potential. Explain.

(b) What do you mean by the terms *atoms per unit cell, coordination number* and *atomic packing factor*?

(c) Show that the packing fraction of FCC is greater than BCC.

(d) The density of BCC iron is 7.9 gm/cc and its atomic weight is 56. Calculate the length of the side of the cubic until cell and nearest neighbouring distance.

(e) The spacing of planes in a crystal is 1.2Å and the angle for the 1st order Bragg's reflection is 30°. Determine the energy of the X-rays beam in eV.

Contents

Preface to the Second Edition xvii
Preface to the First Edition xviii
New Syllabus xix-xx

1. Vibration and Wave—Simple Harmonic Motion | Module 1.1 1-33
1.1 Introduction 1
1.2 Factors Causing a System to Oscillate 2
1.3 Projection of a Uniform Circular Motion on Reference Circle 2
1.3.1 Some definitions 2
1.4 Differential Equation of SHM 5
1.5 Energy of a Linear Harmonic Oscillator 6
1.6 Illustrative Examples of SHM 8
1.6.1 Motion of a simple pendulum 8
1.6.2 Vertical oscillation of an elastic spring supporting a weight at its lower end 9
1.6.3 Oscillation of a liquid in a U-tube 10
1.6.4 Simple harmonic motion of electrical system 10
1.7 Principle of Superposition 11
1.7.1 Superposition of two collinear harmonic vibrations 12
1.7.2 Applications of beats 14
1.8 Superposition of Two Harmonic Vibrations at Right Angles to Each Other 15
1.8.1 Demonstrations of Lissajous figures 18
1.8.2 Uses of Lissajous figures 19
1.9 Answers of Some Important Questions 19
1.10 Solved Problems 22
EXERCISES
1(A)–Objective Type Questions 29
1(B)–Short Answer Type Questions 31
1(C)–Numerical Problems 31
1(D)–Broad Answer Type Questions 33

2. Vibration and Wave—Free and Damped Vibrations | Module 1.2 **34–54**

2.1 Free Vibration 34
2.2 Damped Vibration 34
2.3 Damping Forces 35
2.4 Analysis of Damped Vibration 35
2.4.1 Logarithmic decrement 41
2.4.2 Energy of a weakly damped oscillator 42
2.4.3 Quality factor (or Q-value) 43
2.4.4 Time constant 44
2.4.5 Damping in LCR circuit 44
2.5 Answers of Some Important Questions **45**
2.6 Solved Problems **48**
EXERCISES
2(A)–Objective Type Questions **52**
2(B)–Short Answer Type Questions **53**
2(C)–Numerical Problems **53**
2(D)–Broad Answer Type Questions **54**

3. Forced Vibration | Module 1.3 **55-76**

3.1 Introduction 55
3.2 Analytical Treatment of Forced Vibration 55
3.3 Early Stage of Forced Vibration 56
3.4 Steady State of Forced Vibration 56
3.5 Mechanical Impedance 58
3.6 Applied Force—Frequency Dependence of the Forced Oscillator : Resonance 59
3.7 Amplitude Resonance 60
3.8 Variation of phase of the forced oscillator 61
3.9 Velocity Resonance 61
3.10 Phase Variation of V_0 with ω 63
3.11 Power Supply and Loss of Energy 63
3.11.1 Power supply by the forcing system 63
3.12 Rate of Work Done by the Dissipative Force 64
3.13 Sharpness of Resonance 64
3.14 Expression for Sharpness of Resonance 66
3.15 Q in terms of Dissipation (Decay) of Energy 66
3.16 Analogy with Electric Circuit 67
3.17 Answers of Some Important Questions **69**
3.18 Solved Problems **71**
EXERCISES
3(A)–Objective Type Questions **74**
3(B)–Short Answer Type Questions **75**
3(C)–Numerical Problems **75**
3(D)–Broad Answer Type Questions **76**

4. Interference of Electromagnetic Waves | Module 2.1 77–121

4.1 Nature of Light 77

4.2 Corpuscular Theory of Light 77

4.3 Huygens' Wave Theory 78

4.4 Wavefront 78

4.5 Huygens' Principle of Wave Propagation 79

4.6 Interference of Light (Electromagnetic Waves) 80

4.7 Young's Double Slit Experiment 80

4.8 Analytical Treatment of Interference 81

4.8.1 The Condition for Permanent (or Sustained) Interference 82

4.9 Two Real Sources of Light Cannot Produce Interference Fringes 83

4.10 The Intensity Distribution 83

4.11 Conservation of Energy 85

4.12 Coherence 85

4.13 The Fringe Width 86

4.14 Production of Coherent Sources 88

4.15 Fresnel's Bi-prism Experiment 88

4.16 Displacement of Fringes due to Introduction of a Thin Transparent Plate (Determination of the Thickness of a Thin Transparent Plate) 90

4.17 Phase Change on Reflection: Stokes' Law 92

4.18 Interference by Division of Amplitude: Colour of Thin Films 93

4.19 Classification of Fringes Exhibited by Thin Films 95

4.20 Explanation of Colour Effect 96

4.21 Non-Reflecting Films 96

4.22 Interference Due to Wedge-Shaped Film 97

4.23 Necessity of an Extended Source 99

4.24 Newton's Rings 99

4.25 Determination of the Wavelength of Monochromatic Light 102

4.26 Determination of Refractive Index of a Liquid 103

4.27 Newton's Rings by Transmitted Light 103

4.28 Newton's Ring with White Light 104

4.29 Applications of Newton's Rings 105

4.30 Answers of Some Important Questions 105

4.31 Solved Problems 107

EXERCISES

4(A)–Objective Type Questions 115

4(B)–Short Answer Type Questions 117

4(C)–Numerical Problems 119

4(D)–Broad Answer Type Questions 120

5. Diffraction of Light | Module 2.2 122–157

5.1 Introduction 122

5.2 Distinction between Interference and Diffraction 122
5.3 Classification of Diffraction 123
5.4 Fraunhofer Diffraction at a Single-Slit 124
5.5 Fraunhofer Diffraction Due to a Double-Slit 128
5.6 Fraunhofer Diffraction Due to N-slits (Theory of a Diffraction Grating) 132
5.6.1 Spectrum Produced by a Grating 137
5.6.2 Angular Dispersive Power of a Grating 138
5.6.3 Determination of Wavelength of Monochromatic Light by Grating 139
5.6.4 Rayleigh Criterion of Resolution 139
5.6.5 Resolving Power of a Grating 140
5.6.6 Resolving Power of a Microscope 141
5.7 **Answers of Some Important Questions** **142**
5.8 **Solved Problems** **145**
EXERCISES
5(A)–Objective Type Questions **152**
5(B)–Short Answer Type Questions **154**
5(C)–Numerical Problems **155**
5(D)–Broad Answer Type Questions **156**

6. Polarization of Light | Module 3.1 **158–191**
6.1 Introduction 158
6.2 General Concept of Polarization 158
6.3 Polarization of Light Waves 159
6.4 Plane of Vibration and Plane of Polarization 160
6.5 Pictorial Representation of Light Vibration: Unpolarized Light 160
6.6 Polarization by Reflection: Brewster's Law 162
6.7 Double Refraction (or Birefringence): Ordinary Ray (O-ray) and Extraordinary Ray (E-ray) 164
6.8 Geometry of Calcite: Optic Axis and Principal Section 166
6.9 Nicol Prism 167
6.10 Malus Law 168
6.11 Polarization by Scattering 169
6.12 Polarization by Selective Absorption 169
6.13 Huygens' Theory of Double Refraction 170
6.14 Theory of Production of Different Types of Polarized Light 171
6.15 Left-handed and Right-handed Polarized Light 175
6.16 Retardation Plates 176
6.17 Production of Plane, Circularly and Elliptically Polarized Light 177
6.17.1 Plane Polarized Light 177
6.17.2 Circularly Polarized Light 177
6.17.3 Elliptically Polarized Light 178
6.18 Analysis of Polarized Light 179
6.19 **Answers of Some Important Questions** **181**

	6.20	**Solved Problems**	**183**
		EXERCISES	
		6(A)–Objective Type Questions	**187**
		6(B)–Short Answer Type Questions	**188**
		6(C)–Numerical Problems	**189**
		6(D)–Broad Answer Type Questions	**190**
7.	**Laser**	**Module 3.2**	**192–214**
	7.1	Introduction	192
	7.2	Properties of Laser Light	192
	7.3	The Basic Principle of Laser	193
	7.4	Spontaneous Emission Versus Stimulated Emission of Radiation	194
	7.5	Comparison between Laser and Ordinary Light	194
	7.6	Relation between Spontaneous and Stimulated Emission Probabilities —Einstein's Theory	195
	7.7	Important Discussion	196
	7.8	Theory of Amplification of Light	197
		7.8.1 Gain Coefficient (β)	197
	7.9	Metastable State	198
	7.10	Active Medium	198
	7.11	Pumping	199
	7.12	Population Inversion	199
	7.13	Threshold Condition for Light Amplification	201
	7.14	Main Components of the Laser	201
	7.15	Cavity Resonance	202
		7.15.1 Types of Laser Systems	202
	7.16	Pulsed Ruby Laser	203
		7.16.1 Working	204
		7.16.2 Applications of Ruby Laser	204
	7.17	Continuous He-Ne Visible Laser	205
		7.17.1 Working	205
		7.17.2 Applications of He-Ne Laser	206
	7.18	Applications of Laser	207
	7.19	**Answers of Some Important Questions**	**207**
	7.20	**Solved Problems**	**209**
		EXERCISES	
		7(A)–Objective Type Questions	**211**
		7(B)–Short Answer Type Questions	**212**
		7(C)–Numerical Problems	**213**
		7(D)–Broad Answer Type Questions	**213**
8.	**Holography**	**Module 3.3**	**215–222**
	8.1	Introduction	215

8.2	Principle of Holography	215
8.3	Characteristics of Holograms	217
8.4	Application of Holography	219
8.5	Distinction Between Holography and Photography	220
8.6	**Answers of Some Important Questions**	**221**
	EXERCISES	
	8(A)–Objective Type Questions	**221**
	8(B)–Short Answer Type Questions	**222**
	8(C)–Broad Answer Type Questions	**222**
9.	**Special Theory of Relativity | Module 4.1**	**223–260**
9.1	Introduction	223
9.2	Frame of Reference	223
9.3	Galilean Transformation	224
	9.3.1 Galilean Coordinate Transformation	224
	9.3.2 Galilean Velocity Transformation	225
	9.3.3 Galilean Acceleration Transformations	225
	9.3.4 Galilean Transformation of Force	226
9.4	Michelson-Morley Experiment	227
9.5	Postulates of Special Theory of Relativity	230
9.6	Insufficiency of Galilean Transformation	230
9.7	Lorentz Transformation of Space and Time	231
9.8	Consequences of Lorentz Transformation Equations	233
	9.8.1 Length Contraction (or Lorentz-Fitzgerald Contraction)	233
	9.8.2 Two Events are said to be Simultaneous if They Occur at the Same Time in a Frame of Reference	235
	9.8.3 Einstein's Time Dilation (or Apparent Retardation of Clocks)	235
	9.8.4 Experimental Verification of Time Dilation	237
9.9	Einstein's Theorem of Velocity Addition	237
9.10	Relativistic Variation of Mass	238
	9.10.1 Derivation of Mass Variation Relation	239
9.11	Einstein's Mass-Energy Relation	241
9.12	Energy and Momentum of Relativistic Particle	242
9.13	**Answers of Some Important Questions**	**242**
9.14	**Solved Problems**	**245**
	EXERCISES	
	9(A)–Objective Type Questions	**254**
	9(B)–Short Answer Type Questions	**255**
	9(C)–Numerical Problems	**257**
	9(D)–Broad Answer Type Questions	**258**
10.	**Quantum Physics–I | Module 4.2**	**261–291**
10.1	Introduction	261

10.2	Black Body Radiation	261
10.3	Distribution of Energy in the Spectrum of Black Body Radiation	262
	10.3.1 Wien's Law of Energy Distribution	262
	10.3.2 Rayleigh-Jeans Law	263
	10.3.3 Failure of Classical Theory to explain the Spectral Distribution of Energy	264
	10.3.4 Planck's Radiation Law	264
10.4	Photoelectric Effect	267
	10.4.1 Experimental Study of Photoelectric Effect	267
	10.4.2 Experimental Observations	268
	10.4.3 Characteristics (Laws) of Photoelectric Effect	269
	10.4.4 Failure of Classical Theory to explain Photoelectric Effect	269
	10.4.5 Einstein's Photoelectric Effect	270
10.5	The Compton Effect	270
	10.5.1 The Direction of the Recoil Electron	273
	10.5.2 Kinetic Energy of the Recoil Electron	274
10.6	Failure of Classical Theory to explain Compton Effect	275
10.7	Explanation of Compton Effect on the Basis of Quantum Theory	275
10.8	Dual Nature of Light (Wave-Particle Dualism)	276
10.9	Bohr's Principle of Complementarity	276
10.10	**Answers of Some Important Questions**	**276**
10.11	**Solved Problems**	**279**
	EXERCISES	
	10(A)–Objective Type Questions	**286**
	10(B)–Short Answer Type Questions	**288**
	10(C)–Numerical Problems	**289**
	10(D)–Broad Answer Type Questions	**290**
11.	**Quantum Physics–II | Module 4.3**	**292–318**
11.1	Introduction	292
11.2	Phase Velocity and Group Velocity	292
11.3	de Broglie Hypothesis	295
	11.3.1 de Broglie Wavelength of a Charged Particle Accelerated by an Electric Field V volt/m	295
	11.3.2 Distinction between Matter Waves and Electromagnetic Waves	296
	11.3.3 Bohr's Quantization Condition in the Light of de Broglie's Hypothesis	296
	11.3.4 Davisson and Germer Experiment	297
11.4	Heisenberg's Uncertainty Relation	298
	11.4.1 Statement of Uncertainty Principle	299
	11.4.2 Simple Proof of Heisenberg's Uncertainty Relation	299
	11.4.3 Simple Proof of Uncertainty Relation between Energy and Time	300
11.5	Illustration of Heisenberg's Uncertainty Principle by Thought Experiments	300

11.5.1	Diffraction of Electrons through Single-Slit	300
11.5.2	Gamma Ray Microscope (Thought) Experiment	301
11.6	Applications of Uncertainty Relation	302
11.7	Heisenberg's Uncertainty Principle from de Broglie Wave Concept	305
11.8	**Answers of Some Important Questions**	**306**
11.9	**Solved Problems**	**308**
	EXERCISES	
	11(A)–Objective Type Questions	**313**
	11(B)–Short Answer Type Questions	**315**
	11(C)–Numerical Problems	**317**
	11(D)–Broad Answer Type Questions	**318**
12.	**Crystallography | Module 5.1**	**319–349**
12.1	Introduction	319
12.1.1	Properties of Crystalline Solids	320
12.1.2	Properties of Amorphous Solids	320
12.2	Space Lattice (or Point Lattice)	320
12.3	Crystal Lattice	320
12.4	Basis	320
12.5	Some Crystallographic Terms	321
12.5.1	Unit Cell	321
12.5.2	Wigner-Seitz Primitive Cell	321
12.5.3	Basis Vectors	322
12.5.4	Primitive Unit Cell	322
12.5.5	Primitive Basis Vectors	322
12.6	Bravais Lattice	322
12.7	Space Lattices of Cubic Systems	323
12.7.1	Calculation of Number of Atoms per Unit Cell	324
12.7.2	Coordination Number	324
12.7.3	Nearest Neighbour Distance	324
12.7.4	Atomic Radius (r)	324
12.7.5	Atomic Packing Factor	325
12.8	Calculation of Lattice Constant	326
12.9	Lattice Planes and Miller Indices	327
12.10	Crystallographic Directions	327
12.10.1	Rules to find the Miller Indices of a Lattice Plane	328
12.10.2	Important Features of Miller Indices	328
12.11	Inter-planar Spacing	328
12.11.1	Angle between Two Planes having Miller Indices ($h_1k_1l_1$) and ($h_2k_2l_2$)	329
12.11.2	Inter-planar Spacing between Lattice Planes of Different Cubic Lattice	330
12.11.3	Comparison of SC, BCC, FCC Structure	331
12.12	**Answers of Some Important Questions**	**331**

12.13	**Solved Problems**	**333**
	EXERCISES	
	12(A)–Objective Type Questions	**344**
	12(B)–Short Answer Type Questions	**346**
	12(C)–Numerical Problems	**348**
	12(D)–Broad Answer Type Questions	**348**
13.	**X-rays | Module 5.2**	**350–374**
13.1	Introduction	350
13.2	Production of X-rays	350
13.3	Production of X-rays by Coolidge Tube	350
13.4	Quality of X-rays	351
13.5	Intensity of X-rays	351
13.6	Interaction of X-rays with Matter	351
13.7	Origin of X-rays	352
13.8	X-rays Spectrum	352
	13.8.1 Continuous X-rays	352
	13.8.2 Quantum Theory and Duane-Hunt Law	353
	13.8.3 Origin of Continuous X-rays	354
13.9	Characteristic X-rays	354
	13.9.1 Origin of Characteristic X-rays	354
13.10	Applications of X-rays	355
13.11	Absorption of X-rays	356
13.12	Variation of Intensity with Thickness	356
13.13	Properties of X-rays	357
13.14	Differences between Visible Light and X-rays	358
13.15	Comparison between Cathode-rays and X-rays	358
13.16	X-ray Diffraction: Laue Method	359
13.17	The Laue Photographs	359
13.18	Bragg's Laws	359
13.19	Determination of Lattice Constant	360
13.20	**Answers of Some Important Questions**	**362**
13.21	**Solved Problems**	**364**
	EXERCISES	
	13(A)–Objective Type Questions	**369**
	13(B)–Short Answer Type Questions	**371**
	13(C)–Numerical Problems	**372**
	13(D)–Broad Answer Type Questions	**373**
14.	**Fibre Optics | Module 6**	**375–389**
14.1	Introduction	375
14.2	What is an Optical Fibre?	375
14.3	Structure of an Optical Fibre	375

14.4	Working Principle of Optical Fibres	376
14.5	Classification of Optical Fibres	376
	14.5.1 Comparison of Step-Index and Graded-Index Optical Fibres	378
14.6	Acceptance Angle and Acceptance Cone	379
	14.6.1 Numerical Aperture	380
14.7	Attenuation or Transmission Losses	380
14.8	Advantages of the Fibre Optic Communication	381
14.9	Applications of Fibre Optics	382
14.10	**Answers of Some Important Questions**	**382**
14.11	**Solved Problems**	**384**
	EXERCISES	
	14(A)–Objective Type Questions	**386**
	14(B)–Short Answer Type Questions	**387**
	14(C)–Numerical Problems	**388**
	14(D)–Broad Answer Type Questions	**388**

Appendices	**391–402**
Appendix A: Some Fundamental Constants	391
Appendix B: Miscellaneous Conversion Factors	392
Appendix C: SI Prefixes	393
Appendix D: Some Derived Units	394
Appendix E: Some Useful Mathematical Formulae	395
Appendix F: Some Useful Trigonometric Formulae	396–397
Appendix G: (a) Some Indefinite Integrals, (b) Some Definite Integrals	398–399
Appendix H: Some Important Hyperbolic Functions	400
Appendix I: Solutions of Some Differential Equations	401
Appendix J: Gamma Functions	402
Bibliography	**403–404**
Index	**405–406**
Engineering and Management Examinations	**1–**

Preface to the Second Edition

The second edition of the book, *Principles of Engineering Physics,* Volume I has been designed strictly in accordance with the New Syllabus (2010) of Engineering Physics–I meant for the First Year students pursuing BTech course under West Bengal University of Technology.

The salient features of the present edition of the book include:

- Concise and lucid coherent text
- A new chapter on *Holography*
- Neat and illustrative diagrams
- Answers to some important questions
- Diverse numerical worked-out problems
- The EXERCISES are split up into four sections: (A) *Objective Type Questions,* (B) *Short Answer Type Questions,* (C) *Numerical Problems,* and (D) *Broad Answer Type Questions*—to test the depth of understanding of the subject.
- An exhaustive bibliography for further reading.
- I take this opportunity to express my thanks and gratitude to my numerous students, colleagues and friends working in different parts of West Bengal and outside for their constant encouragement and constructive suggestions for the improvement of the book. Special thanks are due to Mr U K Ghosh, Assoc. Professor of Engg. Physics, BIET, Suri, Birbhum.

I would also like to thank Mr Amitabha Sen, Mr R M Nath, Mr Soumen Paul, Mr Dipan Ray and Mr Prabhat Jas of New Central Book Agency (P) Ltd for bringing out the book in its present form.

23 May 2011 S P Kuila
Department of Applied Physical Sciences
Bengal Institute of Technology and Management
Santiniketan, WB

Preface to the First Edition

Technology is the applied aspect of pure science which provides concepts, theories, and formulae. All technological progresses depend on scientific understanding of the working of nature; pure science and technology, therefore, are closely interlinked. It is transparent that a learner must have appropriate knowledge of advanced level topics of pure science for facing challenges in future as a technologist.

Though many books have been published on the subject, a major part of them are either extremely insufficient or highly exaggerated, thereby unsuitable for the understanding of the students. Keeping this point in mind, a long experience spanning over more than three decades has been exploited to present a suitable textbook on the subject for the students—specially of WBUT. The book has been written primarily on the basis of class lecture notes in the most coherent and exhaustive manner so that the students can grasp the subject with minimum labour and time. The critical points have been discussed with utmost lucidity and clarity in a self-developed technique of expression.

Apart from appropriate illustrations and examples, the book is enriched with a large number of solved problems and a section entitled "*Answers of Some Important Questions*" at the end of each chapter to provide a comprehensive and thorough concept of the subject. Moreover, the EXERCISE of each chapter comprises of four different types of questions: (a) Objective Type, (b) Short Answer Type, (c) Numerical Problems, and (d) Broad Answer Type—to test the depth of understanding of the subject. I shall deem my labour amply rewarded if the book meets all the requirements of the students for whom it is intended.

I take this opportunity to express my sincere thanks to numerous students and colleagues working in different colleges and institutions for their encouragement and constructive suggestions during the preparation of the manuscript. I would like to thank Mr Amitabha Sen, Mr R M Nath, Mr Soumen Paul, and Mr Prabhat Jas of New Central Book Agency (P) Ltd, Kolkata, for bringing out the book in due time. Thanks are also due to my wife Mrs Jyotsna Kuila and son Mr Sandip Kuila for their help and encouragement during the preparation of the manuscript.

29 November 2008

S P Kuila

Department of Applied Physical Sciences
Bengal Institute of Technology and Management
Santiniketan, WB

NEW SYLLABUS

WBUT 2010

Physics/Paper I

Module 1

Oscillation

1.1 Simple harmonic motion: Preliminary concepts, Superposition of S.H.Ms in two mutually perpendicular directions: Lissajous figure. **2L**

1.2 Damped vibration: Differential equation and its solution, Logarithmic decrement, Quality factor. **3L**

1.3 Forced vibration: Differential equation and its solution, Amplitude and Velocity resonance, Sharpness of resonance. Application in LCR Circuit. **3L**

Module 2

Optics 1

2.1 Interference of electromagnetic waves: Conditions for sustained interference, double slit as an example, Qualitative idea of Spatial and Temporal Coherence, Conservation of energy and intensity distribution, Newton's ring. **3L**

2.2 Diffraction of light: Fresnel and Fraunhofer class, Fraunhofer diffraction for single slit and double slits. Intensity distribution of N-slits and plane transmission grating (No deduction of the intensity distributions for N-slits is necessary), Missing orders. Rayleigh criterion, Resolving power of grating and microscope. (Definition and formulae). **5L**

Module 3

Optics 2

3.1 Polarization: General concept of Polarization, Plane of vibration and plane of polarization, Qualitative discussion on Plane, Circularly and Elliptically polarized light, Polarization through reflection and Brewster's law, Double refraction (birefringence) Ordinary and Extraordinary rays. Nicol's Prism, Polaroid. Half wave plate and Quarter wave plate. **4L**

3.2 Laser: Spontaneous and Stimulated emission of radiation, Population inversion, Einstein's A and B coefficient (derivation of the mutual relation), Optical resonator and Condition necessary for active Laser action, Ruby Laser, He-Ne-Laser-applications of laser. **4L**

3.3 Holography: Theory of holography, viewing the hologram, Applications. **3L**

Module 4

Quantum Physics

4.1 Concept of dependence of mass with velocity, mass energy equivalence, energy-momentum relation (no deduction required). Blackbody radiation: Rayleigh-Jeans' law (derivation without the calculation of number of states), Ultraviolet catastrophe, Wien's law, Planck's radiation law (Calculation of the average energy of the oscillator), Derivation of Wien's displacement law and Stephan's law from Planck's radiation law. Rayleigh-Jean's law and Wien's law as limiting cases of Planck's law. Compton Effect (calculation of Compton wavelength is required). **5L**

4.2 Wave-particle duality and de Broglie's hypothesis, Concept of matter waves, Davisson-Germer experiment, Concept of wave packets and Heisenberg's uncertainty principle. **4L**

Module 5

Crystallography

5.1 Elementary ideas of crystal structure: lattice, basis, unit cell, Fundamental types of lattices—Bravais lattice, Simple cubic, f.c.c. and b.c.c. lattices (use of models in the class during teaching is desirable), Miller indices and miller planes, Co-ordination number and Atomic packing factor. **4L**

5.2 X-rays: Origin of Characteristic and Continuous X-ray, Bragg's law (No derivation), Determination of lattice constant. **2L**

"All progress in technology depends on scientific understanding of the way in which nature works, in order that we may make her work for our ends. Pure science and technology are not independent fields of study. Agriculture cannot prosper without the study of genetics, light metals cannot be made without research in chemistry and automation depends on electronic devices that need a detailed understanding of physics of matter and of the logic of control systems."

J. Bronowski, G. Barry, J. Fisher, and J. Huxley

Chapter 1 | Module 1.1

Vibration and Wave—Simple Harmonic Motion

1.1 Introduction

In our everyday life we generally come across two types of motion in physical systems: *translation* and *vibration.* Translational motion is that in which a body moves such that the position of the body varies linearly with time, e.g., the motion of a train on a straight track. On the other hand, if the motion of a body repeats itself in equal interval of time on the same path, then it is *vibration* or *oscillation.* The motion of a pendulum bob, the motion of the prongs of a tuning fork, the motion of the arms of a walking person, etc. are the examples of vibratory (or oscillatory) motion.

Vibratory motion is of the following two types:

(1) Periodic motion

Any motion that repeats itself at a regular interval of time on the same path is known as periodic motion. The motion of the earth round the sun, the motions of the hour-hand and the minute-hand of a clock are periodic motion.

(2) Simple harmonic motion (SHM)

Simple harmonic motion is a periodic motion in which the acceleration of the particle is directly proportional to its displacement from the mean position of rest and is always directed towards it.

Simple harmonic motion, depending on the nature of the applied force, are of two types:

(a) *Linear simple harmonic motion*: If a particle moves in a straight line under the constant constraint force, then the resulting simple harmonic motion is a *linear simple harmonic motion.*

(b) *Angular simple harmonic motion*: If a particle executes rotational motion about an axis under a constant constraint torque, then the resulting SHM is called *angular simple harmonic motion.*

Characteristics of SHM

(i) The motion is periodic and oscillatory.

(ii) The force and hence the acceleration of the particle executing SHM at any instant is proportional to its displacement measured from the mean position of the path.

(iii) The force and hence acceleration of the particle executing SHM at any instant is directed towards the mean position of the path.

Note that all SHM are periodic but all periodic motions are not SHM.

1.2 Factors Causing a System to Oscillate

A physical system oscillates due to two basic properties:

(i) *Elasticity of the system and*

(ii) *Inertia of the system.*

The elastic property of the system determines the magnitude of the restoring force, which brings back the displaced system to its mean position of rest and inertia, on the other hand, tries to oppose any change in its velocity. The restoring force tries to bring back, i.e., *restore* the body by imparting a velocity, and inertia preserves the velocity. As a result the displacement of the body overshoots, repeats itself and oscillation continues.

1.3 Projection of a Uniform Circular Motion on Reference Circle

We consider a particle moving along the circumference of a circle of radius A with a constant velocity v in the anti-clockwise direction [Fig. 1.1]. A perpendicular is drawn from the position of the particle P on XX' and let M be the foot of the perpendicular at any instant t. This circle is known as the *circle of reference* and the particle P is the *reference particle.* N is the foot of the perpendicular from P on YY'.

From [Fig. 1.1], it is evident that when the particle P is at X, the point N is at O. When P is at Y, the projection N is at Y, i.e., when P moves from X to Y, N moves from O to Y. When P is at X', N is at O and when P reaches to Y', N comes to Y'. When P comes to X, the projection N finally comes back to O. Thus, as the particle P moves along the reference circle the projection executes motion along the diameter YY' through O. Similarly, the projection M on XX' executes motion along XX' as the particle P moves along the reference circle.

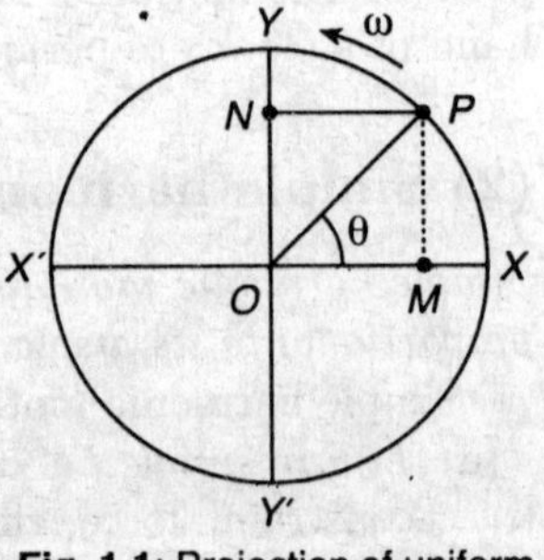

Fig. 1.1: Projection of uniform circular motion.

If instead of the projections M or N, a particle moves along XX' or YY' under the action of some force, then the particle is said to execute a linear SHM.

1.3.1 Some Definitions

(1) Displacement. The distance described by the particle along the path from its mean position at a given instant of time is called *displacement.*

The displacement of the particle at M in [Fig. 1.1] is $OM = x$ given by

$$OM = OP\cos\theta \text{ or, } x = A\cos\theta. \tag{1.1}$$

Since ω is the angular velocity of the particle along the reference circle and θ is the angle described by OP in time t, we have $\theta = \omega t$.

$$\therefore\ x = A\cos\omega t. \tag{1.2}$$

The *time-displacement curve* of the particle is simply a graph obtained by plotting the displacement (x) of the particle (M) against time (t) and is shown in [Fig. 1.2]. The position of the particle M at any instant of time can be easily found from the time-displacement curve of the particle. Table 1.1 shows the angular and linear displacement of the particle at different values of time t.

Table 1.1

Time	0	$T/4$	$T/2$	$3T/4$	T
Angle	0	90°	180°	270°	360°
Displacement	0	A	0	$-A$	0

Time is measured in terms of the total time taken by the particle for one complete vibration (T), T is also called the *time-period* of vibration. For drawing the time-displacement curve of the particle, T is divided into equal eight parts each being equal to $T/8$ as indicated in [Fig. 1.2].

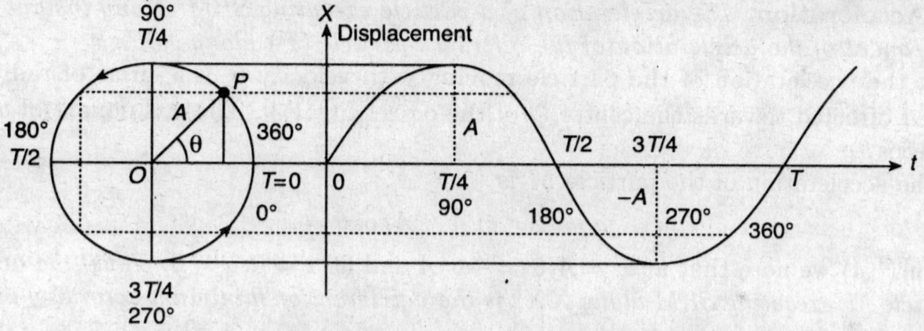

Fig. 1.2: Time-displacement graph of a particle executing SHM.

(2) Amplitude. *The maximum displacement described by the particle on either side of the mean position is called the amplitude of vibration.* It is denoted by A and is equal to the radius of the *circle of reference.*

(3) Velocity. *The velocity of the particle M is equal to the component of velocity of P along XOX', which is equal to $v\sin\theta$, where θ is the angle as shown in* Fig. 1.3 *and v is the velocity of P at any instant t.*

So the velocity of M is

$$V = v\sin\theta = v\sin\omega t$$

$$= \omega A\sqrt{1-\cos^2\omega t} \qquad (\because\ v = \omega A)$$

$$= \omega A\sqrt{1-\left(\frac{OM}{OP}\right)^2}$$

$$= \omega A\sqrt{1-\frac{x^2}{A^2}}$$

$$= \omega\sqrt{A^2-x^2}$$

$$\text{or, } V = \omega\sqrt{A^2-x^2}. \tag{1.3}$$

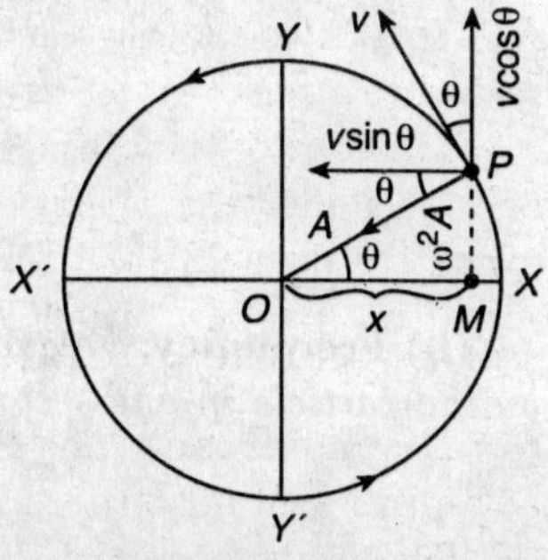

Fig. 1.3: Dependence of velocity on displacement *x*.

Since ω and A are constant quantities, the velocity V depends only on the displacement x. When

$$x = 0, \quad V = \omega A = V_{\max}, \text{ the maximum velocity}$$
$$x = A, \quad V = 0 = V_{\min}, \text{ the minimum velocity.}$$

Thus, the velocity of the particle is the maximum (ωA) at the mean position O and minimum (0) at the extreme positions X or X'. The *time-velocity curve* of the particle is shown in Fig. 1.4. From Fig. 1.4, it is evident that the particle assumes the maximum velocity $V_{\max} = \omega A$, when it passes through the mean position O and minimum velocity $V_{\min} = 0$ at the extremities.

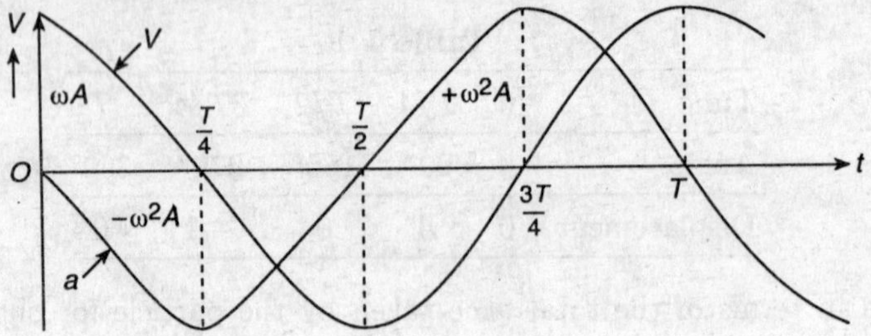

Fig. 1.4: Variation of V and a with t of a particle executing SHM.

(4) Acceleration. *The acceleration of a particle executing SHM at any instant t is equal to the component of the acceleration of the reference particle (P) along XX'.*

Now, the acceleration of the particle moving with velocity v in a circle of radius A is $f = \frac{v^2}{A} = \omega^2 A$ directed towards the centre O of the circle [Fig. 1.3]. The component of a along XX' is $-\omega^2 A \cos\theta$.

Or the acceleration of the particle M is

$$a = -\omega^2 x \quad [\because \; A\cos\theta = x]. \tag{1.4}$$

From eqn (1.4) we note that at $x = A, a = -\omega^2 A$ and at $x = 0, a = 0$. *Thus, the acceleration of the particle M executing SHM along XX' is the maximum or minimum according as the particle is at the extremities or mean position.*

The variation of a with t of the particle M is represented in Fig. 1.4.

(5) Time period. *The time-period of a particle executing SHM is defined as the time required to complete one vibration and is denoted by T.*

Here the time required by the particle P to complete one revolution is the same as the time required by M to move from O to X, X to X' and then X' to O.

Now, angular velocity of the particle is

$$\omega = \frac{2\pi}{T} \text{ or, } T = \frac{2\pi}{\omega} \tag{1.5}$$

$$\text{or, } T = \frac{2\pi}{\sqrt{\frac{a}{x}}} \quad [\because \; |a| = \omega^2 x, \text{ from eqn (1.4)}]$$

$$\text{or, } T = 2\pi\sqrt{\frac{x}{a}} = 2\pi\sqrt{\frac{\text{displacement}}{\text{acceleration}}}. \tag{1.6}$$

(6) Frequency. *Frequency of a vibrating particle is defined as the number of vibrations made by the particle in one second and is denoted by n.*

$$\text{or, } n = \frac{1}{T}. \tag{1.7}$$

(7) Phase. *The phase of a vibrating particle speaks of its state of displacement and motion at any instant t.* It is measured either in terms of angle described by the reference particle, or in terms of fraction of time-period, which has elapsed since the particle last passed through its mean position in the positive direction.

If the particle starts its motion from X and comes to P at time t, then $\angle XOP = \theta = \omega t$ is the phase of the particle given by

$$x = A\cos\omega t. \tag{1.8}$$

If, however, time is measured from the instant when P was at P_0, then $\angle P_0OP = \omega t$, hence

$$x = (\omega t + \phi) \quad [\text{where } \angle XOP_0 = \phi]. \tag{1.9}$$

At $t = 0$, the phase $(\omega t + \phi)$ reduces to ϕ, that is why ϕ is called the *initial phase* or *epoch.* If the particle starts motion from a point P_0' at $t = 0$, which is below the XX' line, then

$$x = A\cos(\omega t - \phi). \tag{1.10}$$

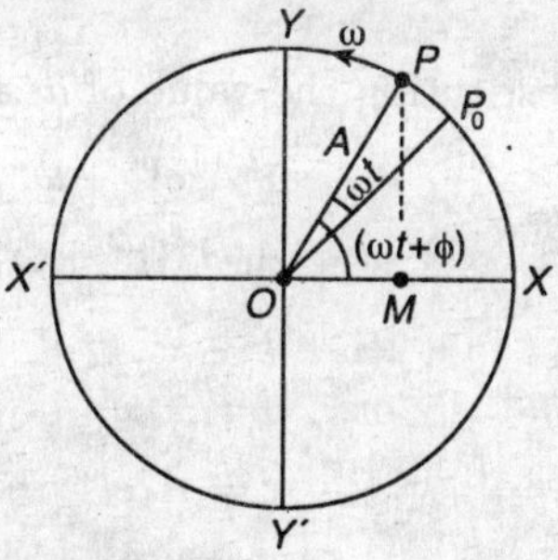

Fig. 1.5: Phase of a particle executing SHM.

In general,

$$x = A\cos(\omega t \pm \phi). \tag{1.11}$$

If $(\omega t + \phi)$ be the phase of a particle at any time t, then

$$\omega t + \phi = \omega\left(t + \frac{\phi}{2\pi}\right)$$

and the displacement is

$$x = A\cos\omega\left(t + \frac{\phi}{2\pi}\right). \tag{1.12}$$

Hence, the displacement is repeated after a time $\frac{\phi}{2\pi}$, which is the time-period T of SHM.

1.4 Differential Equation of SHM

If a particle be displaced from its equilibrium position and caused to move uniformly along the X-axis, then it is under the action of the inertial force,

$$m\frac{d^2x}{dt^2},$$

where m, the mass of the particle and $\dfrac{d^2x}{dt^2}$ is the acceleration at a distance x from the mean position.

Over and above, a *restoring force or the force of restitution* acts on the particle, given by

$$F = -sx \text{ (when the displacement } x \text{ is small)},$$

which tends to bring back the particle to its position of equilibrium. Here s is the restoring force per unit displacement and is called the *stiffness constant* and $1/s$ is called the *compliance.* Under the dynamical equilibrium of the system,

$$m\frac{d^2x}{dt^2} = -sx \text{ or, } \frac{d^2x}{dt^2} + \frac{s}{m}x = 0.$$

Putting $s/m = \omega^2$, we have

$$\frac{d^2x}{dt^2} + \omega^2 x = 0. \tag{1.13}$$

Eqn (1.13) is called the *differential equation of motion of the particle executing SHM* (the particle being called the linear harmonic oscillator). To investigate the nature of motion, we have to solve eqn (1.13).

As a trial solution, let $x = A_0 e^{pt}$.

$$\therefore \quad \frac{dx}{dt} = pA_0 e^{pt} \text{ and } \frac{d^2x}{dt^2} = p^2 A_0 e^{pt}.$$

Substituting the value of d^2x/dt^2 in eqn (1.13), we get

$$\begin{aligned}
& p^2 A_0 e^{pt} + \omega^2 A_0 e^{pt} = 0 \\
\text{or,} \quad & A_0 e^{pt}(p^2 + \omega^2) = 0 \\
\text{or,} \quad & p^2 + \omega^2 = 0 \quad (\because A_0 e^{pt} \neq 0) \\
\text{or,} \quad & p = \pm i\omega. \\
\therefore \quad & x = A_1 e^{i\omega t} + A_2 e^{-i\omega t} \\
& = (A_1 + A_2)\cos\omega t + i(A_1 - A_2)\sin\omega t \\
& = C_1 \cos\omega t + C_2 \sin\omega t, \text{ where } C_1 = A_1 + A_2 \text{ and } C_2 = i(A_1 - A_2).
\end{aligned}$$

Putting

$$C_1 = A\sin\phi \text{ and } C_2 = A\cos\phi,$$

we get

$$\phi = \tan^{-1}(C_1/C_2) \text{ and } A = \sqrt{C_1^2 + C_2^2}.$$

$$\therefore \; x = A\sin\phi\cos\omega t + A\cos\phi\sin\omega t$$

$$x = A\sin(\omega t + \phi). \tag{1.14}$$

If $\phi = \theta + \pi/2$, then

$$x = A\cos(\omega t + \theta). \tag{1.15}$$

Eqns (1.14) and (1.15) are the alternative forms of the general solution of eqn (1.13). Eqns (1.14) and (1.15) represent the displacement of the particle at any time t where A is the maximum displacement and is called the *amplitude* of oscillation. The displacement is repeated if ωt increases by 2π. Again,

$$\omega t + 2\pi = \omega\left(t + \frac{2\pi}{\omega}\right) = \omega(t + T),$$

where $T = 2\pi/\omega$ is the *time-period of oscillation.*

The frequency of oscillation $n = 1/T = \omega/2\pi$. The quantity $\omega = \sqrt{s/m}$, the restoring force per unit displacement per unit mass, denotes the angular velocity (or circular frequency) of the particle. The angles $(\omega t + \phi)$ and $(\omega t + \theta)$ are known as the *phase angles.* At $t = 0$, the phase angle is ϕ or θ, which are also called the *initial phase* or *epoch.*

1.5 Energy of a Linear Harmonic Oscillator

If x be the displacement of the particle executing SHM at time t, then the restoring force acting on it is $F = -sx$, which opposes the displacement. If the particle is displaced a further distance dx, then the work

$$dW = Fdx = sxdx \text{ [taking the magnitude of } F]. \quad [\because |F| = sx]$$

This work remains stored in the particle in the form of its PE. Hence, the PE for a displacement x is

$$E_p = \int_0^x sxdx = \frac{1}{2}sx^2. \tag{1.16}$$

The KE of the particle at time t is

$$E_k = \frac{1}{2}m\left(\frac{dx}{dt}\right)^2. \tag{1.17}$$

The total energy of the particle is given by

$$E = E_k + E_p = \frac{1}{2}m\left(\frac{dx}{dt}\right)^2 + \frac{1}{2}sx^2. \tag{1.18}$$

In absence of any dissipation, the total energy E remains constant and the system is called a *conservative system.* So for a conservative system, $dE/dt = 0$

$$\text{or,} \quad \frac{d}{dt}\left[\frac{1}{2}m\left(\frac{dx}{dt}\right)^2 + \frac{1}{2}sx^2\right] = 0$$

$$\text{or,} \quad m\left(\frac{dx}{dt}\right)\frac{d^2x}{dt^2} + sx\frac{dx}{dt} = 0$$

$$\text{or,} \quad m\frac{d^2x}{dt^2} + sx = 0 \quad \left[\because \frac{dx}{dt} \neq 0\right]$$

$$\text{or,} \quad \frac{d^2x}{dt^2} + \omega^2 x = 0 \ [\omega^2 = s/m]. \tag{1.19}$$

Thus, we obtain the differential equation of motion of the particle executing SHM.

Average Kinetic Energy

Let the displacement of the particle at any instant t is

$$x = A\cos(\omega t + \theta)$$

and the velocity

$$V = \frac{dx}{dt} = -A\omega\sin(\omega t + \theta).$$

$$\therefore \ E_k = \frac{1}{2}mV^2 = \frac{1}{2}m\omega^2A^2\sin^2(\omega t + \theta). \tag{1.20}$$

The maximum KE is

$$E_{km} = \frac{1}{2}m\omega^2A^2. \tag{1.21}$$

The instantaneous potential energy is

$$E_p = \frac{1}{2}sA^2\cos^2(\omega t + \theta) = \frac{1}{2}m\omega^2A^2\cos^2(\omega t + \phi) \ [\because \ s = m\omega^2]. \tag{1.22}$$

The maximum PE is

$$E_{pm} = \frac{1}{2}m\omega^2A^2. \tag{1.23}$$

The average KE of the particle is

$$\begin{aligned}\langle E_k \rangle &= \frac{1}{T}\left[\frac{1}{2}m\int_0^T \left(\frac{dx}{dt}\right)^2 dt\right] \\ &= \frac{m}{2T}\omega^2A^2\int_0^T \sin^2(\omega t + \theta)dt \\ &= \frac{1}{4}m\omega^2A^2 \quad \left[\because \int_0^T \sin^2(\omega t + \theta)dt = \frac{T}{2}\right] \\ &= \frac{1}{2}E_{km}.\end{aligned} \tag{1.24}$$

The average PE of the particle is

$$\langle E_p \rangle = \frac{1}{T}\left[\frac{1}{2}s\int_0^T x^2 dt\right]$$
$$= \frac{sA^2}{2T}\int_0^T \cos^2(\omega t + \theta)dt$$
$$= \frac{1}{4}sA^2 = \frac{1}{4}m\omega^2 A^2$$
$$= \frac{1}{2}E_{pm}. \quad (1.25)$$

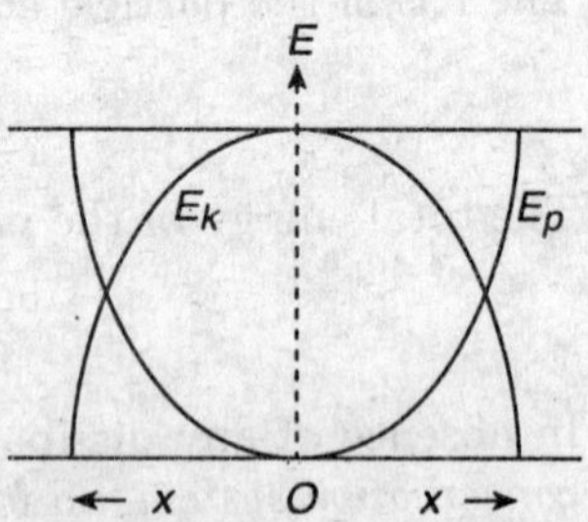

Fig. 1.6: Variation of E_k and E_p with displacement.

Thus

$$\langle E_k \rangle = \langle E_p \rangle = \frac{1}{2}E_{pm}.$$

i.e., *the average KE and average PE of the particle are half the corresponding maximum energy E_{km} and E_{pm} respectively.*

Thus

$$E = E_k + E_p = \frac{1}{2}m\omega^2 A^2$$

is a constant and is equal to 1/2 the maximum KE or PE. During oscillation, the KE (E_k) changes to PE (E_p) and vice versa so that at any instant ($E_k + E_p$) is equal to a constant. The variation of E, E_k and E_p with x is represented in Fig. 1.6.

1.6 Illustrative Examples of SHM

1.6.1 Motion of a Simple Pendulum

A simple pendulum of effective length l is suspended from the point O and the weight of the bob (mg) is balanced by the tension (T) of the string [Fig. 1.7]. Let the bob be displaced to the position B so that $\angle AOB = \theta(\theta < 4°)$ and the weight acts vertically downward through B. Resolving mg along and perpendicular to OB, we note that the component $mg\cos\theta$ along OB balances the tension (T) acting along BO whereas the perpendicular component $mg\sin\theta$ tries to bring back the bob to A.

Thus, the effective force on the bob is

$$F = mg\sin\theta$$
$$= mg\theta \ [\because \ \sin\theta \simeq \theta \text{ as } \theta \text{ is small}]$$
$$= mg\frac{x}{l} \ [\text{where } AB = x].$$

This force is equal to the displacing force on the particle, $m(d^2x/dt^2)$ in magnitude but opposite in sign to the restoring force F.

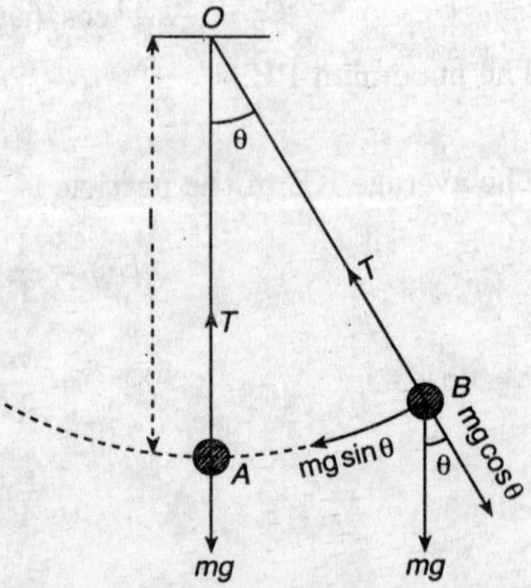

Fig. 1.7: Motion of a simple pendulum.

In equilibrium condition,

$$m\frac{d^2x}{dt^2} = -mg\frac{x}{l}$$

$$\text{or,} \quad \frac{d^2x}{dt^2} + \frac{g}{l}x = 0$$

$$\text{or,} \quad \frac{d^2x}{dt^2} + \omega^2 x = 0 \text{ [where } \omega^2 = g/l\text{].} \tag{1.26}$$

Eqn (1.26) is the *differential equation of motion of the simple pendulum* whose time-period is given by

$$T = \frac{2\pi}{\omega} = 2\pi\sqrt{\frac{l}{g}}. \tag{1.27}$$

1.6.2 Vertical Oscillation of an Elastic Spring Supporting a Weight at Its Lower End

We consider a helical spring of mass m carrying a concentrated mass M at its lower end, suspended from the point O. By lowering the mass M and then releasing, vertical oscillations may be set in the spring.

If x be the displacement of M from the position of rest at any time t, then its KE is

$$= \frac{1}{2}M\left(\frac{dx}{dt}\right)^2 = \frac{1}{2}M\dot{x}^2.$$

If l be the length of the spring, then its mass per unit length is m/l and the mass of element dy at a distance y from the support is $\frac{m}{l}dy$ and velocity of the element will be $\frac{\dot{x}}{l}y$. So the KE of the spring is

$$= \int_0^l \frac{1}{2}\left(\frac{m}{l}dy\right)\left(\frac{y}{l}\dot{x}\right)^2 \quad \left[\text{where } \dot{x} = \frac{dx}{dt}\right]$$

$$= \frac{1}{2}\frac{m}{l}\left(\frac{\dot{x}}{l}\right)^2 \int_0^l y^2 dy$$

$$= \frac{1}{2}\frac{m}{l^3}\dot{x}^2 \times \frac{l^3}{3}$$

$$= \frac{1}{6}m\dot{x}^2.$$

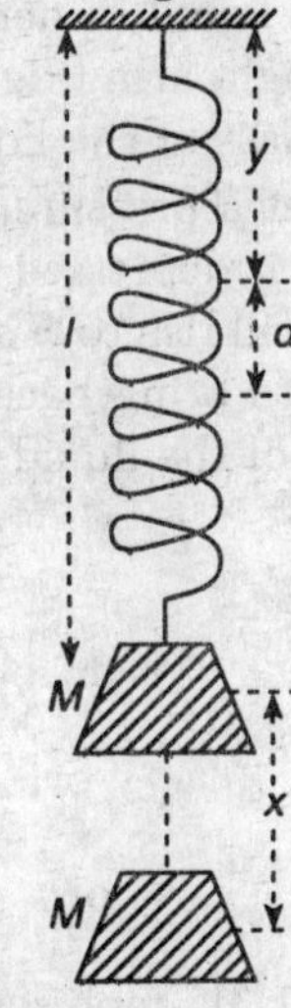

Fig. 1.8: Vertical oscillation of an elastic spring.

Therefore, the total KE of the spring and mass system is

$$\frac{1}{2}M\dot{x}^2 + \frac{1}{6}m\dot{x}^2 = \frac{1}{2}\left(M + \frac{m}{3}\right)\dot{x}^2.$$

The PE of the system is $\frac{1}{2}sx^2$, where s is the restoring force per unit displacement.

Since the system is conservative,

$$\frac{1}{2}\left(M + \frac{m}{3}\right)\dot{x}^2 + \frac{1}{2}sx^2 = \text{constant}.$$

Differentiating w.r.t. t, we get

$$\frac{1}{2}\left(M+\frac{m}{3}\right)2\dot{x}\ddot{x}+\frac{1}{2}s2x\dot{x}=0$$

$$\text{or,}\quad \left(M+\frac{m}{3}\right)\ddot{x}+sx=0\ [\because\ \dot{x}\neq 0]$$

$$\text{or,}\quad \ddot{x}+\frac{s}{\left(M+\frac{m}{3}\right)}x=0$$

$$\text{or,}\quad \ddot{x}+\omega^2x=0\ \left[\omega^2=\frac{s}{M+\frac{m}{3}}\right]. \tag{1.28}$$

Eqn (1.28) is the differential equation of motion of the system, which is simple harmonic in nature. The spring-mass system is thus equivalent to a single mass $M'=M+\frac{m}{3}$. The time-period of the system is

$$T=\frac{2\pi}{\omega}=2\pi\sqrt{\frac{M+\frac{m}{3}}{s}}. \tag{1.29}$$

If $m \ll M$, then

$$T=2\pi\sqrt{\frac{M}{s}}. \tag{1.30}$$

1.6.3 Oscillation of a Liquid in a U-tube

Let a liquid of density ρ be poured in a U-tube and PQ represents the liquid levels in the left and right arm [Fig. 1.9]. If l be the height of the liquid levels and α, the cross sectional area, then the mass of the liquid $M=2l\alpha\rho$.

Let a pressure be now applied on the liquid in the left arm when it depressed to R so that $PR=x$. The liquid in the right arm will rise to S so that $R'S=2x$. The liquid column $R'S$ will tend to bring back the two levels to PQ. The force due to the mass of the liquid column RS is

$$F=2x\alpha\rho g.$$

$$\therefore\quad 2l\alpha\rho\frac{d^2x}{dt^2}=-2x\alpha\rho g$$

$$\text{or,}\quad \ddot{x}+\frac{g}{l}x=0$$

$$\text{or,}\quad \ddot{x}+\omega^2x=0\ \left(\text{where } \omega^2=\frac{g}{l}\right). \tag{1.31}$$

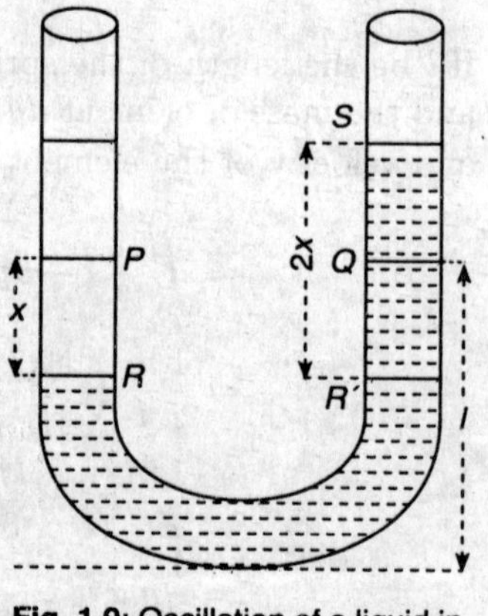

Fig. 1.9: Oscillation of a liquid in a U-tube.

Eqn (1.31) represents a SHM of time-period,

$$T=\frac{2\pi}{\omega}=2\pi\sqrt{\frac{l}{g}}. \tag{1.32}$$

1.6.4 Simple Harmonic Motion of Electrical System

Let us consider an electric circuit containing a capacitor C and an inductance L as shown in Fig. 1.10. When the capacitor is uncharged, the system is in equilibrium state and no current passes through the circuit. If the key K is closed, then the capacitor begins to charge and the state of equilibrium is disturbed.

If q be the charge on the capacitor at any time t, then the voltage across the plates of C is $V = q/C$. When the key K is opened, the capacitor discharges through L producing a current $i = dq/dt$.

The repulsive force between the electrons gives rise to the restoring force distributing the electrons equally on the plates of the capacitor, there being no net charge. The inductance L tends to oppose the distribution of charge and the current through it.

The voltage across the inductor at any time t is given by

$$V = -L\frac{di}{dt} = -L\frac{d^2q}{dt^2} \quad \left[\because\ i = \frac{dq}{dt}\right].$$

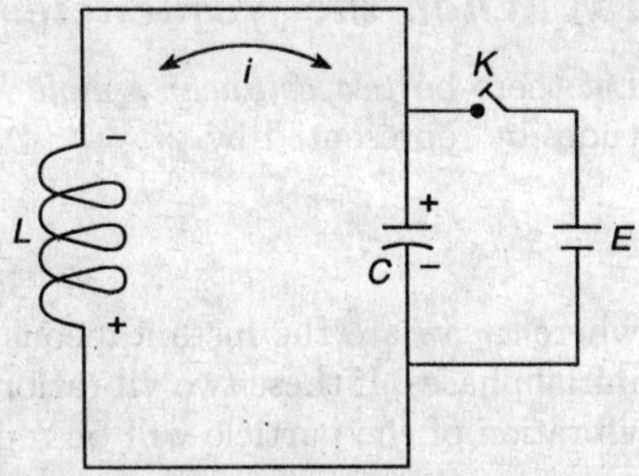

Fig. 1.10: Oscillation of charge in an *L*-*C* circuit.

Here, the significance of the negative sign is that the voltage V opposes the increase of current in the circuit. According to Kirchhoff's law,

$$-L\frac{d^2q}{dt^2} = \frac{q}{C}$$

$$\text{or,}\quad \frac{d^2q}{dt^2} + \frac{1}{LC}q = 0$$

$$\text{or,}\quad \frac{d^2q}{dt^2} + \omega^2 q = 0, \qquad (1.33)$$

where

$$\omega = \frac{1}{\sqrt{LC}}.$$

In the electric circuit comprising of L and C, the charge q thus executes SHM with angular frequency $\omega = 1/\sqrt{LC}$. The time-period of oscillation is

$$T = \frac{2\pi}{\omega} = 2\pi\sqrt{LC} \qquad (1.34)$$

1.7 Principle of Superposition

Statement 1.1 *The principle states that the resultant of two or more harmonic displacements is given by the algebraic sum of the individual displacements. This is true only when the equation of motion is linear.*

If two harmonic motions represented by

$$y_1 = A_1\cos(\omega t + \phi_1) \text{ and } y_2 = A_2\cos(\omega t + \phi_2),$$

where A_1, A_2 are the amplitudes and ϕ_1 and ϕ_2, the initial phases of two vibrations of angular frequency ω, superpose simultaneously on a medium particle, then according to the principle of superposition the resultant displacement of the particle will be

$$y = y_1 + y_2.$$

The principle was first stated by *Thomas Young* in 1802 while studying the interference phenomenon of light waves.

1.7.1 Superposition of Two Collinear Harmonic Vibrations

(a) *When the frequencies are equal*

Let there be *two collinear simple harmonic vibrations* of same frequency ω but different amplitudes be represented by

$$x_1 = A_1 \cos(\omega t + \phi_1) \tag{1.35}$$

$$\text{and } x_2 = A_2 \cos(\omega t + \phi_2), \tag{1.36}$$

where x_1, x_2 are the instantaneous displacements, A_1, A_2 are the amplitudes and ϕ_1, ϕ_2 are the initial phases. If these two vibrations act simultaneously on a medium particle, then the resultant vibration of the particle will be represented by

$$\begin{aligned} x &= x_1 + x_2 \\ &= A_1 \cos(\omega t + \phi_1) + A_2 \cos(\omega t + \phi_2) \\ &= (A_1 \cos\phi_1 + A_2 \cos\phi_2)\cos\omega t - (A_1 \sin\phi_1 + A_2 \sin\phi_2)\sin\omega t. \end{aligned} \tag{1.37}$$

Putting $(A_1 \cos\phi_1 + A_2 \cos\phi_2) = A\cos\phi$ and $(A_1 \sin\phi_1 + A_2 \sin\phi_2) = A\sin\phi$, we get

$$\begin{aligned} A^2 &= (A_1 \cos\phi_1 + A_2 \cos\phi_2)^2 + (A_1 \sin\phi_1 + A_2 \sin\phi_2)^2 \\ &= A_1^2 + A_2^2 + 2A_1A_2\cos(\phi_1 - \phi_2) \\ &= A_1^2 + A_2^2 + 2A_1A_2\cos\phi. \end{aligned}$$

$$\text{where} \quad \phi = (\phi_1 - \phi_2) \tag{1.38}$$

$$\text{and } \tan\phi = \frac{A_1 \sin\phi_1 + A_2 \sin\phi_2}{A_1 \cos\phi_1 + A_2 \cos\phi_2}. \tag{1.39}$$

Using these values eqn (1.39) can be written as

$$x = A\cos(\omega t + \phi). \tag{1.40}$$

Hence, *the resultant motion is also simple harmonic having the same angular frequency ω as that of the component vibrations.*

Special cases

(i) When the phase difference $\phi = 0$ or π, the resultant amplitude is $A = (A_1 + A_2)$ or $(A_1 - A_2)$ and the resultant displacement

$$x = (A_1 + A_2)\cos\omega t \text{ or, } (A_1 - A_2)\cos\omega t. \tag{1.41}$$

If $A_1 = A_2$, then the resultant amplitude is 0 for $\phi = \pi$.

(ii) When $\phi = \pi/2$, the resultant amplitude is

$$A = \sqrt{A_1^2 + A_2^2} \text{ and } \phi = \tan^{-1}\left(\frac{A_2}{A_1}\right)$$

and the resultant displacement is

$$x = \sqrt{A_1^2 + A_2^2}\cos(\omega t + \phi) \tag{1.42}$$

(b) *When the frequencies are slightly different (Beats)*

When two simple harmonic vibrations of slightly different frequencies superpose, the amplitude of the resultant vibration changes regularly with time between maximum and minimum values. This periodic changes of amplitude of the resultant vibration is known as the phenomenon of beats.

To analyze the phenomenon of beats we consider two harmonic vibrations of frequencies ω_1 and ω_2 and amplitudes A_1 and A_2 respectively as below

$$x_1 = A_1 \cos \omega_1 t \tag{1.43}$$

$$x_2 = A_2 \cos \omega_2 t. \tag{1.44}$$

The phase constant of the two vibrations is assumed to be zero for simplicity.
By the *principle of superposition,* the resultant vibration is given by

$$x = x_1 + x_2 = A_1 \cos \omega_1 t + A_2 \cos \omega_2 t. \tag{1.45}$$

Now,

$$\tfrac{1}{2}(\omega_1 + \omega_2) = \omega_a \text{ is the average frequency}$$

$$\text{and } \tfrac{1}{2}(\omega_2 - \omega_1) = \omega_m \text{ is the modulation frequency with } \omega_2 > \omega_1.$$

Eqn (1.45) then becomes

$$x = A_1 \cos(\omega_a - \omega_m)t + A_2 \cos(\omega_a + \omega_m)t$$

$$\text{or, } x = (A_1 + A_2) \cos \omega_m t \cos \omega_a t + (A_1 - A_2) \sin \omega_m t \sin \omega_a t. \tag{1.46}$$

$$[\because \omega_1 = (\omega_a - \omega_m) \text{ and } \omega_2 = (\omega_a + \omega_m)]$$

Let

$$(A_1 + A_2) \cos \omega_m t = A_m \cos \theta_m \tag{1.47}$$

$$\text{and } (A_1 - A_2) \sin \omega_m t = A_m \sin \theta_m. \tag{1.48}$$

Then

$$x = (A_m \cos \theta_m) \cos \omega_a t + (A_m \sin \theta_m) \sin \omega_a t$$

$$= A_m[\cos \omega_a t \cos \theta_m + \sin \omega_a t \sin \theta_m] = A_m \cos(\omega_a t - \theta_m), \tag{1.49}$$

where

$$A_m^2 = (A_1 + A_2)^2 \cos^2 \omega_m t + (A_1 - A_2)^2 \sin^2 \omega_m t$$

$$= A_1^2 + A_2^2 + 2A_1 A_2 (\cos^2 \omega_m t - \sin^2 \omega_m t)$$

$$\text{or, } A_m = [A_1^2 + A_2^2 + 2A_1 A_2 \cos(2\omega_m t)]^{1/2} \tag{1.50}$$

$$\text{and } \theta_m = \frac{(A_1 - A_2) \sin \omega_m t}{(A_1 + A_2) \cos \omega_m t}. \tag{1.51}$$

Since A_m and θ_m are both functions of time, the eqn (1.49) does not represent a SHM. However, it can be a periodic motion with angular frequency ω_m.

Beats

If $\omega_2 \simeq \omega_1$, then $\omega_a \gg \omega_m$. In that case, A_m and θ_m are almost constant w.r.t. time and eqn (1.49) may be regarded as a harmonic vibration of frequency ω_a and period $T = 2\pi/\omega_a$. *The resultant vibration of two harmonic vibrations of nearly equal frequency thus produces beats.*

From eqn (1.50), the resultant vibration will have maximum amplitude, when

$$\cos(2\omega_m t) = +1$$

$$\text{or, } \quad 2\omega_m t = 2p\pi \; [p = 0, 1, 2, 3, \cdots]$$

$$\text{or, } \quad (\omega_2 - \omega_1)t = 2p\pi \text{ or, } 2\pi(n_2 - n_1)t = 2p\pi$$

$$\text{or, } \quad t = \frac{p}{n_2 - n_1} = 0, \frac{1}{n_2 - n_1}, \frac{2}{n_2 - n_1}, \cdots,$$

where $n_1 = \omega_1/2\pi$ and $n_2 = \omega_2/2\pi$ are the *cyclic frequencies* of the two SHMs. The time interval between any two consecutive maxima $t_b = 1/(n_2 - n_1)$, is also called the *beat period* and $N = 1/t_b = (n_2 - n_1)$ is the *beat frequency*.

Again, A_m is the minimum $(A_2 - A_1)$, when

$$\cos(2\omega_m t) = -1$$
$$\text{or,}\quad 2\omega_m t = (2p+1)\pi \;[p = 0, 1, 2, 3, \cdots]$$
$$\text{or,}\quad 2\pi(n_2 - n_1)t = (2p+1)\pi$$
$$\text{or,}\quad t = \frac{2p+1}{2(n_2 - n_1)} = \frac{1}{2(n_2 - n_1)}, \frac{3}{2(n_2 - n_1)}, \frac{5}{2(n_2 - n_1)}, \cdots.$$

Thus, $t_b = 1/(n_2 - n_1)$ and $N = n_2 - n_1$. Therefore, the time interval between two consecutive mimima is also $t_b = 1/(n_2 - n_1)$.

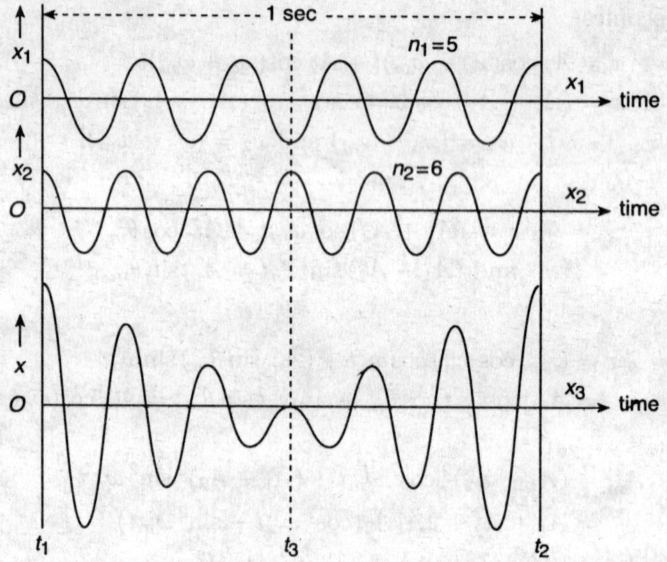

Fig. 1.11: Formation of beats.

Fig. 1.11 illustrates the phenomenon of beats. The component vibrations represented by curves x_1 and x_2 have the same amplitude and initial phase. x_1 and x_2 vibrations are of frequency 5 and 6 respectively. The resultant vibration is represented by x. Fig. 1.11 shows that at time t_1 and t_2, x_1 and x_2 vibrations are in phase causing the resultant amplitude the maximum. But at t_3, the resultant amplitude is the minimum where x_1 and x_2 vibrations are in opposite phase. A beat comprises of two consecutive maxima or minima. Fig. 1.11 shows one beat per second as $N = n_2 - n_1 = 6 - 5 = 1$.

1.7.2 Applications of Beats

Beats have the following important practical applications:

(i) To determine the small difference between the frequencies of two sources of sound,

(ii) To produce a pleasing tone to the resulting sound of an orchestra,

(iii) To transit a signal from one place to another through the formation of wave packets.

1.8 Superposition of Two Harmonic Vibrations at Right Angles to Each Other

(a) Oscillations having equal frequencies

Let two SHMs of equal frequency ω and phase difference ϕ be represented by

$$x = A_1 \cos \omega t \tag{1.52}$$

$$\text{and } y = A_2 \cos(\omega t + \phi), \tag{1.53}$$

where A_1 and A_2 are the amplitude of x and y vibrations respectively. The resultant vibration due to the simultaneous action of the two vibrations on a particle is obtained by the method of superposition as below:

From eqn (1.53),

$$\frac{y}{A_2} = \cos \omega t \cos \phi - \sin \omega t \sin \phi.$$

From eqn (1.52),

$$\cos \omega t = \frac{x}{A_1}$$

$$\text{and} \quad \sin \omega t = \sqrt{1 - \frac{x^2}{A_1^2}}.$$

$$\therefore \quad \frac{y}{A_2} = \frac{x}{A_1} \cos \phi - \left(1 - \frac{x^2}{A_1^2}\right)^{1/2} \sin \phi$$

$$\text{or,} \quad \left(\frac{x}{A_1} \cos \phi - \frac{y}{A_2}\right) = \left(1 - \frac{x^2}{A_1^2}\right) \sin \phi.$$

Squaring both sides, we get

$$\frac{x^2}{A_1^2} + \frac{y^2}{A_2^2} - \frac{2xy}{A_1 A_2} \cos \phi = \sin^2 \phi. \tag{1.54}$$

Eqn (1.54) represents a *general ellipse* whose axes are inclined to the X- and Y-axes. So the resultant vibration causes the particle, on which it acts, to move in general in an elliptical path.

Special cases

(i) When $\phi = 0$, eqn (1.54) reduces to

$$\frac{x^2}{A_1^2} + \frac{y^2}{A_2^2} - \frac{2xy}{A_1 A_2} = 0$$

$$\text{or,} \quad \left(y - \frac{A_2}{A_1} x\right)^2 = 0. \tag{1.55}$$

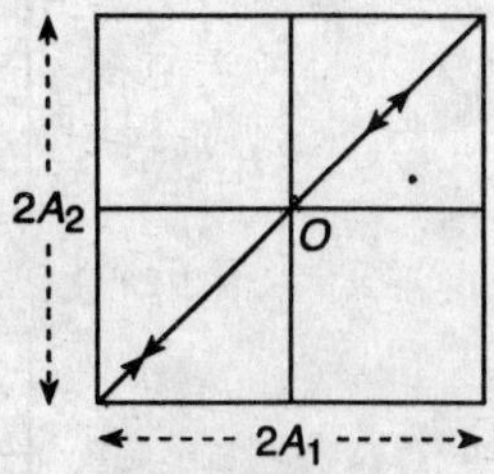

Fig. 1.12: ($\phi = 0$).

Eqn (1.55) represents a pair of coincident straight lines whose equation is $y = (A_2/A_1)x$ and slope A_2/A_1, which is positive passing through the origin O [Fig. 1.12].

(ii) When $\phi = \pi/2$, the eqn (1.54) reduces to

$$\frac{x^2}{A_1^2} + \frac{y^2}{A_2^2} = 1. \tag{1.56}$$

Eqn (1.56) represents an ellipse whose axes are parallel to X- and Y-axes as shown in Fig. 1.13. The particle moves in an elliptical path. It can be shown that the motion of the particle is in the clockwise direction as indicated in Fig. 1.13. In optics this represents *a right-handed elliptically polarised light.*

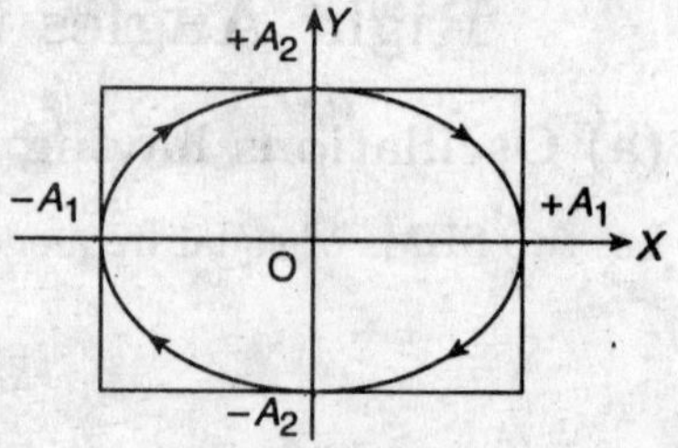

Fig. 1.13: ($\phi = \pi/2$).

In addition to $\phi = \pi/2$, if $A_1 = A_2 = A$, then the ellipse degenerates into a circle

$$x^2 + y^2 = A^2 \tag{1.57}$$

as shown in Fig. 1.14.

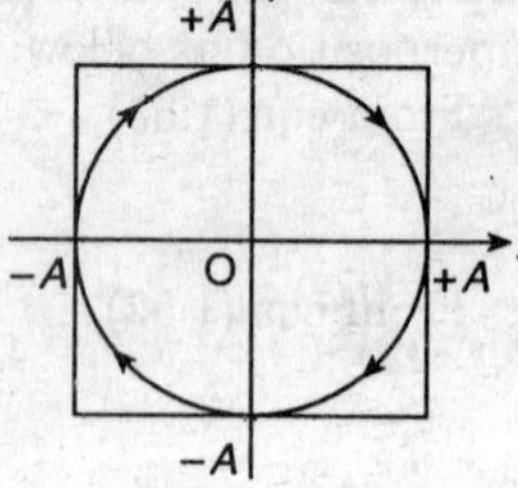

Fig. 1.14: ($\phi = \pi/2$, $A_2 = A$).

(iii) When $\phi = \pi$, the eqn (1.54) becomes

$$\frac{x^2}{A_1^2} + \frac{y^2}{A_2^2} + \frac{2xy}{A_1 A_2} = 0 \text{ or, } \left(y + \frac{A_2}{A_1}x\right)^2 = 0. \tag{1.58}$$

Eqn (1.58) represents *a pair of coincident straight lines*, $y = -(A_2/A_1)x$, with negative slope $(-A_2/A_1)$ and passing through the origin. The ellipse under such condition degenerates into a straight line with negative slope [Fig. 1.15].

(iv) When $\phi = 3\pi/2$,

$$x = A_1 \cos \omega t$$

and $\quad y = A_2 \cos(\omega t + 3\pi/2) = A_2 \sin \omega t.$

$$\therefore \quad \frac{x^2}{A_1^2} + \frac{y^2}{A_2^2} = 1. \tag{1.59}$$

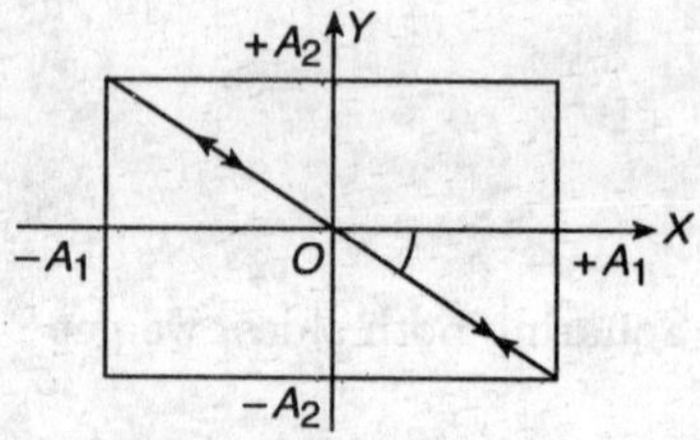

Fig. 1.15: ($\phi = \pi$).

Eqn (1.59) represents an ellipse described in the anticlockwise direction. This represents a *left-handed elliptically polarised light* in optics.

The sequence of pattern for a few values of $\phi = 0$ to 2π is shown in Fig. 1.16.

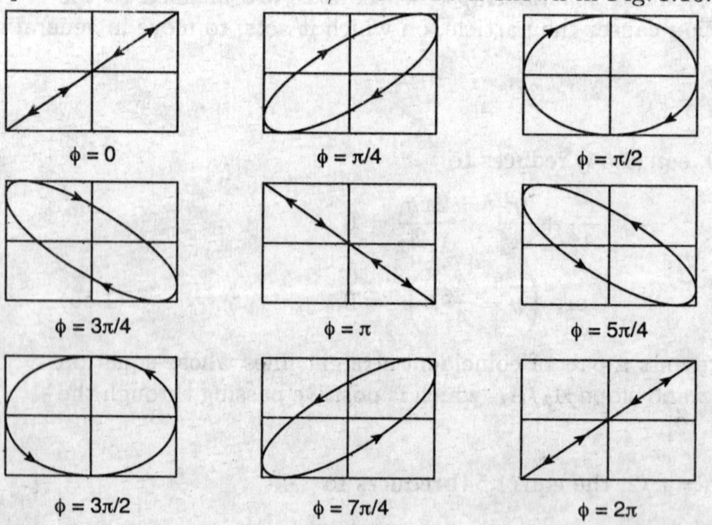

Fig. 1.16

(b) Oscillations having different frequencies (Lissajous figures)

When two simple harmonic vibrations of different frequency as well as phase at right angles to each other act simultaneously on a particle, the patterns traced out are known as *Lissajous figures after* J.A. Lissajous (1822–1880).

Frequencies in the ratio 1 : 2. Let two perpendicular simple harmonic vibrations of frequencies ω and 2ω acting simultaneously on a medium particle be represented by

$$x = A_1 \cos \omega t \tag{1.60}$$

$$\text{and } y = A_2 \cos(2\omega t + \phi), \tag{1.61}$$

where ϕ is the phase difference between them and A_1 and A_2 are respectively the amplitudes of the vibrations.

Then

$$\frac{y}{A_2} = \cos 2\omega t \cos\phi - \sin 2\omega t \sin\phi$$

$$= (2\cos^2 \omega t - 1)\cos\phi - 2\sin\omega t \cos\omega t \sin\phi.$$

Now, from eqn (1.60),

$$\cos\omega t = \frac{x}{A_1} \text{ and } \sin\omega t = \left(1 - \frac{x^2}{A_1^2}\right)^{1/2}$$

$$\therefore \quad \frac{y}{A_2} = \left(\frac{2x^2}{A_1^2} - 1\right)\cos\phi - \frac{2x}{A_1}\left(1 - \frac{x^2}{A_1^2}\right)^{1/2} \sin\phi$$

$$\text{or,} \quad \frac{y}{A_2} - \left(2\frac{x^2}{A_1^2} - 1\right)\cos\phi = -2\frac{x}{A_1}\left(1 - \frac{x^2}{A_1^2}\right)^{1/2} \sin\phi$$

Squaring both sides, we get

$$\left[\frac{y}{A_2} - \left(\frac{2x^2}{A_1^2} - 1\right)\cos\phi\right]^2 = 4\frac{x^2}{A_1^2}\left(1 - \frac{x^2}{A_1^2}\right)\sin^2\phi. \tag{1.62}$$

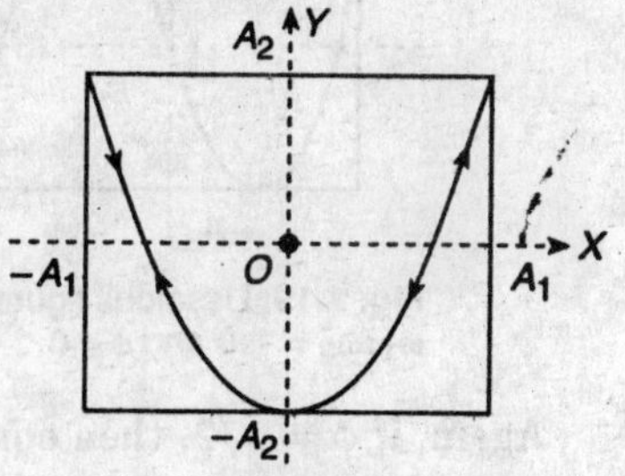

Fig. 1.17: Lissajous figure with $\omega_1 : \omega_2 = 1:2$ and $\phi = 0$.

If $\phi = 0$, then eqn (1.62) becomes

$$\left(\frac{y}{A_2} - \frac{2x^2}{A_1^2} + 1\right)^2 = 0. \tag{1.63}$$

Eqn (1.63) represents *two coincident parabolas* [Fig. 1.17] each of which is represented by the equation,

$$x^2 = \frac{A_1^2}{2A_2}(y + A_2).$$

If $\phi = \pi/2$, then eqn (1.62) reduces to

$$\frac{y^2}{A_2^2} + \frac{4x^2}{A_1^2}\left(\frac{x^2}{A_1^2} - 1\right) = 0. \tag{1.64}$$

Since eqn (1.64) is of *fourth degree* in x, it represents a curve containing two loops [Fig. 1.18]. The shape of the loops changes with the change of phase difference ϕ between the two perpendicular vibrations.

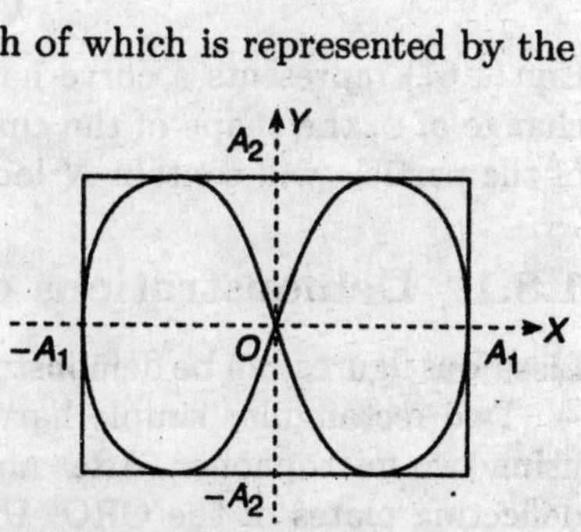

Fig. 1.18: Lissajous figure with $\omega_1 : \omega_2 = 1:2$ and $\phi = \pi/2$.

(c) Frequencies in the ratio 1 : 3

Let two simple harmonic vibrations at right angles to each other having frequencies ω and 3ω respectively act simultaneously on a particle. If ϕ be the phase difference between them, then vibrations are represented by

$$x = A_1 \cos\omega t \text{ and } y = A_2 \cos(3\omega t + \phi).$$

$$\therefore \quad \frac{y}{A_2} = \cos 3\omega t \cos\phi - \sin 3\omega t \sin\phi$$

$$= (4\cos^3\omega t - 3\cos\omega t)\cos\phi - (3\sin\omega t - 4\sin^3\omega t)\sin\phi$$

$$= \left(4\frac{x^3}{A_1^3} - 3\frac{x}{A_1}\right)\cos\phi - \left[3\left(1-\frac{x^2}{A_1^2}\right)^{1/2} - 4\left(1-\frac{x^2}{A_1^2}\right)^{3/2}\right]\sin\phi$$

$$= \left(4\frac{x^3}{A_1^3} - 3\frac{x}{A_1}\right)\cos\phi - \left(1-\frac{x^2}{A_1^2}\right)^{1/2}\left(4\frac{x^2}{A_1^2} - 1\right)\sin\phi$$

$$\text{or,} \quad \left[\frac{y}{A_2} - \left(\frac{4x^3}{A_1^3} - \frac{3x}{A_1}\right)\cos\phi\right]^2 = \left(1-\frac{x^2}{A_1^2}\right)\left(4\frac{x^2}{A_1^2} - 1\right)^2 \sin^2\phi. \tag{1.65}$$

If $\phi = 0$, then the above equation reduces to

$$\left(\frac{y}{A_2} + \frac{3x}{A_1} - \frac{4x^3}{A_1^3}\right)^2 = 0. \tag{1.66}$$

This equation represents *two coincident cubic curves* as shown in Fig. 1.19.

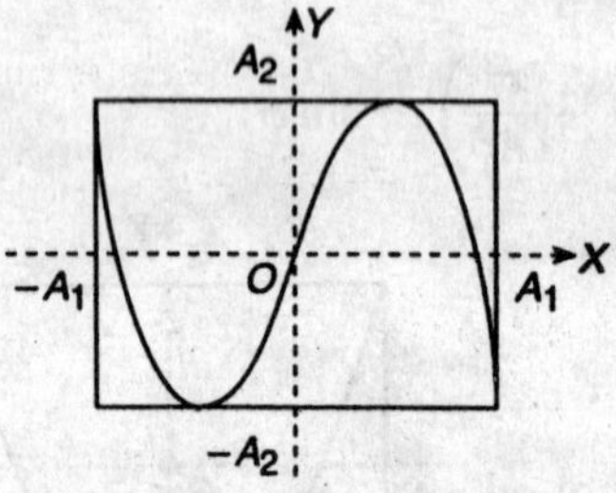

Fig. 1.19: Lissajous figure with $\omega_1 : \omega_2 = 1:3$ and $\phi = 0$.

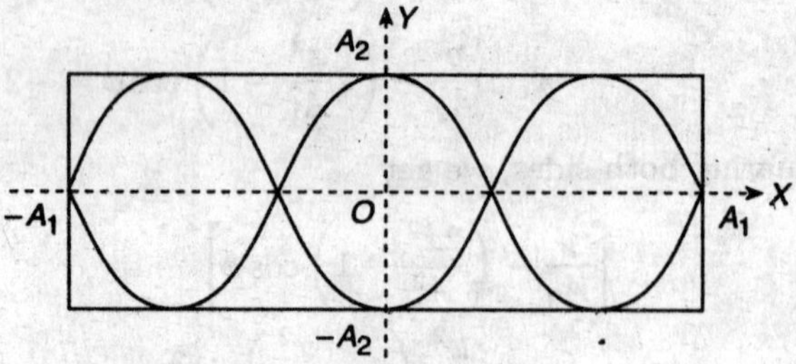

Fig. 1.20: Lissajous figure with $\omega_1 : \omega_2 = 1:3$ and $\phi = \pi/2$.

Again, if $\phi = \pi/2$, then eqn (1.65) reduces to.

$$\left(1-\frac{x^2}{A_1^2}\right)\left(\frac{4x^2}{A_1^2} - 1\right)^2 - \frac{y^2}{A_2^2} = 0. \tag{1.67}$$

Eqn (1.67) represents a curve having three loops [Fig. 1.20] as it is of degree six in x. With the change of ϕ, the shape of the curve changes. If $\omega_1 : \omega_2 = 1 : N$, then curve representing the path of the particle will contain N loops.

1.8.1 Demonstrations of Lissajous Figures

Lissajous figures can be demonstrated on the screen of a cathode ray oscillograph (CRO) [Fig. 1.21].

Two rectangular simple harmonic vibrations are converted into two sinusoidal voltages by using two microphones. After amplification these voltages are fed to the horizontal and vertical deflecting plates of the CRO. Under the simultaneous action of the two sinusoidal voltages at right angles, the cathode ray beam traces out the Lissajous figures, which can be viewed on the fluorescent screen of the CRO. If the frequencies of the electrical oscillations are not exactly in a simple ratio, the figure changes its form slowly. Very beautiful patterns are obtained for more complicated frequency ratios.

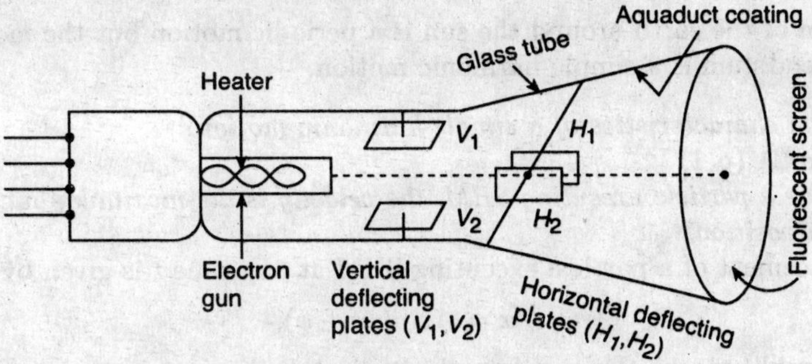

Fig. 1.21: Production of Lissajous figures by CRO.

1.8.2 Uses of Lissajous Figures

Lissajous figures have important practical applications:

(i) The unknown frequency of a tuning fork can be determined using Lissajous figures.

(ii) The ratio of the two exactly commensurate frequencies of two unknown vibrations can be determined.

(iii) The shape and nature of the signal waveform in electronic circuits can be determined.

(iv) The amplitude of the signal wave can be determined.

1.9 Answers of Some Important Questions

1.1. *What are the factors responsible for the oscillatory motion of a physical system?*
Ans. Following two factors are responsible for the oscillatory motion of a physical system:

(a) Elasticity of the system and

(b) Inertia of the system.

The elastic properties of the system determines the magnitude of the restoring force, which brings back the displaced system to its *mean position.* Inertia, on the other hand, tries to oppose the change in its velocity. The restoring force tries to bring back, i.e., *restore* the body by imparting a velocity and inertia preserves the velocity. Due to this reason the displacement of the body overshoots and repeats itself and oscillation continues.
1.2. *How does periodic motion differs from simple harmonic motion?*
Ans.

(a) **Periodic motion**: Any motion that repeats itself at a regular interval of time on the same path is called a *periodic motion.*

Simple harmonic motion: Simple harmonic motion is a periodic motion in which the acceleration of the particle is directly proportional to displacement from the mean position of rest and is always directed towards it.

(b) A simple harmonic motion must be a periodic motion but a periodic motion may not be a simple harmonic motion.

(c) The motion of the earth around the sun is a periodic motion but the motion of the bob of a simple pendulum is a simple harmonic motion.

1.3. *What are the characteristics of a simple harmonic motion?*
Ans. See Section 1.1 (p. 1)
1.4. *Show that for a particle executing SHM, the velocity is the maximum but the acceleration is zero at the mean position.*
Ans. The displacement of a particle executing SHM at any time t is given by

$$x = A\sin(\omega t + \phi),$$

where A is the amplitude, ω, the angular velocity and ϕ, the initial phase.
The velocity

$$v = \frac{dx}{dt} = A\omega\cos(\omega t + \phi).$$

But

$$\cos(\omega t + \phi) = \sqrt{1 - \frac{x^2}{A^2}} = \frac{\sqrt{A^2 - x^2}}{A}.$$

$\therefore\ v = \omega\sqrt{A^2 - x^2}$, at mean position $x = 0$, $\therefore\ v = v_{\text{max}} = \omega A$.
Again, acceleration

$$f = \frac{dv}{dt} = -A\omega^2\sin(\omega t + \phi) = -\omega^2 x$$

$\because$ at mean position $x = 0$, $\therefore\ f = f_{\text{min}} = 0$.
1.5. *Show that for a body executing SHM, the acceleration leads the velocity by $\pi/2$ and displacement by π.*
Ans. If the SHM of a body be represented by

$$x = A\sin(\omega t + \phi), \tag{1.68}$$

then velocity

$$v = A\omega\cos(\omega t + \phi) = A\omega\sin\left(\omega t + \phi + \frac{\pi}{2}\right). \tag{1.69}$$

Comparing eqns (1.68) and (1.69) we note that velocity leads the displacement by $\pi/2$.
Again, acceleration

$$f = \frac{dv}{dt} = -A\omega^2\sin(\omega t + \phi) = A\omega^2\sin(\omega t + \phi + \pi). \tag{1.70}$$

Comparing eqns (1.69) and (1.70), we see that acceleration of a partial executing SHM leads the velocity by $\pi/2$. From eqns (1.68) and (1.69) it is evident that the acceleration leads the displacement by π.
1.6. *A hollow sphere is filled with water, used as pendulum bob. If water trickles out slowly through the whole made at the bottom, how will the time-period be affected?*
Ans. The time-period of a simple pendulum of effective length l is

$$T = 2\pi\sqrt{\frac{l}{g}},$$

where g is the acceleration due to gravity. When water trickles out slowly through the hole made at the bottom of the sphere, mass of the bob comes down gradually. The time-period T being independent on the mass of the bob will remain unchanged if l is assumed to remain unchanged.
1.7. *What does actually oscillates in an oscillatory L-C circuit?*

Ans. In an oscillatory electric circuit containing an inductance L and a capacitance C, the charge on the capacitor is the harmonically varying quantity, which is responsible for electrical oscillations. The frequency of oscillation of such a circuit is

$$f = \frac{1}{2\pi\sqrt{LC}}.$$

1.8. *What are Lissajous figures? What are the factors that determine the nature of the figures?*
Ans. When two simple harmonic vibrations of slightly different frequencies as well as phases at right angles to each other act simultaneously on a particle, the patterns traced out are known as *Lissajous figures* after J.A. Lissajous (1822–1880).

The factors, which determine the shape of the figures are as follows:

(a) the amplitudes of vibration of the two SHMs,

(b) the frequencies of the two SHMs,

(c) the phase difference between the two SHMs.

1.9. *What are beats?*
Ans. When two SHMs of slightly different frequencies superpose, the amplitude of the resultant vibration changes regularly with time between maximum and minimum values. This periodic changes of amplitude of the resultant vibration is known as the *phenomenon of beat.*

If n_1 and n_2 be the frequencies of the two component SHMs, then the beat period is

$$t_b = \frac{1}{n_1 - n_2} \cdot (n_1 > n_2)$$

1.10. *What do you mean by the term phase of a particle executing SHM?*
Ans. See Section 1.3.1 (p. 2)
1.11. *Show that for a particle executing SHM*

$$\begin{aligned}(\textit{average KE}) &= (\textit{average PE}) \\ &= (\textit{half the corresponding maximum energy}).\end{aligned}$$

Ans. See Section 1.5 (p. 6)
1.12. *What are the applications of beats?*
Ans. See Section 1.7.2 (p. 14)
1.13. *Mention some practical applications of Lissajous figures.*
Ans. See Section 1.8.2 (p. 19)
1.14. *Is* $x = \sin kt + \cos kt$ *an equation of SHM?*
Ans. We have velocity

$$v = \frac{dx}{dt} = k\cos kt - k\sin kt$$

and acceleration

$$f = \frac{dv}{dt} = -k^2 \sin kt - k^2 \cos kt = -k^2 x.$$

$\therefore$ $f \propto -x$, so, the equation represents a SHM.
1.15. *What are the dimensions of force constant of a vibrating spring?*
Ans. Force F on the mass of the spring is $F = -kx$, x = displacement of the mass from the equilibrium position at any time t.

$$\therefore \text{dimension of } k = \frac{[F]}{[x]} = \frac{[MLT^{-2}]}{[L]} = [MT^{-2}].$$

1.10 Solved Problems

Ex. 1 *Calculate the time-period of the liquid column of length l in a U-tube, if it is depressed in one arm by x, p is the density of liquid and A is the cross sectional area of each arm of U-tube.*

[WBUT 2007]

Soln. The liquid column in the U-tube, when depressed in one arm and then released will execute SHM [Fig. 1.22].

The length of the liquid column in the U-tube is l. When it is depressed by a length x in the left arm, the difference in levels in the two arms will be $2x$. The mass of this column of liquid is $m = 2A\rho x$, where A is the cross sectional area of the U-tube and ρ, the density of the liquid.

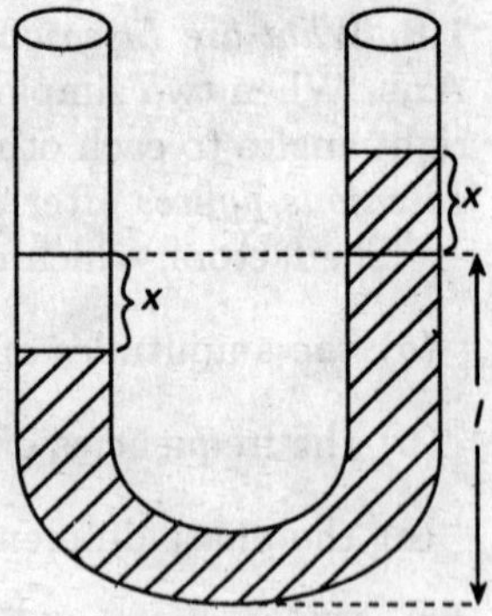

Fig. 1.22: Oscillation of a liquid in a U-tube.

The restoring force on the liquid of mass m' is

$$F = -m'g = -2A\rho xg$$
$$= -2A\rho gx = -kx,$$

where the force constant $k = 2A\rho g$.

If l be the length of the level from the bottom of the U-tube, then mass of the liquid $m = 2lA\rho$.

$\therefore$ the equation of motion of the liquid column is

$$m\frac{d^2x}{dt^2} = -kx \text{ or, } \frac{d^2x}{dt^2} + \left(\frac{k}{m}\right)x = 0$$

$$\text{or, } \frac{d^2x}{dt^2} + \left(\frac{g}{l}\right)x = 0$$

$$\text{or, } \frac{d^2x}{dt^2} + \omega^2 x = 0 \ [\because\ m = 2lA\rho \text{ and } k = 2A\rho g]. \quad (1.71)$$

Eqn (1.71) represents a SHM whose angular frequency of oscillation is $\omega = \sqrt{g/l}$.

$\therefore$ time-period of oscillation is

$$T = \frac{2\pi}{\omega} = 2\pi\sqrt{\frac{l}{g}}.$$

Ex. 2 *A particle moving simple harmonically in a straight line, has velocities u_1 and u_2 when it is at distances x_1 and x_2 respectively from the mean position. Show that the time-period of the motion is given by*

$$T = 2\pi\sqrt{\frac{x_2^2 - x_1^2}{u_1^2 - u_2^2}}.$$

Find also the amplitude and the maximum velocity of the particle.

Soln. If the displacement of a particle executing SHM along X-direction be

$$x_1 = A\sin(\omega t + \phi),$$

then the velocity at that instant t is

$$u_1 = \frac{dx_1}{dt} = A\omega\cos(\omega t + \phi) = A\omega\sqrt{1 - \frac{x_1^2}{A^2}} = \frac{2\pi}{T}\sqrt{A^2 - x_1^2}.$$

Similarly,

$$u_2 = \frac{2\pi}{T}\sqrt{A^2 - x_2^2}.$$

$$\therefore\ u_1^2 = \frac{4\pi^2}{T^2}(A^2 - x_1^2) \text{ and } u_2^2 = \frac{4\pi^2}{T^2}(A^2 - x_2^2).$$

$$\text{or, } u_1^2 - u_2^2 = \frac{4\pi^2}{T^2}[A^2 - x_1^2 - A^2 + x_2^2] = \frac{4\pi^2}{T^2}(x_2^2 - x_1^2)$$

$$\text{or, } T^2 = 4\pi^2\frac{(x_2^2 - x_1^2)}{(u_1^2 - u_2^2)} \text{ or, } T = 2\pi\sqrt{\frac{x_2^2 - x_1^2}{u_1^2 - u_2^2}} \textbf{ (Proved)}.$$

Again,

$$u_1^2 = \frac{4\pi^2(A^2 - x_1^2)}{4\pi^2(x_2^2 - x_1^2)} \times (u_1^2 - u_2^2) = \frac{(u_1^2 - u_2^2)(A^2 - x_1^2)}{(x_2^2 - x_1^2)} \quad \left[\because T = 2\pi\sqrt{\frac{x_2^2 - x_1^2}{u_1^2 - u_2^2}}\right]$$

$$\text{or, } A^2 - x_1^2 = \frac{u_1^2(x_2^2 - x_1^2)}{(u_1^2 - u_2^2)}$$

$$\text{or, } A = \left(\frac{u_1^2 x_2^2 - u_2^2 x_1^2}{u_1^2 - u_2^2}\right)^{1/2}.$$

And the maximum velocity of the particle is

$$u_{\max} = A\omega = \frac{2\pi}{T}A = \left(\frac{u_1^2 - u_2^2}{x_2^2 - x_1^2}\right)^{1/2} \times \left(\frac{u_1^2 x_2^2 - u_2^2 x_1^2}{u_1^2 - u_2^2}\right)^{1/2}$$

$$= \left(\frac{u_1^2 x_2^2 - u_2^2 x_1^2}{x_2^2 - x_1^2}\right)^{1/2}.$$

Ex. 3 *A solid glass cylinder heavier at the bottom is floating vertically in a liquid with* 20 cm *of its length within the liquid. If the cylinder be slightly depressed and then released, find the time-period of its oscillation.*

Soln. If a cylinder of mass m, cross section A floats vertically in a liquid of density ρ, then according to the law of floatation,

weight of the cylinder = weight of displaced liquid.

$$\therefore\ mg = hA\rho g,$$

where h is the immersed length of the cylinder in the liquid and g is the acceleration due to gravity [Fig. 1.23] or, $m = hA\rho$.

If the cylinder be further immersed by a length x by depressing a little, then the weight of the additional displaced liquid is $xA\rho g$.
$\therefore$ the force due to buoyancy acting on the cylinder is $F = -A\rho g x$ in the upward direction.

The −ve sign represents that F and x are opposite in direction.
$\therefore$ acceleration of the cylinder is

$$f = \frac{F}{m} = -\frac{A\rho g x}{hA\rho} = -\frac{g}{h}x$$

$$= -\omega^2 x \quad \left[\because \omega = \sqrt{g/h}\right].$$

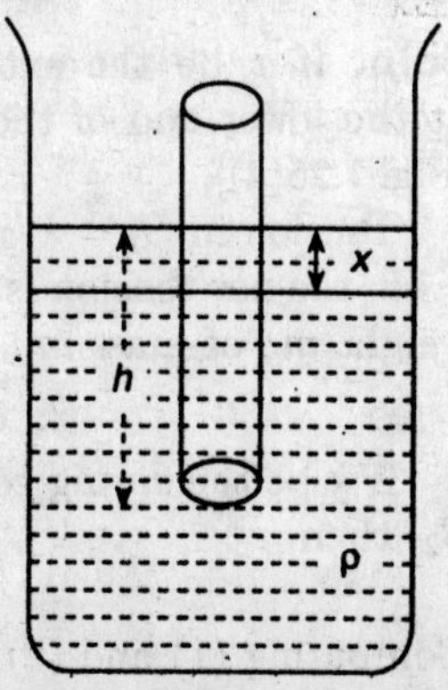

Fig. 1.23: Floating of a glass cylinder in a liquid.

So the motion of the cylinder is SHM.

Time-period $T = \frac{2\pi}{\omega} = 2\pi\sqrt{\frac{h}{g}}$.

Here $h = 20$ cm, and $g = 980$ cm/s^2.

$$\therefore\ T = 2\pi\sqrt{\frac{20}{980}} = \frac{2\pi}{7}\ \text{s} = \mathbf{0.9\ s.}$$

Ex. 4 *A cubical block of side L cm and density d is floating in water of density $\rho(\rho > d)$. The block is slightly depressed and released. Show that it will execute simple harmonic motion. Determine the frequency of oscillation.* **[WBUT 2001]**

Soln. The mass of the block = L^3d and mass of displaced water $m = xA\rho$, where x is the length of the block in water [Fig. 1.24].

According to the *Archimedes principle*

$$xA\rho g = L^3\rho g$$

$$\text{or, } x\rho = Ld\ (\because\ A = L^2)$$

$$\text{or, } x = \frac{Ld}{\rho}.$$

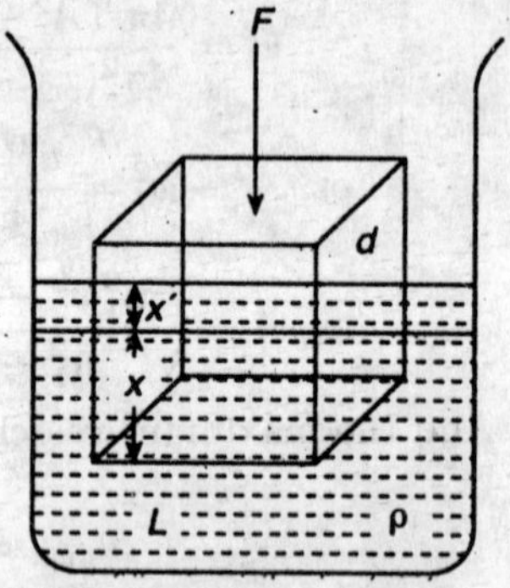

Fig. 1.24: Floating wooden cube in water.

The additional upward thrust on the block when it is depressed a further length of x' cm,

$$F = -x'A\rho g$$

and the acceleration of the block is

$$f = \frac{F}{m} = -x'\frac{A\rho g}{m} = -x'\omega^2 \left[\text{where } \omega^2 = \frac{A\rho g}{m}\right].$$

Thus, $f \propto -x'$ and hence the motion is SHM.

$$\omega = \sqrt{\frac{A\rho g}{m}} = \sqrt{\frac{L^2\rho g}{xL^2\rho}} = \sqrt{\frac{\rho g}{Ld}}\ \left[\because\ m = xL^2\rho \text{ and } x = \frac{Ld}{\rho}\right].$$

$\therefore$ the frequency of oscillation is

$$n = \frac{\omega}{2\pi} = \frac{1}{2\pi}\sqrt{\frac{\rho g}{Ld}}.$$

Ex. 5 *Two springs S_1 and S_2 each of length l, have force constants k_1 and k_2. Find the force constant k of the spring system connected as shown in figure below.*

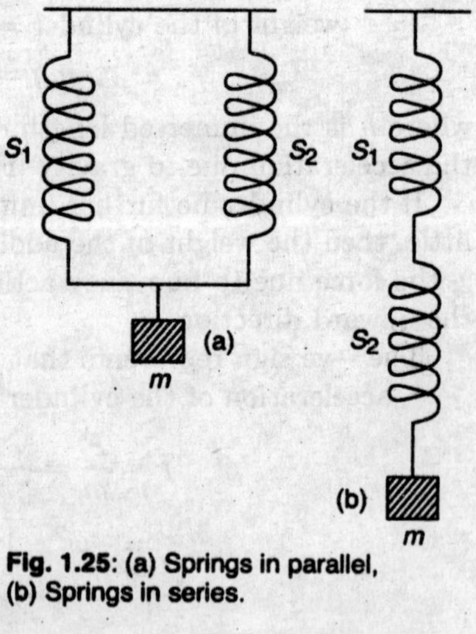

Fig. 1.25: (a) Springs in parallel, (b) Springs in series.

Soln. If x be the extension when a mass m is attached at the lower end of the combination of S_1 and S_2, then in [Fig. 1.25(a)],

Tension in $S_1 = k_1x$ and that in $S_2 = k_2x$.

$\therefore$ the net tension is $(k_1 + k_2)x$, which is balanced by the weight mg of mass m.

$$\therefore\ mg = (k_1 + k_2)x. \tag{1.72}$$

If k be the spring constant of the combination of S_1 and S_2, then

$$mg = kx. \tag{1.73}$$

Comparing (1) and (2), we get

$$\mathbf{k = k_1 + k_2}. \tag{1.74}$$

If x_1 and x_2 be the extension of S_1 and S_2 in Fig. 1.25(b), then total extension $= x_1 + x_2$. Again,

$$\because\ mg = k_1x_1,\ \therefore\ x_1 = \frac{mg}{k_1}$$

$$\text{and}\ \because\ mg = k_2x_2,\ \therefore\ x_2 = \frac{mg}{k_2}.$$

$$\therefore\ x_1 + x_2 = mg\left(\frac{1}{k_1} + \frac{1}{k_2}\right). \tag{1.75}$$

But

$$mg = (x_1 + x_2)k \text{ or, } k = \frac{mg}{(x_1 + x_2)} \text{ or, } x_1 + x_2 = \frac{mg}{k}. \tag{1.76}$$

From (4) and (5), we get

$$\frac{1}{k} = \frac{1}{k_1} + \frac{1}{k_2} = \frac{k_1 + k_2}{k_1k_2} \text{ or, } k = \frac{\mathbf{k_1k_2}}{\mathbf{k_1 + k_2}}. \tag{1.77}$$

Ex. 6 *A small spherical steel ball is placed a little away from the centre of a concave mirror whose radius of curvature is* 2.5 m. *When the ball is released, it begins to oscillate about the centre. What is the period of oscillations?* [$g = 10\ \text{m/s}^2$, *neglect the friction*].

Soln. Let a small steel ball is placed at A, a little away from the centre O of the concave mirror. The radius of curvature of the mirror is $OC = AC = R$ as shown in Fig. 1.26.

The weight mg acts vertically downwards at A. This force is resolved into two rectangular components, viz., $mg\cos\theta$ and $mg\sin\theta$, where $\angle AOC = \theta$. The restoring force is provided by $mg\sin\theta$.

$$\begin{aligned}\therefore\ F &= -mg\sin\theta \\ &= -mg\theta\ (\because\ \theta \text{ is small}) \\ &= -mg\frac{x}{R}\ \ [\because\ OA = x] \\ &= -kx\ \left[\text{where } k = \frac{mg}{R}\right].\end{aligned}$$

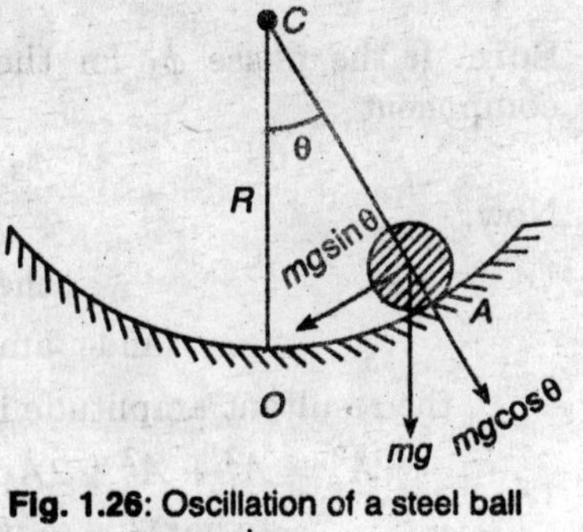

Fig. 1.26: Oscillation of a steel ball on a concave mirror.

$\because\ F \propto -x$, the motion of the ball is SHM with frequency

$$\omega = \sqrt{\frac{k}{m}} = \sqrt{\frac{g}{R}}.$$

$\therefore$ time-period of oscillation,

$$T = 2\pi\sqrt{\frac{R}{g}}$$

$$\text{or, } T = 2 \times 3.142 \times \sqrt{\frac{2.5}{10}} = \mathbf{3.14\ s}.$$

Ex. 7 *A particle in SHM is at its mean position at* $t = 0$. *At time* $t = 0.1$ s *and* $t = 0.2$ s *the displacements are* 1 cm *and* 1.5 cm *respectively. Determine the frequency and amplitude of vibration of the particle.*

Soln. Let the displacement be represented by

$$x = A\cos\omega t + B\sin\omega t,$$

where A and B are constants and ω is the angular frequency.

Now, at $t = 0$, $x = 0$, $\therefore\ 0 = A + B \cdot 0$, or, $A = 0$.
So $x = B \sin \omega t$.
Again, since at $t = 0.1$ s, $x = 1$ cm and $t = 0.2$ s, $x = 1.5$ cm.

$$\therefore \quad 1 = B\sin(0.1\omega) \text{ and } 1.5 = B\sin(0.2\omega).$$

$$\text{or,} \quad 1.5 = \frac{\sin(\omega/5)}{\sin(\omega/10)} = \frac{2\sin(\omega/10)\cos(\omega/10)}{\sin(\omega/10)} = 2\cos\left(\frac{\omega}{10}\right)$$

$$\text{or,} \quad \cos\left(\frac{\omega}{10}\right) = \frac{1.5}{2} = 0.75.$$

$$\therefore \quad \frac{\omega}{10} = 41.4° = 0.723 \text{ radian.}$$

$$\therefore \quad \omega = 7.23 \text{ radian.}$$

The frequency of oscillation,

$$n = \omega/2\pi = 1.15 \text{ Hz}$$

and the amplitude of vibration

$$B = \frac{1}{\sin(0.1\omega)} = \frac{1}{\sin 41.4°} = \mathbf{1.51\ cm.}$$

Ex. 8 *A particle is simultaneously subjected to two simple harmonic motions in the same direction, each of frequency* 5 Hz. *If the amplitudes are* 0.005 m *and* 0.002 m *respectively and phase difference between them is* 45°, *find the amplitude of the resultant displacement and its phase relative to the first component.*

Write down the expression for the resultant displacement as a function of time.

[BUH 1997]

Soln. If the phase ϕ_1 for the first component be taken to be zero, then that of the second component

$$\phi_2 = 45° \ (\because \text{ the phase difference is } 45°).$$

Now,

the amplitude of the first SHM, $A_1 = 0.005$ m
and amplitude of the 2nd SHM, $A_2 = 0.002$ m.

$\therefore$ the resultant amplitude is given by

$$A^2 = A_1^2 + A_2^2 + 2A_1A_2\cos(\phi_2 - \phi_1)$$
$$= (0.005)^2 + (0.002)^2 + 2 \times (0.005) \times (0.002) \times \cos 45° = 43.14 \times 10^{-6} \text{ m}^2.$$

$$\therefore\ A = \mathbf{6.57 \times 10^{-3}\ m.}$$

The phase constant of the resultant motion is given by

$$\tan\delta = \frac{A_1 \sin\phi_1 + A_2 \sin\phi_2}{A_1\cos\phi_1 + A_2\cos\phi_2}$$
$$= \frac{(0.005) \times \sin 0° + (0.002) \times \sin 45°}{(0.005) \times \cos 0° + (0.002) \times \cos 45°} = 0.2205.$$

$$\therefore\ \delta = \tan^{-1}(0.2205) = \frac{\boldsymbol{\pi}}{\mathbf{14.4}}.$$

Again, frequency of each SHM is $\nu = 5$ Hz.
$\therefore$ angular frequency $\omega = 2\pi\nu = 10\pi$ radians/s.
Hence, the resultant displacement at any time t is

$$x = \mathbf{6.57 \times 10^{-3} \cos\left(10\pi t + \frac{\pi}{14.4}\right)} \text{ m.}$$

Ex. 9 9 kg *of mercury is poured into a U-tube of uniform internal diameter of* 1.2 cm. *It oscillates freely about its equilibrium position. Calculate the period of oscillation.*

Soln. If m and l be respectively the mass and length of the liquid column in the U-tube, then

$$m = lA\rho \text{ or, } l = \frac{m}{\rho A},$$

where A is the cross sectional area of the tube.

According to the problem,

$$m = 9 \text{ kg}$$

$$\rho = 13.6 \times 10^3 \text{ kg/m}^3$$

$$A = 3.14 \times \left(\frac{1.2}{2}\right)^2 \text{ cm}^2 = 1.1304 \times 10^{-4} \text{ m}^2.$$

$$\therefore\ l = \frac{9}{(13.6 \times 10^3) \times (1.1304 \times 10^{-4})} = 5.854 \text{ m}^2.$$

Hence, the time-period of oscillation of the mercury column is

$$T = 2\pi\sqrt{\frac{l}{2g}} = 2\pi\sqrt{\frac{5.854}{2 \times 9.8}} = \mathbf{3.43\ s.}$$

Ex. 10 *If a mass M be suspended at the lower end of a helical spring of mass m and length l, then find the time-period of oscillation of the system when it is set into oscillation by lowering the mass M.*

Soln. Let us consider any small length δy at a distance y from the fixed end, then its mass is $m\delta y/l$ and its velocity is $v = \frac{\dot{x}y}{l}$ [cf. Fig. 1.27].

$\therefore$ KE of the spring is

$$E_{ks} = \int_0^l \frac{1}{2}\frac{m}{l}\left(\frac{y}{l}\dot{x}\right)^2 dy$$

$$= \frac{1}{2}\frac{m}{l}\left(\frac{\dot{x}}{l}\right)^2 \int_0^l y^2 dy = \frac{1}{6}m\dot{x}^2.$$

KE of the mass M is $E_{kM} = \frac{1}{2}M\dot{x}^2$.

$\therefore$ total KE of the system is

$$E_k = \frac{1}{2}M\dot{x}^2 + \frac{1}{6}m\dot{x}^2$$

$$= \frac{1}{2}\left(M + \frac{1}{3}m\right)\dot{x}^2 = \frac{1}{2}M'\dot{x}^2.$$

Thus, the composed system appears to have a mass

$$M' = M + \frac{1}{3}m.$$

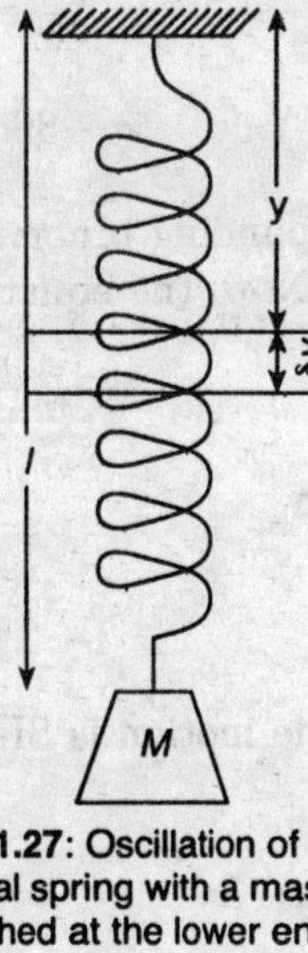

Fig. 1.27: Oscillation of a helical spring with a mass M attached at the lower end.

$\therefore$ the time-period of the system is

$$T = \frac{2\pi}{\sqrt{k/M'}} = \mathbf{2\pi}\sqrt{\frac{\boldsymbol{M + \frac{1}{3}m}}{\boldsymbol{k}}},$$

where k is the force constant.

If the mass m is negligible, then $T = \mathbf{2\pi\sqrt{M/k}}$.

Ex. 11 *The PE of a diatomic molecule is given by*

$$E_p = -\frac{\alpha}{x^6} + \frac{\beta}{x^{12}},$$

where x is the distance between the constituent atoms and α and β are two positive quantities. Determine the equilibrium distance x_0 and the time-period of oscillation. [Assume that one atom is heavy and at rest.]

Soln. Since at equilibrium E_p is the minimum.

$$\therefore \quad \frac{dE_p}{dx} = 0.$$

But

$$\frac{dE_p}{dx} = \frac{6\alpha}{x^7} - \frac{12\beta}{x^{13}} = 0 \text{ or, } x = \left(\frac{2\beta}{\alpha}\right)^{1/6} = x_0 \text{ (say),}$$

the equilibrium separation.

The restoring force on the vibrating atom of mass m is given by

$$F = -\frac{dE_p}{dx} = -\frac{6\alpha}{x^7} + \frac{12\beta}{x^{13}}.$$

If $x = x_0 + x_1$, $(x_1 \geq x_0)$, then

$$F = -\frac{6\alpha}{x_0^7}\left(1 - \frac{7x_1}{x_0}\right) + \frac{12\beta}{x_0^{13}}\left(1 - \frac{13x_1}{x_0}\right)$$

$$\left[\because \frac{1}{(x_1 + x_0)^7} \simeq \frac{1}{x_0^7}\left(1 - \frac{7x_1}{x_0}\right) \text{ and } \frac{1}{(x_1 + x_0)^{13}} \simeq \frac{1}{x_0^{13}}\left(1 - \frac{13x_1}{x_0}\right)\right]$$

$$= \left(\frac{-12 \times 13}{x_0^{14}}\beta + \frac{6 \times 7}{x_0^8}\alpha\right) x_1$$

$$= -36\alpha\left(\frac{\alpha}{2\beta}\right)^{4/3} x_1 \quad \left[\because -\frac{6\alpha}{x_0^7} = \frac{12\beta}{x_0^{13}}\right].$$

[Expanding binomially and retaining terms up to first power in (x_1/x_0).]

Now, the equation of motion of the atom of mass m is

$$m\frac{d^2x_1}{dt^2} = F \text{ or, } m\frac{d^2x_1}{dt^2} = -36\alpha\left(\frac{\alpha}{2\beta}\right)^{4/3} x_1 \text{ or, } \frac{d^2x_1}{dt^2} + \omega^2 x_1 = 0$$

with

$$\omega = \left[\left(\frac{36\alpha}{m}\right)\left(\frac{\alpha}{2\beta}\right)^{4/3}\right]^{1/2} = 3\left(\frac{4\alpha^7}{m^3\beta^4}\right)^{1/6}$$

$\therefore$ the motion is SHM with time-period,

$$T = \frac{2\pi}{\omega} = \frac{2\pi}{3}\left(\frac{m^3\beta^4}{4\alpha^7}\right)^{1/6}$$

Ex. 12 *A particle is executing SHM along a straight line and has velocity $\sqrt{3}b\omega$, when at a distance b from O. If the particle is moving away from O at that instant, show that it will be at a further distance b in time $\pi/3\omega$ before coming to rest.*

Soln. If v and ω be the linear and angular velocity of the particle respectively, then

$$v = \omega\sqrt{a^2 - x^2},$$

where a is the amplitude.

$$\text{or,}\quad \sqrt{3}b\cdot\omega = \omega\sqrt{a^2-b^2}$$

$$\text{or,}\quad 3b^2 = a^2 - b^2$$

$$\text{or,}\quad a = 2b.$$

So the amplitude of vibration of the SHM is $a = 2b$.

$\therefore$ the displacement of the particle executing SHM at time t is

$$x = a\sin\omega t \text{ or, } b = 2b\sin\omega t.$$

$$\therefore\quad \sin\omega t = \frac{1}{2} = \sin\frac{\pi}{6}.$$

$$\therefore\quad \omega t = \frac{\pi}{6} \text{ or, } t = \frac{\pi}{6\omega}.$$

So the time required by the particle to travel a further distance b from the mean position is $\pi/6\omega$. As the periodic time $T = \frac{2\pi}{\omega}$,

$$\therefore\quad \frac{T}{4} = \frac{2\pi}{\omega}\times\frac{1}{4} = \frac{\pi}{2\omega}.$$

But the particle has taken a time $\pi/6\omega$ to travel a distance b. Hence, it will take

$$\frac{\pi}{2\omega} - \frac{\pi}{6\omega} = \frac{\pi}{3\omega}$$

to travel a further distance b.

EXERCISES

1(A)–Objective Type Questions

Select the correct answer:

1.1. For a particle executing SHM, the acceleration is (a) directly proportional to the displacement from the mean position (b) constant (c) inversely proportional to the displacement from the mean position (d) inversely proportional to the square of the displacement from the mean position.

1.2. The maximum velocity v_m of a particle executing SHM represented by $x = A\sin\omega t$ at time t occurs at (a) $x = 0$ (b) $x = A$ (c) $x = -A$ (d) $x = A/2$.

1.3. In the above problem the acceleration is the maximum at (a) $x = 0$ (b) $x = A$ (c) $x = -A$ (d) $x = A/3$.

1.4. The time-period of a particle executing SHM is (a) $T = 2\pi\sqrt{x/f}$ (b) $T = 2\pi\sqrt{xf}$ (c) $T = 2\pi\sqrt{f/x}$ (d) $T = \sqrt{x/f}$.

1.5. If the acceleration of a particle is given by $\alpha = -\beta x$, where x is the distance from the mean position and β is a constant, then its time-period is (a) $2\pi/\beta$ (b) $2\sqrt{\pi}/\beta$ (c) $2\pi/\sqrt{\beta}$ (d) $2\pi\sqrt{\beta}$.

1.6. If the amplitude of a SHM is 4 cm, then its energy will be half-potential and half-kinetic for displacement x from the mean position, when (a) $x = 1$ cm (b) 2 cm (c) $2\sqrt{2}$ cm (d) $\sqrt{2}$ cm.

1.7. Masses M_1 and M_2 are suspended from a spring of negligible mass and force constant k. When the masses are in equilibrium, M_1 is removed without disturbing the system. Then the angular frequency of M_2 is (a) $\sqrt{\frac{k}{M_2}}$ (b) $\sqrt{\frac{k}{M_1+M_2}}$ (c) $\sqrt{\frac{k}{M_1M_2}}$ (d) $\sqrt{\frac{k}{M_1}}$.

1.8. A uniform spring of mass constant k is cut into two pieces whose lengths are in the ratio 1: 2. The ratio of force constant of the pieces is (a) $k: 3k$ (b) $k: 2k$ (c) $3k: \frac{3}{2}k$ (d) $3k: 2k$.

1.9. For the production of the phenomenon of beats, the two collinear SHM must superpose on a medium particle with angular frequencies ω_1 and ω_2 such that (a) $\omega_1 = \omega_2$ (b) $\omega_1 = 2\omega_2$ (c) $\omega_1 \simeq \omega_2$ (d) $\omega_1 \neq \omega_2$.

1.10. To get circular Lissajous figure, the phase difference ϕ and the amplitudes (a and b) of two superposing mutually perpendicular SHMs respectively are (a) $\phi = 0;\ a = b$ (b) $\phi = \frac{\pi}{2};\ a = b$ (c) $\phi = \frac{\pi}{2};\ a \neq b$ (d) $\phi = 0;\ a \neq b$. **[WBUT 2006]**

1.11. If the effective length of a simple pendulum is increased to infinity, then its time-period T will be (a) infinite (b) finite (c) 0 (d) none of these.

1.12. The differential equation of motion of a particle executing SHM along the X-axis is (a) $\frac{d^2x}{dt^2} + \omega^2 x = 0$ (b) $m\frac{d^2x}{dt^2} + \omega^2 x = 0$ (c) $\frac{d^2x}{dt^2} - \omega^2 x = 0$ (d) $\frac{dx}{dt} + \frac{x}{\omega} = 0$.

1.13. The time-period of the SHM in problem 1.12 is (a) $\frac{2\pi}{\omega}$ (b) $\frac{\omega}{2\pi}$ (c) $2\pi\omega$ (d) none of these.

1.14. If the amplitudes of two rectangular SHMs of equal amplitude and differing in phase by $\pi/2$, superpose at a point in the medium, then the resultant motion will be (a) parabolic (b) elliptic (c) circular (d) none of these.

1.15. The average KE of a particle executing SHM is (a) $4ma^2\omega^2$ (b) $\frac{1}{4}m\omega^2a^2$ (c) $ma^2\omega^2$ (d) $\frac{4}{m\omega^2a^2}$.

1.16. The time-period T of torsional oscillation of a cylinder is (a) $2\pi\sqrt{\frac{C}{I}}$ (b) $2\pi\sqrt{IC}$ (c) $2\pi\sqrt{\frac{I}{C}}$ (d) $\sqrt{\frac{I}{2\pi C}}$.

1.17. Lissajous figures are produced when two rectangular SHMs act simultaneously on a medium particle with (a) same frequency (b) commensurate frequencies (c) large frequencies (d) none of these.

1.18. SHM is basically a (a) linear motion (b) circular motion (c) periodic motion (d) none of these.

1.19. Two mutually perpendicular oscillations with same frequency, amplitude but phase difference δ, which produce closed curve with non-zero area enclosed (a) for all values of δ except $\delta = 0$ (b) only for $\delta = \frac{\pi}{2}$ (c) for all values of δ except $\delta = 0$ and $\delta = \pi$ (d) for all values of $\delta \geq \frac{\pi}{2}$. **[WBUT 2007]**

1.20. Waves originating from a point source and travelling in any isotropic medium is described as (a) $\phi = \phi_0 \exp i(\vec{k}\cdot\vec{r} - \omega t)$ (b) $\phi = \phi_0 \exp i(\vec{k}\cdot\vec{r} - \omega t)/r$ (c) $\phi = \phi_0 \exp i(\vec{k}\cdot\vec{r} - \omega t)/r^2$. (d) $\phi = \phi_0 \exp i(\vec{k}\cdot\vec{r} + \omega t)/r$. **[WBUT 2007]**

1.21. Example of weakly damped harmonic oscillator is (a) Dead-beat galvanometer (b) Tangent galvanometer (c) Ballistic galvanometer (d) discharge of a charged capacitor through a resistance. **[WBUT 2007]**

1.22. If a particle is executing simple harmonic motion with frequency ν, then its PE (a) remains constant over time (b) is oscillating with frequency ν (c) is oscillating with frequency $\nu/2$ (d) is oscillating with a frequency 2ν. **[WBUT 2010]**

ANSWERS

1.1. (a) **1.2.** (b) **1.3.** (b) **1.4.** (a) **1.5.** (c) **1.6.** (c) **1.7.** (a) **1.8.** (c) **1.9.** (c) **1.10.** (b) **1.11.** (a) **1.12.** (a) **1.13.** (a) **1.14.** (c) **1.15.** (b) **1.16.** (c) **1.17.** (b) **1.18.** (b) **1.19.** (b) **1.20.** (a) **1.21.** (c) **1.22.** (b).

1(B)–Short Answer Type Questions

1.1. State the characteristic of SHM. Define phase. Prove the principle of conservation of energy. [WBUT 2001]

1.2. Explain the terms *phase* and *phase difference* as applied to SHM.

1.3. What properties of the medium are essential for the production of SHM?

1.4. Show that for a particle executing SHM, the average KE is equal to the average PE and each is half of the maximum energy at any instant of time.

1.5. Prove that the total mechanical energy of a particle executing SHM is constant.

1.6. Obtain the differential equation of motion of a particle executing SHM.

1.7. Deduce the equation of motion of a floating stick in a liquid.

1.8. Write the forms of the SHM for the following boundary conditions: (a) at $t = 0$, $x = a$ and $\dot{x} = 0$ (b) at $t = 0$, $x = 0$ and $\dot{x} = v$ (c) at $t = 0$, $x = x_r$ and $\dot{x} = \dot{x}_r$.

1.9. What are Lissajous figures? How can they be demonstrated?

1.10. A mass is attached at the free end of a vertically fixed spring. What is the time-period of oscillation, if the mass be depressed a little and then released?

1.11. Show that the equation $y = mx + ct + k$ cannot represent a SHM.

1(C)–Numerical Problems

1.1. If $y_1 = 4\sin(10t + \phi)$ and $y_2 = 5\cos(10t)$ be the displacements of two particles at time t, then find the phase difference between the velocities of the particles. [*Ans.* $(\phi - \pi/2)$]

1.2. The displacement of a particle executing SHM is given by $y = \sin kt + \cos kt$. Find (a) time-period, (b) amplitude of vibration. [*Ans.* $T = 2\pi/k, \sqrt{2}$]

1.3. N number of springs of spring constants $k_1, k_2, \ldots, k_N$ are attached one after another and the combination hangs vertically. Find the time-period of oscillation of a mass m attached at the free end of the spring. $\left[\textit{Ans. } 2\pi\sqrt{m \times \sum_{i=1}^{M}\left(\frac{1}{k_i}\right)}\right]$

[**Hints.** The effective spring constant is given by

$$\frac{1}{k} = \frac{1}{k_1} + \frac{1}{k_2} + \cdots + \frac{1}{k_N} = \sum_{i=1}^{N}\left(\frac{1}{k_i}\right).$$

$$\therefore\ k = \frac{1}{\sum_{i=1}^{N}\left(\frac{1}{k_i}\right)}.$$

So time-period,

$$T = 2\pi\sqrt{\frac{m}{k}} = 2\pi\sqrt{\frac{m}{\left\{1\Big/\sum_{i=1}^{N}\left(\frac{1}{k_i}\right)\right\}}} \quad \text{or,}\quad T = 2\pi\sqrt{m \times \sum_{i=1}^{M}\left(\frac{1}{k_i}\right)}.]$$

1.4. A particle of mass 0.1 kg is held between two rigid supports by two springs of force constants $8N/m$ and $2N/m$. If the particle is displaced along the direction of the length of the springs, calculate the frequency of vibration. [*Ans.* $5/\pi$ Hz]

1.5. A point describes SHM in a line 4 cm long. Its velocity, when passes through the centre of the line, is 12 cm/s. Find the time-period of oscillation. [*Ans.* 1.5 s]

1.6. A body executing SHM has an amplitude of 10 cm and time-period 1.5 s. Calculate the time taken by the body to travel a distance $5\sqrt{3}$ from its mean position. [*Ans.* 0.25 s]

1.7. When a man, weighing 60 kg on earth, gets into a car, the centre of gravity of the car is depressed by 0.3 cm. Assuming the mass of the car to be 1000 kg, find the frequency of vibration of the car when it is empty. [*Ans.* 2.2 Hz]

1.8. A point mass m is suspended at the end of a massless wire of length l and cross section A. If Y be the Young's modulus of elasticity of the wire, obtain the frequency of oscillation for the SHM along the vertical line. $\left[Ans.\ \frac{1}{2\pi}\sqrt{\frac{AY}{ml}}\right]$

[IIT 1978]

1.9. A vertical U-tube of uniform cross section contains water up to a height of 30 cm. Show that, if the water of one side is depressed and then released, its motion up and down the two sides of the tube is simple harmonic. Calculate its frequency.

1.10. If the mass and diameter of a planet are twice that of the earth, what will be the period of oscillation of a pendulum on that planet, if it is a seconds pendulum on earth.

1.11. A solid cylinder of mass M is attached to a massless horizontal spring of spring constant k and it executes SHM, when rolls without slipping. Show that, its period of oscillation is

$$T = 2\pi\sqrt{\frac{3M}{2k}}.$$

1.12. A particle of mass 10 gm is describing SHM along a straight line with a period of 2*s* and amplitude 10 cm. What is the KE when it is (a) 2 cm away from its position of equilibrium (b) 5 cm away from its position of equilibrium.

1.13. Two vibrations along the same line are described by the equations $x_1 = 0.05\cos 8\pi t$ and $x_2 = 0.03\cos 10\pi t$, where x is in metre and t in second. Obtain the equation of the resultant motion and hence beat period. **[BUH 1993]**

[*Hints.* $[x = \sqrt{2}\times 10^{-2}\{17 + 15\cos(2\pi t)\}^{1/2}]\cos(9\pi t + \delta)$, $\delta = \tan^{-1}(0.25\tan \pi t)$, beat period = 1 s.]

1.14. Two vibrations at right angles to each other are described by the equations $x = 3\cos 4\pi t$, $y = 3\cos\left(8\pi t + \frac{\pi}{3}\right)$, where x and y are expressed in cm and t in s. Construct the Lissajous figures of the combined motion. **[BUH 1994]**

1.15. A certain fork is found to give 2 beats per second when sounded in conjunction with a stretched string vibrating transversely under a tension of either 10.2 kg-wt or 9.9 kg-wt. Calculate the frequency of the form. [*Ans.* 268.6 Hz]

1.16. A steel wire 50 cm long and 5 mm in diameter gives a note of 240 vibrations when stretched with a certain weight. A second steel wire bear the same weight as the first, but is 40 cm long and 6 mm diameter. Find the periodic time of its fundamental note. [*Ans.* 0.004 s]

1(D)–Broad Answer Type Questions

1.1. (a) What are the characteristics of a simple harmonic motion? Define phase. Prove the principle of conservation of energy in this case. (b) A cubical box of side L and density d is floating on water of density $\rho(\rho > d)$. The block is slightly depressed and released. Show that it will execute simple harmonic motion. Determine the frequency of oscillations. **[WBUT 2001]**

1.2. Establish the differential equation of simple harmonic motion and solve the equation. **[WBUT 2004]**

1.3. (a) Prove that the motion of a simple pendulum of small oscillation is simple harmonic. Hence, find the expression for the period of oscillation. (b) Prove that the vertical oscillation of a loaded spring is simple harmonic and hence find an expression for the time-period of its oscillation.

1.4. Two SHMs having the same period but differing in phase and amplitude are acting in the same direction on a particle. Deduce the expression for the amplitude and phase of the resulting motion. Discuss the special cases, when the phase difference is $0, \pi/2$ and π respectively.

1.5. Two vibrations at right angles to each other are described by the equations:

$$x = A_1 \cos \omega t$$

$$y = A_2 \cos(2\omega t + \delta).$$

Superpose these two vibrations to get the resulting equation. If $\delta = 0$, show that the resulting path represents two coincident parabolas with their vertices $(0, -A_2)$. **[BUH 1996]**

1.6. What are beats? Give an analytical description of the phenomenon of beats and show that the beat frequency is equal to the difference between the frequencies of the component oscillations. **[CUH 1999; BUH 1998, 2000]**

1.7. (a) Prove that the motion of a simple pendulum of small oscillation is simple harmonic. Hence, find an expression for the time-period of its oscillation. (b) Prove that the vertical oscillations of a loaded spring are simple harmonic and hence find out an expression for the time-period of its oscillation.

1.8. Two simple harmonic motions having the same period but differing in phase and amplitude are acting in the same direction on a particle. Deduce the expression for the amplitude and phase of the resulting motion. Discuss the special cases when the phase difference is $0, \pi/2$ and π respectively.

1.9. Two simple harmonic vibrations of same frequency at right angles to each other superpose at a point. Discuss the nature of resultant motion when their phase difference (i) $\phi = 0$ and (ii) $\phi = \pi$.

1.10. (a) Two SHM of different frequencies superpose at a point. Discuss the nature of resultant vibration. (b) Describe an experiment to demonstrate Lissajous figures. (c) What are the uses of Lissajous figures?

Chapter 2 | Module 1.2

Vibration and Wave—Free and Damped Vibrations

2.1 Free Vibration

If a body is set into vibration by applying a force and left to itself, then after removing the force, the body continues to oscillate for ever with constant frequency and constant amplitude. The frequency of the body is determined by the inertial and elastic properties of the body and is called its *natural frequency.* This type of simple harmonic motion, which persists indefinitely without loss of amplitude is called *free (or undamped) vibration.* This type of vibration is, however, an idealized conception.

If x be the displacement of a particle of mass m from the mean position, then the differential equation of the vibration is

$$\frac{d^2x}{dt^2} + \omega_0^2 x = 0,$$

where $\omega_0 = \sqrt{k/m}$, the natural frequency of vibration and k, the restoring force per unit displacement.

The above equation represents a differential equation of motion whose solution is

$$x = A\sin(\omega t + \phi),$$

here ϕ is the initial phase and A, the amplitude of vibration. Since A is constant w.r.t. time, the total energy of the particle is constant. That is why such a system is called a *conservative system.*

Such a vibration is an ideal one and in practical situation does not occur.

Example. The oscillation of the bob of a simple pendulum in vacuum, the vibrations of the prong of a tuning fork, the motion of the coil of a suspended coil type galvanometer (in vacuum), etc. are the examples of free motion.

2.2 Damped Vibration

In practical world, however, the amplitude decays with time and ultimately the body comes to rest. This decay of vibrational amplitude with time is due to the presence of frictional resistance (known as damping). To maintain its motion the vibrating body loses energy continuously so that the amplitude decreases with time. This type of vibration whose amplitude decays continuously with time in presence of damping is known as *damped vibration.*

Example. The motion of a mass attached to one end of a spring, the other end being attached to a rigid support, the motion of the coil of a suspended coil galvanometer in air are the examples of damped vibrations. The damping is caused by the friction of air and at the point of suspension. It is to be noted that the frequency of damped vibration is slightly less than the natural frequency. The time-displacement curves for free and damped vibrations are indicated by curves (a) and (b) respectively [Fig. 2.1].

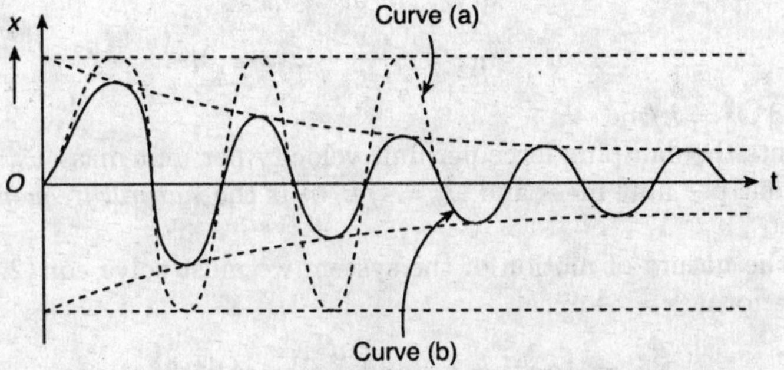

Fig. 2.1: Time-displacement curve of a free and damped oscillator.

2.3 Damping Forces

The damping force acting on a real system arises due to the following:

(a) **Viscosity.** When an oscillating system executes motion in a fluid, the viscous retarding force offered by the fluid in contact with the system causes damping of the system. So here the damping force is the *viscous force.*

(b) **Coulomb friction force.** An oscillating body executing motion in a horizontal plane and connected to a spring, is always in contact with the horizontal surface. The resulting frictional force opposes the motion. The damping force here is thus *Coulomb friction force.*

(c) **Structural damping.** In a solid, a part of energy may be dissipated due to imperfect elasticity or internal friction of the material. The damping force arising due to above is proportional to the amplitude of vibration but independent of frequency. This type of damping is due to *structural properties.*

The damping force F depending on the velocity of the particle in fluid has magnitude given by

$$F = p_1 v + p_2 v^2.$$

If $v < (p_1/p_2)$, then $F = -pv$, where p is the viscous damping force per unit velocity and is called *viscous damping coefficient.* The negative sign indicates that the force opposes the motion and causes damping of the system.

2.4 Analysis of Damped Vibration

Let a damped harmonic oscillator having *one degree of freedom* be at a distance x from the equilibrium position during motion at any instant of time.

The forces acting on the system are as follows:

(a) **Force of restitution.** This force tends to restore the system to its original equilibrium position and is proportional to displacement x and is given by $-kx$, where k is the restoring force per unit displacement.

(b) **Damping force.** This force acting on the oscillating system causes dissipation of energy and is given by $-p\frac{dx}{dt}$, where p is the damping force per unit velocity.

(c) **Force of inertia.** This force is given by $m\frac{d^2x}{dt^2}$, where m is the mass of the oscillator.

According to *Newton's second law of motion*, the equation of motion of the oscillating particle is

$$m\frac{d^2x}{dt^2} = -p\frac{dx}{dt} - kx$$

$$\text{or,}\quad \frac{d^2x}{dt^2} + \frac{p}{m}\frac{dx}{dt} + \frac{k}{m}x = 0$$

$$\text{or,}\quad \frac{d^2x}{dt^2} + \gamma\frac{dx}{dt} + \omega_0^2 x = 0, \tag{2.1}$$

where $\gamma = p/m$ and $\omega_0^2 = k/m$.

Here, γ represents the damping force per unit velocity per unit mass, ω_0^2 is the restoring force per unit displacement per unit mass and $\omega_0 = \sqrt{k/m}$ is the *natural frequency* of the undamped system.

To investigate the nature of motion of the system we must solve eqn (2.1). Let us suppose a trial solution of the form, $x = Ae^{\alpha t}$.

$$\therefore\quad \frac{dx}{dt} = A\alpha e^{\alpha t} = \alpha x \text{ and } \frac{d^2x}{dt^2} = A\alpha^2 e^{\alpha t} = \alpha^2 x.$$

Substituting these values of $\frac{dx}{dt}$ and $\frac{d^2x}{dt^2}$ in eqn (2.1), we get

$$(\alpha^2 + \gamma\alpha + \omega_0^2)x = 0$$

$$\text{or,}\quad \alpha^2 + \gamma\alpha + \omega_0^2 = 0 \;[\because\; x \neq 0].$$

This is a quadratic equation in α and so α has the following two values:

$$\alpha_1 = -\frac{\gamma}{2} + \frac{1}{2}(\gamma^2 - 4\omega_0^2)^{1/2} \text{and } \alpha_2 = -\frac{\gamma}{2} - \frac{1}{2}(\gamma^2 - 4\omega_0^2)^{1/2}.$$

Thus, the two possible solutions of eqn (2.1) are

$$x_1 = A_1 e^{\alpha_1 t} = A_1 e^{\left[-\frac{\gamma}{2} + \frac{1}{2}(\gamma^2 - 4\omega_0^2)^{1/2}\right]t}$$

$$\text{and } x_2 = A_2 e^{\alpha_2 t} = A_2 e^{\left[-\frac{\gamma}{2} + \frac{1}{2}(\gamma^2 - 4\omega_0^2)^{1/2}\right]t}$$

$$\text{or, } x = x_1 + x_2 = A_1 e^{\left[-\frac{\gamma}{2} + \left(\frac{\gamma^2}{4} - \omega_0^2\right)^{1/2}\right]t} + A_2 e^{\left[-\frac{\gamma}{2} - \left(\frac{\gamma^2}{4} - \omega_0^2\right)^{1/2}\right]t}, \tag{2.2}$$

where A_1 and A_2 are two arbitrary constants to be determined from the initial conditions.

Case I *If $\gamma > 2\omega_0$, i.e., for large damping*

$$\left(\frac{\gamma^2}{4} - \omega_0^2\right) > 0.$$

So

$$\left(\frac{\gamma^2}{4} - \omega_0^2\right)^{1/2} = q$$

is *real and positive*. From eqn (2.2), we get

$$x = A_1 e^{(-\frac{\gamma}{2} + q)t} + A_2 e^{(-\frac{\gamma}{2} - q)t} \tag{2.3}$$

and the velocity V of the particle is

$$\frac{dx}{dt} = \left(-\frac{\gamma}{2} + q\right) A_1 e^{(-\frac{\gamma}{2} + q)t} - \left(\frac{\gamma}{2} + q\right) A_2 e^{(-\frac{\gamma}{2} - q)t}. \tag{2.4}$$

Let the oscillator be in equilibrium position at $t = 0$. If a kick be given to it at this position, then $V = V_0$ (say). Here the initial conditions are:

(i) $x = 0$ at $t = 0$

(ii) $V = V_0$ at $t = 0$.

Now, applying the first boundary condition to eqn (2.3), we get

$$0 = A_1 + A_2, \therefore\ A_2 = -A_1$$

and applying second condition to eqn (2.4),

$$\begin{aligned} V_0 &= \left(-\frac{\gamma}{2} + q\right) A_1 - \left(\frac{\gamma}{2} + q\right) A_2 \\ &= \left(-\frac{\gamma}{2} + q\right) A_1 + \left(\frac{\gamma}{2} + q\right) A_1 \quad [\because\ A_2 = -A_1] \\ &= A_1 \left[\frac{\gamma}{2} + q - \frac{\gamma}{2} + q\right] = 2qA_1 \\ \text{or, } A_1 &= \frac{V_0}{2q} = -A_2. \end{aligned}$$

Hence, from eqn (2.3),

$$\begin{aligned} x &= \frac{V_0}{2q} e^{-\frac{\gamma}{2}t} \left(e^{qt} - e^{-qt}\right) \\ &= \frac{V_0}{q} e^{-\frac{\gamma t}{2}} \sinh(qt) \quad \left[\because\ \frac{e^{qt} - e^{-qt}}{2} = \sinh(qt)\right] \qquad (2.5) \\ \text{and } \frac{dx}{dt} &= \frac{V_0}{2} e^{-\frac{\gamma t}{2}} \left[\left(e^{qt} + e^{-qt}\right) - \frac{\gamma}{2q}\left(e^{qt} - e^{-qt}\right)\right] \\ &= V_0 e^{-\frac{\gamma t}{2}} \left[\cos h(qt) - \frac{\gamma}{2q} \sinh(qt)\right]. \qquad (2.6) \end{aligned}$$

From eqn (2.4) it is clear that for $t \to 0$, $e^{-\gamma t/2} \to 1$ and $\sin(qt)$ increases with the increase of t. So x increases with t initially. But with the increase of t, $e^{-\gamma t/2}$ decreases exponentially with t and at a particular value of t, $e^{-\gamma t/2}$ reduces to zero and hence x reduces to zero. At $t = t_0$ (say), the turning point occurs when $dx/dt = 0$.

From eqn (2.6),

$$\begin{aligned} \cosh(qt_0) - \frac{\gamma}{2q} \sinh(qt_0) &= 0 \text{ at } \frac{dx}{dt} = 0 \\ \text{or, } \tanh(qt_0) &= \frac{2q}{\gamma}. \qquad (2.7) \end{aligned}$$

The time-displacement curve of the oscillator for large damping is shown in Fig. 2.2, which shows that x increases up to $t = t_0$ and then decreases to zero but x never becomes negative. This type of motion of the damped oscillator is called *dead beat.* The motion of a pendulum bob through a thick viscous liquid and the motion of the coil of an over-damped dead beat galvanometer are the examples of *dead beat* motion. This type of vibration has important practical applications in the *electrical recording of sound by means of an oscillograph and in the design of certain types of sound receivers.*

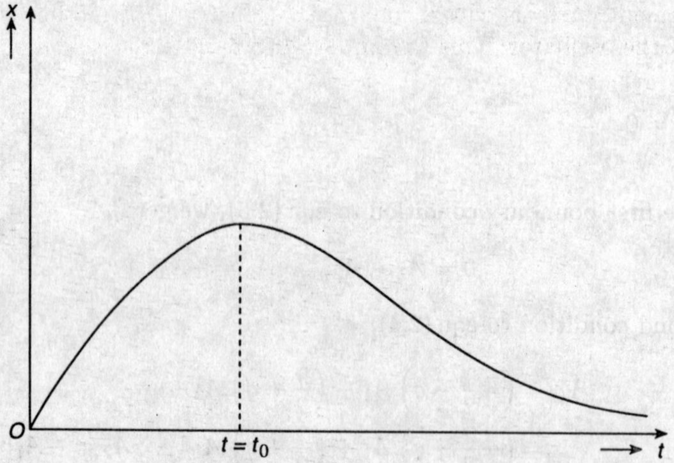

Fig. 2.2: Time displacement curve of an oscillator of large damping.

Case II *If* $\gamma = 2\omega_0$ (*critical damping*), eqn (2.2) *reduces to*

$$x = A_1 e^{-\gamma t/2} + A_2 e^{-\gamma t/2}$$

$$\left[\because q = \left(\frac{\gamma^2}{4} - \omega_0^2\right)^{1/2} = 0, \text{ when } \frac{\gamma}{2} = \omega_0\right]$$

$$= (A_1 + A_2)e^{-\frac{\gamma t}{2}}$$

$$= Be^{-\frac{\gamma t}{2}}, \text{ where } B = A_1 + A_2. \tag{2.8}$$

But according to *the theory of differential equation*, the solution of a second order differential equation must involve *two constants* and as such eqn (2.8) cannot be an accepted solution of eqn (2.1). The solution can, however, be expressed as

$$x = (B + Ct)e^{-\gamma t/2}, \tag{2.9}$$

where both B and C are constants whose values are to be found out from the initial conditions. Since $x = 0$ and $dx/dt = V_0$ at $t = 0$, we get from eqn (2.9), $B = 0$ and $C = V_0$.

$$\therefore\ x = V_0 t e^{-\gamma t/2} \tag{2.10}$$

$$\text{and } \frac{dx}{dt} = = C\left(e^{-\gamma t/2} - t\frac{\gamma}{2}e^{-\gamma t/2}\right) = C\left(1 - \frac{\gamma t}{2}\right)e^{-\gamma t/2}. \tag{2.11}$$

Eqn (2.10) shows as $t \to 0$, $e^{-\gamma t/2} \to 1$ and x increases with the increase of t. After sometime, $e^{-\gamma t/2}$ starts contributing reducing x eventually ~~to~~ zero. At $t = t_0^*$, x becomes maximum and $dx/dt = 0$.

From eqn (2.11) at $t = t_0^*$,

$$0 = C\left(1 - \frac{\gamma t_0^*}{2}\right)e^{-\gamma t_0^*/2}.$$

$$\because Ce^{-\gamma t_0^*/2} \neq 0, \therefore 1 - \frac{\gamma t_0^*}{2} = 0 \text{ or, } t_0^* = \frac{2}{\gamma}$$

$$\text{or, } t_0^* = \frac{1}{\omega_0}\left[\because \gamma = 2\omega_0, \therefore \frac{2}{\gamma} = \frac{1}{\omega_0}\right]$$

The time-displacement curve is shown in Fig. 2.3. This type of motion is called the *critically damped* motion of the oscillator. This has important application in *pointer type galvanometer.*

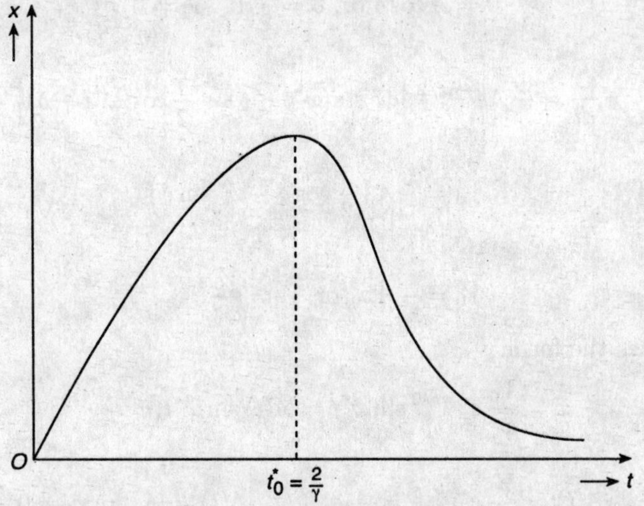

Fig. 2.3: Time displacement curve of a critically damped oscillator.

Case III $\gamma < 2\omega_0$ (*Weak damping*).

The case of *weak damping* is most important as it leads to *oscillatory damped harmonic motion.*

$$\because\ \frac{\gamma}{2} < \omega_0,\ \left(\frac{\gamma^2}{4} - \omega_0^2\right)^{1/2} = \sqrt{-1}\left(\omega_0^2 - \frac{\gamma^2}{4}\right)^{1/2} = j\omega^*,$$

where

$$j = \sqrt{-1} \text{ and } \omega^* = \left(\omega_0^2 - \frac{\gamma^2}{4}\right)^{1/2}$$

is a *real positive quantity.*

Now, from eqn (2.2),

$$x = e^{-\gamma t/2}\left[A_1 e^{j\omega^* t} + A_2 e^{-j\omega^* t}\right]$$
$$= e^{-\gamma t/2}[(A_1 + A_2)\cos\omega^* t + j(A_1 - A_2)\sin\omega^* t]. \tag{2.12}$$

Put $A_1 + A_2 = A\cos\delta$ and $j(A_1 - A_2) = A\sin\delta$, where A and δ are new constants to be found out from the initial condition of the problem.

$$\therefore\ x = Ae^{-\gamma t/2}\cos(\omega^* t - \delta). \tag{2.13}$$

Eqn (2.13) shows that the motion is oscillatory but not simple harmonic as the amplitude is not constant but decays exponentially with time due to the term $e^{-\gamma t/2}$. The motion is not even strictly periodic since it never repeats itself as the successive amplitudes decay exponentially with time. But if $\gamma \to 0$, then

$$\omega^* = \left(\omega_0^2 - \frac{\gamma^2}{4}\right)^{1/2} \to \omega_0 \text{ and } e^{-\gamma t/2} \to 1$$

and the *amplitude remains almost constant* over a large number of cycles. Under this condition the motion is *almost periodic and simple harmonic* of natural frequency ω_0.

If we assume $x = 0$ and $dx/dt = V_0$ at $t = 0$, then from eqn (2.13), we get

$$0 = A\cos\delta \text{ or, } \delta = \frac{\pi}{2} \; (\because \; A \neq 0).$$

Also

$$\frac{dx}{dt} = -Ae^{-\gamma t/2}\left[\omega^* \sin(\omega^* t - \delta) + \frac{\gamma}{2}\cos(\omega^* t - \delta)\right] \tag{2.14}$$

reduces to

$$V_0 = -A\left(\frac{\gamma}{2}\cos\delta - \omega^* \sin\delta\right).$$

Putting $\delta = \pi/2$,

$$V_0 = +A\omega^* \text{ or, } A = \frac{V_0}{\omega^*}.$$

Then eqn (2.13) takes the form

$$x = \frac{V_0}{\omega^*}e^{-\gamma t/2}\sin\omega^* t = A(t)\sin\omega^* t, \tag{2.15}$$

where

$$A(t) = \frac{V_0}{\omega^*}e^{-\gamma t/2} = A_0 e^{-\gamma t/2}, \; A(t) = A_0 \text{ for } \gamma = 0.$$

From eqn (2.14),

$$\frac{dx}{dt} = V_0 e^{-\gamma t/2}\left[\cos\omega^* t - \frac{\gamma}{2\omega^*}\sin\omega^* t\right] \; \left[\because \; A = \frac{V_0}{\omega^*}, \delta = \frac{\pi}{2}\right]. \tag{2.16}$$

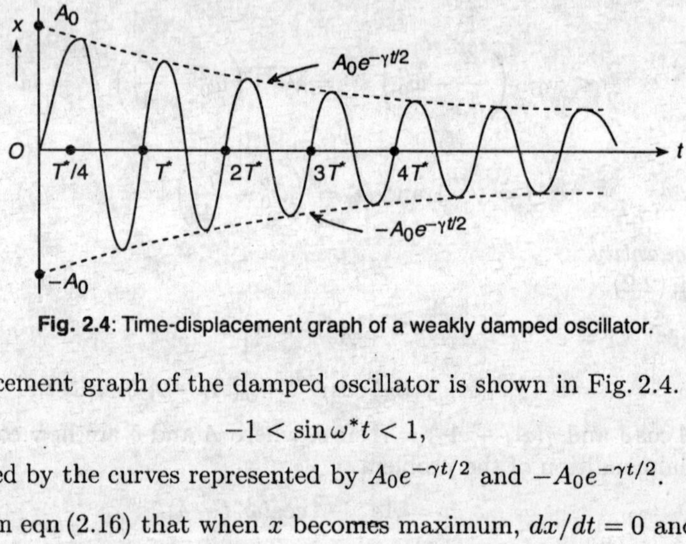

Fig. 2.4: Time-displacement graph of a weakly damped oscillator.

The time-displacement graph of the damped oscillator is shown in Fig. 2.4. Since

$$-1 < \sin\omega^* t < 1,$$

the graph is bounded by the curves represented by $A_0e^{-\gamma t/2}$ and $-A_0e^{-\gamma t/2}$.

It is evident from eqn (2.16) that when x becomes maximum, $dx/dt = 0$ and so

$$0 = \cos\omega^* t - \frac{\gamma}{2\omega^*}\sin\omega^* t = 0 \text{ or, } \tan\omega^* t = \frac{2\omega^*}{\gamma}.$$

When $\gamma \ll 2\omega_0$, i.e., $\frac{2\omega_0}{\gamma} \to \infty$,

$$\tan\omega^* t = \frac{2\omega^*}{\gamma} = \frac{2\omega_0}{\gamma} \to \infty \left[\because \; \omega^* = \left(\omega_0^2 - \frac{\gamma^2}{4}\right)^{1/2} \to \omega_0\right].$$

$$\therefore \; \omega^* t = (2n+1)\frac{\pi}{2} = \frac{\pi}{2}, \frac{3\pi}{2}, \frac{5\pi}{2}, \cdots,$$

x becomes maximum, when $t = t_1$ given by

$$\omega^* t_1 = \frac{\pi}{2} \text{ or, } t_1 = \frac{\pi}{2\omega^*} = \frac{T^*}{4} \left[\text{as } T^* = \frac{2\pi}{\omega^*}\right],$$

where T^* is the time-interval between two successive zeros of displacement and is called the *time-period* of the damped oscillation. The maximum displacements are also midway between two zeros of x.

Hence, just as in case of SHM, *the maxima and minima are midway between the zeros of displacement.*

Effect of the damping force

The effect of the damping force on the oscillator is of *two-fold*:

(i) The loss of amplitude from A_0 to $A_0 e^{-\gamma t/2}$ and

(ii) Fall of angular frequency from ω_0 to $\omega^* = \left(\omega_0^2 - \frac{\gamma^2}{4}\right)^{1/2} = \omega_0 \left(1 - \frac{\gamma^2}{4\omega_0^2}\right)^{1/2}$.

2.4.1 Logarithmic Decrement

In case of damped oscillation, *logarithmic decrement measures the rate of decrease of amplitude.* The displacement of a damped oscillator at any time t is

$$x = A_0 e^{-\gamma t/2} \sin \omega^* t = A(t) \sin\left(\frac{2\pi}{T^*}\right) t, \tag{2.17}$$

where $A(t) = A_0 e^{-\gamma t/2}$, A_0 is the amplitude of the undamped oscillator and $T^* = \frac{2\pi}{\omega^*}$ is its time-period. Fig. 2.5 represents the time-displacement graph of a damped oscillator. The displacement is zero at $t = 0, \frac{T^*}{2}, \frac{2T^*}{2}$ and maximum at $t = \frac{T^*}{4}, \frac{3T^*}{4}, \frac{5T^*}{4}, \cdots$ as evident from eqn (2.17).

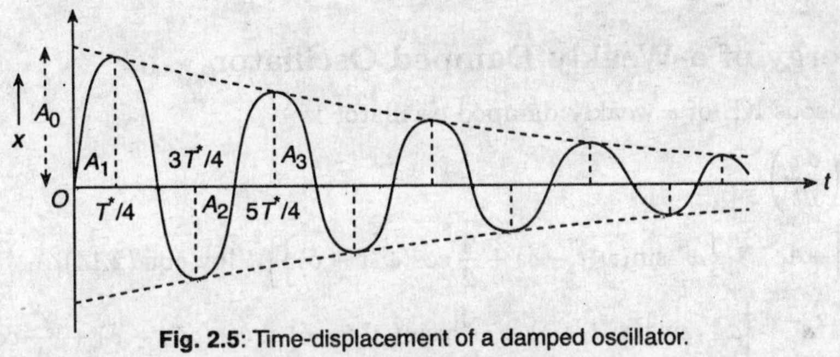

Fig. 2.5: Time-displacement of a damped oscillator.

Eqn (2.17) shows that at $t = \frac{T^*}{4}, \frac{3T^*}{4}, \frac{5T^*}{4}$, the amplitudes are

$$A_1 = A_0 e^{-\frac{\gamma T^*}{8}}$$

$$A_2 = A_0 e^{-\frac{3\gamma T^*}{8}}$$

$$A_3 = A_0 e^{-\frac{5\gamma T^*}{8}} \text{ and so on.}$$

$$\therefore \frac{A_1}{A_2} = \frac{A_2}{A_3} = \cdots = \frac{A_{n-1}}{A_n} = e^{\frac{\gamma T^*}{4}} = d \text{ (say)},$$

where d is a constant. Thus, the ratio of the two successive amplitudes of the damped oscillator is a constant (d) and is termed as the *decrement of the motion.* The logarithm of d is called the *logarithmic decrement* denoted by λ, i.e.,

$$\lambda = \ln d = \ln e^{\frac{\gamma T^*}{4}} = \frac{\gamma T^*}{4} = \frac{\gamma}{2} \cdot \frac{T^*}{2}.$$

Hence, *logarithmic decrement is the natural logarithm of the ratio of successive amplitudes separated by a time* $t = \frac{T^*}{2}$.

Now,

$$\frac{A_1}{A_2} = d = e^{\lambda} \; [\because \; \lambda = \ln d].$$

$$\therefore \; \frac{A_1}{A_3} = \frac{A_1}{A_2} \times \frac{A_2}{A_3} = d^2 = e^{2\lambda} = e^{(3-1)\lambda}.$$

Similarly,

$$\frac{A_1}{A_n} = e^{(n-1)\lambda}$$

$$\text{or, } \lambda = \frac{1}{(n-1)} \ln\left(\frac{A_1}{A_n}\right). \tag{2.18}$$

Determining A_1, A_n and noting the value of n, we can determine the value of λ.

If θ_0 and θ_1 be the amplitudes of oscillation of the coil of a *Ballistic galvanometer* in absence and in presence of damping respectively, then

$$\theta_1 = \theta_0 e^{-\frac{\gamma T^*}{8}}$$

$$\text{or, } \theta_0 = \theta_1 e^{\frac{\gamma T^*}{8}} = \theta_1 e^{\lambda/2} \left[\text{where } \lambda = \frac{\gamma T^*}{4}\right]$$

$$\text{or, } \theta_0 \simeq \theta_1 \left(1 + \frac{\lambda}{2}\right) \left[\text{for } \lambda \to 0, e^{\lambda/2} \simeq \left(1 + \frac{\lambda}{2}\right)\right]. \tag{2.19}$$

By noting the *observed amplitude of oscillation through* θ_1 in presence of damping, we can determine the *actual decrement through* θ_0, if λ be measured experimentally.

2.4.2 Energy of a Weakly Damped Oscillator

The instantaneous KE of a weakly damped oscillator is

$$\begin{aligned} E_k &= \frac{1}{2} m \left(\frac{dx}{dt}\right)^2 \\ &= \frac{1}{2} m \left[-Ae^{-\frac{\gamma t}{2}} \left\{\omega^* \sin(\omega^* t - \delta) + \frac{\gamma}{2} \cos(\omega^* t - \delta)\right\}\right]^2 \quad \text{[by eqn (2.14)]} \\ &= \frac{1}{2} m A^2 e^{-\gamma t} \left[{\omega^*}^2 \sin^2(\omega^* t - \delta) + \omega^* \gamma \sin(\omega^* t - \delta) \times \cos(\omega^* t - \delta) + \frac{\gamma^2}{4} \cos^2(\omega^* t - \delta)\right]. \end{aligned}$$

The instantaneous PE is

$$E_p = \frac{1}{2} kx^2 = \frac{1}{2} m\omega_0^2 A^2 e^{-\gamma t} \cos^2(\omega^* t - \delta). \text{ [using } k = m\omega_0^2 \text{ and eqn (2.13)]}$$

So the total mechanical energy of the oscillator at any instant is

$$\begin{aligned} E(t) &= E_k + E_p \\ &= \frac{1}{2} m A^2 e^{-\gamma t} \left[{\omega^*}^2 \sin^2(\omega^* t - \delta) + \frac{\omega^* \gamma}{2} \sin 2(\omega^* t - \delta) \right. \\ &\quad \left. + \left(\frac{\gamma^2}{4} + \omega_0^2\right) \cos^2(\omega^* t - \delta)\right]. \end{aligned}$$

Now, the average value of $E(t)$ at any instant t

$$\langle E(t)\rangle = \frac{1}{2}mA^2e^{-\gamma t}\left[\omega^{*^2}\left\langle \sin^2(\omega^* t-\delta)\right\rangle + \frac{\omega^*\gamma}{2}\langle \sin 2(\omega^* t-\delta)\rangle + \left(\frac{\gamma^2}{4}+\omega_0^2\right)\left\langle \cos^2(\omega^* t-\delta)\right\rangle\right]$$

$$= \frac{1}{2}mA^2e^{-\gamma t}\left[\omega^{*^2}\times\frac{1}{2}+\frac{\omega^*\gamma}{2}\times 0+\left(\frac{\gamma^2}{4}+\omega_0^2\right)\times\frac{1}{2}\right]$$

$$= \frac{1}{4}mA^2e^{-\gamma t}\left(\omega^{*^2}+\frac{\gamma^2}{4}+\omega_0^2\right).$$

$$\because \left\langle \sin^2(\omega^* t-\delta)\right\rangle = \left\langle \cos^2(\omega^* t-\delta)\right\rangle = \frac{1}{2} \text{ and } \langle \sin 2(\omega^* t-\delta)\rangle = 0,$$

$$\therefore \langle E(t)\rangle = \frac{1}{4}mA^2e^{-\gamma t}\left[\left(\omega_0^2-\frac{\gamma^2}{4}\right)+\frac{\gamma^2}{4}+\omega_0^2\right]$$

$$\left[\because \omega^* = \left(\omega_0^2-\frac{\gamma^2}{4}\right)^{1/2}\right]$$

$$\text{or, } \langle E(t)\rangle = \frac{1}{2}mA^2e^{-\gamma t}\omega_0^2 = E_0e^{-\gamma t}, \tag{2.20}$$

where $E_0 = \frac{1}{2}mA^2\omega_0^2$ is the total mechanical energy of an undamped oscillator. Eqn (2.20) shows that the energy of an undamped oscillator decays *exponentially* with time [Fig. 2.6].

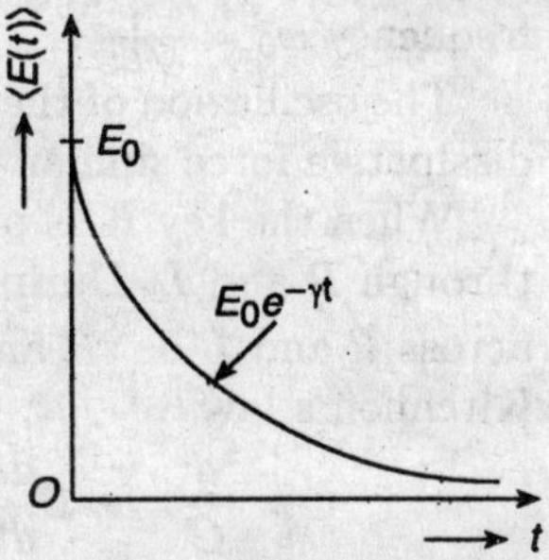

Fig. 2.6: Energy decay of a weakly damped oscillator.

Now, the average power dissipated during one time-period T^* is the average rate of loss of energy $\langle E(t)\rangle$.

$$\therefore \langle P(t)\rangle = \frac{d}{dt}\langle E(t)\rangle$$

$$= \gamma E_0 e^{-\gamma t}$$

$$= \gamma\langle E(t)\rangle. \tag{2.21}$$

2.4.3 Quality Factor (or Q-value)

Quality factor of an oscillator expresses its efficiency of performance. Average energy stored in the oscillator is $\langle E(t)\rangle$ and the average rate of energy loss is

$$\frac{d}{dt}\langle E(t)\rangle = \langle P(t)\rangle = \gamma\langle E(t)\rangle.$$

In one time-period T^*, the average rate of loss of energy is

$$T^*\gamma\langle E(t)\rangle = \frac{2\pi}{\omega^*}\gamma\langle E(t)\rangle$$

$$= \frac{2\pi}{Q}\langle E(t)\rangle \quad \left[\text{where } Q = \frac{\omega^*}{\gamma}\right]$$

$$= \frac{2\pi}{Q}\times \text{ [average energy stored]}.$$

$$\therefore Q = 2\pi \times \frac{\text{[average energy stored in one period]}}{\text{[average energy lost in one period]}}.$$

Here Q is known as the *quality factor* or *Q-value* of the damped oscillator.

2.4.4 Time Constant

The time in which the amplitude of the damped oscillator decreases to $\frac{1}{e}$ of its initial value is called time constant (τ).

Time constant of a damped oscillator is given by

$$\frac{1}{e} = \frac{A_0 e^{-\frac{\gamma(t+\tau)}{2}}}{A_0 e^{-\gamma t/2}} = e^{-\frac{\gamma}{2}\tau} \text{ or, } e^{-1} = e^{-\frac{\gamma\tau}{2}}.$$

$$\therefore \frac{\gamma\tau}{2} = 1 \text{ or, } \tau = \frac{2}{\gamma}$$

$$\text{or, } \tau = \frac{2Q}{\omega_0} \left[\because Q = \frac{\omega_0}{\gamma}\right]. \quad (2.22)$$

τ is also called the *relaxation time* or *modulus of decay.*

For smaller damping ($\gamma \to 0$), Q, $\tau \to \infty$, i.e., the time required for the oscillator to damp out is large.

2.4.5 Damping in *LCR* Circuit [cf. WBUT 2007]

We consider an electrical circuit containing inductance L, capacitance C and a resistance R [Fig. 2.7]. We have seen that, if $R = 0$, then the oscillation of the circuit is undamped whose frequency $\omega_0 = \frac{1}{\sqrt{LC}}$.

The oscillation of the LCR circuit is damped due to the presence of R, which offers a resistive dissipative force similar to friction or viscosity in case of a mechanical oscillator.

When the key K is pressed, the capacitor C is charged and when K is released, C discharges through R and L. During discharge, let at any time t, the voltage across C be $V = \frac{q}{C}$ and that across R and L be iR and $L\frac{di}{dt}$ respectively. We have from Kirchhoff's law

$$\frac{q}{C} = -L\frac{di}{dt} - Ri$$

$$\text{or, } L\frac{di}{dt} + iR + \frac{q}{C} = 0$$

$$\text{or, } L\frac{d^2q}{dt^2} + R\frac{dq}{dt} + \frac{q}{C} = 0 \left[\because i = \frac{dq}{dt}\right]$$

$$\text{or, } \frac{d^2q}{dt^2} + \frac{R}{L}\frac{dq}{dt} + \frac{1}{LC}q = 0$$

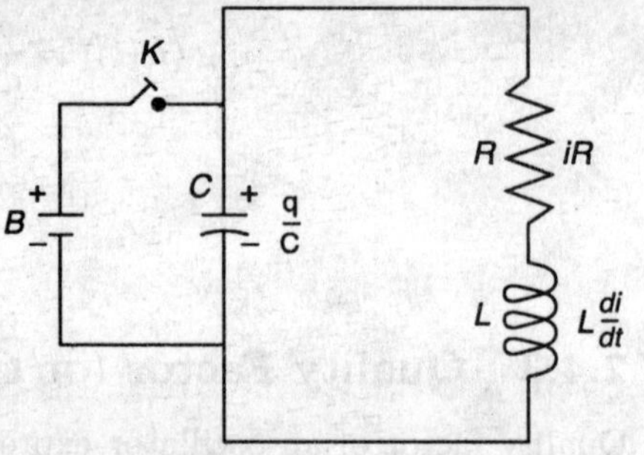

Fig. 2.7: Damping in *LCR* circuit.

Putting

$$\gamma = \frac{R}{L} \text{ and } \omega_0^2 = \frac{1}{LC},$$

we get

$$\frac{d^2q}{dt^2} + \gamma\frac{dq}{dt} + \omega_0^2 q = 0. \quad (2.23)$$

Eqn (2.23) is analogous to that of a *damped mechanical oscillator* [eqn (2.1)]. In this equation q, L and $\frac{1}{C}$ are analogous to x, m and k respectively in eqn (2.1).

Case I *When $\gamma > 2\omega_0$ or $R > 2\sqrt{\frac{L}{C}}$, the damping is large and the charge q decays gradually until C is completely discharged. This discharge is dead beat or non-oscillatory.*

Case II *When $\gamma = 2\omega_0$ or $R = 2\sqrt{\frac{L}{C}}$, the discharge is critically damped and q dies out rapidly.*

Case III *When $\gamma < 2\omega_0$ or $R < 2\sqrt{\frac{L}{C}}$, the oscillatory discharge of the charge on C takes place. The charge q repeatedly assumes +ve and −ve values and ultimately reduces to zero.*

The charge at any time t on the capacitor is given by

$$q(t) = q_0 e^{-\frac{R}{2L}t}\cos(\omega^* t - \delta), \text{ where } \omega^* = \sqrt{\frac{1}{LC} - \frac{R^2}{4L^2}}.$$

If $R = 0$,

$$\omega^* = \sqrt{\frac{1}{LC}} = \omega_0,$$

the frequency of the undamped oscillation.

The Q-value of the circuit is

$$Q = \frac{\omega^*}{\gamma} = \frac{\omega^* L}{R} \quad \left[\because \frac{R}{L} = \gamma\right].$$

So when R is small, the Q-value of the circuit is large. For $R = 0$, $\omega^* = \omega_0$ and hence

$$Q = \frac{\omega_0 L}{R} = \frac{1}{R}\sqrt{\frac{L}{C}} \quad \left[\because \omega_0 = \sqrt{\frac{1}{LC}}\right].$$

If R is absent, then the circuit is purely inductive and Q-value of an inductive circuit is

$$Q = \infty.$$

2.5 Answers of Some Important Questions

2.1. *Distinguish between free and damped vibrations.*
Ans.

Free vibration	*Damped vibration*
(i) SHM's, which persists indefinitely without loss of energy are called *free* or *undamped* vibrations.	(i) The vibration in which the amplitude (or energy) and frequency are reduced gradually with time are called *damped* vibrations.
(ii) There is no effect of dissipative forces due to radiation by the oscillator, mechanical friction, etc.	(ii) Dissipative forces due to radiation by the oscillator, friction, etc. are operative on the oscillator.
(iii) Once started, the oscillation continues for ever with constant amplitude and frequency.	(iii) The amplitude and frequency dies out with time.
(iv) This is an ideal conservative system.	(iv) This is real and non-conservative.
(v) *Example.* Oscillation of a simple pendulum in absolute vacuum.	(v) *Example.* Oscillation of a simple pendulum in air, etc.

2.2. *What are the sources of damping force?*
Ans. The different sources of damping are:

(a) the loss of energy by radiation of an oscillator

(b) the loss of energy by mechanical friction (i.e., viscosity, etc.)

(c) the transference of heat from a layer at higher temperature to a layer at lower temperature

(d) inter-molecular exchange of energy.

2.3. *What are the effects of dissipative forces on a simple harmonic oscillator?*
Ans. The effect of the dissipative force on a simple harmonic oscillator is two-fold:

(a) the loss of amplitude with time and

(b) fall of frequency.

2.4. *Define logarithmic decrement. Derive a relation for it.*
Ans. If A_0 be the amplitude of a damped oscillator, then its amplitude at any time t is

$$A = A_0 e^{-\frac{\gamma t}{2}},$$

where

$$\gamma = \frac{p}{m},$$

p being the damping constant and m, the mass of the oscillator.

If the oscillator starts from mean position, then at $t = \frac{T}{4}$ (T = time-period) it goes to the maximum displacement. Let this displacement be denoted by A_1, then

$$A_1 = A_0 e^{-\frac{\gamma T}{2 \cdot 4}}.$$

The oscillator goes to the maximum displacement (A_2) on the same side after a time $\left(T + \frac{T}{4}\right)$, where

$$A_2 = A_0 e^{-\frac{\gamma}{2}\left(T+\frac{T}{4}\right)}.$$

Similarly, the successive amplitudes $A_3, A_4, \cdots, A_n$ on the same side are

$$A_3 = A_0 e^{-\frac{\gamma}{2}\left(2T+\frac{T}{4}\right)}$$

$$A_4 = A_0 e^{-\frac{\gamma}{2}\left(3T+\frac{T}{4}\right)}$$

$$\cdots\cdots\cdots\cdots\cdots\cdots\cdots\cdots\cdots$$

$$A_n = A_0 e^{-\frac{\gamma}{2}\left[(n-1)T+\frac{T}{4}\right]}.$$

$$\therefore \frac{A_1}{A_2} = \frac{A_2}{A_3} = \frac{A_3}{A_4} = \cdots = \frac{A_{n-1}}{A_n}$$

$$= e^{\frac{\gamma T}{2 \cdot 2}} = d \text{ (a constant) } d \text{ is called the } \textit{decrement}.$$

$$\text{or, } \ln\left(\frac{A_{n-1}}{A_n}\right) = \frac{\gamma T}{2 \cdot 2} = \lambda \text{ (say)}.$$

λ is called the *logarithmic decrement.*

So the logarithmic decrement is defined as *the natural logarithm of the ratio between the two successive amplitudes on the same side of the mean position of a damped oscillation separated by one full period.*

2.5. *What is critical damping?*

Ans. When $\gamma = 2\omega_0$, where $\gamma = \frac{p}{m}$ and $\omega_0 = \sqrt{\frac{k}{m}}$, the displacement of the oscillator at any time t is

$$x = (B + Ct)e^{-\frac{\gamma t}{2}},$$

where B and C are two constants. This type of motion where the amplitude dies out very quickly to zero is called *critically damped* motion and the corresponding damping is known as the *critical damping.* The motion is neither oscillatory nor non-oscillatory. It has important application in pointer type galvanometer.

2.6. *Discuss the physical significance of logarithmic decrement.*

Ans. The logarithmic decrement of a damped oscillator is

$$\lambda = \frac{\gamma T}{4} = \frac{p}{4m}T,$$

where p is the damping coefficient; m, the mass and T, the periodic time. The above relation provides a method of measuring p.

The significance of logarithmic decrement of a damped oscillator represents the time rate of decay of amplitude of the oscillator.

2.7. *What is relaxation time and how does it vary with damping coefficient γ?*

Ans. Definition: *The time in which the amplitude of a damped oscillator decays to $\frac{1}{e}$ times of its original amplitude is called the relaxation time.*

If A and A_0 be the amplitudes of the oscillator at $t = 0$ and $t = t$ respectively, then

$$A = \frac{1}{e}A_0.$$

But

$$\frac{1}{e} = 0.368, \therefore\ A = 36.8\% \text{ of } A_0.$$

So higher the relaxation time (τ) slower is the rate of decay of amplitude and hence slower is the dissipative rate of energy of the damped oscillator.

Now,

$$A = A_0 e^{-\frac{\gamma t}{2}} = \frac{1}{e}A_0 \ [\text{for } t = \tau].$$

$$\therefore\ e^{-\frac{\gamma t}{2}} = e^{-1} \text{ or, } \frac{\gamma\tau}{2} = 1 \text{ or, } \tau = \frac{2}{\gamma}.$$

So *the relaxation time of the damped oscillator varies inversely as the damping coefficient* (γ).

2.8. *Define quality factor of damped oscillator?*

Ans. Quality factor of a damped oscillator gives a *measure of the rate of decay of energy of the damped oscillator.* It may be defined in various ways:

(a) *The quality factor of a damped oscillator is defined as the ratio of average energy stored in one period to the average energy lost in one period,* i.e.,

$$Q = 2\pi \times \frac{[\text{average energy stored in one period}]}{[\text{average energy lost in one period}]}.$$

Now, average energy stored in one time-period $= \langle E(t)\rangle$ and that lost in one time-period is $T^*\gamma\langle E(t)\rangle$.

$$\therefore\ Q = \frac{2\pi \times \langle E(t)\rangle}{T^*\gamma\langle E(t)\rangle} = \frac{2\pi}{T*}\frac{1}{\gamma}$$

$$= \frac{\omega^*}{\gamma} \quad \left[\because \frac{2\pi}{T*} = \omega^*\right]$$

$$\text{or, } Q = \frac{\omega^*}{2/\tau} = \frac{\omega^*\tau}{2} \quad \left[\because \tau = \frac{2}{\gamma}\right].$$

(b) *The quality factor of a damped oscillator may also be defined as the ratio of maximum value of restoring force to the maximum value of damping force,* i.e.,

$$Q = \frac{\text{maximum value of restoring force}}{\text{maximum value of damping force}}.$$

Now, the maximum value of restoring force $= m\omega^{*2}A$ and the maximum value of restoring force

$$= p\left(\frac{dx}{dt}\right)_{\max} = pv_{\max}$$

$$= pA\omega^* = m\gamma A\omega^* \; [\because \; p = m\gamma].$$

$$\therefore \; Q = \frac{m\omega^{*2}A}{m\gamma A\omega^*} = \frac{\omega^*}{\gamma}$$

$$\text{or, } Q = \frac{\omega^*}{2/\tau} = \frac{\omega^*\tau}{2} \left[\because \; \gamma = \frac{2}{\tau}\right],$$

where τ is the *relaxation time* of the damped oscillator.

(c) *The quality factor of the damped oscillator may also be defined as*

$$Q = 2\pi \frac{\text{maximum KE at resonance}}{\text{energy dissipated per cycle at resonance}}$$

2.6 Solved Problems

Ex. 1 *A massless spring suspended from a rigid support, carries a flat disc of mass* 100 gm *at its lower end. It is observed that the system oscillates with a frequency of* 10 Hz *and the amplitude of damped oscillations reduces to half its undamped value in one minute. Calculate.* (a) *the resistance force constant* (b) *the relaxation time* (c) *the quality factor and* (d) *the force constant of the spring.* **[BUH 1995]**

Soln.

(a) If A_0 and A be the amplitudes at $t = 0$ and $t = t$ of the damped oscillation, then

$$A = A_0 e^{-\frac{\gamma t}{2}} \text{ [where } \gamma \text{ is the damping coefficient].}$$

But $A/A_0 = 1/2$, $t = 1\,\text{min} = 60$ s.

$$\therefore \; e^{-\frac{\gamma t}{2}\times 60} = \frac{1}{2} \text{ or, } 30\gamma = \ln 2$$

$$\text{or, } \gamma = \frac{2.303 \times 0.301}{30} = 0.023/\text{s}$$

$\therefore$ the resistive force per unit velocity is

$$p = m\gamma = 0.1 \times 0.023 = \mathbf{0.0023 \; N\text{-}s/m} \; [\because \; m = 0.1 \text{ kg}].$$

(b) The relaxation time

$$\tau = \frac{2}{\gamma} = \mathbf{86.96 \; s.}$$

(c) The qualitv factor of the damped oscillator is

$$Q = \frac{\omega^*}{\gamma} = \frac{2\pi\nu^*}{\gamma} = \frac{2\pi \times 10}{\gamma} = \frac{20\pi}{\gamma},$$

where ν^* is the cyclic frequency of the damped oscillator.

$$\therefore \; Q = \frac{20 \times 3.14}{0.023} = \mathbf{2730.}$$

(d) The force constant k is given by

$$\omega_0 = \sqrt{\frac{k}{m}} \text{ or, } k = m\omega_0^2 \ [\omega_0 = \text{undamped frequency}].$$

Now,

$$\omega^{*2} = \omega_0^2 - \frac{\gamma^2}{4}$$

$$\text{or, } \omega_0^2 = \omega^{*2} + \frac{\gamma^2}{4} = (20\pi)^2 + \frac{(0.023)^2}{4} = 3949.$$

$$\therefore \ k = m\omega_0^2 = 0.1 \times 3949 = \mathbf{394.9 \ N/m}.$$

Ex. 2 *In an LCR series circuit* $L = 1$ mH, $C = 5$ μF *and* $R = 0.5\ \Omega$. *Calculate the frequency and quality factor.*

Soln. The angular frequency of the damped oscillation is

$$\omega^* = \left(\frac{1}{LC} - \frac{R^2}{4L^2}\right)^{1/2}$$

$$= \left[\frac{1}{10^{-3} \times 5 \times 10^{-6}} - \frac{(0.5)^2}{4(10^{-3})^2}\right]^{1/2}$$

$$= \mathbf{1.414 \times 10^4 \ rad/s}$$

The cyclic frequency

$$\nu^* = \frac{\omega^*}{2\pi} = \mathbf{0.225 \times 10^4 \ Hz.}$$

Relaxation time

$$\tau = \frac{2}{\gamma} = \frac{2L}{R} = \frac{2 \times 10^{-3}}{0.5} = \mathbf{4 \times 10^{-3} \ s}$$

The Q-factor of the circuit is

$$Q = \frac{\omega^*}{\gamma} = \frac{L\omega^*}{R} = \frac{10^{-3} \times 1.414 \times 10^4}{0.5} = \mathbf{28.3.}$$

Ex. 3 *A massless spring of spring constant* 10 N/m *is suspended from a rigid support and carries a mass of* 0.1 kg *at its lower end. The system is subjected to a resistive force* $-pv$ *where* p *is a constant and* v *is the velocity. It is observed that, the system performs damped oscillatory motion and its energy decays to* $1/e$ *of its initial value in* 50 s. (a) *What is the value of* p? (b) *What is the* Q*-value of the oscillator?* (c) *Show that the fractional change in frequency of the damped oscillation is* $\simeq (8Q^2)^{-1}$. *What is the percentage change in frequency due to damping?*

Soln.

(a) The energy decay of the damped oscillator is given by $E(t) = E_0 e^{-\gamma t}$, where $E(t)$ and E_0 are the energies at time $t = t$ and $t = 0$ respectively. $\gamma = p/m$, where p is the resistive force constant per unit velocity.

Now,

$$\frac{E(t)}{E_0} = \frac{1}{e} = e^{-1}$$

$$\text{or,} \quad \frac{E_0 e^{-50t}}{E_0} = e^{-1} \; [\because t = 50]$$

$$\text{or,} \; \gamma = \frac{1}{50} = 2 \times 10^{-2}/\text{s}.$$

$$\therefore \; p = m\gamma = 0.1 \times 2 \times 10^{-2} = 2 \times 10^{-3} \text{ kg/s}$$

$$= \mathbf{2 \times 10^{-3} \; N.s/m.}$$

(b) The frequency of the undamped oscillator is

$$\omega_0 = \sqrt{\frac{k}{m}} = \sqrt{\frac{10}{0.1}} = 10 \text{ rad/s}.$$

The frequency of the damped oscillator

$$\omega^{*^2} = \omega_0^2 - \frac{\gamma^2}{4} = 10^2 - 10^{-4} \simeq 10^2.$$

$$\therefore \; \omega^* = 10.$$

$\therefore$ the Q-factor,

$$Q = \frac{\omega^*}{\gamma} = \frac{10}{2 \times 10^{-2}} = \mathbf{500.}$$

(c) The fractional change in frequency is given by

$$\frac{\omega_0 - \omega^*}{\omega_0} = \frac{\omega_0 - \omega_0\left(1 - \frac{1}{8Q^2}\right)}{\omega_0} \simeq \frac{1}{8Q^2}.$$

$$\left[\because \; \omega^* = \omega_0\left(1 - \frac{\gamma^2}{4\omega_0^2}\right)^{1/2} = \omega_0\left(1 - \frac{1}{8Q^2}\right)\right].$$

$\therefore$ percentage change in frequency is

$$\frac{\omega_0 - \omega^*}{\omega_0} \times 100 = \frac{100}{8Q^2} = \frac{100}{8 \times (500)^2} = \mathbf{0.5 \times 10^{-5}}.$$

Ex. 4 *In a one-dimensional motion, a mass of* 10 gm *is acted on by a restoring force of* 10 dynes/cm *and a damping force of* 2 dyne-s/cm. *Find* (a) *whether the motion is periodic or oscillatory* (b) *the resisting force per unit velocity, which will make the motion critically damped* (c) *the mass for which the given forces will make the motion critically damped.* **[CU 1992]**

Soln.

(a) In this problem,

$$m = 10 \text{ gm}, \; k = 10 \text{ dyne/cm}, \; p = 2 \text{ dyne-s/cm}$$

$$\therefore \; \gamma = \frac{p}{m} = \frac{2}{10} = 0.2 \text{ N/s}$$

$$\omega_0^2 = \frac{k}{m} = 1 \text{ rad}^{n^2}/\text{s}^2$$

$$\frac{\gamma^2}{4\omega_0^2} = \frac{(0.2)^2}{4 \times 1} = \frac{0.04}{4} = 0.01 < 1$$

$\therefore$ the motion is **oscillatory**.

(b) For initial damping of the motion,

$$\gamma = 2\omega_0 \text{ or, } \gamma^2 = 4\omega_0^2 \text{ or, } \frac{p^2}{m^2} = 4\frac{k}{m}.$$

$$\therefore\ p = 2\sqrt{mk} = 2\sqrt{10 \times 10}$$
$$= 20 \text{ dyne-s/cm}$$

So for critical damping, $p = $ **20 dyne-s/cm.**

(c) Again,

$$\because\ \frac{p^2}{m^2} = 4\frac{k}{m},$$

$$\therefore\ m = \frac{p^2}{4k} = \frac{20 \times 20}{4 \times 10} = \mathbf{10\ gm.}$$

[$\because$ for critical damping $p = 20$ dyne-s/cm.]

Ex. 5 *Calculate the frequency, the relaxation time and the quality factor of an LCR circuit with* $L = 10\,\text{mH}, c = 0.05\,\mu\text{F}$ *and* $R = 100\,\Omega$.

Soln. Here $L = 10\,\text{mH} = 10 \times 10^{-3}\,\text{H} = 10^{-2}\,\text{H}$

$C = 0.05\,\mu\text{F} = 5 \times 10^{-8}\,\text{F}$

and $R = 100\,\Omega$

The angular frequency of the damped oscillator is

$$\omega^* = \left(\frac{1}{LC} - \frac{R^2}{4L^2}\right)^{\frac{1}{2}}$$
$$= \left(\frac{1}{10^{-2} \times 5 \times 10^8} - \frac{100^2}{4 \times (10^{-2})^2}\right)^{\frac{1}{2}}$$
$$= (2 \times 10^9 - 2.5 \times 10^7)^{\frac{1}{2}} = 10^4\sqrt{19.75}$$
$$= 10^4 \times 4.44 = 4.44 \times 10^4 \text{ rad/s.}$$

$$\therefore \text{ frequency } \nu^* = \frac{\omega^*}{2\pi} = \frac{4.44 \times 10^4}{2\pi} = \mathbf{7.078 \times 10^3\ Hz.}$$

The relaxation time,

$$\tau = \frac{2}{\gamma} = \frac{2L}{R} = \frac{2 \times 10^{-2}}{100} = \mathbf{2 \times 10^{-4}\ s.}$$

The quality factor,

$$Q = \frac{\omega^*}{\gamma} = \frac{L\omega^*}{R} = \frac{10^{-2} \times 4.44 \times 10^4}{100} = \mathbf{4.44.}$$

Ex. 6 *A mechanical oscillator has initial energy* $E_0 = 50$ J *having damping coefficient* $\gamma = 2/\text{s}$. *Calculate the time required to decrease to* $E = E_0/e$.

Soln. If E_0 and E be the energies of the oscillator at $t = 0$ and $t = t$ respectively, then

$$E = E_0 e^{-\gamma t} \text{ or, } \frac{E}{E_0} = e^{-\gamma t}.$$

But according to the problem, $E = E_0/e$.

$$\therefore\ E/E_0 = e^{-1}.$$

So

$$e^{-1} = e^{-2t} \; [\because \gamma = 2]$$

$$\text{or,} \;\; 2t = 1, \; \therefore \; t = \frac{1}{2} \text{ s.}$$

$\therefore$ the energy of the oscillator will decrease to E_0/e after $\frac{1}{2}$ s.

EXERCISES

2(A)–Objective Type Questions

Select the correct answer:

2.1. If ω_0 is the natural frequency of a body and k is the damping constant, then its quality factor is (a) ω_0/k (b) $\omega_0/2k$ (c) $2\omega_0/k$. **[WBUT 2004]**

2.2. The effect of the damping force on a harmonic oscillator is to (a) reduce angular frequency and amplitude of vibration (b) increase amplitude and angular frequency of vibration (c) reduce angular frequency and increase amplitude of vibration (d) increase angular frequency and reduce amplitude of vibration.

2.3. Relaxation time is the time in which the amplitude A_0 of a damped oscillator falls to (a) $\frac{A_0}{3}$ (b) $\frac{A_0}{e}$ (c) $\frac{A_0}{\gamma}$ (d) $A_0 e$.

2.4. If τ be the relaxation time and γ, the damping coefficient, then (a) $\tau = \frac{2}{\gamma}$ (b) $\tau = \frac{\gamma}{2}$ (c) $\gamma\tau = 1$ (d) $\gamma\tau =$ constant.

2.5. The quality factor of a series LCR circuit is (a) $\frac{1}{R}\sqrt{\frac{1}{LC}}$ (b) $\frac{1}{R}\sqrt{\frac{L}{C}}$ (c) $\sqrt{\frac{RL}{C}}$ (d) $\sqrt{RLC}$.

2.6. If γ be the damping constant and T^*, the periodic time of a damped oscillator, then logarithmic decrement of motion is (a) $\frac{\gamma}{2}\frac{T^*}{2}$ (b) γT^* (c) $\frac{T^*}{\gamma}$ (d) $\frac{\gamma}{4T^*}$.

2.7. If the damping constant λ is small, the resonance curve is (a) flat (b) sharp (c) unchanged (d) neither of these.

2.8. The amplitude of a damped oscillator with damping factor γ varies as (a) $e^{\gamma t}$ (b) $e^{\frac{\gamma t}{2}}$ (c) $e^{-\frac{\gamma t}{2}}$ (d) $e^{\frac{1}{\gamma t}}$.

2.9. In terms of Q, the logarithmic decrement λ is given by (a) $\lambda = \frac{\pi}{2Q}$ (b) $\lambda = \frac{Q}{2\pi}$ (c) $2\pi Q$ (d) $\frac{\omega_0}{Q}$.

2.10. The relaxation time of a damped oscillator is given by (a) $\tau = \frac{\omega_0}{2Q}$ (b) $\frac{2}{\omega_0 Q}$ (c) $\frac{2Q}{\omega_0}$ (d) $\frac{1}{2\omega_0 Q}$.

2.11. The energy decay time Γ of a damped oscillator is (a) $\Gamma = \frac{2}{\tau}$ (b) $\Gamma = 2\tau$ (c) $\Gamma = \frac{\tau}{2}$ (d) none of these.

2.12. In terms of Q, the energy decay time is given by (a) $\Gamma = \frac{T_0 Q}{2\pi}$ (b) $\Gamma = \frac{2\pi}{T_0 Q}$ (c) $\Gamma = \frac{2\pi T_0}{Q}$ (d) $\Gamma = 2\pi T_0 Q$.

2.13. The number of complete oscillations executed by a damped oscillator in time Γ (energy decay time) is (a) $n = \frac{\omega_0 \Gamma}{2\pi}$ (b) $n = \frac{\omega_0}{2\pi\Gamma}$ (c) $n = 2\pi\omega_0\Gamma$ (d) $\frac{\Gamma}{2\pi\omega_0}$.

2.14. If γ be the damping constant and ω_0 the natural frequency of the damped oscillator, then for oscillatory motion (a) $\gamma > 2\omega_0$ (b) $\gamma = 2\omega_0$ (c) $\gamma < 2\omega_0$ (d) $\gamma = \frac{\omega_0}{2}$.

2.15. The frequency ω^* of a damped oscillator is (a) $\omega_0\left(1 - \frac{\gamma^2}{4\omega_0^2}\right)^{1/2}$ (b) $\omega_0\left(1 - \frac{\gamma^2}{4\omega_0^2}\right)$ (c) $\left(1 - \frac{\gamma^2}{4\omega_0^2}\right)$ (d) $\frac{\omega_0}{\left(1 - \frac{\gamma^2}{4\omega_0^2}\right)^{1/2}}$.

2.16. The quality factor Q for an LCR circuit is (a) $\frac{\omega R}{L}$ (b) $\frac{\omega L}{R}$ (c) $\frac{\omega}{LR}$ (d) $\frac{R}{\omega L}$. **[WBUT 2010]**

ANSWERS

2.1. (a) **2.2.** (a) **2.3.** (b) **2.4.** (a) **2.5.** (b) **2.6.** (a) **2.7.** (c) **2.8.** (b) **2.9.** (d) **2.10.** (c) **2.11.** (c) **2.12.** (a) **2.13.** (a) **2.14.** (c) **2.15.** (a) **2.16.** (b).

2(B)–Short Answer Type Questions

2.1. What are free and damped vibrations? Does the principle of conservation of energy hold good in case of damped vibration? Explain.

2.2. Show that the average energy of a damped oscillator dies down according to the relation $\langle E(t)\rangle = E_0 e^{-\gamma t}$ where E_0 is the energy at $t = 0$, and γ is the damping constant.

2.3. What do you understand by *logarithmic decrement, relaxation time* and *quality factor* of a weakly damped oscillator? What are the relationships between them?

2.4. Establish the equation of motion of a damped harmonic motion subjected to a resistive force proportional to velocity.

2.5. Show that the amplitude of a weakly damped oscillator reduces to half its initial value in time $t = \tau \ln 2$, where τ is the relaxation time.

2(C)–Numerical Problems

2.1. A mass of 0.1 kg is attached at the free end of a massless spring hanging from a rigid support. The system oscillates in a liquid that exerts a viscous force $-pv$, where p is a constant and v is the velocity. It is observed that the energy deceays to half its original value in 5 s.

(a) What is the value of p?

(b) What is the Q-value of the oscillator?

(c) What is the change in frequency due to damping?

[*Ans.* Spring constant = 9 N/m (a) 0.014 N-s/m, (b) ~ 68, (c) 4×10^{-5} Hz]

2.2. In an LCR series circuit, $L = 0.5$ H, and $C = 5\ \mu$F. Find the maximum value of R for the discharge to be oscillatory. [*Ans.* 632 Ω]

2.3. An object of mass 0.1 kg hangs from a massless spring of spring constant 10 N/m. The top of the spring is oscillated harmonically with an amplitude of 1 cm and an angular frequency ω. The Q-value of the system is 50.

(a) What is the amplitude of the oscillator in the steady state?

(b) What is the mean power input to maintain forced oscillation at $\omega = \omega_0$?

[*Ans.* (a) 0.5 m, (b) 0.25ω rad]

2.4. A vibrator of mass 10 gm is acted on by a restoring force of 5 dyne/cm and a damping force of 2 dyne-sec/cm. Find whether the motion is over damped or oscillatory. If $t = 0$, the vibrator was at position $x = 0$, when a velocity of 1 cm/sec is imparted to it, then calculate the maximum derivation along positive X-axis. **[WBUT 2010]**

2(D)–Broad Answer Type Questions

2.1. (a) What are the characteristics of simple harmonic motion? Define phase. Prove the principle of conservation of energy in this case.

(b) A cubical block of side L cm and density d cc is floating in water of density ρ cc ($\rho > d$). The block is slightly depressed and released. Show that it will execute simple harmonic motion. Determine the frequency of oscillations. **[WBUT 2001]**

2.2. (a) Establish the differential equation of a simple harmonic motion and solve the equation.

(b) Establish the differential equation of damped harmonic motion and explain the different terms in the expression.

(c) What is meant by critical damping? **[WBUT 2004]**

2.3. (a) Show that the intensity of sound wave is proportional to the mean square of the excess pressure.

(b) Two vibrations at right angles to each other are described by the equations:

$$x = A_1 \cos \omega t$$

$$y = A_2 \cos(2\omega t + \delta).$$

Superpose these two vibrations to get the resulting equation. If $\delta = 0$, show that the resulting path represents two coincident parabolas with their vertices $(0, -A_2)$. **[BUH 1996]**

2.4. A mass m is suspended by two similar springs each of stiffness constant k in series. The amplitude of vertical oscillations of the system decreases to 20% of the initial value after five consecutive cycles of oscillations. Determine the damping coefficient b(frictional force $= -bv$), if $k = 80$ N/m and $m = 2.5$ kg. **[BUH 1998]**

Chapter 3 | Module 1.3

Forced Vibration

3.1 Introduction

If a dissipative system be given an impulse and let go, the system oscillates with gradually decaying amplitude. But if the same system be acted upon by a periodic force of constant amplitude and frequency the state of things differ. Initially the system tends to vibrate with its own frequency (ω^*) but the applied periodic force tries to impress its own natural frequency (ω) on the system. The system then vibrates with both frequencies and if $\omega \simeq \omega^*$, then beats are formed, known as *transient beats.* Due to the action of the resistive forces the vibration with frequency ω^* dies out and ultimately in the steady state the oscillator vibrates with the frequency (ω) of the applied force with constant amplitude.

Such a vibration, in which *a system oscillates with a frequency of the externally applied force (which is other than its own) is called forced vibration.*

3.2 Analytical Treatment of Forced Vibration

We consider a damped oscillator of mass m subjected to a simple harmonic force of constant amplitude and frequency. If x be its displacement from the mean position at time t, then the different forces acting on the oscillator are as follows:

(i) A restoring force $-kx$, where k is the stiffness factor, tending to bring back the displaced particle to its mean position.

(ii) The resisting force $-p\frac{dx}{dt}$, where p is the resisting force per unit velocity.

(iii) The periodic force $F \cos \omega t$, where F and ω are the *amplitude* and *frequency* of the force.

The net force acting on the damped oscillator is

$$\left(F \cos \omega t - kx - p\frac{dx}{dt}\right),$$

which must be equal to the force due to *inertia.* The equation of motion of the forced oscillator is given by

$$m\frac{d^2x}{dt^2} = F \cos \omega t - kx - p\frac{dx}{dt}$$

$$\text{or,} \quad \frac{d^2x}{dt^2} + \frac{p}{m}\frac{dx}{dt} + \frac{k}{m}x = \frac{F}{m}\cos \omega t$$

$$\text{or,} \quad \frac{d^2x}{dt^2} + \gamma\frac{dx}{dt} + \omega_0^2 x = f \cos \omega t, \tag{3.1}$$

where $\gamma = p/m$, the damping force per unit velocity per unit mass, $\omega_0 = \sqrt{k/m}$, the natural frequency of the oscillator and $f = F/m$, the force amplitude per unit mass. $\gamma/2$ is the decay constant of the damped oscillator.

The general solution of eqn (3.1)

Eqn (3.1) is an *inhomogeneous second order linear differential equation.* The solution of such a differential equation consists of two parts, (i) the particular solution (PI) and (ii) the complimentary function (CF). The complete solution (CS) is the superposition of PI and CF, i.e., CS = PI + CF.

Now, the particular solution (PI) is the solution (x_1) of the differential equation,

$$\frac{d^2x_1}{dx^2} + \gamma\frac{dx_1}{dt} + \omega_0^2 x_1 = f\cos\omega t \tag{3.2}$$

and the complementary function (CF) is the solution (x_2) of the equation,

$$\frac{d^2x_2}{dt^2} + \gamma\frac{dx_2}{dt} + \omega_0^2 x_2 = 0. \tag{3.3}$$

Then

$$\frac{d^2}{dt^2}(x_1 + x_2) + \gamma\frac{d}{dt}(x_1 + x_2) + \omega_0^2(x_1 + x_2) = f\cos\omega t.$$

The solution x_2 is the same as that of a damped oscillation and is given by

$$x_2 = A_1 e^{-\frac{\gamma t}{2}}\cos(\omega^* t - \delta), \tag{3.4}$$

where

$$\omega^* = \omega_0\left(1 - \frac{\gamma^2}{4\omega_0^2}\right)^{1/2}$$

is the *angular frequency* of the damped oscillator and δ is the *phase constant.* The amplitude of the oscillator decays exponentially according to the factor $A_1 e^{-\gamma t/2}$ and reduces to zero after some time and the oscillation x_1 persists with frequency ω, the frequency of the applied force. This state is called the *steady state* and the corresponding solution is

$$x_1 = B\cos(\omega t - \phi), \tag{3.5}$$

where B is the amplitude of vibration of the forced oscillator and ϕ is the phase constant relative to that of the applied force.

The complete solution is given by

$$x = x_1 + x_2 = B\cos(\omega t - \phi) + Ae^{-\frac{\gamma t}{2}}\cos(\omega^* t - \delta). \tag{3.6}$$

3.3 Early Stage of Forced Vibration

In the *early stage* of forced vibration there is superposition of vibrations of frequencies ω^* and ω. Since ω^* and ω do not differ much in the unsteady state, beats may be produced. This is the phenomenon of *transient beats.* The duration of the transient beats is determined by the *damping coefficient* $\gamma/2$.

3.4 Steady State of Forced Vibration

The steady state solution of the oscillator is represented by

$$x = B\cos(\omega t - \phi) = \text{(Real part of) } Be^{j(\omega t - \phi)} \tag{3.6a}$$

Now, eqn (3.6a) can be written as $x = Be^{j(\omega t - \phi)}$, where the term '*real part of*' has been dropped.

$$\therefore \ \frac{dx}{dt} = j\omega Be^{j(\omega t-\phi)} \text{ and } \frac{d^2x}{dt^2} = -\omega^2 Be^{j(\omega t-\phi)}.$$

Putting these values of $\frac{dx}{dt}$, $\frac{d^2x}{dt^2}$ in equation,

$$\frac{d^2x}{dt^2} + \gamma\frac{dx}{dt} + \omega_0^2 x = fe^{j\omega t},$$

[where $fe^{j\omega t}$ is written in place of $f\cos\omega t$],

$$\text{i.e., } \frac{d^2x}{dt^2} + \gamma\frac{dx}{dt} + \omega_0^2 x = fe^{j(\omega t-\phi+\phi)} = fe^{j(\omega t-\phi)}\cdot e^{j\phi}, \tag{3.7}$$

we get

$$\left[B\left(\omega_0^2 - \omega^2 + j\omega\gamma\right) - fe^{j\phi}\right]e^{j(\omega t-\phi)} = 0.$$

This equation is true for all values of t. Since $e^{j(\omega t-\phi)} \neq 0$,

$$B(\omega_0^2 - \omega^2 + i\omega\gamma) - fe^{j\phi} = 0$$

$$\text{or, } B = \frac{fe^{j\phi}}{\omega_0^2 - \omega^2 + j\omega\gamma}$$

$$\text{or, } Be^{-j\phi} = \frac{f}{\omega_0^2 - \omega^2 + j\omega\gamma} \tag{3.8}$$

$$\text{or, } B(\cos\phi - j\sin\phi) = \frac{f}{\omega_0^2 - \omega^2 + j\omega\gamma}. \tag{3.9}$$

The complex conjugate of the above equation is

$$B(\cos\phi + j\sin\phi) = \frac{f}{\omega_0^2 - \omega^2 - j\omega\gamma}. \tag{3.10}$$

From eqns (3.9) and (3.10), we get

$$B\cos\phi = \frac{f(\omega_0^2 - \omega^2)}{\left(\omega_0^2 - \omega^2\right)^2 + \omega^2\gamma^2} \tag{3.11}$$

$$\text{and } B\sin\phi = \frac{f\omega\gamma}{\left(\omega_0^2 - \omega^2\right)^2 + \omega^2\gamma^2}. \tag{3.12}$$

Now, squaring eqns (3.11) and (3.12) and then adding, we get

$$B^2 = \frac{f^2}{\left(\omega_0^2 - \omega^2\right)^2 + \omega^2\gamma^2}$$

$$\text{or, } B = \frac{f}{\sqrt{\left(\omega_0^2 - \omega^2\right)^2 + \omega^2\gamma^2}}$$

$$= \frac{F/m}{\sqrt{\left(\omega_0^2 - \omega^2\right)^2 + \omega^2\gamma^2}} \tag{3.13}$$

$$\text{and } \tan\phi = \frac{\omega\gamma}{(\omega_0^2 - \omega^2)}$$

$$\text{or, } \phi = \tan^{-1}\left(\frac{\omega\gamma}{\omega_0^2 - \omega^2}\right). \tag{3.14}$$

So the steady state solution of the forced oscillator is

$$x = \frac{F/m}{\sqrt{\left(\omega_0^2 - \omega^2\right)^2 + \omega^2\gamma^2}} \cos\left[\omega t - \tan^{-1}\left(\frac{\omega\gamma}{\omega_0^2 - \omega^2}\right)\right]. \tag{3.15}$$

The following interesting points are evident from the analysis of the *steady state* behaviour of the forced oscillator:

(a) The phase difference ϕ between the displacement and the applied force depends on ω, γ and ω_0.

(b) The amplitude B depends on F and ω of the applied force and also on γ, ω_0 of the oscillator.

(c) The motion of the oscillator does not depend on the initial conditions, i.e., how the oscillator is set into motion. The amplitude, phase and frequency of the forced oscillator depend only on F and ω of the applied force and the constants of the oscillator m, ω_0 and γ. Hence, *in the steady state, the oscillator forgets how the motion has been started.*

3.5 Mechanical Impedance

The force required to produce unit velocity is called the *mechanical impedance of the oscillator.* If F be the force required to produce velocity V, then the mechanical impedance is given by $Z = \frac{F}{V}$. Since the displacement of the forced oscillator at any time is

$$x = \text{(real part of) } Be^{j(\omega t-\phi)} = Be^{j(\omega t-\phi)},$$

where the term (real part of) has been omitted.

$\therefore$ velocity of the oscillator at t is $V = \frac{dx}{dt} = j\omega Be^{j(\omega t-\phi)}$.

So,

$$Z = \frac{Fe^{j\omega t}}{j\omega Be^{j(\omega t-\phi)}} = \frac{F}{j\omega Be^{-j\phi}} = \frac{mf}{j\omega Be^{-j\phi}}$$

$$= \frac{mf}{j\omega \frac{f}{(\omega_0^2-\omega^2)^2+j\gamma\omega}} \quad \left[\because\ Be^{-j\phi} = \frac{f}{(\omega_0^2-\omega^2)+j\omega\gamma}\right]$$

$$= \frac{m[(\omega_0^2-\omega^2)+j\omega\gamma]}{j\omega} = m\gamma - \frac{jm}{\omega}(\omega_0^2-\omega^2).$$

$$\therefore\ Z = m\gamma - \frac{jm(\omega_0^2-\omega^2)}{\omega}. \tag{3.16}$$

Thus, Z is a *complex quantity* having a real part $m\gamma = p$, the *damping coefficient*, and an imaginary part $\dfrac{m(\omega_0^2-\omega^2)}{\omega}$, which is due to the reaction of the oscillator to F.

$$\therefore\ Z^* = \text{the complex conjugate of } Z = m\gamma + \frac{jm}{\omega}(\omega_0^2-\omega^2).$$

Now,

$$|Z|^2 = Z^*Z = \left[m\gamma + \frac{jm}{\omega}(\omega_0^2-\omega^2)\right]\left[m\gamma - \frac{jm}{\omega}(\omega_0^2-\omega^2)\right]$$

$$= m^2\gamma^2 + \frac{m^2}{\omega^2}(\omega_0^2-\omega^2)^2$$

$$\text{or, } |Z| = m\left[\gamma^2 + \frac{1}{\omega^2}(\omega_0^2-\omega^2)^2\right]^{1/2} \tag{3.17}$$

So,

$$B = \frac{f}{\left[\left(\omega_0^2 - \omega^2\right)^2 + \omega^2\gamma^2\right]^{1/2}} = \frac{F}{m\left[\left(\omega_0^2 - \omega^2\right)^2 + \omega^2\gamma^2\right]^{1/2}}$$

$$= \frac{F}{m\omega\left[\gamma^2 + \frac{\left(\omega_0^2 - \omega^2\right)^2}{\omega^2}\right]^{1/2}} = \frac{F}{\omega|Z|}. \tag{3.18}$$

Eqn (3.18) shows that for a particular ω, $B \propto \frac{1}{|Z|}$.

3.6 Applied Force-Frequency Dependence of the Forced Oscillator : Resonance

The amplitude of the forced oscillator is

$$B = \frac{f}{\left[\left(\omega_0^2 - \omega^2\right)^2 + \omega^2\gamma^2\right]^{1/2}} = \frac{f}{D^{1/2}},$$

where $D = \left(\omega_0^2 - \omega^2\right)^2 + \omega^2\gamma^2$.

$$\therefore \quad \frac{dD}{d\omega} = \frac{d}{d\omega}\left[\left(\omega_0^2 - \omega^2\right)^2 + \omega^2\gamma^2\right] = 2\left(\omega_0^2 - \omega^2\right) \times (-2\omega) + 2\gamma^2\omega.$$

For D to be minimum so that B is maximum, we have

$$\frac{dD}{d\omega} = 0 \text{ and } \frac{d^2D}{d\omega^2} \text{ positive}$$

$$\text{or,} \quad 2(\omega_0^2 - \omega^2) = \gamma^2 \text{ or, } \omega = \left(\omega_0^2 - \frac{\gamma^2}{2}\right)^{1/2}$$

We can easily verify that $\frac{d^2D}{d\omega^2}$ is positive for $\omega_0 > \frac{\gamma}{2}$. So, B is maximum, when

$$\omega = \left(\omega_0^2 - \frac{\gamma^2}{2}\right)^{1/2} = \omega_r \text{ (say)}. \tag{3.19}$$

For low damping, i.e., $\gamma < \sqrt{2}\omega_0$, $\omega \simeq \omega_0 = \omega_r$.

Thus, *the amplitude of the forced oscillator* (B) *assumes the maximum value when the frequency of the applied force is nearly equal to the frequency of natural vibrations.* This is the condition of *resonance* and

$$\omega_r \doteq \left(\omega_0^2 - \frac{\gamma^2}{2}\right)^{1/2}$$

is called the *resonant frequency.*

Thus, the amplitude of the forced oscillator at $\omega = \omega_r$ is

$$\frac{f}{\left[\left(\omega_0^2 - \omega_r^2\right)^2 + \gamma^2\omega_r^2\right]^{1/2}}.$$

The maximum value of B is

$$B_0 = \frac{f}{\left[\left\{\omega_0^2 - \left(\omega_0^2 - \frac{\gamma^2}{2}\right)\right\}^2 + \gamma^2\left(\omega_0^2 - \frac{\gamma^2}{2}\right)\right]^{1/2}}$$

$$\text{or, } B_0 = \frac{F/m}{\gamma\left(\omega_0^2 - \frac{\gamma^2}{4}\right)^{1/2}} \left[\because \omega = \omega_r = \left(\omega_0^2 - \frac{\gamma^2}{2}\right)^{1/2}\right]. \tag{3.20}$$

The motion of the forced oscillator is of the following three types:
(a) **Resistance controlled motion.** For small damping ($\gamma/2 \le \omega_0$)

$$B_0 = \frac{F}{m\gamma\omega_0} \text{ or, } B_0 = \frac{F}{p\omega_0} \text{ [where } p = m\gamma]. \tag{3.21}$$

Thus, near resonant frequency under small damping (i.e., $\omega = \omega_r \simeq \omega_0$ for $\gamma/2 \le \omega_0$), the maximum amplitude of the forced oscillator is inversely proportional to the resistive term p and hence the motion of the forced oscillator is called *resistance controlled motion.*

(b) **Stiffness controlled motion.** At very low frequency side ($\omega \ll \omega_0$), $\omega_0^2 - \omega^2 \simeq \omega_0^2$.

$$\therefore B = \frac{F/m}{[\omega_0^4 + \gamma^2\omega^2]^{1/2}} \left[\because B = \frac{f}{\left[\left(\omega_0^2 - \omega^2\right)^2 + \gamma^2\omega^2\right]^{1/2}}\right]$$

$$= \frac{F/m}{\omega_0^2\left(1 + \frac{\omega^2}{\omega_0^2} \cdot \frac{\gamma^2}{\omega_0^2}\right)} = \frac{F}{m\omega_0^2} \left[\because \frac{\gamma^2}{\omega_0^2} \to 0, \text{ when } \gamma < \omega_0\right]$$

$$\text{or, } B = \frac{F}{k} \left[\because \omega_0^2 = \frac{k}{m}\right]. \tag{3.22}$$

The amplitude of the forced oscillator in the region $\omega \ll \omega_0$ and $\gamma < \omega_0$ is inversely proportional to the stiffness constant (k) and hence the motion is called the *stiffness controlled motion.*

(c) **Mass controlled motion.** In the high frequency side ($\omega \gg \omega_0$) and small damping ($\gamma < \omega_0$), the amplitude of the forced oscillator is

$$B = \frac{F/m}{(\omega^4 + \gamma^2\omega^2)^{1/2}} \left[\because \left(\omega_0^2 - \omega^2\right)^2 = \omega^4, \text{ when } \omega \gg \omega_0\right]$$

$$= \frac{F}{m\omega^2\left(1 + \frac{\gamma^2}{\omega_0^2} \cdot \frac{\omega_0^2}{\omega^2}\right)^{1/2}} = \frac{F}{m\omega^2} \text{ [when } \gamma < \omega_0]. \tag{3.23}$$

Hence, the amplitude of the forced oscillator is inversely proportional to the mass m of the oscillator and hence the motion is *mass controlled.*

3.7 Amplitude Resonance

The amplitude of the forced oscillator is

$$B = \frac{F/m}{\left[\left(\omega_0^2 - \omega^2\right)^2 + \gamma^2\omega^2\right]^{1/2}}.$$

When $\omega = 0$,

$$B = \frac{F}{m\omega_0^2} = \frac{F}{k}\left[\because \omega_0^2 = \frac{k}{m}\right].$$

Keeping m, k and F constant the plot of the amplitude of the forced oscillator with frequency ω is shown in Fig. 3.1. For weak damping (i.e., $\frac{\gamma}{2} < \omega_0$), B becomes equal to B_0 at $\omega = \omega_r$. This is known as the *condition of amplitude resonance.* Also for weak damping, B is steeper on either side of $\omega = \omega_r$.

At $\omega = 0$, $B = \frac{F}{k}$ for all values of γ.

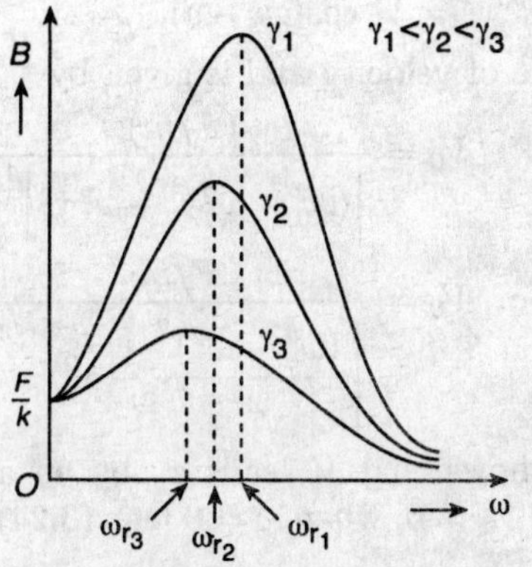

Fig. 3.1: Variation of amplitude with frequency.

3.8 Variation of Phase of the Forced Oscillator

The steady state solution of the forced oscillator is $x_s = B\cos(\omega t - \phi)$, where

$$\phi = \tan^{-1}\left[\frac{\omega\gamma}{(\omega_0^2 - \omega^2)}\right].$$

Thus, x_s lags behind the driving force $F(t) = F\cos\omega t$ by an angle ϕ, which depends on γ and the relative values of ω and ω_0.

(i) If the frequency of the applied force (ω) is very small compared to the natural frequency (ω_0), i.e., $\omega \ll \omega_0$, then

$$\tan\phi \simeq \frac{\gamma}{\omega_0}\cdot\frac{\omega}{\omega_0} \to 0, \text{ as } \frac{\omega}{\omega_0} \ll 1 \text{ for } \omega \to 0.$$

Hence, $\phi \to 0$ as $\omega \to 0$. The *oscillator in this case oscillates approximately in the same phase with the driving force.*

(ii) When $\omega \gg \omega_0$,

$$\tan\phi \simeq -\frac{\gamma}{\omega} = -\frac{\gamma}{\omega_0}\cdot\frac{\omega_0}{\omega} \to 0.$$

Thus, ϕ has a value such that $\phi \simeq \pi$. Thus, *when* $\omega \gg \omega_0$, *the oscillator oscillates* 180° *out of phase w.r.t. the driving force.*

(iii) At resonance, i.e., $\omega = \omega_0(\gamma \to 0)$, $\tan\phi = \infty$, i.e., $\phi = 90°$. The *displacement of the forced oscillator differs* 90° *in phase w.r.t. the driving force.*

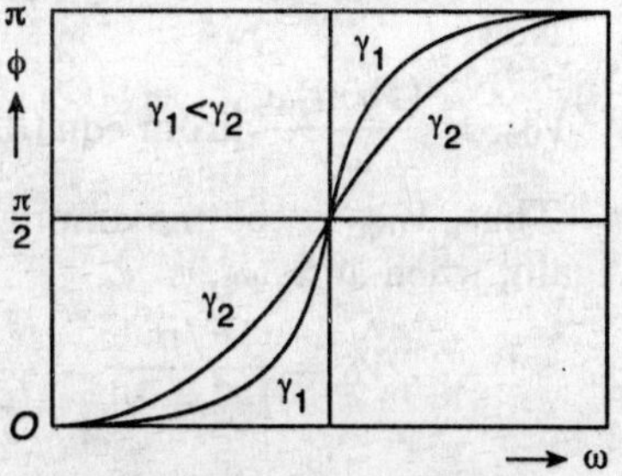

Fig. 3.2: Variation of phase of the forced oscillator with frequency.

The variation of ϕ with ω is shown in Fig. 3.9 for different values of the damping coefficient γ.

3.9 Velocity Resonance

The displacement of the driven oscillator in the steady state is given by $x_s = B\cos(\omega t - \phi)$, where B is the amplitude and ϕ, the phase angle.

The velocity of the oscillator at t is given by

$$V = \frac{dx_s}{dt} = -B\omega \sin(\omega t - \phi)$$
$$= B\omega \cos\left(\omega t - \phi + \frac{\pi}{2}\right)$$
$$= V_0 \cos(\omega t - \delta), \tag{3.24}$$

where $V_0 = B\omega$ is the amplitude of velocity and is given by

$$V_0 = \frac{(F/m)\omega}{\left[\left(\omega_0^2 - \omega^2\right)^2 + \omega^2\gamma^2\right]^{1/2}} \tag{3.25}$$

$$\text{or, } V_0 = \frac{F/m}{\left[\frac{\left(\omega_0^2 - \omega^2\right)^2}{\omega^2} + \gamma^2\right]^{1/2}} \tag{3.26}$$

and $\delta = \phi - \pi/2$. Eqn (3.24) shows that V leads x_s by an angle $\pi/2$. It is clear that V_0 is a function of the frequency ω and $V_0 = 0$, when $\omega = 0$ [eqn (3.24)]. V_0 assumes the maximum value V_{0m}, when

$$\left[\frac{\left(\omega_0^2 - \omega^2\right)^2}{\omega^2} + \gamma^2\right]$$

is the minimum. For this,

$$\frac{d}{d\omega}\left[\frac{\left(\omega_0^2 - \omega^2\right)^2}{\omega^2} + \gamma^2\right] = 0$$

$$\text{or, } \frac{[2(\omega_0^2 - \omega^2) \times (-2\omega)]\omega^2 + \left(\omega_0^2 - \omega^2\right)^2 \times 2\omega}{\omega^4} = 0$$

$$\text{or, } 2\omega(\omega_0^2 - \omega^2)(\omega_0^2 - \omega^2 - 2\omega^2) = 0$$

$$\text{or, } 2\omega(\omega_0^2 - \omega^2)(\omega_0^2 - 3\omega^2) = 0$$

$$\text{since } 2\omega(\omega_0^2 - 3\omega^2) \neq 0,$$

$$\therefore \omega_0^2 = \omega^2 \text{ or, } \omega = \omega_0.$$

Hence, $V_0 = V_{0m}$ at $\omega = \omega_0$, which is the condition of *velocity resonance*.

Now,

$$V_{0m} = \frac{(F/m)\omega}{\omega\gamma} \text{ [from eqn (3.25)]} = \frac{F}{m\gamma} = \frac{F}{p}. \tag{3.27}$$

Thus, V_{0m} is the maximum, when p is the minimum.

Again, when $\omega \gg \omega_0$,

$$V_0 = \frac{F/m}{\left[\frac{\omega^4}{\omega^2} + \gamma\right]} = \frac{F/m}{(\omega^2 + \gamma)}$$

$$\text{or, } V_0 = \frac{F}{m\omega^2} \to 0 \text{ as } \omega \to \infty \ (\gamma \ll \omega^2).$$

Thus, we have

$$V_0 = \begin{cases} 0, & \text{when } \omega = 0 \\ V_{0m} = \frac{F}{p}, & \text{when } \omega = \omega_0 \\ 0, & \text{when } \omega = \infty. \end{cases}$$

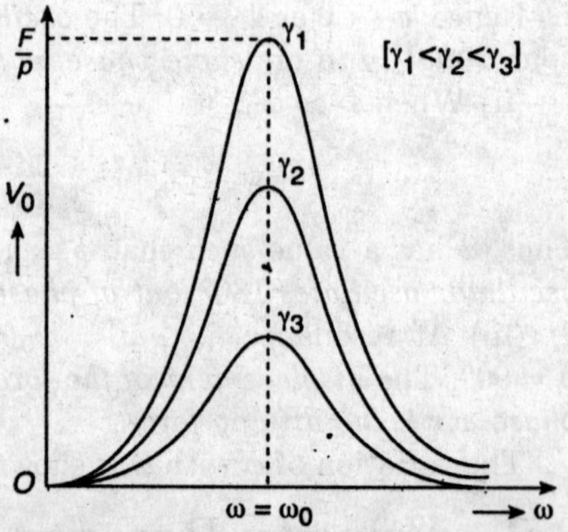

Fig. 3.3: Variation of velocity amplitude with frequency.

Fig. 3.3 shows the variation of V_0 as a function of ω in the steady state.

3.10 Phase Variation of V_0 with ω

The phase of V_0 relative to the applied force is $\delta = \phi - \frac{\pi}{2}$, where

$$\phi = \tan^{-1}\frac{\omega\gamma}{(\omega_0^2 - \omega^2)}.$$

Now, for $\omega \ll \omega_0$, $\phi = 0$, $\therefore\ \delta = -\frac{\pi}{2}$.

For $\omega = \omega_0$, $\phi = \tan^{-1}\infty$ or $\tan\phi = \infty$.

$$\therefore\ \phi = \frac{\pi}{2} \text{ and } \delta = \frac{\pi}{2} - \frac{\pi}{2} = 0$$

and for $\omega \gg \omega_0, \tan\phi \to -\infty$.

$$\therefore\ \phi = \pi \text{ and } \delta = \pi - \frac{\pi}{2} = \frac{\pi}{2}.$$

The variation of δ with ω is shown in Fig. 3.4.

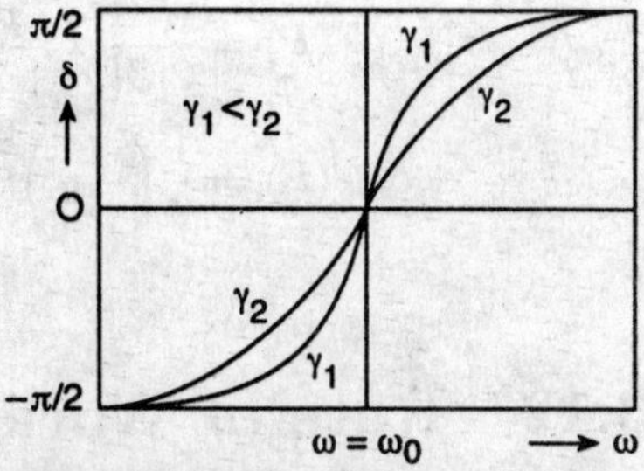

Fig. 3.4: Variation of phase of v_0 with frequency.

3.11 Power Supply and Loss of Energy

3.11.1 Power Supply by the Forcing System

The necessary power for the oscillation of the forced system is supplied by the forcing system. In the steady state the instantaneous power of the driver in forced vibration is

$$\begin{aligned} P &= \text{instantaneous applied force} \times \text{instantaneous velocity} \\ &= F\cos\omega t \times \frac{F}{|Z|}\cos\left(\omega t - \phi + \frac{\pi}{2}\right) \\ &\left[\because \text{ velocity } \frac{dx}{dt} = \frac{d}{dt}[B\cos(\omega t - \phi)] = \frac{F}{|Z|}\cos\left(\omega t - \phi + \frac{\pi}{2}\right)\right] \\ &= \frac{F^2}{2|Z|}\left[\cos\left(2\omega t - \phi + \frac{\pi}{2}\right) + \cos\left(\frac{\pi}{2} - \phi\right)\right]. \end{aligned}$$

The average power $\langle P \rangle$ is

$$\begin{aligned} \langle P \rangle &= \frac{1}{T}\int_0^T P dt, \quad \left[\text{where } T = \frac{2\pi}{\omega} \text{ is the period of the applied force}\right] \\ &= \frac{F^2}{2|Z|}\frac{1}{T}\left[\int_0^T \cos\left(2\omega t - \phi + \frac{\pi}{2}\right)dt + \int_0^T \cos\left(\frac{\pi}{2} - \phi\right)dt\right] \\ &= \frac{F^2}{2|Z|}\frac{1}{T}\left[\int_0^T \cos\left(2\omega t - \phi + \frac{\pi}{2}\right)dt + \int_0^T \sin\phi dt\right] \\ &= \frac{F^2}{2|Z|}\frac{1}{T}[0 + T\sin\phi] \\ &= \frac{F^2}{2|Z|}\sin\phi \quad \left[\because \int_0^T \cos\left(2\omega t - \phi + \frac{\pi}{2}\right)dt = 0\right]. \end{aligned}$$

Again, since

$$\tan\phi = \frac{\omega\gamma}{(\omega_0^2 - \omega^2)},$$

$$\therefore \sin\phi = \frac{\omega\gamma}{\sqrt{(\omega_0^2 - \omega^2)^2 + \omega^2\gamma^2}} = \frac{\omega\gamma}{|Z|(\omega/m)} = \frac{\gamma m}{|Z|} = \frac{p}{|Z|} \left[\because \gamma = \frac{p}{m}\right].$$

$$\therefore \langle P\rangle = \frac{F^2}{2|Z|} \times \frac{p}{|Z|}$$

$$\text{or, } \langle P\rangle = \frac{1}{2}\left(\frac{F}{|Z|}\right)^2 p. \tag{3.28}$$

3.12 Rate of Work Done by the Dissipative Force

Since the dissipative force (damping force) acting on the forced oscillation is $p\frac{dx}{dt}$, the rate of work done,

$$\frac{dW}{dt} = \text{instantaneous frictional force} \times \text{instantaneous velocity}$$

$$\text{or, } \frac{dW}{dt} = p\frac{dx}{dt}\cdot\frac{dx}{dt} = p\left(\frac{dx}{dt}\right)^2$$

$$= \frac{pF^2}{|Z|^2}\cos^2\left(\omega t - \phi + \frac{\pi}{2}\right) \left[\because \frac{dx}{dt} = \frac{F}{|Z|}\cos\left(\omega t - \phi + \frac{\pi}{2}\right)\right].$$

Now, the average rate of work done is

$$\left\langle\frac{dW}{dt}\right\rangle = \frac{1}{T}\int_0^T \left(\frac{dW}{dt}\right) dt$$

$$= \frac{1}{T}\int_0^T p\frac{F^2}{|Z|^2}\cos^2\left(\omega t - \phi + \frac{\pi}{2}\right) dt = p\left(\frac{F}{|Z|}\right)^2 \frac{1}{T}\cdot\frac{T}{2}.$$

$$\therefore \left\langle\frac{dW}{dt}\right\rangle = \frac{1}{2}\left(\frac{F}{|Z|}\right)^2 p. \tag{3.29}$$

From eqns (3.28) and (3.29) we note that

$$\langle P\rangle = \left\langle\frac{dW}{dt}\right\rangle.$$

That is, in the steady state of forced vibration, *the average rate of power supplied by the forcing system is equal to the average rate of work done by the forced system against the damping force.*

3.13 Sharpness of Resonance

The *sharpness of resonance* represents the rapidity with which $\langle P\rangle$ diminishes with the increase of ω from ω_0.

We have the average rate of power, i.e., $\langle P\rangle$ supplied by the driving system is

$$\langle P\rangle = \frac{1}{2}\left(\frac{F}{|Z|}\right)^2 p$$

$$= \frac{F^2 p}{2\left[m^2\gamma^2 + m^2\frac{\left(\omega_0^2-\omega^2\right)^2}{\omega^2}\right]} \qquad (3.30)$$

$$= \frac{F^2 p}{2\left[p^2 + \left(\frac{k}{\omega} - m\omega\right)^2\right]}. \qquad (3.31)$$

Eqn (3.30) shows that $\langle P\rangle$ at resonance, i.e., $\omega = \omega_0$ is

$$\langle P_0\rangle = \frac{F^2 p}{2p^2} = \frac{F^2}{2p}. \qquad (3.32)$$

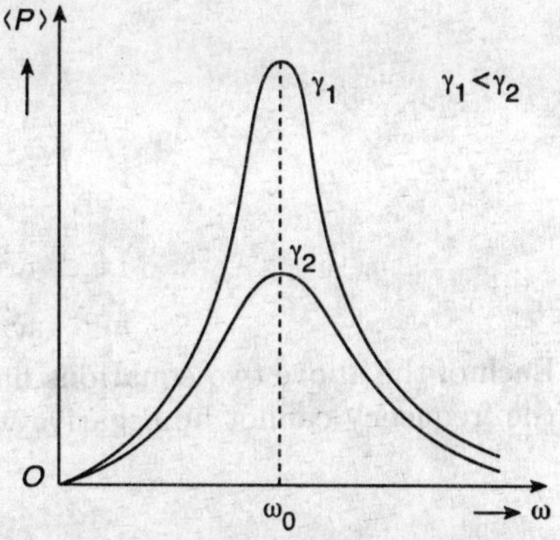

Fig. 3.5: Variation of $\langle P\rangle$ with ω.

When p is small, $\langle P_0\rangle$ is large and vice versa. *Thus, for smaller value of p, the resonance is sharp.* The plot of $\langle P\rangle$ with the frequency of the driving force is shown in Fig. 3.5.

Now,

$$\frac{\langle P\rangle}{\langle P_0\rangle} = \frac{F^2 p}{2\left[p^2 + \frac{m^2}{\omega^2}\left(\omega_0^2-\omega^2\right)^2\right]} \times \frac{2p}{F^2}$$

$$= \frac{p^2}{p^2 + m^2\left(\frac{\omega_0^2}{\omega} - \omega\right)^2} = \frac{m^2\gamma^2}{m^2\gamma^2 + m^2\omega_0^2\left(\frac{\omega_0}{\omega} - \frac{\omega}{\omega_0}\right)^2}$$

$$= \frac{\gamma^2}{\gamma^2 + \omega_0^2\left(\frac{\omega}{\omega_0} - \frac{\omega_0}{\omega}\right)^2}$$

$$\text{or,}\quad \frac{\langle P\rangle}{\langle P_0\rangle} = \frac{\gamma^2}{\gamma^2 + \omega_0^2\Delta^2}, \qquad (3.33)$$

where

$$\Delta = \frac{\omega}{\omega_0} - \frac{\omega_0}{\omega}. \qquad (3.34)$$

For $\omega = \omega_0$,

$$\Delta = 0 \text{ and } \frac{\langle P\rangle}{\langle P_0\rangle} = 1.$$

But for $\omega < \omega_0$ or $\omega > \omega_0$, $\Delta \neq 0$,

$$\therefore\ \frac{\langle P\rangle}{\langle P_0\rangle} < 1.$$

The plot of $\frac{\langle P\rangle}{\langle P_0\rangle}$ against ω for different values of γ is shown in Fig. 3.6.

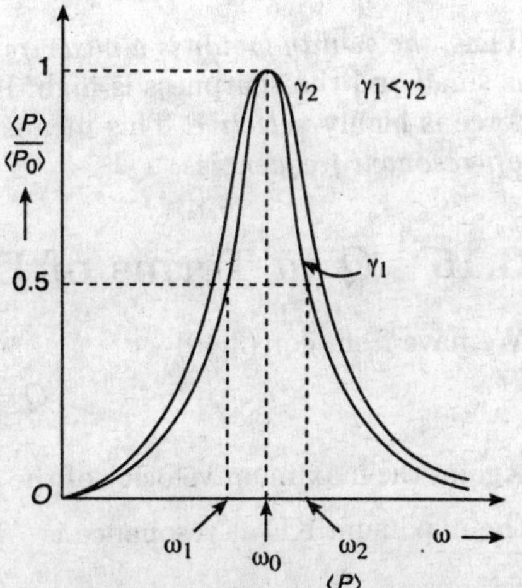

Fig. 3.6: Variation of $\frac{\langle P\rangle}{\langle P_0\rangle}$ with ω

For

$$\frac{\langle P\rangle}{\langle P_0\rangle} = \frac{1}{2},$$

we have

$$\frac{1}{2} = \frac{\gamma^2}{\gamma^2 + \omega_0^2\Delta^2} \text{ or, } \omega_0^2\Delta^2 = \gamma^2 \tag{3.35}$$

$$\text{or,} \quad \omega_0\Delta = \pm\gamma \text{ or, } \omega_0\left(\frac{\omega}{\omega_0} - \frac{\omega_0}{\omega}\right) = \pm\gamma$$

$$\text{or,} \quad \omega - \frac{\omega_0^2}{\omega} = \pm\gamma \text{ or, } \omega^2 \mp \gamma\omega - \omega_0^2 = 0,$$

$$\text{i.e.,} \quad \omega^2 + \gamma\omega - \omega_0^2 = 0 \tag{3.36}$$

$$\text{and} \quad \omega^2 - \gamma\omega - \omega_0^2 = 0. \tag{3.37}$$

Each of the above two equations have two roots — one positive and the other negative. Since ω, the frequency cannot be negative we accept only the positive roots as

$$\omega_1 = -\frac{\gamma}{2} + \left(\omega_0^2 + \frac{\gamma^2}{4}\right)^{1/2} \tag{3.38}$$

$$\text{and } \omega_2 = +\frac{\gamma}{2} + \left(\omega_0^2 + \frac{\gamma^2}{4}\right)^{1/2}. \tag{3.39}$$

ω_1 and ω_2 are known as the *low half-power point* and *high half-power point* respectively. The difference between the two half-points is called the *bandwidth*, i.e., $\omega_2 - \omega_1 = \Delta\omega = \gamma$ is the bandwidth.

3.14 Expression for Sharpness of Resonance

The *sharpness of resonance* is given by

$$S_r = \frac{1}{|\Delta|} = \frac{\omega_0}{\gamma} \quad [\because \omega_0\Delta = \pm\gamma]. \tag{3.40}$$

From eqns (3.38) and (3.39), we have

$$\gamma = \omega_2 - \omega_1, \quad \therefore S_r = \frac{\omega_0}{\omega_2 - \omega_1}. \tag{3.41}$$

But $\dfrac{\omega_0}{\omega_2 - \omega_1}$ is the *quality factor* and is denoted by Q.

$$\therefore Q = S_r = \frac{\omega_0}{\omega_2 - \omega_1}. \tag{3.42}$$

Thus, *the quality factor is a measure of sharpness of resonance.* For small damping the bandwidth is small and the sharpness is high. In this case, the response of the driven system to the driving force is highly *selective.* This means *the response is very high at resonant frequency and poor at off-resonant frequencies.*

3.15 Q in Terms of Dissipation (Decay) of Energy

We have from eqn (3.40)

$$Q = \frac{\omega_0}{\gamma} = \frac{\omega_0 m}{p} \quad \left[\because \gamma = \frac{p}{m}\right]. \tag{3.43}$$

Again, the maximum velocity of the oscillator at resonance is $V_{0m} = \dfrac{F}{p}$ [cf. eqn (3.27)] and hence the maximum KE at resonance is

$$E_m = \frac{1}{2}mV_{0m}^2 = \frac{1}{2}m\left(\frac{F}{p}\right)^2.$$

Also the energy dissipated per unit time at resonance is $\langle P_0 \rangle = \dfrac{F^2}{2p}$ [cf. eqn (3.32)].

$\therefore$ the energy dissipated per cycle is

$$E_d = \langle P_0 \rangle T = \frac{F^2 T}{2p} = \frac{\pi F^2}{\omega_0 p} \quad \left[\because \ T = \frac{2\pi}{\omega_0}\right].$$

$$\therefore \ \frac{E_m}{E_d} = \frac{1}{2} m \left(\frac{F}{p}\right)^2 \times \frac{\omega_0 p}{\pi F^2} = \frac{\omega_0 m}{2\pi p}$$

$$= \frac{1}{2\pi} \frac{\omega_0}{(p/m)} = \frac{1}{2\pi}\left(\frac{\omega_0}{\gamma}\right) = \frac{Q}{2\pi} \quad \left[\because \ \gamma = \frac{p}{m} \text{ and } Q = \frac{\omega_0}{\gamma}\right].$$

$$\therefore \ Q = 2\pi \frac{E_m}{E_d}$$

$$\text{or, } Q = 2\pi \frac{\text{maximum KE at resonance}}{\text{energy dissipated per cycle at resonance}}. \tag{3.44}$$

3.16 Analogy with Electric Circuit

Let a sinusoidal voltage $V = V_0 \sin \omega t$ be applied to a circuit containing a resistance (R), an inductance (L) and a capacitance (C) in series [Fig. 3.7]. Then at any instant t, if I be the current, q the charge on the condenser plates, then the voltage induced across the inductance L is $L\frac{dI}{dt}$ and the voltage across the condenser is q/C and the potential drop across R is RI.

The emf equation is

$$L\frac{dI}{dt} + RI + \frac{q}{C} = V_0 \sin \omega t$$

$$\text{or, } L\frac{d^2 I}{dt^2} + R\frac{dI}{dt} + \frac{1}{C}\frac{dq}{dt} = V_0 \omega \cos \omega t.$$

$$\text{or, } L\frac{d^2 I}{dt^2} + R\frac{dI}{dt} + \frac{1}{C} I = V_0 \omega \cos \omega t. \tag{3.45}$$

$$\therefore \ I = \frac{dq}{dt}.$$

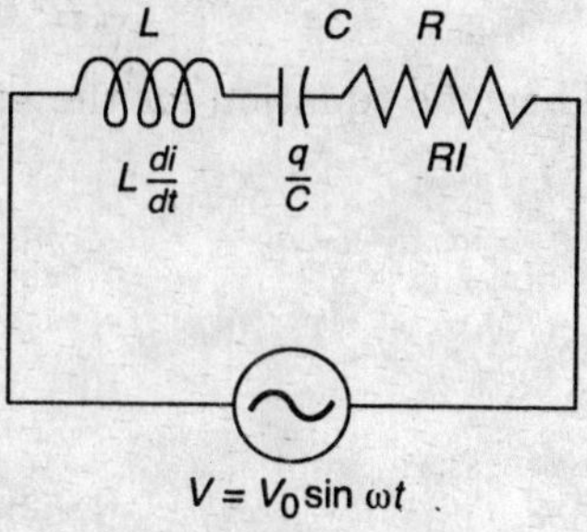

Fig. 3.7: *LCR* resonant circuit.

At steady state the alternating current will be of the same frequency (ω) as the applied voltage. So the most general solution for current may be written as

$$I = I_0 \sin(\omega t - \phi). \tag{3.46}$$

$$\therefore \ \frac{dI}{dt} = \omega I_0 \cos(\omega t - \phi) \tag{3.47}$$

$$\text{and } \frac{d^2 I}{dt^2} = -\omega^2 I_0 \sin(\omega t - \phi). \tag{3.48}$$

Using the values of I, $\dfrac{dI}{dt}$, $\dfrac{d^2 I}{dt^2}$ in eqn (3.45), we get

$$-L\omega^2 I_0 \sin(\omega t - \phi) + R\omega I_0 \cos(\omega t - \phi) + \frac{I_0}{C}\sin(\omega t - \phi) = \omega V_0 \cos \omega t$$

$$\text{or, } I_0 \left(\frac{1}{C} - L\omega^2\right)(\sin \omega t \cos \phi - \cos \omega t \sin \phi)$$
$$+ R\omega I_0 (\cos \omega t \cos \phi + \sin \omega t \sin \phi) = \omega V_0 \cos \omega t. \tag{3.49}$$

Eqn (3.49) holds for all values of t. For $t = \pi/2\omega$,

$$I_0\left(\frac{1}{C} - L\omega^2\right)\cos\phi + R\omega I_0 \sin\phi = 0$$

$$\text{or,} \quad \tan\phi = \frac{(\omega L - \frac{1}{\omega C})}{R}. \tag{3.50}$$

$$\therefore \quad \sin\phi = \frac{(\omega L - \frac{1}{\omega C})}{\sqrt{R^2 + (\omega L - \frac{1}{\omega C})^2}}, \quad \cos\phi = \frac{R}{\sqrt{R^2 + (\omega L - \frac{1}{\omega C})^2}}. \tag{3.51}$$

At $t = 0$, eqn (3.49) reduces to

$$I_0\left(\frac{1}{C} - \omega^2 L\right) \times (-\sin\phi) + R\omega I_0 \cos\phi = \omega V_0$$

$$\text{or,} \quad I_0.\left[R \cdot \frac{R}{\sqrt{R^2 + (\omega L - \frac{1}{\omega C})^2}} + \left(\omega L - \frac{1}{\omega C}\right) \cdot \frac{(\omega L - \frac{1}{\omega C})}{\sqrt{R^2 + (\omega L - \frac{1}{\omega C})^2}}\right]$$

$$= V_0 \quad \text{[using eqn (3.51)]}$$

$$\text{or,} \quad I_0\left[\sqrt{R^2 + \left(\omega L - \frac{1}{\omega C}\right)^2}\right] = V_0$$

$$\text{or,} \quad I_0 = \frac{V_0}{\sqrt{R^2 + (\omega L - \frac{1}{\omega C})^2}}. \tag{3.52}$$

Then the equation for the instantaneous current becomes

$$I = \frac{V_0}{\sqrt{R^2 + (\omega L - \frac{1}{\omega C})^2}} \sin(\omega t - \phi) \tag{3.53}$$

$$\text{or,} \ I = \frac{V_0}{Z} \sin(\omega t - \phi), \tag{3.54}$$

$$\text{[where} \quad Z = \sqrt{R^2 + \left(\omega L - \frac{1}{\omega C}\right)^2}$$

is the impedance of the LCR circuit]

$$\text{or,} \ I = I_0 \sin(\omega t - \phi), \tag{3.55}$$

where

$$I_0 = \frac{V_0}{Z} = \frac{V_0}{\sqrt{R^2 + (\omega L - \frac{1}{\omega C})^2}}. \tag{3.56}$$

Since Z depends on ω, I_0 also depends on ω. If the frequency of the ac voltage be gradually increased, then at a certain value of $\omega = \omega_0$ (say),

$$\omega_0 L = \frac{1}{\omega_{0C}} \ \text{or,} \ \omega_0 = \frac{1}{\sqrt{LC}}. \tag{3.57}$$

Then $Z = Z_0 = R$ and the current

$$I_0 = \frac{V_0}{R}$$

is *the maximum value of current*, say, I_{0m}, i.e.,

$$I_{0m} = \frac{V_0}{R}.. \tag{3.58}$$

This is known as the *phenomenon of resonance* in *LCR* series circuit. The variation of *current amplitude* as well as *impedance with the frequency* ω of the applied ac voltage frequency is shown in Fig. 3.15. It is evident that as the resistance in the circuit is reduced, the resonance curve becomes sharper. The more quickly the current amplitude falls for changes of frequency from ω_0, the sharper is the resonance.

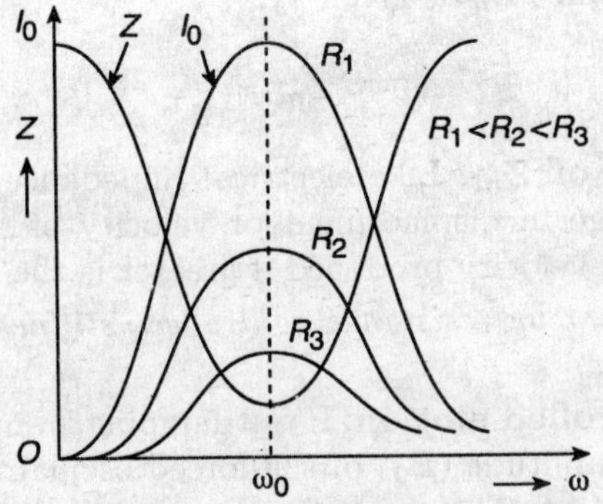

Fig. 3.8: Variation of I_0, Z with frequency in an *LCR* circuit.

The *sharpness of resonance* is described by the quality factor Q, defined by

$$Q = \frac{\omega_0}{\omega_2 - \omega_1},$$

where ω_1 and ω_2 represent respectively *the lower- and upper-half power frequencies,* i.e., ω_1 and ω_2 are the frequencies at which the power dissipation in the circuit drops to one-half its resonance value.

3.17 Answers of Some Important Questions

3.1. *What do you mean by forced vibration of a damped oscillator?*
Ans. See Section 3.2 (p. 55)
3.2. *Distinguish between amplitude resonance and velocity resonance of a force oscillator.*
Ans. The main difference between *amplitude resonance* and *velocity resonance* of a force oscillator are as follows:

Amplitude resonance	*Velocity resonance*
(i) In amplitude resonance the amplitude of forced oscillator is the maximum for a particular frequency of the applied force.	(i) In velocity resonance, the velocity amplitude of the forced oscillator is the maximum at a particular frequency of the applied force.
(ii) Amplitude resonance occurs at resonant frequency $\omega_r = \left(\omega_0^2 - \frac{\gamma^2}{2}\right)^{1/2}$, γ is the damping coefficient.	(ii) Velocity resonance occurs at frequency $\omega = \omega_0$, ω_0 is the natural frequency of the oscillator.
(iii) At applied frequency $\omega = 0$, the amplitude of the forced oscillator is F/k, where F is the force amplitude and k the force constant.	(iii) At applied frequency $\omega = 0$, the velocity amplitude is zero.
(iv) At amplitude resonance the phase of the forced oscillator w.r.t. that of the applied force is $\pi/2$.	(iv) At velocity resonance, the phase of the forced oscillator w.r.t. that of the applied force is zero.

3.3. *Define mechanical impedance of a forced oscillator. What is its physical significance?*

Ans. *The mechanical impedance of a forced oscillator is defined as the ratio of the instantaneous force exerted by the external periodic force to the instantaneous velocity of the oscillator.*

If F be the instantaneous force acting on the damped oscillator produces a velocity v in it, then the mechanical impedance Z_m is given by

$$Z_m = \frac{F}{v}.$$

Physical significance of Z_m. Like electrical impedance in an ac circuit, the mechanical impedance opposes a change in displacement or velocity of the forced oscillator. For a given applied force, smaller is the velocity produced if greater is the impedance of the oscillator.

3.4. *Explain what are meant by resistance controlled, stiffness controlled and mass controlled motion of a forced oscillator?*

Ans. (a) **Resistance controlled motion.** If the damping is small $\left(\frac{\gamma}{2} \ll \omega_0\right)$, then near resonant frequency, the maximum amplitude (B_0) of the forced oscillator is inversely proportional to the resistance term (p), i.e., $B_0 \propto \frac{1}{p}$. This type of motion is called *resistance controlled motion* of the forced oscillator.

(b) **Stiffness controlled motion.** In the low frequency side (i.e., $\omega \ll \omega_0$) and small damping (i.e., $\gamma < \omega_0$), the amplitude of the forced oscillator is inversely proportional to the stiffness constant (k). Hence, the motion of the forced oscillator under such condition is called the *stiffness controlled motion.*

(c) **Mass controlled motion.** In the high frequency region ($\omega \gg \omega_0$) and small damping ($\gamma < \omega_0$), the amplitude of the forced oscillator is inversely proportional to the mass m of the oscillator and hence the motion is called the *mass controlled.*

3.5. *Characterize early stage and steady state of the forced vibration.*

Ans. In the early stage the oscillator vibrates under the action of two vibrations, viz., the damped vibration of frequency ω^* and vibration with forcing frequency ω. Since $\omega^* \simeq \omega$, *beats* are formed. But the damped vibration dies away quickly and hence the oscillator vibrates only with frequency ω. Due to the short duration of the beats, these are called *transient beats* and the early stage is characterized by the *phenomenon of transient beats.*

After the early stage, the oscillator vibrates with the frequency of the applied force and the state of the oscillator is called the *steady state.* Steady state of the forced oscillator is characterized by the fact that, *power supplied by the forcing system is equal to the rate of work done by the dissipative force.*

3.6. *Define sharpness of resonance. How is it related to the quality factor?*

Ans.

Definition of sharpness of resonance. It is defined as *the rate of fall of amplitude with the change in frequency of the applied force on each side of the resonant frequency.*

Effect of damping on the sharpness of resonance

(a) If the damping is negligible (i.e., $p = 0$), then the sharpness is high, the amplitude of vibration $B \to \infty$. The resonance at this condition is called *sharp resonance.*

(b) For small damping (i.e., $p \to$ small) the rate of fall of amplitude with frequency of the applied force is large and the sharpness is high on both sides of the resonant frequency (ω_r).

(c) If damping is large (i.e., $k \to \infty$), then sharpness is low and the corresponding resonance is called *flat response.* The amplitude of vibration remains practically constant over a wide range of frequency near resonant frequency.

If $\omega_2 - \omega_1 = 4\omega = \gamma$ be the *bandwidth* of the resonance curve of the forced oscillator, then sharpness S_r is inversely proportional to $4\omega = \gamma$, i.e.,

$$S_r \propto \frac{1}{|\Delta\omega|}, \quad [\Delta\omega = \omega_2 - \omega_1]$$

$$\text{or, } S_r = \frac{\omega_0}{\gamma} \quad [\because \omega_0 \Delta\omega = \pm\gamma]$$

$$\text{or, } S_r = \frac{\omega_0}{\omega_2 - \omega_1} = Q,$$

where Q is the *quality factor* of the oscillator.

Hence, *quality factor gives a measure of sharpness of resonance.* This is used to increase the *selectivity* of radio sets. Sharpness in radio sets helps to reproduce signals free from interference with signals of frequency close to it.

3.7. *If the frequency of the forcing system becomes equal to the natural frequency of the oscillator, the amplitude of vibration is large. Explain why.*

Ans. We know that the phase of the applied force becomes equal to that of the velocity of vibration, if the frequency of the former is the same as the natural frequency of the oscillator. Under this condition the applied force helps to increase the amplitude of the body in each step. A new impulse adds to all the previous ones and the effect is to increase the amplitude of vibration to a large value.

3.8. *Define figure of merit in terms of bandwidth.*

Ans. *The figure of merit of a forced oscillator is defined as the ratio of the frequency at velocity resonance to the full bandwidth at half maximum power.*

Hence,

$$\text{figure of merit} = \frac{\text{frequency at velocity resonance}}{\text{full bandwidth at half maximum power}}$$

$$= \frac{\omega_0}{\omega_2 - \omega_1} = Q, \text{ the quality factor.}$$

3.9. *Why a thin wire rather than a thick wire is preferred in sonometer?*

Ans. For the thin wire the mass per unit length is smaller than that of the thick wire. So, the *damping constant* of the thin wire is less as compared to the thick wire which helps to create *sharp resonance* in a thin wire of the sonometer.

If the lengths of the thin and thick wires are adjusted so as to be in unison with the fork, it will be observed that, the paper rider on the thin wire is thrown off only at the resonant length while that on the thick wire it is thrown off when the length of the wire is much away from the strictly resonant length. So, for greater accuracy in the measurement of frequency of the fork the thiner wire is preferred in the construction of the sonometer.

3.10. *Why are soldiers asked to stop marching while crossing a wooden bridge?*

Ans. A troop of soldiers while crossing a wooden bridge can set forced oscillations in the bridge. If the frequency of their foot steps happens to match with one of the natural frequencies of the bridge, resonance occurs. Then the bridge starts oscillations with destructively large amplitude and the bridge may break down. To protect the bridge from the hand of damage, soldiers are ordered to stop marching while passing over a wooden bridge.

3.18 Solved Problems

Ex. 1 *An object of mass* 2 kg *hangs from a spring of negligible mass. The spring is extended by* 2.5 cm, *when the object is attached. The top end of the spring is oscillating up and down in* SHM *with an amplitude of* 2 mm. *The Q-value of the system is* 20. *If* $g = 10\ \text{m/s}^2$: (a) *What is the angular frequency* ω_0 *of the free undamped oscillations?* (b) *What is the amplitude of forced* *oscillations at $\omega = \omega_0$? (c) What is the mean power input to maintain the oscillations at an angular frequency ω, one per cent greater than ω_0?* **[BUH 1994]**

Soln. If k be the spring constant, then

$$kx = mg \text{ at equilibrium of the spring.}$$

$$\therefore \quad k \times 2.5 \times 10^{-2} = 2 \times \text{g}$$

$$\text{or,} \quad k = \frac{2 \times 10}{2.5 \times 10^{-2}} = 800 \text{ N/m.}$$

(a) The angular frequency,

$$\omega_0 = \sqrt{\frac{k}{m}} = \sqrt{\frac{800}{2}} = \mathbf{20\ rad/s.}$$

(b) The amplitude of the forced oscillation at $\omega = \omega_0$ is

$$B = \frac{F/m}{\sqrt{(\omega_0^2 - \omega^2)^2 + \gamma^2\omega^2}}.$$

For large Q, γ is small. So, B assumes the maximum value B_m given by

$$B_m = \frac{F/m}{\gamma\sqrt{\omega_0^2 - \frac{\gamma^2}{4}}} = \frac{F}{\gamma m \omega_0} = \frac{FQ}{m\omega_0^2} \quad \left[\because \frac{\omega_0}{\gamma} = Q\right].$$

Now,

$$F = k \times 2 \times 10^{-3} \quad \text{[where } F \text{ is the force amplitude]}$$

$$= 800 \times 2 \times 10^{-3} = 1.6 \text{ N.}$$

$$\therefore \quad B_m = \frac{1.6 \times 20}{2 \times 20^2} = 0.04 \text{ m} = \mathbf{4\ cm.}$$

(c) The average input power to the oscillator by the forcing system is

$$\langle P \rangle = \frac{F^2 p}{2|Z_m|^2},$$

where

$$p = \gamma m = \frac{\omega_0}{Q} m = \frac{20}{20} \times 2 = 2 \text{ kg/s.}$$

$$\therefore \quad |Z_m|^2 = p^2 + \left(\omega m - \frac{k}{\omega}\right)^2$$

$$= 4 + \left(20.2 \times 2 - \frac{800}{20.2}\right)^2 \quad [\because \omega = \omega_0 + 0.01\omega_0 = 20.2 \text{ rad/s}]$$

$$= 4 + (40.4 - 39.6)^2 = 4.64 \text{ kg}^2/\text{s}^2.$$

$$\therefore \quad \langle P \rangle = \frac{(1.6)^2 \times 2}{2 \times 4.64} = \mathbf{0.55\ watt.}$$

Ex. 2 *The amplitude of a forced oscillator at frequencies $\omega_1 = 400$/s and $\omega_2 = 600$/s are equal. What is the value of the resonant frequency?*

Soln. The displacement of a forced oscillator at any time t is

$$x = \frac{f\cos(\omega t - \phi)}{\sqrt{\left(\omega_0^2 - \omega^2\right)^2 + \omega^2\gamma^2}} = B\cos(\omega t - \phi),$$

where

$$B = \frac{f}{\sqrt{\left(\omega_0^2 - \omega^2\right)^2 + \omega^2\gamma^2}} \text{ is the amplitude of the oscillator.}$$

Now, for $\omega = \omega_1$, $B_1 = \dfrac{f}{\sqrt{\left(\omega_0^2 - \omega_1^2\right)^2 + \omega_1^2\gamma^2}}$

and for $\omega = \omega_2$, $B_2 = \dfrac{f}{\sqrt{\left(\omega_0^2 - \omega_2^2\right)^2 + \omega_1^2\gamma^2}}$.

By the problem, $B_1 = B_2$,

$$\therefore \quad \left(\omega_0^2 - \omega_1^2\right)^2 + \omega_1^2\gamma^2 = \left(\omega_0^2 - \omega_2^2\right)^2 + \omega_2^2\gamma^2$$

$$\text{or,} \quad \left(\omega_0^2 - \omega_1^2\right)^2 - \left(\omega_0^2 - \omega_2^2\right)^2 = \gamma^2\left(\omega_2^2 - \omega_1^2\right)$$

$$\text{or,} \quad \left(2\omega_0^2 - \omega_1^2 - \omega_2^2\right)\left(\omega_2^2 - \omega_1^2\right) = \gamma^2\left(\omega_2^2 - \omega_1^2\right)$$

$$\text{or,} \quad \gamma^2 = \left(2\omega_0^2 - \omega_1^2 - \omega_2^2\right).$$

The resonant frequency,

$$\omega_r = \sqrt{\omega_0^2 - \frac{\gamma^2}{2}} = \left[\omega_0^2 - \frac{2\omega_0^2 - \omega_1^2 - \omega_2^2}{2}\right]^{1/2}$$

$$= \left(\frac{\omega_1^2 + \omega_2^2}{2}\right)^{1/2} = \left[\frac{(400)^2 + (600)^2}{2}\right]^{1/2}$$

$$= 100\sqrt{26} = \mathbf{510/s}.$$

Ex. 3 *If the displacement at resonance is given by* $A_{\max} = \dfrac{F_0}{\omega' r}$, *where* $\omega' = \sqrt{\dfrac{s}{m} - \dfrac{r^2}{4m^2}}$, *show that, the displacement at low frequencies is amplified by a factor Q at displacement resonance.*

[HPU 1991]

Soln. The amplitude of the forced oscillator at any frequency ω is given by

$$A = \frac{f_0}{\sqrt{4k^2\omega^2 + \left(\omega^2 - \omega_0^2\right)^2}} \quad \left[\text{where } k = \frac{r}{2m}\right].$$

At low frequencies, i.e., $\omega \to 0$, the amplitude,

$$A_0 = \frac{f_0}{\omega_0^2} = \frac{f_0}{(s/m)} = \frac{F_0}{m \times \frac{s}{m}} = \frac{F_0}{s}.$$

$$\therefore \left(\frac{A_{\max}}{A_0}\right)^2 = \frac{F_0^2}{\omega'^2 r^2} \times \frac{s^2}{F_0^2} = \frac{m^2\omega_0^4}{r^2\left(\omega_0^2 - \frac{r^2}{4m^2}\right)} \quad \left[\because \frac{s}{m} = \omega_0^2\right]$$

$$= \frac{m^2\omega_0^2}{r^2\left(1 - \frac{r^2}{4m^2\omega_0^2}\right)} = \frac{Q^2}{1 - \frac{1}{4Q^2}} \quad \left[\text{where } Q = \frac{m\omega_0}{r}\right]$$

$$\text{or, } \frac{A_{max}}{A_0} = \frac{Q}{\left(1-\frac{1}{4Q^2}\right)^{1/2}} = Q\left(1-\frac{1}{4Q^2}\right)^{-1/2}$$

$$= Q\left(1+\frac{1}{8Q^2}+\cdots\right) \simeq Q\left(1+\frac{1}{8Q^2}\right)$$

$$\text{or, } \frac{A_{max}}{A_0} \simeq Q \left[\because \; \frac{1}{8Q^2} \ll Q\right].$$

$\therefore$ amplification factor $\left(\frac{A_{max}}{A_0}\right)$ of the oscillator = Q-factor.

EXERCISES

3(A)–Objective Type Questions

Select the correct answer:

3.1. The resonant frequency of a forced oscillator of natural frequency ω_0 in a medium of damping coefficient γ is (a) $\omega_r = \omega_0$ (b) $\omega_r = \left(\omega_0^2 - \frac{\gamma^2}{2}\right)^{1/2}$ (c) $\omega_r = \left(\omega_0^2 - \frac{\gamma^2}{4}\right)^{1/2}$ (d) $\omega_r = \sqrt{\gamma^2 - \omega_0^2}$.

3.2. If F_0 be the amplitude of the applied force and k the restoring force per unit displacement, then the amplitude of the forced oscillator for $\omega = 0$ is (a) $\frac{k}{F_0}$ (b) kF_0 (c) $\frac{F_0}{k}$ (d) $\frac{1}{kF_0}$.

3.3. If ω and ω_0 be the applied frequency and natural frequency of a driven oscillator, then the velocity resonance occurs at (a) $\omega > \omega_0$ (b) $\omega = \omega_0$ (c) $\omega = \frac{\omega_0}{2}$ (d) $\omega = 2\omega_0$.

3.4. The phase difference between external force and the velocity of the forced oscillator is (a) $\frac{\pi}{2}$ (b) 0 (c) π (d) $\frac{3\pi}{2}$.

3.5. For higher sharpness of the resonance curve, the Q value must be (a) small (b) large (c) medium (d) none of these.

3.6. In case of a forced oscillator, transient beats are observed in (a) steady state (b) early stage (c) middle state (d) none of these.

3.7. In a series L-C-R circuit, the discharge of the capacitor is oscillatory, if (a) $R > 2\sqrt{\frac{L}{C}}$ (b) $R = 2\sqrt{\frac{L}{C}}$ (c)$R < 2\sqrt{\frac{L}{C}}$ (d) none of these.

3.8. For the motion of the forced oscillator to be *resistance controlled*, near $\omega = \omega_0$: (a) $\gamma < \omega_0$ (b) $\gamma = \omega_0$ (c) $\gamma > \omega_0$ (d) $\gamma\omega_0$ = constant.

3.9. For the motion of the forced oscillator to be *stiffness controlled* under the condition, $\gamma < \omega_0$ is (a) $\omega = \omega_0$ (b) $\omega \ll \omega_0$ (c) $\omega \gg \omega_0$ (d) none of these.

3.10. *Mass controlled* motion of the forced oscillator occurs for ($\gamma < \omega_0$), if (a) $\omega = \omega_0$ (b) $\omega < \omega_0$ (c) $\omega > \omega_0$ (d) none of these.

3.11. If P_{input} and $P_{dissipated}$ be the input power to the forced oscillator and power dissipated due to friction, then at resonance: (a) $P_{input} > P_{dissipated}$ (b) $P_{input} = P_{dissipated}$ (c) $P_{input} < P_{dissipated}$ (d) none of these.

3.12. Velocity resonance of the forced oscillator occurs at (a) $\omega = 2\omega_0$ (b) $\omega = \omega_0$ (c) $\omega = \frac{\omega_0}{2}$ (d) none of these.

3.13. At resonance, the maximum amplitude of the forced oscillator is (a) $\frac{F_0}{m\gamma}$ (b) $\frac{F_0 m}{\gamma}$ (c) $\frac{\gamma}{F_0 m}$ (d) $F_0 m\gamma$.

3.14. The velocity of a forced oscillator leads its displacement by (a) 2π (b) π (c) $\frac{\pi}{2}$ (d) $\frac{3\pi}{2}$.

3.15. The sharpness of velocity resonance curve is more, if (a) γ is medium (b) γ is small (c) $\gamma \to \infty$ (d) none of these.

3.16. At resonance, the phase difference between displacement and velocity of a forced oscillator is (a) π (b) $\frac{2\pi}{3}$ (c) 0 (d) $\frac{3\pi}{4}$.

3.17. The dissipated power $P_{dissipated}$ varies as (a) ω (b) ω_0 (c) ω^* (d) $\gamma\omega$.

3.18. An external force $F = F_0 e^{i\omega x}$ is applied to a slightly damped oscillator of natural frequency ω_0, then in the steady state it vibrates with a frequency (a) ω_0 (b) ω (c) $(\omega_0 - \omega)$ (d) $\sqrt{\omega_0^2 - \omega^2}$.
[WBUT 2010]

ANSWERS

3.1. (b) **3.2.** (c) **3.3.** (b) **3.4.** (b) **3.5.** (b) **3.6.** (b) **3.7.** (c) **3.8.** (a) **3.9.** (b) **3.10.** (c) **3.11.** (b) **3.12.** (b) **3.13.** (a) **3.14.** (c) **3.15.** (b) **3.16.** (c) **3.17.** (a) **3.18.** (b).

3(B)–Short Answer Type Questions

3.1. Establish the differential equation of motion of a damped oscillator subjected to a periodic force of constant amplitude and frequency.

3.2. What do you mean by early stage and steady state of forced vibration. Discuss the factors characterizing them.

3.3. Define the term *impedance* of the oscillator. Obtain an expression for it and discuss its variation with the frequency ω of the applied periodic force.

3.4. What is resonance and resonant frequency of a forced oscillator. Discuss amplitude resonance.

3.5. Discuss velocity resonance in forced vibration. Distinguish between amplitude resonance and velocity resonance.

3.6. What are *resistance controlled, stiffness controlled* and *mass controlled* motion of a forced oscillator.

3.7. Show that *in the steady state, the time-averaged input power equals the time-averaged power dissipated through friction.*

3.8. What are sharpness of resonance and Q-factor? Show that the Q-factor is given by $Q = \frac{\omega_0 L}{R}$ in case of a series LCR resonant circuit.

3.9. If a harmonic oscillator of mass m and natural frequency ω_0 is driven by a force $F_0 \sin \omega t$ and the damping is proportional to 2μ times the velocity of the oscillator, then show that the displacement is given by

$$x = \frac{F_0/m}{\sqrt{(\omega_0^2 - \omega^2)^2 + 4\mu^2\omega^2}} \sin(\omega t - \theta),$$

where $\tan\theta = \dfrac{2\mu\omega}{(\omega_0^2 - \omega^2)}$.

Using the above relation (you need not deduce the relation), show that at velocity resonance, the velocity is in phase with the driving force.
[WBUT 2007]

3(C)–Numerical Problems

3.1. A particle of mass 0.01 kg is subjected to a restoring force 0.5 N/m, a damping force 10^{-4} kg/s and an external sinusoidal restoring force of constant amplitude. For what frequency of the driving force is velocity resonance achieved? Calculate the Q-factor and bandwidth at resonance.
[*Ans.* 7.07 rad/s, 707, 10^{-2} rad/s]

3.2. A 50 Hz ac emf of 200 V rms acts on a circuit containing a resistance of 100 Ω and an inductance 2 H in series. Calculate the capacity to be put in series to make the power factor unity? [*Ans.* 5.07 μF]

3.3. Derive an expression for the angular frequency ω_0 at which a series LCR circuit is in resonance. Show that the impedance of such a circuit at angular frequency ω may be given by

$$Z = R\left[1 + Q_0^2 \times \left(\frac{\omega}{\omega_0} - \frac{\omega_0}{\omega}\right)^2\right]^{1/2}$$

where $Q_0 = \omega L/R$ and is called the *figure of merit.*

3(D)–Broad Answer Type Questions

3.1. (a) Write down the differential equation of a forced vibration, explaining each term.

(b) Starting from the solution of the equation for forced vibration, explain the phenomenon of amplitude and velocity resonance deriving the value of the driving frequency in each case.

(c) Starting from the equation of plane progressive wave in one dimension, derive the differential equation of the wave. **[WBUT 2006]**

3.2. Set up the differential equation of motion of a simple harmonic oscillator subjected to a damping force and an external simple harmonic force.

Obtain expressions for the amplitude and the phase angle of the displacement in the steady state. **[cf. CUH 1996]**

3.3. For the forced vibration of a mechanical damped harmonic oscillator driven by a sinusoidal force of angular frequency ω, obtain the value of ω for amplitude resonance. What is the amplitude of the oscillator at resonance? What are the time-averaged kinetic and potential energies of the oscillator? Determine the value of ω for which they are equal. **[BU (HONS) 1992]**

3.4. What is resonance? Draw the plots of the displacement amplitude and velocity amplitude against the frequency of the sinusoidal force driving a mechanical oscillator for different values of damping. What is the power factor at velocity resonance? What is sharpness of resonance? Give a measure for it. **[cf. CUH 1992]**

3.5. (a) Derive an expression for the average power supplied by the sinusoidal force driving a mechanical oscillator over a complete cycle in the steady state. What is power factor and what is its value at resonance?

(b) Show that in the steady state the time-averaged input power equals the time-averaged power dissipated through friction. **[CUH 1996, 2006]**

3.6. (a) Distinguish between amplitude resonance and velocity resonance. Show that at velocity resonance the maximum velocity is inversely proportional to damping constant and the velocity of the oscillator is in phase with the driving force. **[CUH 1996]**

(b) What are half power frequencies? How are they related to sharpness of resonance? Show that the resonant frequency is the geometric mean between the half power frequencies.

3.7. (a) Derive the differential equation of motion of transverse waves generated in a stretched string.

(b) Find out the frequency for the fundamental tone in the above case.

3.8. (a) A series LCR circuit is subjected to a sinusoidal emf $V = V_0 \cos \omega t$. Find an expression for the amplitude of current.

(b) Discuss the phenomenon of resonance in such a circuit.

(c) Discuss the dependence of current amplitude with R. What is value of Q?

3.9. (a) Starting from the equation of motion after solving it, show that, for a forced oscillator in the steady state, the displacement amplitude at low frequencies ($\omega \to 0$), the velocity amplitude at velocity resonance ($\omega = \omega_0$) are independent of the frequency of the driving force.

(b) Explain the terms logarithmic decrement and quality factor of a damped oscillatory system. How are they related? **[WBUT 2010]**

Chapter 4 | Module 2.1

Interference of Electromagnetic Waves

4.1 Nature of Light

Light is a form of energy, which produces the sensation of vision in our eyes. It also produces heat, affects photographic plates, influences the growth and nourishment of plants. The experiment of Russian scientist Lebedev in 1901 proved that the incident light on a surface exerts pressure. This established the fact that *light is a form of energy.*

If light is a form of energy, then the natural question arises: *How does it propagate from one place to another*? Two processes have been proposed: (i) Swarm of *minute particles* travelling through the medium (like bullets of rifle) carry energy from one point to another. They do not require any material medium for propagation. (ii) Energy is transferred from one point to another in the form of *continuous waves.*

Sir Isaac Newton, in 1642, proposed *corpuscular* (or *emission*) *theory* in accordance with the first proposal. According to the second proposal Huygens, Young, Fresnel and others proposed the *wave theory* in 1678. To remove the inconsistency in wave theory James Clerk Maxwell proposed his *electromagnetic theory* in the later years. To explain the complex optical phenomena like photoelectric effect, Raman effect, Compton effect, etc. Albert Einstein put forward the *quantum theory* of light.

4.2 Corpuscular Theory of Light

According to this theory (proposed by Newton) *light consists of a stream of extremely light material particles shot out from a luminous body with tremendous velocity very much like the shots fired from a gun.* The impact of these particles on the *retina* produces the sensation of light. The corpuscular theory successfully explains the phenomena of rectilinear propagation of light, reflection, refraction, total internal reflection, dispersion, etc. The theory, however, fails to explain the phenomena of *interference, diffraction, polarization, double refraction, photoelectric effect, etc.*

The facts, which goes against corpuscular theory are: (a) the mass of a source should decrease when corpuscles are emitted. But *no such change in mass was detected.* (b) according to corpuscular theory the velocity of light in denser mediam is higher than that in rarer medium. But the experimental results of Foucault established the reverse fact. (c) According to this theory the velocity of light should be higher at higher temperature but experiment shows that the velocity of light is independent of temperature.

4.3 Huygens' Wave Theory

The theory states that, *every point on wavefront is the source of new disturbance. It sends small secondary wavelets propagated in all directions. These wavelets are spherical in shape and move forward in a homogeneous medium with the same velocity with which the original wave travels. At any instant the position and shape of the new wavefront can be obtained by drawing a surface enveloping these secondary wavelets.*

Ether pervades all space and is continuous having large elasticity and extremely low density so that the velocity of light in it is $c = \sqrt{E/\rho} \sim 3 \times 10^8$ m/s. According to Huygens' these waves are *mechanical in nature* and *longitudinal in form.*

Huygens' theory could successfully explain the phenomena of reflection, refraction, total internal reflection, dispersion, double refraction, etc. It, however, fails to explain the *rectilinear propagation of light, polarization.* The existence of ether is not yet proved or disproved.

In Maxwel's electromagnetic theory of light, the waves are transverse. The revival of corpuscular theory in some different form known as *quantum theory* of light was proposed by Max Planck. According to this, light consists of *quanta* or *photons* each of energy $h\nu$.

Thus, light sometimes behaves as particle, sometimes wave. There is not a single theory, which account all the optical phenomena. That is why light is called *the darkest part of physics.*

L.de Broglie of France, however, in 1926 reconciled the *particle* and *wave aspect* by proposing his hypothesis of *matter wave, which says, a moving particle of momentum* $p = mv$ *is equivalent to a wave of wavelength* $\lambda = \frac{h}{p}$, *h is the Planck's constant.* In fact, the particle and wave aspects are *complementary conceptions and not rival ones.*

4.4 Wavefront

In Fig. 4.1, let a vibrating particle O be placed in a *homogeneous* and *isotropic medium* extending in all directions. Then the vibration of O will be communicated to the neighbouring particles. The disturbance will travel with same velocity in all directions and at a particular time t it will reach to all the particles, which lie on the surface of a sphere with O as centre and $r = ct$ as the radius. This spherical surface is the wavefront at t.

Definition 4.1. *The wavefront at any instant of time is defined as the locus of all the neighbouring particles in the medium, which are being just disturbed at that instant of time and are consequently in the same phase of vibration.*

The surfaces, which pass through the maxima or minima of waves are called *wave surfaces.* The shape of the wavefront is determined by the source generating the waves:

(a) **Spherical wavefront**: If the source is a point source, then the radiated waves travel in all directions uniformly in a homogeneous, isotropic and continuous medium. The resulting wavefront will be a family of concentric spheres [Fig. 4.2].

(b) **Cylindrical wavefront**: If the source is linear, then it produces a family of coaxial cylindrical wavefronts [Fig. 4.3].

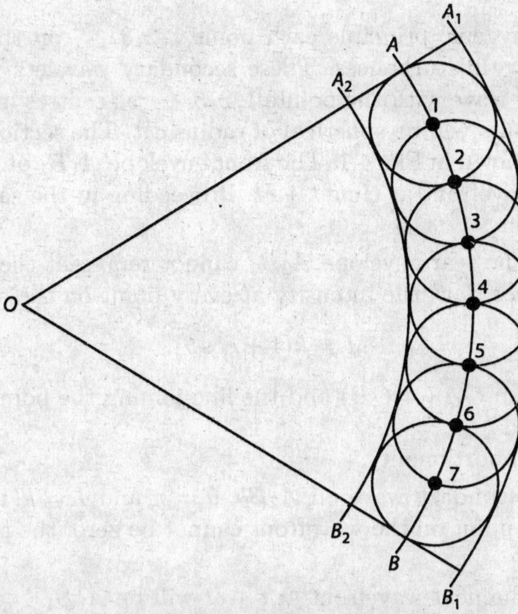

Fig. 4.1: Propagation of wavefront.

(c) **Plane wavefront**: At a large distance from the source both the spherical and cylindrical wavefronts appear as plane wavefront [Fig. 4.4]. The wavefront remains plane if the size of the slit through, which a plane wave passes is much larger than λ, [λ is the wavelength of the light wave]. If, however, the slit is $\sim \lambda$, the wavefront becomes spherical.

Rays: The lines normal to the wavefront are called *rays*, which represent the direction of energy flow. Rays are parallel lines for plane wavefront but for spherical wavefront they are radial lines.

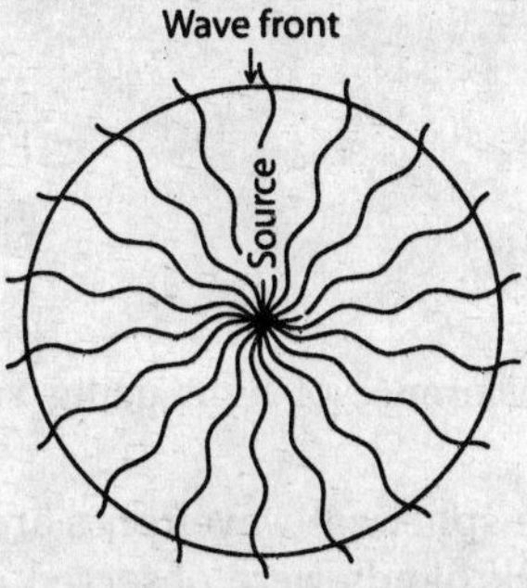

Fig. 4.2: Spherical wavefront.

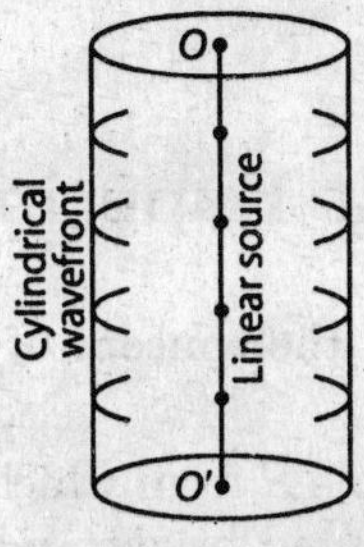

Fig. 4.3: Cylindrical wavefront.

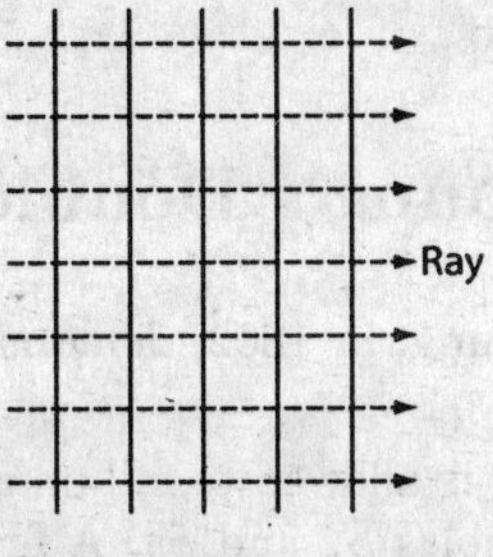

Fig. 4.4: Plane wavefront.

4.5 Huygens' Principle of Wave Propagation

In Fig. 4.1, let O be a point source of light placed in an isotropic medium and AB represents the position of the spherical wavefront at any instant t. The position of the wavefront at a later time $t + \delta t$ can be found out by applying Huygens' principle, which states, *every point on a wavefront is the source of a new disturbance. These wavelets are spherical in shape and move forward in a homogeneous medium with the same velocity with which the original wave travels. At any instant the position and shape of the new wavefront can be obtained by drawing a surface enveloping these secondary wavelets.*

Thus, according to Huygens' principle each point $1, 2, 3, \cdots$ on the wavefront AB at t will be the sources of secondary disturbances. These secondary wavelets will begin to spread with velocity (c) of the primary wave with the points $1, 2, 3, \cdots$ as centres in all directions. The shape of these secondary wavefronts will be spherical of radius $c\delta t$. The sections of these spherical wave surfaces will be circles as shown in Fig. 4.1. The front envelope A_1B_1 of these circles will represent the position of the next wavefront at time $t + \delta t$. Proceeding in the same way we can construct one wavefront from another.

It is to be noted that the rear envelope A_2B_2 cannot represent the next wavefront at $t + \delta t$. Because according to *Stokes' law*, the intensity at every point on the secondary wave is

$$I = k(1 + \cos\theta),$$

where θ = the angle between OA (or OB) and the line joining the point on the secondary wave to its centre, and

k = a constant of proportionality.

For the point on the secondary wave on A_2B_2, $\theta = \pi$ and $I = k(1 + \cos\pi) = k(1 - 1) = 0$. Since the intensity on any point on the wavefront cannot be zero, the rear envelope A_2B_2 can be the wavefront at $t + \delta t$.

Thus, the position of the next wavefront at $t + \delta t$ will be A_1B_1.

4.6 Interference of Light (Electromagnetic Waves)

When light waves from two *monochromatic sources* proceed in the same direction and superpose on a particle of the medium on the screen, the displacement of the particle (and hence the intensity) will be *the maximum or minimum* according as the waves meet the particle in the *same phase or in opposite phase respectively.* This phenomenon is the *interference of light* and provides direct evidence for the *wave nature of light.*

4.7 Young's Double Slit Experiment

Thomas Young, in 1802, demonstrated the concept of interference of light using *sunlight* and a *double slit.*

Sunlight is allowed to fall on a pinhole S from which the spherical wavefronts are made to fall on two pinholes S_1 and S_2. A few coloured bright and dark bands were observed on the screen PQ kept at a certain distance from the two pinholes [Fig. 4.5].

Later on Young replaced white light by monochromatic source of light and the pinholes by narrow slits.

In the improved set up, Young used a small fine slit S in front of a monochromatic source of light of wavelength λ. The wavefront from slit S advances towards a screen having two narrow slits S_1 and S_2, which are symmetrically placed w.r.t. S. A series of bright and dark bands will be observed on the screen PQ placed at a suitable distance from S_1S_2.

Now, the point O is equidistant from S_1 and S_2 and hence the path difference between the light waves reaching O is zero. So the point O will be the maximum intensity. At A and B the path difference is $\lambda/2$ and as such the intensity is the *minimum.* At point P and Q the path difference is λ and the intensity is *maximum.*

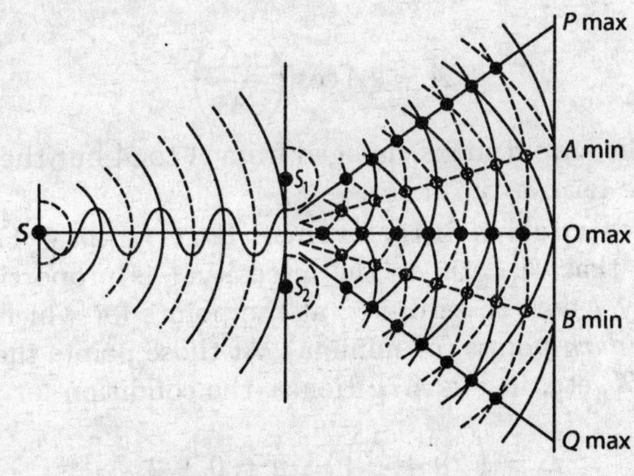

Fig. 4.5: Young's double slit experiment.

In Fig. 4.5 the solid arcs show the positions of *crests* and dotted ones, the positions of *troughs.* The superposition of *crest-crest* or *trough-trough* gives maximum intensity. As S_1 and S_2 are exactly similar, the pattern on the screen PQ is called *interference pattern.*

4.8 Analytical Treatment of Interference

We consider two monochromatic light sources each emitting waves of wavelength λ. The waves emitted from S_1 and S_2 in phase superpose at point P [Fig. 4.6].

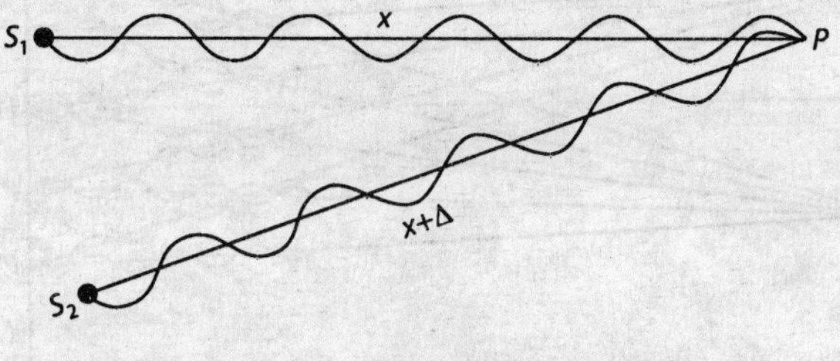

Fig. 4.6: Superposition of light wave at P from S_1 and S_2.

Let the waves coming from S_1 produce a displacement y_1 at P at any time t. Then

$$y_1 = A\sin\frac{2\pi}{\lambda}(ct - x),$$

where A = the amplitude
c = the velocity of the light wave and
x = distance between S_1 and P.

Similarly, if y_2 be the displacement at P due to the waves from S_2, then

$$y_2 = A\sin\frac{2\pi}{\lambda}[ct - (x + \Delta)],$$

where Δ = the path difference of P from S_1 and $S_2 = (S_2P - S_1P)$.

Now, according to the *principle of superposition*, the *resultant displacement at* P is given by

$$y = y_1 + y_2 = A\sin\frac{2\pi}{\lambda}(ct - x) + A\sin\frac{2\pi}{\lambda}[ct - (x + \Delta)]$$

$$= 2A\cos\left(\frac{\pi\Delta}{\lambda}\right)\sin\frac{2\pi}{\lambda}\left[ct - \left(x + \frac{\Delta}{2}\right)\right] = \mathcal{A}\sin\frac{2\pi}{\lambda}\left[ct - \left(x + \frac{\Delta}{2}\right)\right], \qquad (4.1)$$

where

$$\mathcal{A} = 2A\cos\left(\frac{\pi\Delta}{\lambda}\right). \tag{4.2}$$

Eqn (4.2) shows that the amplitude is changed from A to $\mathcal{A}$ but the waveform of the resultant displacement has the same form as y_1 or, y_2.

(a) **Condition for destructive interference.** If $\Delta = (2n+1)\frac{\lambda}{2}$, where $n = 0, 1, 2, 3, \cdots$, then eqn (4.2) shows that $\mathcal{A} = 0$. As the intensity I is proportional to $\mathcal{A}^2$, the points at which $\Delta = 0$, the intensity I will be zero, i.e., at the points for which $\Delta = \frac{\lambda}{2}, \frac{3\lambda}{2}, \frac{5\pi}{2}, \cdots$ (odd multiples of λ) will be the *dark* points (or minima). At those points the waves y_1 and y_2 interfere *destructively* [Points P_1, P_1', etc. in Fig. 4.7]. Hence, the condition for *destructive interference* is

$$\Delta = \left(2n + \frac{\lambda}{2}\right), \quad n = 0, 1, 2, \cdots. \tag{4.3}$$

(b) **Condition for constructive interference.** If $\Delta = 2n\frac{\lambda}{2}$, $n = 0, 1, 2, 3, \cdots$; then $\cos\left(\frac{\pi\Delta}{\lambda}\right) = 1$, then eqn (4.2) shows that the resultant amplitude $\mathcal{A} = 2a$. Hence, at those points for which $\Delta = 2n\frac{\lambda}{2}$, i.e., even multiple of $\frac{\lambda}{2}$, the intensity $I = 4a^2$, which is four times that due to a single source. The waves at those points interfere *constructively* producing *bright points* (or, maxima), points P, P_2, P_2', etc. in Fig. 4.6.

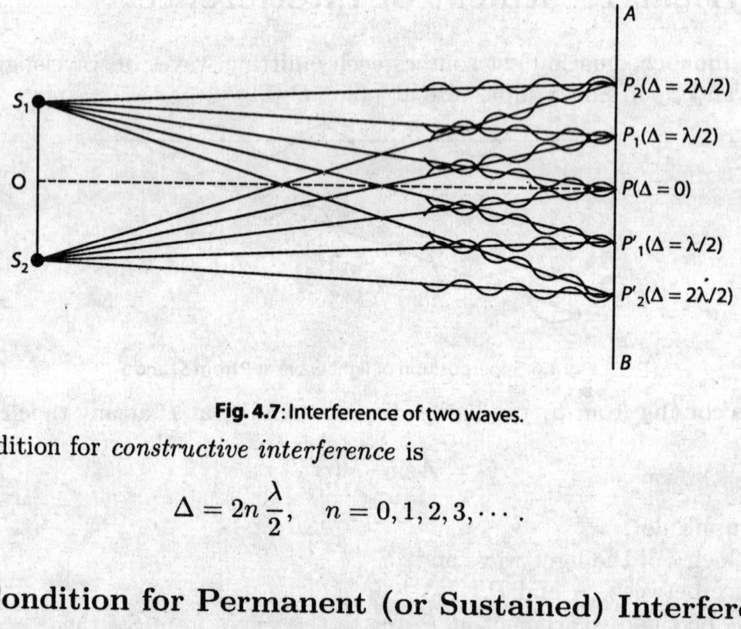

Fig. 4.7: Interference of two waves.

Thus, the condition for *constructive interference* is

$$\Delta = 2n\frac{\lambda}{2}, \quad n = 0, 1, 2, 3, \cdots. \tag{4.4}$$

4.8.1 The Condition for Permanent (or Sustained) Interference

For permanent interference a dark point should remain dark and a bright, bright for all values of time. Now, the intensity at a point dependends on the path differences, $\Delta = \Delta_1 + \Delta_2$, Δ_1 is the path difference due to the *initial phase difference of the two waves originating from S_1 and S_2 respectively* and Δ_2 is the *path difference due to the relative positions of the sources* and the screen. In a particular experiment Δ_2 is constant w.r.t. time. But as the emission of light from a source is an atomic phenomenon, it cannot be controlled from outside. That is why Δ_1 cannot be constant for two different sources. On the other hand, if the two sources S_1 and S_2 are derived from the same source S, then any change in phase in the original source S will produce the same change in S_1 and S_2. As a result the phase difference and hence the path difference Δ_1 will be zero for all time. Such sources are called *coherent sources.*

So the *conditions for permanent interference* are as follows:

(i) *The wavelength and the amplitude of the interfering light waves must be the same.*

(ii) *The two interfering sources should vibrate either in the same phase or have a constant phase difference between them, i.e., the sources must be coherent.*

(iii) *For maxima and minima at a point, the path difference should be even multiple of* $\frac{\lambda}{2}$ *and odd multiple of* $\frac{\lambda}{2}$ *respectively, i.e.,*

$$\Delta = 2n\frac{\lambda}{2} \qquad (n = 0, 1, 2, 3, \cdots) \quad \text{for maxima}$$

$$\text{and} \quad \Delta = (2n+1)\frac{\lambda}{2} \qquad (n = 0, 1, 2, 3, \cdots) \quad \text{for minima.}$$

(iv) *For distinct fringes the sources should be narrow and close to each other.*

4.9 Two Real Sources of Light Cannot Produce Interference Fringes

The two real sources cannot produce permanent interference due to the following two reasons: (i) *They will not emit light waves of the same frequency and same amplitude.* (ii) *The phase of the waves emerging out from the two sources change independently resulting a variation in phase.* Consequently, there will be a *general illumination.*

The fact behind the above reason is that the vibration of a particle in the source is influenced by the random collisions of the surrounding particles, and the phase of the vibration undergoes an *abrupt change.* In a real source there are numerous particles having all possible phases and as such the waves emerging from it cannot maintain a constant phase relationship among them. Hence, *two real sources cannot be coherent.* Due to sudden and continual changes in phase of these waves the positions of maxima and minima in the fringe system change rapidly producing *uniform illumination* on the screen.

4.10 The Intensity Distribution

Let $\vec{E}_1$ and $\vec{E}_2$ be the electric intensities at P due to the sources S_1 and S_2 respectively. If S_1P and S_2P are very large in comparison to S_1S_2, then $\vec{E}_1$ and $\vec{E}_2$ will be in the same direction for practical purposes. Then

$$\vec{E}_1 = \hat{\imath}E_{01}\cos\left(\frac{2\pi}{\lambda}S_1P - \omega t\right) \tag{4.5}$$

$$\text{and} \quad \vec{E}_2 = \hat{\imath}E_{02}\cos\left(\frac{2\pi}{\lambda}S_2P - \omega t\right), \tag{4.6}$$

where $\hat{\imath}$ is the unit vector along the direction of propagation.

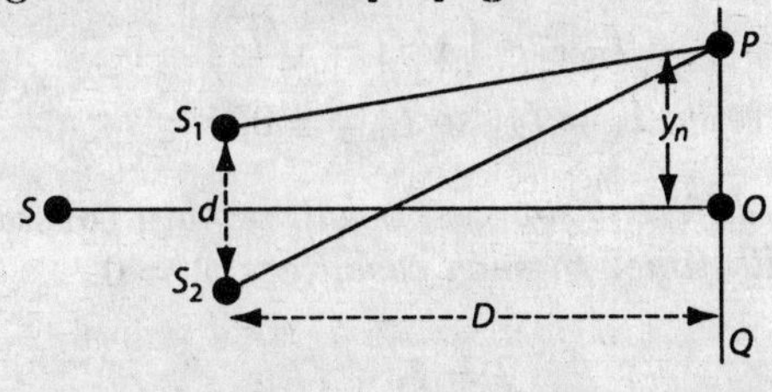

Fig. 4.8: Intensity distribution on the screen *PQ*.

Due to the superposition of $\vec{E}_1$ and $\vec{E}_2$, the resultant field at P is

$$\vec{E} = \vec{E}_1 + \vec{E}_2 = \hat{i}\left[E_{01}\cos\left(\frac{2\pi}{\lambda}S_1P - \omega t\right) + E_{02}\cos\left(\frac{2\pi}{\lambda}S_2P - \omega t\right)\right]. \tag{4.7}$$

The intensity at P is

$$I = kE^2, \text{ where } k = \text{a constant of proportionality}$$

$$= k\left[E_{01}\cos\left(\frac{2\pi}{\lambda}S_1P - \omega t\right) + E_{02}\cos\left(\frac{2\pi}{\lambda}S_2P - \omega t\right)\right]^2$$

$$\text{or,}\quad I = k\left[E_{01}^2\cos^2\left(\frac{2\pi}{\lambda}S_1P - \omega t\right) + E_{02}^2\cos^2\left(\frac{2\pi}{\lambda}S_2P - \omega t\right)\right.$$
$$\left. + E_{01}E_{02}\left\{\cos\left[\frac{2\pi}{\lambda}(S_2P - S_1P)\right] + \cos\left[2\omega t - \frac{2\pi}{\lambda}(S_2P + S_1P)\right]\right\}\right]. \tag{4.8}$$

Now,

$$\left\langle \cos^2\left(\frac{2\pi}{\lambda}S_1P - \omega t\right)\right\rangle = \left\langle \cos^2\left(\frac{2\pi}{\lambda}S_2P - \omega t\right)\right\rangle = \frac{1}{2}$$

and

$$\cos\left\{2\omega t - \frac{2\pi}{\lambda}(S_2P + S_1P)\right\}$$

oscillates between +1 and −1 and, therefore,

$$\left\langle \cos\left\{2\omega t - \frac{2\pi}{\lambda}(S_2P + S_1P)\right\}\right\rangle = 0.$$

∴ eqn (4.8) reduces to

$$I = I_1 + I_2 + 2\sqrt{I_1 I_2}\cos\phi, \tag{4.9}$$

where

$$\phi = \frac{2\pi}{\lambda}(S_2P - S_1P). \tag{4.10}$$

Eqn (4.10) represents the *phase difference between the waves $\vec{E}_1$ and $\vec{E}_2$ at P.*

$I_1 = \frac{1}{2}kE_{01}^2$ is the intensity produced by S_1 only and

$I_2 = \frac{1}{2}kE_{02}^2$ is the intensity at P produced by S_2 only.

Depending on the value of the term $\cos\phi$, eqn (4.9) leads to the following cases.

Case I *Now,* $\cos\phi = 1$, *when* $\frac{2\pi}{\lambda}(S_2P - S_1P) = 2n\pi$, $n = 0, 1, 2, \cdots$ *or,* $\frac{2\pi}{\lambda}\Delta = 2n\pi$ [$\Delta = S_2P - S_1P$, *the path difference of P from* S_1 *and* S_2] *or,* $\Delta = 2n\,\frac{\lambda}{2}$, *then from eqn* (4.9),

$$I_{\max} = \left(\sqrt{I_1} + \sqrt{I_2}\right)^2. \tag{4.11}$$

Similarly, $\cos\phi = 0$, *when* $\frac{2\pi}{\lambda}(S_2P - S_1P) = (2n+1)\pi$, $n = 0, 1, 2, \cdots$ *or,* $\frac{2\pi}{\lambda}\Delta(2n+1)\pi$ *or,* $\Delta = (2n+1)\,\frac{\lambda}{2}$, *then*

$$I_{\min} = \left(\sqrt{I_1} - \sqrt{I_2}\right)^2. \tag{4.12}$$

If $I_1 = I_2$, $I_{\min} = 0$. *Generally,* $I_1 \neq I_2$, *so* $I_{\min} \neq 0$.

Case II *If* S_1 *and* S_2 *be two different sources,* ϕ *will remain constant for about* 10^{-8} s *and then will vary in a random way with time. In such case,* $\langle\cos\phi\rangle = 0$.

Then eqn (4.9) *gives*

$$I = I_1 + I_2. \tag{4.13}$$

No interference pattern will be observed in this case of incoherent sources.

Case III *If S_2P and S_1P are extremely large compared to S_1S_2, then*

$$I_1 = I_2 = I_0 \text{ (say)}$$

and from eqn (4.9),

$$I = 2I_0 + 2I_0 \cos\phi = 4I_0 \cos^2 \frac{\phi}{2}. \qquad (4.14)$$

The intensity distribution pattern is shown in Fig. 4.9, *where the intensity I is plotted against ϕ. The curve is called the* $\cos^2$*-pattern.*

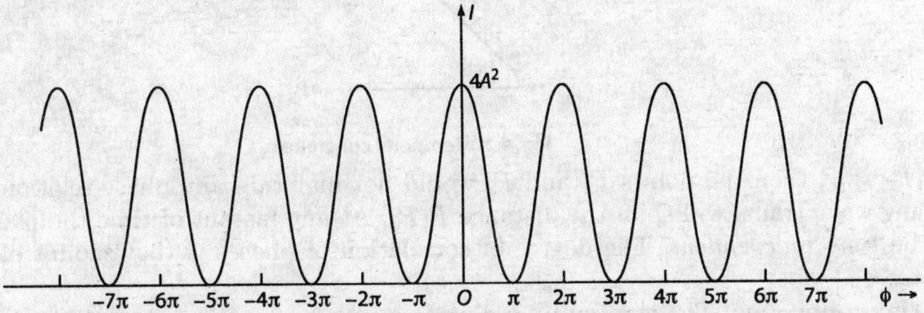

Fig. 4.9: Interference maxima and minima.

4.11 Conservation of Energy

The *principle of conservation of energy* holds good in the phenomenon of interference of light. This can be understood as follows:

We consider four consecutive points, two bright and two dark in the interference pattern. At each bright point the resultant displacement is $2A$, where A represents the amplitude of each interfering light wave at the bright point. The intensity at each bright point is, therefore, $I' = k(2A)^2 = 4kA^2$. Total energy at two bright points is thus $I = 2I' = 8kA^2$, where k is a constant of proportionality.

Again, the total light energy at the two dark points is zero.

Thus, for all the four points the total energy $= 8kA^2 + 0 = 8kA^2$.

If now each of the sources S_1 and S_2 successively illuminated. The light energy due to a single source at all the four points will be $4kA^2$. For the two sources the light energy on all these four points in thus $4kA^2 + 4kA^2 = 8kA^2$.

Hence, we conclude that total light energy remains the same, there is no destruction of light energy. In other words, the interference of light obeys the *principle of conservation of energy.*

[Also see the answer of Q.4.6 (p. 106)]

4.12 Coherence

If a fixed and predictable phase difference between several waves travelling in a particular direction be maintained, then we say the motion is coordinated or *coherence.* The corresponding waves are called *coherent waves* and sources emitting them, the *coherent sources.*

Coherence effects are mainly of two types: (i) *temporal coherence* and (ii) *spatial coherence.*

Temporal coherence: If the phase difference at a single point in the bundle of light waves propagating in space, at the beginning and end of a fixed time interval does not change with time, then the waves are said to have *temporal coherence.* The phase difference between any two fixed points P_1 and P_2 [Fig. 4.10] along any ray will be independent of time but depends on P_1P_2 and the coherent length (l_c) of the light beam. [The coherent length is the product of the number of waves (N) and the wavelength λ, i.e., $l_c = N\lambda$], i.e., the distance $P_1P_2 < l_c$. The waves will correlate in their rise and fall maintaining a *constant phase difference.*

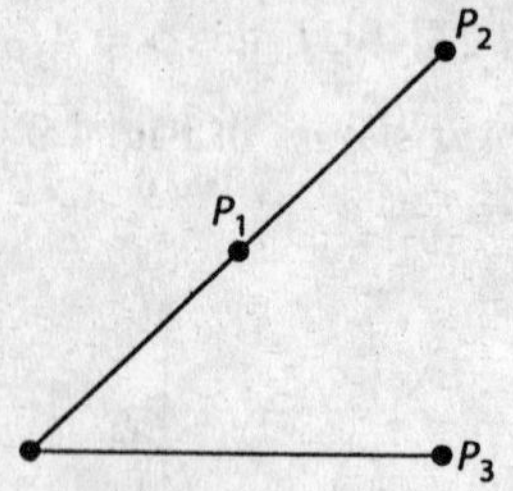

Fig. 4.10: Temporal coherence.

If $P_1P_2 > l_c$, then the points P_1 and P_2 would not maintain any phase relationship. In that case, many wave trains will span the distance P_1P_2. At any instant of time the plane at P_1 and P_2 will be *time independent.* The degree of correlation of phases is the amount of *longitudinal coherence.*

Spatial coherence: The continuity and uniformity of a wave in a direction perpendicular to the direction of propagation refers to *spatial coherence.* The wave is said to have *spatial coherence* if the phase difference for any two fixed points in a plane normal to the wave propagation does not vary with time. In Fig. 4.9, if $SP_1 = SP_3$, then the field points P_1 and P_3 would have the same phase. Since the waves produced by an ideal source exhibit spatial coherence as the phases of the waves at any two points, which are equidistant from the source, are equal. An extended source, however, exhibits less lateral coherence.

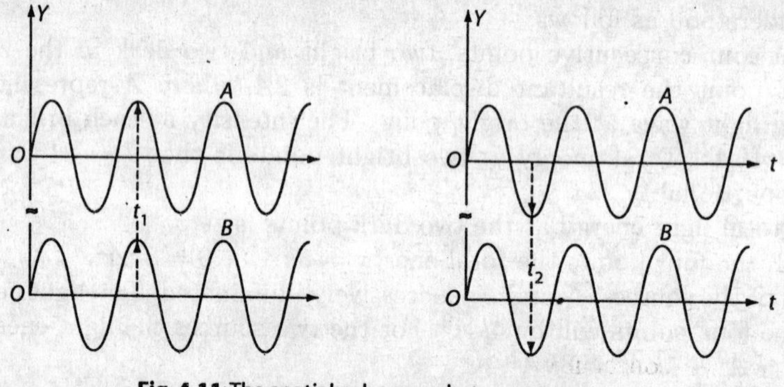

Fig. 4.11: The spatial coherence between two waves.

The *degree of contrast* of the interference fringes is a measure of the degree of spatial coherence of the source emitting the waves. *Spatial coherence is better if the contrast is higher.*

Ex. 1 *The phase difference between two waves at A and B is zero (is constant at t_1 and t_2). So the waves are spatially coherent.*

4.13 The Fringe Width

Let the coherent monochromatic sources of light S_1 and S_2 are placed at a distance $2d$ and a screen PQ be placed at a distance $OC = D$ from S_1S_2 [Fig. 4.10]. The point C being equidistant from S_1 and S_2, the waves at C due to the two sources will be in phase. So the amplitude and hence the intensity at C will be the maximum. The point C is called the *central maximum.* At any point P, at a distance x from C, the intensity will be either maximum or minimum as the path difference $\Delta = S_2P - S_1P$ is *even* or *odd* multiple of $\frac{\lambda}{2}$.

Now, from Fig. 4.12,

$$S_2P^2 = D^2 + (x+d)^2$$
$$S_1P^2 = D^2 + (x-d)^2.$$

$$\therefore \quad S_2P^2 - S_1P^2 = (x+d)^2 - (x-d)^2 = 4dx$$

$$\text{or,} \quad S_2P - S_1P = \frac{4dx}{S_2P + S_1P} = \frac{4dx}{2D} = \frac{2dx}{D}. \tag{4.15}$$

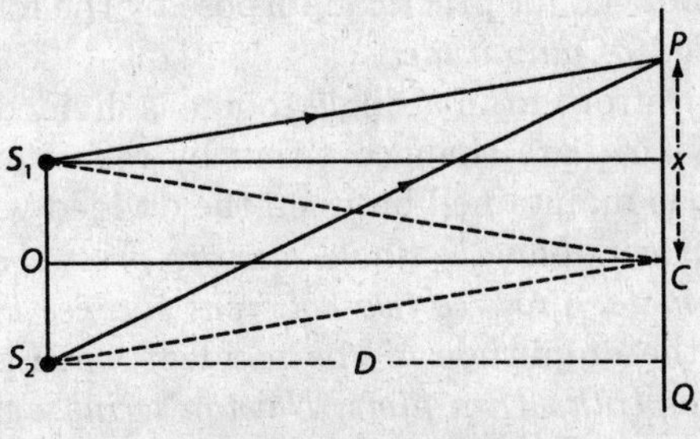

Fig. 4.12: Fringe width.

Since D is very large we have taken $S_2P \simeq S_1P = D$.

So the path difference between the two interfering beams is

$$\Delta = S_2P - S_1P = \frac{2dx}{D}. \tag{4.16}$$

Bright points: If P be the position of the nth order bright fringe, then

$$\Delta = 2n\frac{\lambda}{2} \quad \text{or,} \quad \frac{2dx}{D} = 2n\frac{\lambda}{2} \quad \text{or,} \quad x = \frac{Dn\lambda}{2d}. \tag{4.17}$$

If x_{n+1} be the distance of the $(n+1)$th order bright point from C, then

$$x_{n+1} = \frac{D(n+1)\lambda}{2d}. \tag{4.18}$$

Let the distance of the nth order bright point from O be represented by

$$x_n = \frac{Dn\lambda}{2d}. \tag{4.19}$$

$\therefore$ distance between consecutive bright points is

$$x_{n+1} - x_n = \frac{D(n+1)\lambda}{2d} - \frac{Dn\lambda}{2d} = \frac{D\lambda}{2d}. \tag{4.20}$$

The *fringe width* β *is defined as the distance between two consecutive maxima* (or *minima*) and is given by

$$\beta = x_{n+1} - x_n = \frac{D\lambda}{2d}. \tag{4.21}$$

Dark points: The condition of dark point is $\Delta = (2n+1)\frac{\lambda}{2}$.

If P be the position of the nth dark point, then

$$\Delta = (2n+1)\frac{\lambda}{2} \quad \text{or,} \quad \frac{2dx}{D} = (2n+1)\frac{\lambda}{2} \quad \text{or,} \quad x = \frac{D}{2d}(2n+1)\frac{\lambda}{2}.$$

Taking two consecutive dark points we get the fringe width

$$\beta = \frac{D\lambda}{2d}. \tag{4.22}$$

Since the fringe width β is the same for the bright or dark fringes, so we conclude that *the interference fringes in this case are of equal width.*

From eqn (4.22), the following points are apparent: (i) *the fringe width is directly proportional to the distance between the source and the screen* (ii) *the fringe width is proportional to the wavelength* λ and (iii) *the fringe width is inversely proportional to the distance between the sources* S_1 *and* S_2. Hence, β may be increased by increasing λ and D and decreasing d.

4.14 Production of Coherent Sources

Coherent sources can be produced for practical purpose by the following two methods: (i) *division of wavefront* and (ii) *division of amplitude.*

In the *first method,* a wavefront from a single source is divided laterally into two parts and the directions of these divided parts are changed either by reflection or refraction or diffraction. A constant phase relationship is maintained between the divided wavefronts and thus two coherent sources are produced. *Bi-prism, Lloyd's single mirror, Fresnel's double mirror, Billet's divided lens and Rayleigh interferometer produce two coherent sources by this method.*

In the *second method,* the amplitude of the incident wave is split up into two parts either by *reflection or refraction or both. Thin films, Newton's ring apparatus, Michelson interferometer, Jamin's interferometer and Fabry-Perot interferometer produce two coherent sources by the method of division of amplitude.*

4.15 Fresnel's Bi-prism Experiment

A bi-prism ABC is an obtuse-angled prism having the obtuse angle (179°) at A. The acute angles at B and C are each 30′. The experimental arrangement is shown in Fig. 4.13.

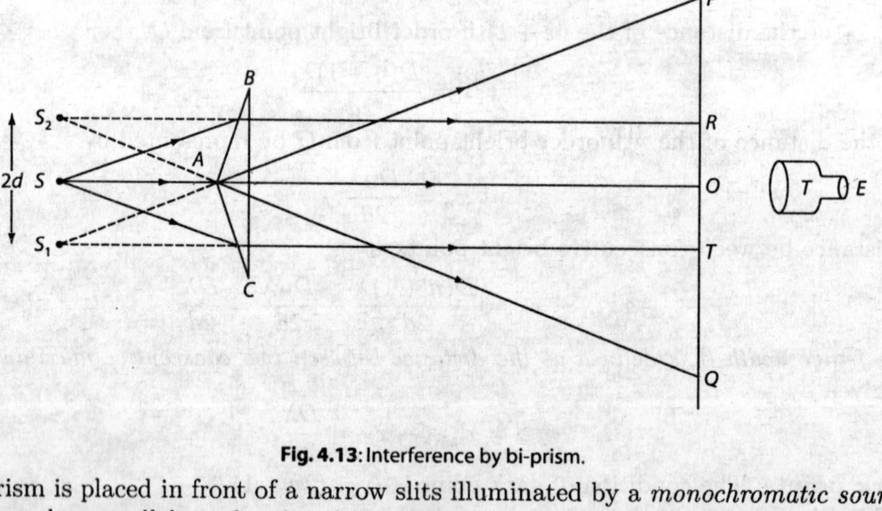

Fig. 4.13: Interference by bi-prism.

The bi-prism is placed in front of a narrow slits illuminated by a *monochromatic source* with the refracting edge parallel to the slit. After refraction through AB and AC the light beam appears to come from S_1 and S_2. S_1 and S_2 are thus a *pair of virtual images* of the slit S and behave as *coherent sources.* S_1 and S_2 are closer together since $\angle A = 179°$ is an obtuse angle. The emergent beams overlap in the regions S_1PT and S_2QR, where interference pattern is produced. Dark and bright fringes formed in the focal plane of the eye-piece, which can be observed by an eye-piece E whose objective is facing towards the bi-prism.

Theory. In Section 4.13 we have seen that the fringe width of the interference pattern is given by

$$\beta = \frac{D\lambda}{2d}.$$

Then the wavelength

$$\lambda = \frac{2d}{D}\beta, \tag{4.23}$$

where $2d$ = the distance between the two *coherent sources*

D = the distance between the focal plane of the eye-piece and the plane of the source.

Measuring β, $2d$ and D, we can find out the wavelength λ of the source S.

Determination of β: Fringes are observed in the field of view of the micrometer eye-piece E. The vertical cross-wire of the eye-piece is made to coincide with one of the fringes and the position is noted, then the micrometer is moved sideways crossing a number of fringes. This number of fringes is counted and the micrometer's vertical cross-wire is made to coincide with the final fringe and the position of the eye-piece is noted. Then the fringe width

$$\beta = \frac{\text{Distance moved}}{\text{No. of fringes moved across the cross-wire}}.$$

Determination of D : The distance between the slit and the focal plane of the eye-piece is measured by noting the scale reading of the optical bench. If x_1 and x_2 be the positions of the slit and that of the focal plane of the eye-piece, then $D = x_2 - x_1$.

Determination of $2d$: Let $S_1'S_2'$ be the distance of the images of the virtual sources S_1 and S_2, when a convex lens is placed at L_1 and $S_1''S_2''$ be the distance of the images of the same two sources when the lens is shifted to the position L_2 [Fig. 4.12].

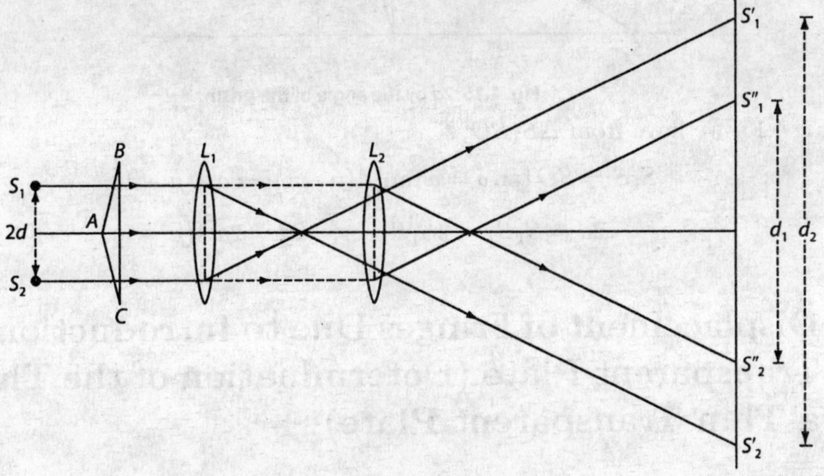

Fig. 4.14: Separation of S_1 and S_2.

Let $S_1'S_2' = d_2$ and $S_1'S_2'' = d_1$, then

$$\text{for } L_1, \quad \frac{d_1}{2d} = \frac{v}{u} = \frac{n}{m}$$

and

$$\text{for } L_2, \quad \frac{d_2}{2d} = \frac{v}{u} = \frac{m}{n}.$$

$$\therefore \quad \frac{d_1 d_2}{(2d)^2} = 1 \quad \text{or,} \quad 2d = \sqrt{d_1 d_2}, \tag{4.24}$$

where u and v are object and image distances respectively.

Thus, measuring d_1 and d_2, we can find out $2d$.

Alternative Method of Measuring 2*d* : Knowing the acute angle A of the bi-prism and the refractive index of the material of the bi-prism, $2d$ can be determined.

If δ be the minimum deviation of the emergent beam from the bi-prism, then

$$\mu = \frac{\sin\left(\frac{A+\delta}{2}\right)}{\sin\frac{A}{2}}.$$

Since A and δ are very small, we have

$$\mu = \frac{\left(\frac{A+\delta}{2}\right)}{\frac{A}{2}} \quad \text{or,} \quad \delta = A\mu - A = A(\mu - 1).$$

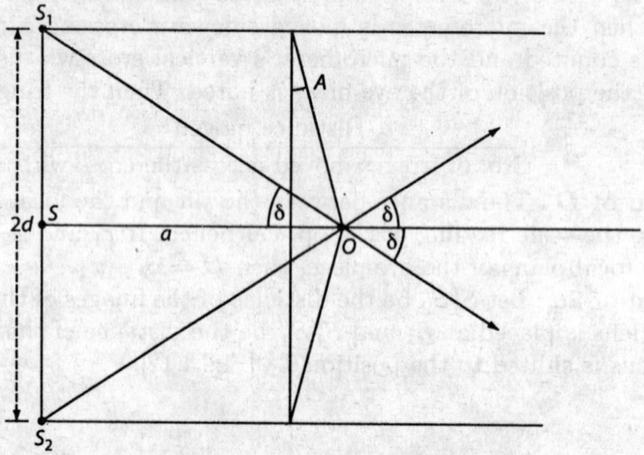

Fig. 4.15: 2*d* by the angle of the prism.

From Fig 4.15, we have from ΔS_1SO,

$$S_1S = SO\tan\delta = a\tan[A(\mu - 1)] = aA(\mu - 1).$$

$$\therefore \quad 2d = S_1S_2 = 2S_1S \quad \text{or,} \quad 2d = 2aA(\mu - 1). \tag{4.25}$$

4.16 Displacement of Fringes Due to Introduction of a Thin Transparent Plate (Determination of the Thickness of a Thin Transparent Plate)

Let S_1 and S_2 be two monochromatic light sources producing interference on the screen PQ [Fig. 4.16]. Since $S_1C = S_2C$, then C will be the position of the *central maxima*.

If now a *thin transparent sheet of thickness* t and *refractive index* μ be inserted in the path of one of the interfering beams, the entire fringe system will be shifted towards the path of the beam in which the plate is inserted. The fringe width β, however, will remain unchanged.

The time taken by the ray S_1P to travel from S_1 to P is

$$\frac{S_1P - t}{c} + \frac{t}{c_m} = \frac{S_1P - t}{c} + \frac{\mu t}{c} = \frac{S_1P + (\mu - 1)t}{c},$$

where $c_m = \frac{c}{\mu}$, the velocity of light in the plate and c is velocity of light in air. Hence, path S_1P is equivalent to an air path $S_1P + (\mu - 1)t$.

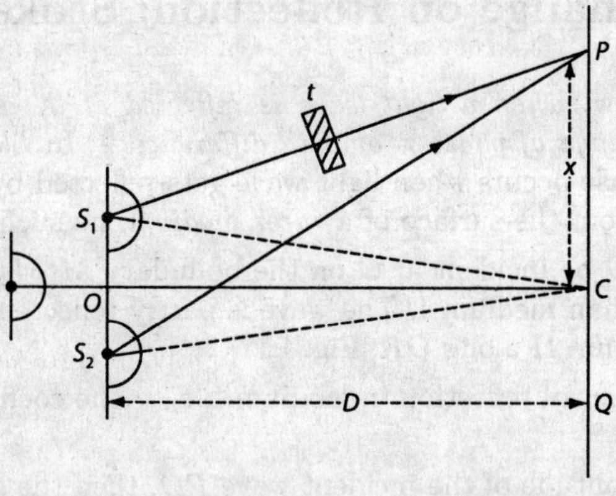

Fig. 4.16: Fringe displacement due to a transparent plate.

The path difference of the rays S_2P and S_1P is

$$\Delta = S_2P - [S_1P + (\mu - 1)t] = (S_2P - S_1P) - (\mu - 1)t = \frac{2xd}{D} - (\mu - 1)t. \qquad (4.26)$$

$\left[\text{since } S_2P - S_1P = \frac{2xd}{D}\right]$

If P be the position of the nth maxima, then

$$\Delta = 2n\frac{\lambda}{2} = n\lambda. \qquad (4.27)$$

From eqns (4.26) and (4.27),

$$\frac{2x_n d}{D} - (\mu - 1)t = n\lambda \quad \text{or,} \quad x_n = \frac{D}{2d}[n\lambda + (\mu - 1)t]. \qquad (4.28)$$

Eqn (4.28) gives the position of the nth maxima.

Now, in absence of the plate $t = 0$, the position of the nth maxima is

$$x_n' = \frac{Dn\lambda}{2d}. \qquad (4.29)$$

If δ be the displacement of the nth maxima, then

$$\delta = \frac{D}{2d}[n\lambda + (\mu - 1)t] - \frac{Dn\lambda}{2d} \quad \text{or,} \quad \delta = \frac{D}{2d}(\mu - 1)t. \qquad (4.30)$$

Since eqn (4.30) does not contain the order number n, δ is the same for all order number. Therefore, δ is also the shift of the central bright fringe (x_0).

$$\therefore \quad x_0 = \frac{D}{2d}(\mu - 1)t. \qquad (4.31)$$

Again, the fringe width $\beta = \frac{D\lambda}{2d}$ or, $\frac{\beta}{\lambda} = \frac{D}{2d}$.

$$\therefore \quad x_0 = \frac{\beta}{\lambda}(\mu - 1)t. \qquad (4.32)$$

Thus, measuring x_0, β and t we can determine the *refractive index* μ *of the material of the film* for the monochromatic light of wavelength λ.

Measurement of thickness of a mica sheet: We have from eqn (4.31) the shift of the central maxima by the introduction of a thin mica sheet of thickness t in the path S_1P,

$$x_0 = \frac{D}{2d}(\mu - 1)t \quad \text{or,} \quad t = \frac{2dx_0}{D(\mu - 1)}. \qquad (4.33)$$

To determine x_0, the monochromatic source S is replaced by a white source when the central maximum will be a bright white fringe. By the introduction of the mica sheet the central bright fringe will be shifted by x_0. By a bi-prism set-up, the value of x_0 can be determined. Measuring $2d$ and D, the thickness of the mica sheet can be determined.

4.17 Phase Change on Reflection: Stokes' Law

According to Stokes' law, *when a light wave is reflected at the surface of an optically denser medium, it suffers a change of phase* π *or path difference* $\frac{\lambda}{2}$. In *Lloyd's mirror experiment,* such an abrupt change of phase occurs when light wave gets reflected by a *denser medium.* When the light wave is reflected from the surface of a *rarer medium,* no such phase change occurs.

Let a ray of light PO be incident at O on the boundary M_1M_2 of media I and II, medium II being *optically denser* than medium I. The wave is partly reflected along OQ and the remaining part refracted into medium II along OR [Fig. 4.17].

Let r_1 = the coefficient of reflection in medium I; t_1 = the coefficient of refraction in medium II.

Then if a be the amplitude of the incident wave PO, then the amplitude of the reflected and refracted waves along OQ and OR respectively will be ar_1 and at_1.

Now, let a ray RO of amplitude at_1 be incident at O. This will give rise to a reflected wave of amplitude at_1r_2 and a transmitted wave OP of amplitude at_1t_2. Here r_2 and t_2 are the reflection and transmission coefficients of media II and I respectively. Similarly, the ray QO of amplitude ar_1 will give rise to a reflected ray OP of amplitude ar_1^2 and a refracted ray OS of amplitude ar_1t_1.

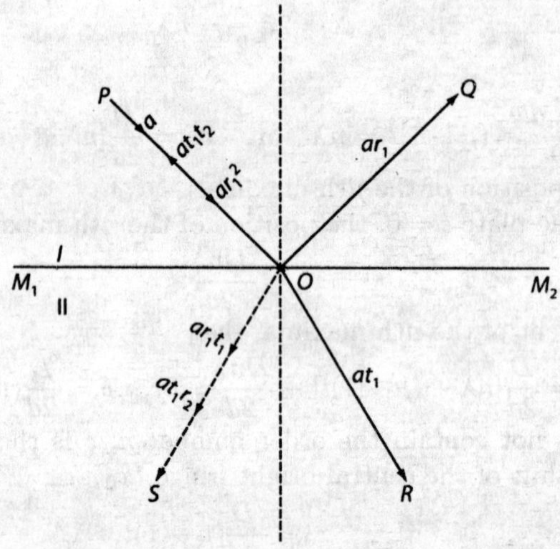

Fig. 4.17: Phase change by reflection.

The two rays of amplitudes ar_1^2 and at_1t_2 will combine to give the wave of amplitude a if the optical reversibility be true. So

$$ar_1^2 + at_1t_2 = a \quad \text{or,} \quad t_1t_2 = 1 - r_1^2. \tag{4.34}$$

Again, since there was no wave along OS, we have

$$at_1r_2 + ar_1t_1 = 0 \quad \text{or,} \quad r_2 = -r_1. \tag{4.35}$$

Eqn (4.35) shows that r_2 and r_1 are numerically equal but of *opposite in sign.* This shows that a relative phase difference π occurs between the rays reflected from the surface of denser medium. We thus conclude that a *phase change* π *occurs when a wave is reflected from the surface of a denser medium.*

4.18 Interference by Division of Amplitude: Colour of Thin Films

If *a thin film of any transparent substance*, e.g., a soap film or thin layer of oil spread on the surface of water be observed in sunlight, we see *brilliant colours.* This is due to the interference of the rays reflected from the upper and lower surfaces of the film. With the increase of thickness of the film, the intensity of colour diminishes and beyond a certain thickness of the film the colour effect disappears completely.

To understand the colour effect by thin films we consider a thin film of thickness t and refractive index μ [Fig. 4.18] with parallel surfaces. Let a ray of monochromatic light of wavelength λ be incident at C, which is partly reflected along CQ and partly refracted along CE inside the film. This will again give rise to a reflected ray ED by division of amplitude and a transmitted ray ES. The ray ED will again suffer reflection at D giving a reflected ray DF and a transmitted ray DR. Finally, we shall get a set of parallel reflected rays each being parallel to CQ and a similar set of transmitted rays parallel to ES.

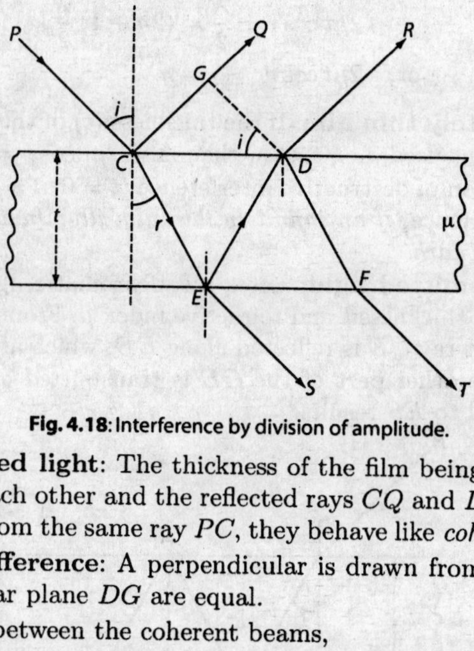

Fig. 4.18: Interference by division of amplitude.

Interference by reflected light: The thickness of the film being very small, the points C and D will be very close to each other and the reflected rays CQ and DR will be parallel. Again, both the rays being derived from the same ray PC, they behave like *coherent waves* and interfere.

Calculation of path difference: A perpendicular is drawn from D on CQ. The paths of the ray from the perpendicular plane DG are equal.

Now, the path difference between the coherent beams,

$$\Delta' = \mu(CE + ED) - CG = 2\mu CE - CG \quad [\because CE = ED]$$

$$= 2\mu CE - CD \sin i, \quad [\angle i = \text{the angle of incidence}]$$

$$= \frac{2\mu t}{\cos r} - 2t \cdot \tan r \sin i, \quad [\angle r = \text{the angle of refraction}]$$

$$= \frac{2\mu t}{\cos r} - \frac{2t \sin r \sin i}{\cos r} = \frac{2\mu t}{\cos r} - \frac{2\mu t \sin^2 r}{\cos r} \quad \left[\because \frac{\sin i}{\sin r} = \mu\right]$$

$$= \frac{2\mu t}{\cos r}\left(1 - \sin^2 r\right).$$

$$\therefore \quad \Delta' = 2\mu t \cos r.$$

Again, the ray CQ is reflected from the surface of a denser medium, a path difference $\frac{\lambda}{2}$ will occur between CQ and DR.

∴ the total path difference,

$$\Delta = \Delta' \pm \frac{\lambda}{2}$$

$$\Delta = 2\mu t \cos r \pm \frac{\lambda}{2}. \tag{4.36}$$

For *constructive interference,* $\Delta = 2n\,\frac{\lambda}{2}$ and the corresponding portion of the film will appear bright. Thus, the condition for *constructive interference*

$$2\mu t \cos r \pm \frac{\lambda}{2} = 2n\,\frac{\lambda}{2}$$

$$\text{or,}\quad 2\mu t \cos r = (2n \pm 1)\,\frac{\lambda}{2}. \tag{4.37}$$

Again, for *destructive interference,* $\Delta = (2n+1)\,\frac{\lambda}{2}$ and the portion of the film, from which these interfering beans come, will appear dark. Thus, the condition for *destructive interference* is

$$2\mu t \cos r \pm \frac{\lambda}{2} = (2n+1)\,\frac{\lambda}{2}$$

$$\text{or,}\quad 2\mu t \cos r = n\lambda. \tag{4.38}$$

Interference by infinitely thin film: If the thickness (t) of the film is very small compared to the wavelength of the light used, i.e., $t \ll \lambda$, then $\Delta = 2\mu t \cos r = 0$ [$\because t$ is very small].

This satisfies the condition of destructive interference ($t = 0, n = 0$). This corresponds to zero intensity in reflected part. Hence, *if an infinitely thin film illuminated by white light is observed by reflected light, it appears dark.*

Interference by transmitted light: Let a monochromatic light ray PC of wavelength λ be incident at C of a film of thickness t and refractive index μ. From Fig. 4.19 it is evident that a portion of the refracted light ray CE is reflected along ED, which after being internally reflected at D, is incident at F. The other part of the CE is transmitted along ES. Similarly, another transmitted ray FT parallel to ES results.

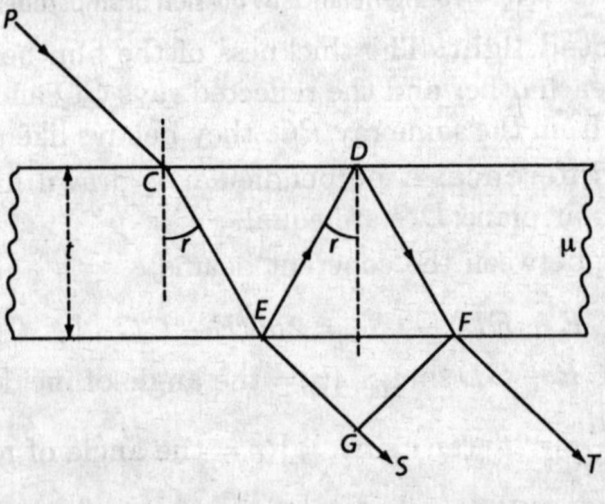

Fig. 4.19: Interference by transmitted light.

To find out the path difference between the parallel transmitted rays ES and FT, FG is drawn perpendicular from F on EG. The path difference is thus

$$\Delta = (ED + DF)\mu - EG = 2\mu t \cos r \quad \text{(as in case of reflected light).}$$

As there is no additional phase difference due to reflected light, the total path difference between ES and FT is $\Delta = 2\mu t \cos r$.

For *constructive interference*,

$$\Delta = 2n\frac{\lambda}{2} \quad \text{or,} \quad 2\mu t \cos r = n\lambda. \tag{4.39}$$

The portion of the film from which these interfering beams originate will appear bright. Again, if

$$\Delta = (2n \pm 1)\frac{\lambda}{2} \quad \text{or,} \quad 2\mu t \cos r = (2n \pm 1)\frac{\lambda}{2}, \tag{4.40}$$

destructive interference occurs and the corresponding portion of the beam appears dark.

It is readily observed that *the conditions of constructive and destructive interference for transmitted light are just the reverse to those for reflected light.* A part of the film appearing bright in reflected light will appear dark in transmitted light. If white light is used, then a particular colour of maximum brightness in reflected light will have its minimum in the transmitted light and vice versa.

So the sensation of colour in fringe system due to reflected and transmitted light are just complementary to each other.

4.19 Classification of Fringes Exhibited by Thin Films

The path difference between two interfering reflected rays of light from a transparent film of thickness t and refractive index μ is

$$\Delta = 2\mu t \cos r + \frac{\lambda}{2}.$$

So the phase difference

$$\phi = \left(\frac{2\pi}{\lambda}\right)\left(2\mu t \cos r + \frac{\lambda}{2}\right). \tag{4.41}$$

Clearly ϕ depends on (i) λ (ii) μt and (iii) r *for a particular film.*

(i) **Fringes of equal inclination (FEI) (or Haidinger fringes)**: We have

$$\phi = \frac{4\pi\mu t \cos r}{\lambda} = (2n+1)\pi \tag{4.42}$$

for the *bright fringes.*

If μt and λ are constant, then eqn (4.40) shows that a particular order number n of bright fringe is governed by the particular angle r, i.e., *the inclination of the rays with the normal to the film.* Such fringes are called the *fringes of equal inclination* (FEI).

Circular fringes in *Michelson interferometer* are the examples of *fringes of equal inclination* (FEI).

(ii) **Fringes of equal thickness (FET) (or Fizeau fringes)**: If λ and r are constant, then from eqn (4.40) we note that a particular order number n of the bright fringe is dependent on a particular value of μt, i.e., the *thickness of the film.* Such fringes are called the *fringes of equal thickness* (FET).

Circular fringes as *Newton's rings* are the example of this class of fringe.

(iii) **Fringes of equal chromatic order (FECO)**: If μt and r are constants, then a particular order number n of a bright fringe is governed by λ, i.e., *particular colour.* A particular colour fringe satisfies the condition for a particular wavelength. These fringes are called *fringes of equal chromatic order* (FECO).

4.20 Explanation of Colour Effect

When a thin film is exposed to white light, e.g., sunlight, beautiful colours appear in the reflected light. The cause of this could be understood from the following explanation.

Soup buble: Let the thickness t of the film of the soap buble is constant. White light has different λ and μ. Due to the spherical nature of the buble the angle of refraction r varies from point to point even for parallel incident beam. Hence, varying values of μ and r can satisfy the condition of *constructive interference,* i.e., $2\mu t\cos r = (2n+1)\frac{\lambda}{2}$ for a particular wavelength λ, i.e., for a particular colour, so that point will be the maximum in that particular colour.

In a similar way the other points may satisfy the condition of *constructive interference* and may appear bright in other colours.

Thin layer of oil film: A thin layer of oil film may be obtained by pouring a little oil on water surface. When parallel sunrays are incident, the angle of refraction r will remain constant. For different values of μ due to different λ the thickness of the film t may not be constant for different points of the film. Thus, the different points of the film satisfy the condition of *constructive interference* for different colours depending on μt values. That is why *the film appears multicoloured.*

If a *monochromatic light* of wavelength λ and refractive index μ be incident on a film, where thickness t is not constant. Depending on the values of λ, μ and t, certain points of the film may satisfy the condition of darkness and this points will appear dark. The other points may appear bright. So the film will have alternate dark and bright bounds.

If t be large for a film, then the condition of brightness, viz., $2\mu t\cos r = (2n\pm 1)\frac{\lambda}{2}$ may be satisfied for a large number of wavelengths. The film under this condition will exhibit *no colour effect.*

4.21 Non-Reflecting Films

Optical instruments in general consist of a system of lenses, prisms, etc. in which a part of the incident light is lost by reflection at various surfaces reducing the quality of performance of the instrument. This can be prevented by the use of *non-reflecting films* on the lens, prism surfaces.

The material to be coated on the glass should have refractive index μ_f in between that of air (μ_a) and glass (μ_g), i.e., $\mu_g > \mu_f > \mu_a$ [Fig. 4.20].

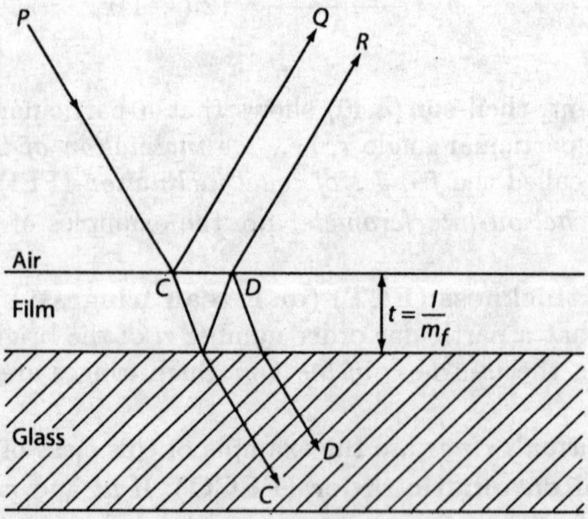

Fig. 4.20: Non-reflecting film.

If the wavelength of the incident light be λ, then the path difference between two reflected rays from the surface of the film is $\Delta = 2\mu_f t \cos r$.

For *normal incidence* the angle of refraction $r = 0$. $\therefore$, $\cos r = 1$.

$\therefore \Delta = 2\mu_f t$.

For *destructive interference*, $\Delta = \frac{\lambda}{2}$,

$$\text{i.e.,}\quad 2\mu_f t = \frac{\lambda}{2} \quad \text{or,}\quad t = \frac{\lambda}{4\mu_f}. \tag{4.43}$$

Thus, the thickness of the non-reflecting film should be $\frac{\lambda}{4\mu_f}$, λ being the wavelength of light, for which λ is usually chosen near the *mean wavelength of the visible spectrum.*

It is to be noted that the light, which would be lost in absence of the non-reflecting film, appears mostly in the transmitted beam enhancing the performance of the instrument if the non-reflecting coating be made.

This technique of reducing the reflectivity is called *blooming.*

4.22 Interference Due to Wedge-Shaped Film

Let AB and CD be the bounding surfaces of a *thin wedge-shaped film,* which enclose an angle α [Fig. 4.21].

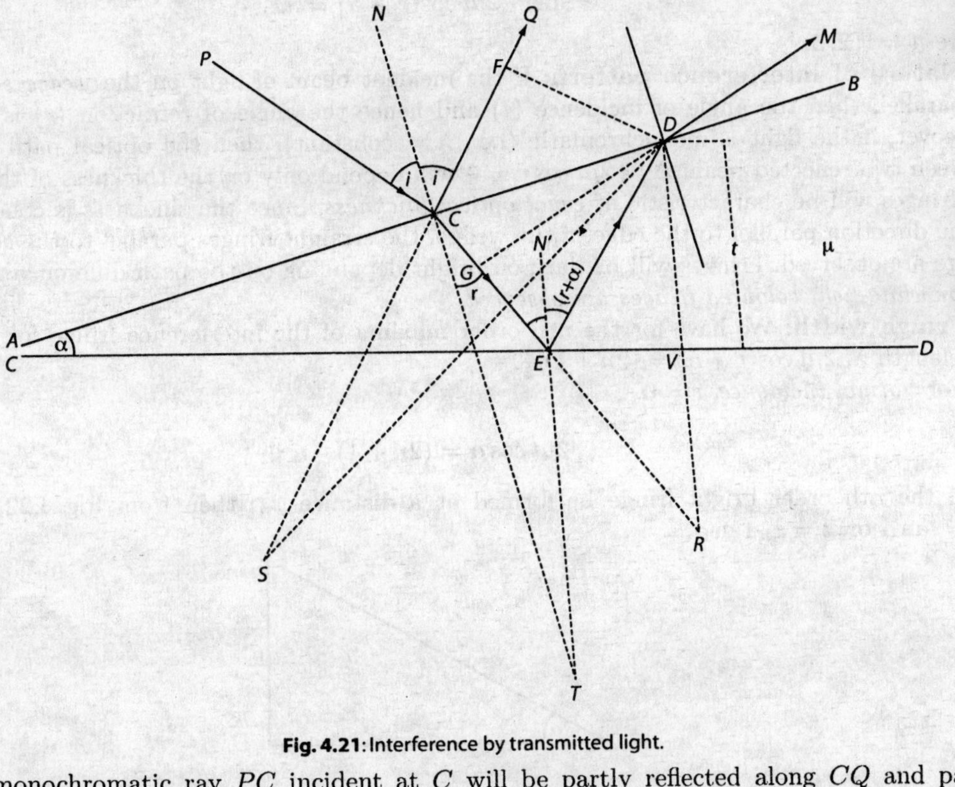

Fig. 4.21: Interference by transmitted light.

The monochromatic ray PC incident at C will be partly reflected along CQ and partly refracted along CE. The ray CE after being partly reflected from E travels along ED and emerges along DM.

The rays CQ and DM are derived from the same incident ray PC and hence are *coherent.* If the incident beam comes from an extended source and the film is thin, then the rays CQ and DM

will produce interference. These two rays will produce constructive or destructive interference depending on (i) *optical path difference* and (ii) *the conditions under which reflection occurs.*

Calculation of optical path difference: The optical path difference between CQ and DM is

$$\Delta = \mu(CE+ED) - CF = \mu(CG+GE+CD) - \mu CG \quad [\because CF = \mu CG]$$
$$= \mu(GE+CD) = \mu(GR) = 2\mu t\cos(r+\alpha) \quad [\because ED = ER]. \tag{4.44}$$

Due to the reflection at a point *backed by denser medium* an additional path difference $\frac{\lambda}{2}$ is introduced.

$$\therefore \Delta = 2\mu t\cos(r+\alpha) \pm \frac{\lambda}{2}. \tag{4.45}$$

For *constructive interference*,

$$\Delta = 2n\frac{\lambda}{2} = n\lambda \quad \text{or,} \quad 2\mu t\cos(r+\alpha) \pm \frac{\lambda}{2} = n\lambda$$
$$\text{or,} \quad 2\mu t\cos(r+\alpha) = (2n\pm 1)\frac{\lambda}{2}. \tag{4.46}$$

For *destructive interference*,

$$\Delta = (2n\pm 1)\frac{\lambda}{2} \quad \text{or,} \quad 2\mu t\cos(r+\alpha) \pm \frac{\lambda}{2} = (2n\pm 1)\frac{\lambda}{2}$$
$$\text{or,} \quad 2\mu t\cos(r+\alpha) = n\lambda, \tag{4.47}$$

where $n = 1, 2, 3, \cdots$.

Nature of interference pattern: If the incident beam of light on the *wedge shaped film* be parallel, then the angle of incidence (i) and hence the angle of refraction (r) is constant. Moreover, if the light is monochromatic (i.e., λ = constant), then the optical path difference between two reflected beam $\Delta = 2\mu t\cos r \pm \frac{\lambda}{2}$ will depend only on the thickness of the film. So the fringes will be characteristic of *equal optical thickness*. Since the thickness is constant only in the direction parallel to the edge of the wedge, the straight fringes parallel to the edge of the wedge are observed. Fringes will be dark on bright depending on the path difference for a fixed t. *For white light coloured fringes are observed.*

Fringe width: We have for the nth order maxima of the interference fringe for a light of wavelength λ, $2\mu t\cos(r+\alpha) = (2n+1)\frac{\lambda}{2}$.

For *normal incidence*, $r = 0$.

$$\therefore 2\mu t\cos\alpha = (2n+1)\frac{\lambda}{2}. \tag{4.48}$$

If the nth order bright fringe be formed at a distance x_n, then from Fig. 4.22, we have $\frac{t}{x_n} = \tan\alpha$ or, $t = x_n\tan\alpha$.

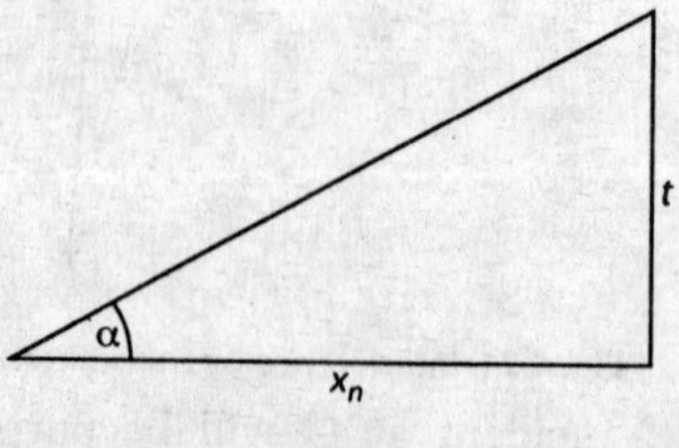

Fig. 4.22: Fringe width in wedge shaped film.

$\therefore$ from eqn (4.46),

$$2\mu x_n\tan\alpha\cos\alpha = (2n+1)\frac{\lambda}{2} \quad \text{or,} \quad 2\mu x_n\sin\alpha = (2n+1)\frac{\lambda}{2}. \tag{4.49}$$

For the $(n+1)$th order maximum,

$$2\mu x_{n+1} \sin\alpha = [2(n+1)+1]\,\frac{\lambda}{2} = (2n+3)\,\frac{\lambda}{2}. \tag{4.50}$$

Subtracting eqn (4.49) from eqn (4.50), we get

$$2\mu(x_{n+1} - x_n)\sin\alpha = \lambda \quad \text{or,} \quad x_{n+1} - x_n = \frac{\lambda}{2\mu\sin\alpha}.$$

But $x_{n+1} - x_n = \beta$ is the fringe width.

$$\therefore \beta = \frac{\lambda}{2\mu\sin\alpha}. \tag{4.51}$$

Since $\sin\alpha \simeq \alpha$, when α is small.

$$\therefore \beta = \frac{\lambda}{2\mu\alpha}. \tag{4.52}$$

For air film, $\mu = 1$.

$$\therefore \beta = \frac{\lambda}{2\alpha}. \tag{4.53}$$

4.23 Necessity of an Extended Source

In Fig. 4.23 a thin uniform film is illuminated by a *point source* S. For different angles of incidence different pair of interfering rays are obtained [Fig. 4.23]. If the film is viewed by keeping the eye at a fixed position, then the whole interference pattern will not be observed. To observe the whole interference pattern eye is to be shifted along the film surface from one position to another.

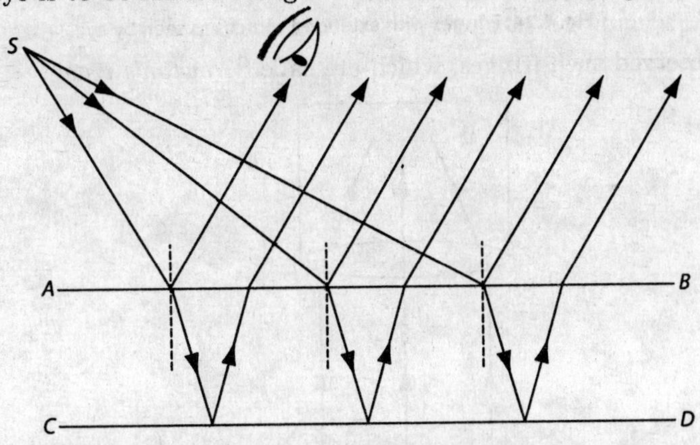

Fig. 4.23: Necessity of extended source.

An extended source of light is assumed to be made up of a large number of point sources S_1, S_2, S_3, $\cdots$ [Fig. 4.24]. If this extended source is used to illuminate the film surface, then light reflected from every point of the film reaches the eye E and one can see the entire interference pattern keeping the eye E at a suitable position.

For viewing interference fringes in a thin film, we use an extended (and not a point) source of light.

4.24 Newton's Rings

A symmetrical air film of varying thickness can be obtained by placing a plano-convex lens on a plane glass plate. The thickness of the air film at the point of contact is zero and gradually increases with the distance from the contact point. If a monochromatic light is allowed to fall normally, then concentric bright and dark circular fringes are observed with reflected light with the point of contact as the centre.

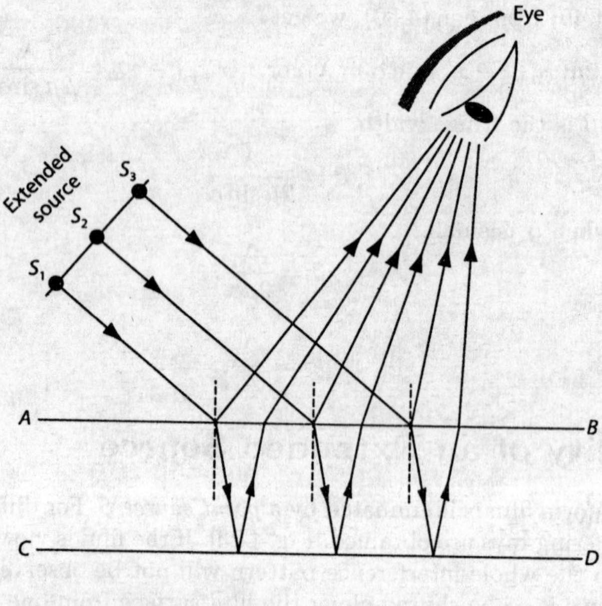

Fig. 4.24: Fringes with extended sources as seen by eye.

Newton first observed such fringes, which are called *Newton's rings.*

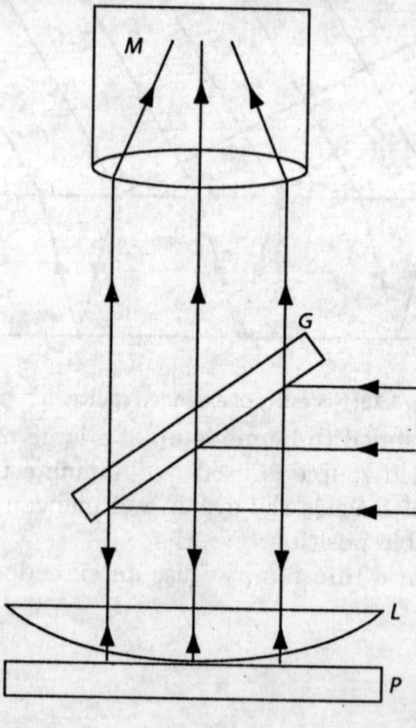

Fig. 4.25: Newton's rings.

The experiment: A parallel beam of monochromatic light is made to be incident on a glass plate G inclined at 45° to the incident beam. The normally reflected light beam is then made to be incident on the plano-convex lens L placed on the glass plate P. The incident light beam is partly reflected from the curved surface of the lens and partly from the plane surface of the glass plate. These two reflected beams of light interfere producing concentric circular bright and dark rings, which can be viewed by a microscope focussed on the film [Fig. 4.25]. These are *fringes of equal thickness and localized in the air film between L and P.* The central fringe in reflected light will be dark.

Theory. In Fig. 4.26, R is the radius of curvature of the lens L, and O is the centre of curvature. Let $BG = t$ be the thickness of the air film at G, where $AB = r_n$, the radius of the nth ring.

From the triangle OAB of Fig. 4.26,

$$R^2 = r_n^2 + (R-t)^2 = r_n^2 + R^2 + t^2 - 2Rt$$

$$\text{or,} \quad r_n^2 = 2Rt - t^2$$

$$\text{or,} \quad r_n^2 \simeq 2Rt.$$

t^2 is neglected due to smallness.

$$\therefore t = \frac{r_n^2}{2R}. \tag{4.54}$$

Again, for a wedge shaped film, the optical path difference is given by $\Delta = 2\mu t\cos(r+\alpha)$, where α is the angle of the wedge and r is the angle of refraction.

For *normal incidence*, $r = 0$ and α is very small. $\therefore \Delta = 2\mu t$.

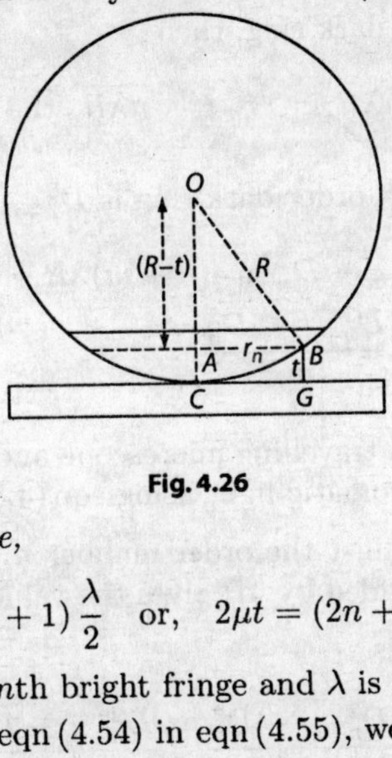

Fig. 4.26

Now, for *constructive interference*,

$$\Delta = (2n+1)\frac{\lambda}{2} \quad \text{or,} \quad 2\mu t = (2n+1)\frac{\lambda}{2}, \tag{4.55}$$

where n is the order number of the nth bright fringe and λ is the wavelength of the light used.

Substituting the value of t from eqn (4.54) in eqn (4.55), we get

$$2\mu\frac{r_n^2}{2R} = (2n+1)\frac{\lambda}{2} \quad \text{or,} \quad r_n^2 = (2n+1)\frac{\lambda R}{2} \quad [\because \mu = 1 \text{ for air film}]. \tag{4.56}$$

Putting $n = 0, 1, 2, 3$, etc. in eqn (4.56), we get

$$r_1 = \sqrt{\frac{\lambda R}{2}}, \quad r_2 = \sqrt{\frac{3\lambda R}{2}}, \quad r_3 = \sqrt{\frac{5\lambda R}{2}}, \quad \text{etc.}$$

Thus, *the radii of the bright rings are proportional to the square root of odd natural numbers.*

For *destructive interference*, we have

$$\Delta = 2n\frac{\lambda}{2} = n\lambda \quad \text{or,} \quad 2\mu t = n\lambda$$

$$\text{or,} \quad 2\mu\frac{r_n^2}{2R} = n\lambda \quad \left[\because t = \frac{r_n^2}{2R}\right]$$

$$\text{or,} \quad r_n^2 = Rn\lambda \qquad (\because \mu = 1 \text{ for air film}). \tag{4.57}$$

Here r_n represents the radius of the nth order dark ring.

For $n = 1, 2, 3, \cdots$,

$$r_1 = \sqrt{\lambda R}, \quad r_2 = \sqrt{2\lambda R}, \quad r_3 = \sqrt{3\lambda R}, \quad \text{etc.}$$

Thus, *the radii of the dark rings are proportional to the square root of the natural numbers.*

4.25 Determination of the Wavelength of Monochromatic Light

If D_n be the diameter of the nth dark ring, then

$$D_n = 2r_n \quad \text{or,} \quad D_n^2 = 4n\lambda R \quad [\because r_n = \sqrt{n\lambda R}, \text{ the radius of the } n\text{th dark ring}].$$

The diameter of the $(n+m)$th order dark ring is $D_{n+m}^2 = 4(n+m)\lambda R$.

$$\therefore \quad D_{n+m}^2 - D_n^2 = 4(n+m)\lambda R - 4n\lambda R = 4m\lambda R$$

$$\text{or,} \quad \lambda = \frac{D_{n+m}^2 - D_n^2}{4mR}. \tag{4.58}$$

Measuring D_{n+m} and D_n by a travelling microscope and R by a spherometer we can find out λ, the wavelength of the monochromatic light using eqn (4.58).

Alternatively, plotting D_n^2 against the order number n, we get a straight line. The slope of this straight line ($4\lambda R$), when divided by $4R$ gives the value of λ.

Slope of the straight line OB is

$$\frac{D_{n+m}^2 - D_n^2}{(n+m) - n} = \frac{D_{n+m}^2 - D_n^2}{m}.$$

$\therefore$ the wavelength λ is given by

$$4\lambda R = \frac{D_{n+m}^2 - D_n^2}{m}$$

$$\text{or,} \quad \lambda = \frac{D_{n+m}^2 - D_n^2}{4mR}.$$

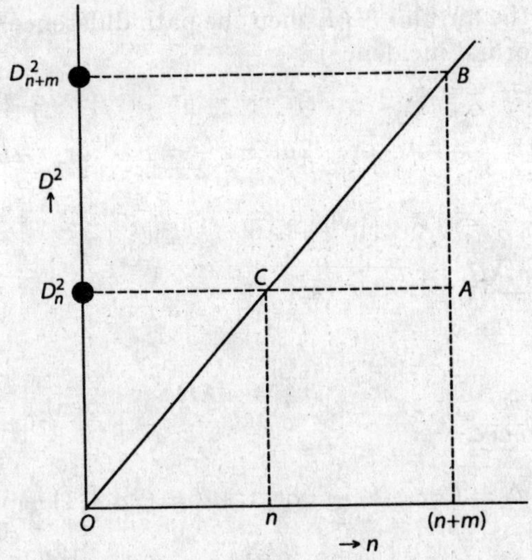

Fig. 4.27: Plot of D_n^2 against order number.

4.26 Determination of Refractive Index of a Liquid

A liquid film is formed by putting a few drops of liquid between the glass plate and the convex lens whose retractive index is to be determined. Using monochromatic light of known wavelength the diameter of nth and $(n+m)$th order dark rings are measured.

If D_n and D_{n+m} be diameters of the nth and $(n+m)$th order dark rings, then

$$D_n^2 = \frac{4n\lambda R}{\mu} \quad \text{and} \quad D_{n+m}^2 = \frac{4(n+m)\lambda R}{\mu}.$$

$$\therefore D_{n+m}^2 - D_n^2 = \frac{4m\lambda R}{\mu} \quad \text{or,} \quad \mu = \frac{4m\lambda R}{D_{n+m}^2 - D_n^2}. \tag{4.59}$$

Measuring D_{n+m} and D_n by a travelling microscope and R by a spherometer, the value of refractive index μ of the liquid can be found out using eqn (4.59).

If R is not known, then D'_n and D'_{n+m}, i.e., the diameters of the nth order and $(n+m)$th order dark fringes are to be measured first in air, when $D'^2_{n+m} - D'^2_n = 4m\lambda R$.

For liquid film, $D_{n+m}^2 - D_n^2 = \dfrac{4m\lambda R}{\mu}$.

$$\therefore \mu = \frac{D'^2_{n+m} - D'^2_n}{D^2_{n+m} - D^2_n}. \tag{4.60}$$

Measuring D'_n and D'_{n+m} we can find out μ of the liquid using eqn (4.60).

4.27 Newton's Rings by Transmitted Light

In Fig. 4.28 the ray after passing through the lens is partly reflected at M on the upper surface of the glass plate G and partly transmitted through the glass plate. The reflected beam on reaching N on the curved surface of the lens is again partly reflected back, which is then transmitted through G. These two transmitted part interfere in the focal plane of the eye-piece of the microscope to produce interference.

If t be the thickness of the air film NM, then the path difference between the two transmitted rays is $\Delta = 2\mu t \pm \lambda$ (for normal incidence).

For *constructive interference,*

$$\Delta = 2n\frac{\lambda}{2} = n\lambda \quad \text{or,} \quad 2\mu t \pm 2\frac{\lambda}{2} = n\lambda \quad \text{or,} \quad 2\mu t = (n-1)\lambda$$

$$\text{or,} \quad 2\mu \times \frac{r_n^2}{2R} = n'\lambda, \quad [\text{where } (n-1) = n']$$

$$\text{or,} \quad r_n^2 = \frac{n'\lambda R}{\mu}.$$

For air film, $\mu = 1$.

$$\therefore r_n^2 = n'\lambda R. \tag{4.61}$$

For *destructive interference,*

$$\Delta = (2n+1)\frac{\lambda}{2} \quad \text{or,} \quad 2\mu t = (2n+1)\frac{\lambda}{2}$$

$$\text{or,} \quad 2\mu \times \frac{r_n^2}{2R} = (2n+1)\frac{\lambda}{2} \quad \text{or,} \quad r_n^2 = \frac{(2n+1)\lambda R}{2}. \tag{4.62}$$

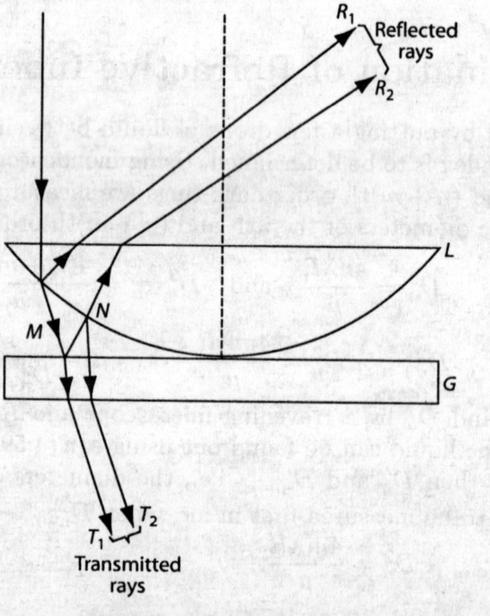

Fig. 4.28: Diagram showing production of coherent rays.

Thus, the conditions of constructive and destructive interference are *reversed* in case of interference by *transmitted* light. So a *bright fringe in reflected light will appear dark in transmitted light.* The central fringe will be *bright* in this case.

4.28 Newton's Ring with White Light

The square of diameter of a ring is proportional to the wavelength (λ). The wavelength of red light (λ_r) is the greatest and that of the violet light (λ_v) is the smallest. So for the same order, the diameter of the red ring will be the greatest and that of the violet ring will be the smallest. The diameters of the rings of other colours for the same order will occupy the intermediate positions.

Thus, with reflected white light the Newton's ring system will have central dark ring encircled by mixed colours from violet to red. For large order number, overlapping of successive order number will lead to uniform illumination.

4.29 Applications of Newton's Rings

Newton's ring has the following important applications: (i) *To measure wavelength λ of monochromatic light* (ii) *To measure refractive index μ of a liquid* (iii) *To check planeness of a glass surface* (iv) *To measure radius of curvature of a lens, when λ is known* (v) *To measure the wavelength difference ($\Delta\lambda$) of two closely spaced wavelength* (iv) *Knowing $\Delta\lambda$, the coherent length $L \sim \frac{\lambda^2}{\Delta\lambda}$ can be found out.*

4.30 Answers of Some Important Questions

4.1. *What is a wavefront? What are different forms of wavefronts?*
Ans. The wavefront of a wave at any instant of time is defined as the locus of all the neighbouring particles in the medium, which are being just distributed at that instant of time and are consequently in the same phase of vibration.

Depending on the geometrical shape of the source of light, wavefronts are of the following three types: (i) *Spherical wavefront,* if the source is a point (ii) *Cylindrical wavefront,* if the source is linear (iii) *Plane wavefront,* if the source is at an infinite distance.
4.2. *State Huygens' principle and point out its success and failure.*
Ans. To overcome the difficulties in Newton's corpuscular theory regarding the nature of light, Huygens put forward his wave theory as: *Every point on a wavefront is the source of a new disturbance. It sends small secondary wavelets propagated in all directions. These wavelets are spherical in shape and move forward in a homogeneous medium with same velocity with which the original wave travels. At any instant the position and shape of the new wavefront can be obtained by drawing a surface enveloping these secondary wavelets.*

The success of the theory is that it can explain the phenomena of *reflection, refraction, total internal reflection, interference, diffraction, dispersion, double refraction,* etc.

The theory, however, fails to explain the phenomena of *rectilinear propagation of light and polarization.*
4.3. *What is interference of light?*
Ans. If two monochromatic light waves having same wavelength and constant phase difference proceeding simultaneously in the same direction produce alternate bright and dark fringes by superposition on a screen, then the phenomenon is called *interference of light.*

For *bright fringe,* the optical path difference between the waves is even multiples of half-wavelength, i.e., $\Delta = 2n \cdot \frac{\lambda}{2}$, where $n = 0, 1, 2, \cdots$ and λ is the wavelength of light. This is called *constructive interference.*

The condition for dark fringe is that, the optical path difference between the waves is odd multiples of half-wavelength, i.e., $\Delta = (2n \pm 1)\,\frac{\lambda}{2}$.

This is called *destructive interference.*
4.4. *State the essential conditions for sustained, i.e., permanent interference.*
[BUH 2003, CUH 1995, 2000]

Ans. See Section 4.8.1 (p. 82).
4.5. *Why two different real sources cannot produce permanent interference?*
Ans. See Section 4.9 (p. 81).
4.6. *Show how energy is conserved in interference phenomena?* **[WBUT 2002, '08]**

Ans. If two light waves are represented by

$$y_1 = a_1 e^{i(\omega t - kx_1)}$$

$$\text{and} \quad y_2 = a_2 e^{i(\omega t - kx_2)},$$

where a_1 and a_2 are the amplitudes of the two waves and x_1 and x_2 are the respective distances from the sources to the point of superposition, then the resultant displacement at P is

$$y = y_1 + y_2 = Ae^{i(\omega t - \phi)},$$

where $A = \sqrt{a_1^2 + a_2^2}$ and $\phi = \tan^{-1}(b/a)$, $a = a_1 \cos kx_1 + a_2 \cos kx_2$; $b = a_1 \sin kx_1 + a_2 \sin kx_2$.

The resultant intensity at P is $I = a_1^2 + a_2^2 + 2a_1a_2 \cos\delta$, where $\delta = k(x_2 - x_1)$, the phase difference between the two waves at P. Thus, it is clear from the above equation that *I depends on δ.*

Intensity at P in absence of interference is $a^2 + a^2 = 2a^2$ [when $a_1 = a_2 = a$ (say)]. But due to interference the intensity at maxima points is

$$I_{\max} = a^2 + a^2 + 2a^2 \cos 0 = 4a^2 \quad [\because \delta = (2m+1)\pi \text{ for maxima}].$$

The average value of intensity in the range from 0 to 2π is

$$\langle I \rangle = \frac{\int_0^{2\pi} I\,d\delta}{\int_0^{2\pi} d\delta} = \frac{\int_0^{2\pi} (a^2 + a^2 + 2a^2 \cos\delta)\,d\delta}{2\pi} = \frac{\left[a^2\delta + a^2\delta + 2a^2 \sin\delta\right]_0^{2\pi}}{2\pi}$$

$$= 2a^2. \quad [\text{where } a_1 = a_2 = a, \text{ say}]$$

This shows that average intensity at a point is equal to the sum of the intensities of the two real sources. So the energy, which appears to be absent at minima has actually been transferred to the maxima so that the intensity at the maxima is $2a^2 + 2a^2 = 4a^2$. Hence, in the phenomenon of interference, energy is *not destructed,* rather it is *redistributed.* Thus, *the laws of conservation of energy holds good in interference phenomena.*

4.7. *What are coherent and non-coherent sources of light?* **[WBUT 2003]**

Ans. If two light sources be such that (i) they emit light waves of the same frequency and nearly the same amplitude and (ii) the phase difference of the waves emitted is constant of time, then the sources are called *coherent sources.*

On the other hand, (i) if the two sources do not emit light of same frequency, nor of same amplitude (ii) the phase difference of waves emitted changes continuously with time, then the sources are called *incoherent sources.*

4.8. *In Young double slit experiment, if one slit is covered with blue transparent paper and the other with red, then can you expect interference of light?*

Ans. No, because the wavelengths of light from the two sources will be different, i.e., ($\lambda_b \neq \lambda_r$). As the sources are not coherent they cannot produce interference pattern.

4.9. *If the slit separation d is less than wavelength of light λ, then can you expect interference pattern?*

Ans. We have the fringe width, $\beta = \dfrac{D}{2d}\lambda$. If $2d < \lambda$, then $\beta \to$ an extremely large value. In this case uniform illumination and not interference fringes will be observed.

4.10. *What are different types of interference fringes?*

Ans. Interference fringes are of the following two types:

(i) *Interference by division of wavefront*: Here, the two coherent sources are produced by *dividing laterally* the incident wave into two parts through reflection, refraction, etc.

In *Fresnel's bi-prism, Lloyd's mirror experiment,* pair of coherent sources are obtained by this technique.

(ii) *Interference by division of amplitude*: Here two coherent sources are produced by dividing the incident wavefront into two or more parts through *partial reflection or refraction.*

In Newton ring, Michelson interferometer experiment coherent sources are produced by this technique.

4.11. *Can two real sources of light behave as coherent sources?*

Ans. Two real sources of light cannot behave as coherent sources due to the following reasons: (i) They cannot emit light waves of *same frequency and nearly of same amplitude.* (ii) The *phase of emitted waves will not be constant of time.* They will produce general illumination on a screen.

4.12. *What do you mean by Temporal coherence and Special coherence?*

Ans. See Section 4.12 (p. 85).

4.13. *An infinitely thin film illuminated by white light appears dark when observed by reflected light. Why?*

Ans. In case of interference of light, the path difference between two successive reflected rays is $\Delta = 2\mu t \cos r \pm \frac{\lambda}{2}$, where t = thickness, μ = refractive index, r = the angle of refraction and λ = the wavelength of light.

Now, for infinitely thin film, $t \ll \lambda$. $\therefore$ $\Delta = 2\mu t \cos r = 0$.

This satisfies the condition of *destructive interference* (as $t = 0$, $\Delta = 0$) and corresponds to zero intensity in reflected part. Hence, the result follows.

4.14. *Explain what do you mean by fringes of equal thickness. Are Newton's rings fringes of equal thickness?*

Ans. If the phase difference of two interfering light waves is governed by nt, then each fringe is a locus of all the points corresponding to the same thickness of the air film [n = r.i. and t the thickness of the film]. Such fringes are called the *fringes of equal thickness.*

Each bright or dark fringe in Newton's ring is a locus of points corresponding to constant thickness of the air film and hence are fringes of equal thickness.

4.15. *Discuss the colour of Newton's rings with white light.*

Ans. See Section 4.28 (p. 104).

4.16. *Why two wide slits cannot produce interference?*

Ans. A wide source can be imagined to be equivalent to a large number of narrow slits placed side by side. Each pair of narrow sources in two different slits will produce interference pattern of their own, which will overlap producing general illumination on the screen.

4.17. *Why the central fringe of Newton's ring pattern is dark? How can it be made bright?*

Ans. At the contact point O of the glass plate and the lens the thickness of the air film is zero. Due to Stokes' law the two reflected rays from the point O suffer a phase change π and hence produce *destructive interference.* That is why the central ring is dark.

The central fringe can, however, be made *bright by introducing a liquid between the convex surface of the lens and the glass plate whose r.i. n is greater than that of the lens but less than that of the glass plate.*

4.31 Solved Problems

Ex. 1 *In an experiment using Young's double slits, the distance between the centre of the interference pattern and tenth bright fringe on either side is* 3.44 cm *and the distance between the slits and the screen is* 200 cm. *If the wavelength of light used is* 5.89×10^{-5} cm, *determine the separation between the slits.* **[WBUT (QB)]**

Soln. Here the wavelength of light used $\lambda = 5.89 \times 10^{-5}$ cm.

Distance between the slits and screen $D = 200$ cm.

The distance between the nth bright fringe from the centre of interference pattern is $x_n = \dfrac{Dn\lambda}{2d}$.

$\therefore$ the separation between the slits is

$$2d = \frac{Dn\lambda}{x_n} = \frac{200 \times 10 \times 5.89 \times 10^{-5}}{3.44} = \frac{11780 \times 10^{-5}}{3.44} = \mathbf{0.0342\ cm.}$$

Ex. 2 *In a Young's double slit experiment the separation of four bright fringes is* 2.5 mm, *when the wavelength used is* 6.25×10^{-5} cm. *The distance from the slits to the screen is* 80 cm. *Calculate the separation between the two slits.*

Soln. Fringe width $\beta = \dfrac{2.5}{4} = 0.625$ mm $= 0.0625$ cm.

Now, the expression for fringe width

$$\beta = \frac{D\lambda}{2d} \quad \text{or,} \quad 2d = \frac{D\lambda}{\beta} = \frac{80 \times 6.25 \times 10^{-5}}{0.0625} = \mathbf{0.08\ cm.}$$

Ex. 3 *A double slit of* 0.5 mm *separation is illuminated by light of blue cadmium light of* $\lambda =$ 5000 Å. *How far behind the slits must go to obtain fringes that are* 0.1 cm *apart? The wavelength of blue cadmium light is* 4800 Å.

Soln. Here $2d = 0.5$ mm $= 0.05$ cm, $\beta = 0.1$ cm, $\lambda = 4800 \times 10^{-8}$ cm, $D = ?$

We have $\beta = \dfrac{D\lambda}{2d}$ or, $D = \dfrac{2d\beta}{\lambda}$ or, $D = \dfrac{0.05 \times 0.1}{4800 \times 10^{-8}} = \mathbf{104.17\ cm.}$

Ex. 4 *In an experiment of Young's double slit, the slits are* 2 mm *apart and are illuminated with a mixture of two wavelengths* 5000 Å *and* 6000 Å. *At what minimum distance from the common central bright fringe on a screen,* 2 m *away from the slits, will a bright fringe of one interference pattern coincide with a bright fringe from the other?*

Soln. The distance of the nth bright fringe from the central fringe of wavelength λ_1,

$$x_{n_1} = \frac{Dn_1\lambda_1}{2d} \tag{1}$$

and that for λ_2,

$$x_{n_2} = \frac{Dn_2\lambda_2}{2d}. \tag{2}$$

Now, from eqns (1) and (2),

$$\frac{Dn_1\lambda_1}{2d} = \frac{Dn_2\lambda_2}{2d} \quad \text{or,} \quad \frac{n_1}{n_2} = \frac{\lambda_2}{\lambda_1} = \frac{6000}{5000} = \frac{6}{5}.$$

Thus, in the first position, $n_1 = 6$ and $n_2 = 5$, where overlapping takes place.

$$\therefore x_{n_1} = \frac{D\lambda_1.6}{2d} = \frac{2 \times 100 \times 5000 \times 10^{-8} \times 6}{0.2} = \mathbf{0.30\ cm.}$$

For the 2nd overlapping, $n_1 = 12$, $n_2 = 10$ and so on.

Ex. 5 *In a Young's double slit experiment, the slits are* 0.2 mm *apart and the screen is* 1.5 m *away. It is observed that the distance between the central bright fringe and the fourth dark fringe is* 1.8 cm. *Find the wavelength of the light.* **[WBUT (QB)]**

Soln. Distance of the 4th dark fringe from the central fringe is

$$x = (2n+1)\frac{D\lambda}{2d}.$$

Here distance between the slit and source, $D = 1.5$ m $= 150$ cm.
Distance between the slits $2d = 0.02$ cm, $x = 1.8$ cm.

$$\therefore \lambda = \frac{2dx}{(2n+1)D} = \frac{0.02 \times 1.8}{9 \times 150} \text{ cm} = 0.000026667 \text{ cm} = \mathbf{2666\ \text{Å}}.$$

Ex. 6 *Two coherent sources whose intensity ratio is* 100 : 1 *produce interference fringes. Deduce the ratio of maximum to minimum intensity.*

Soln. If a_1 and a_2 be the amplitudes of the coherent sources, then the maximum and minimum intensities are given by $I_{\text{max}} = (a_1 + a_2)^2$; $I_{\text{min}} = (a_1 - a_2)^2$.

$$\therefore \frac{I_{\text{max}}}{I_{\text{min}}} = \frac{(a_1+a_2)^2}{(a_1-a_2)^2}.$$

By the problem, $\dfrac{a_1^2}{a_2^2} = \dfrac{I_1}{I_2} = \dfrac{100}{1}$. $\quad \therefore a_1 = 10a_2$.

$$\therefore \frac{I_{\text{max}}}{I_{\text{min}}} = \frac{(10a_2+a_2)^2}{(10a_2-a_2)^2} = \left(\frac{11}{9}\right)^2 = \frac{121}{81}.$$

$$\therefore I_{\text{max}} : I_{\text{min}} = \mathbf{121 : 81}.$$

Ex. 7 *Calculate the fringe width of interference pattern in Young's double slit experiment with two slits* 10^{-3} m *apart on a screen* 1 m *away.* [$\lambda = 5893 \times 10^{-8}$ cm]

Soln. Fringe width $\beta = \dfrac{D\lambda}{2d}$.

Hence, $2d = 10^{-3}$ m $= 0.1$ cm, $D = 1$ m $= 100$ cm and $\lambda = 5893 \times 10^{-8}$ cm

$$\therefore \beta = \frac{100 \times 5893 \times 10^{-8}}{0.1} = 5893 \times 10^{-5} \text{ cm} = \mathbf{5893 \times 10^{-7}\ m}.$$

Ex. 8 *In the Fresnel's bi-prism arrangement show that* $d = 2x(\mu - 1)\alpha$, *where* α *and* x *are the refracting angle and distance of the source from the base of the bi-prism and* μ *is the refractive index of the material of the bi-prism.*

Soln. As the bi-prism has a very small refracting angle ($\alpha < 1°$) the deviation of the refracted ray by this bi-prism is $\delta = (\mu - 1)\alpha$.

If θ be the total deviation as shown in Fig. 4.29, we have

$$\frac{\theta}{2} = (\mu - 1)\alpha \quad \text{or,} \quad \theta = 2(\mu - 1)\alpha.$$

If x is the distance of the bi-prism from the source, then $d = x\theta$ or, $\mathbf{d = 2x(\mu - 1)\alpha}$.

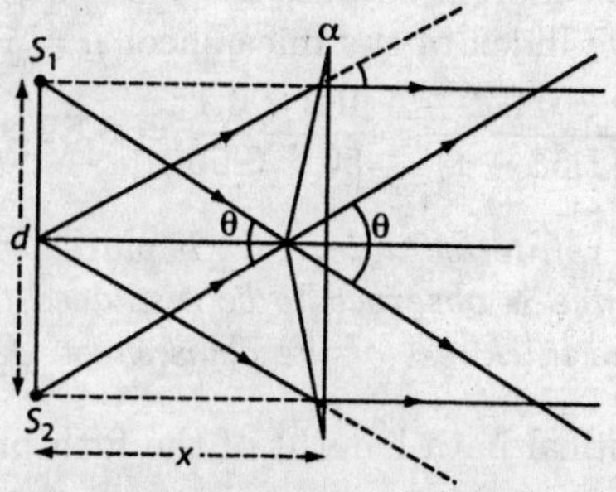

Fig. 4.29

Ex. 9 *A Fresnel's bi-prism having an angle* 1° *and* $\mu = 1.5$ *forms interference fringes on a screen distant* 80 cm *from the bi-prism. If the distance between the source and the bi-prism is* 20 cm, *find the fringe separation when the light used is* 6900 Å.

Soln. Here the distance between the bi-prism and screen = 80 cm.

$\therefore D = 20 + 80 = 100$ cm.

$\mu = 1.5$ and $\alpha = 1° = \dfrac{2\pi}{360}$ radians $= \dfrac{\pi}{180}$ radians.

$\therefore$ the distance between the sources is

$$2d = 2x(\mu - 1)\alpha \quad [x \text{ is the distance of the bi-prism from } S_1S_2]$$

$$= 2 \times 20(1.15 - 1) \times \frac{\pi}{180} = \frac{\pi}{9} \text{ cm.}$$

Now, the fringe width $\beta = \dfrac{D\lambda}{2d}$.

$\lambda = 6900$ Å $= 6900 \times 10^{-8}$ cm.

$$\therefore \beta = \frac{100 \times 6900 \times 10^{-8}}{\pi/9} = \mathbf{1.977 \times 10^{-2}\ cm.}$$

Ex. 10 *In a bi-prism experiment, the successive bright bands are* 0.201 apart mm *and the separation of the coherent sources are* 3.732 mm *and* 1.242 mm *respectively for two positions of the convex lens. Determine the wavelength of the light used, if the distance between the source and the screen is* 73.47 cm.

Soln. Here, $d_1 = 3.732$ mm $= 0.3732$ cm, $d_2 = 1.242$ mm $= 0.1242$ cm.

$\therefore 2d = \sqrt{d_1 d_2} = \sqrt{0.3732 \times 0.1242} = 0.2153$ cm. $D = 73.47$ cm, $\beta = 0.201$ mm $= 0.0201$ cm.

$\therefore$ the wavelength of light used

$$\lambda = \frac{2d\beta}{D} = \frac{0.2153 \times 0.0201}{73.47} = \mathbf{5889 \times 10^{-8}\ cm.}$$

Ex. 11 *A mica sheet of refractive index* 1.58 *is introduced in one of the interfering beams and the central fringe gets shifted by* 0.2 cm. *Distance between the coherent sources is* 0.1 cm *and the screen is placed at a distance of* 50 cm *from the source. Determine the thickness of the mica sheet.* **[WBUT (QB)]**

Soln. If x be the displacement of the central fringe and t the thickness of the mica sheet, then

$$t = \frac{x2d}{D(\mu - 1)},$$

where the distance between the coherent sources $2d = 0.1$ cm, distance between the sources and the screen $D = 50$ cm, refractive index of the mica-sheet $\mu = 1.58$ and $x = 0.2$ cm.

$$\therefore t = \frac{0.2 \times 0.1}{50(1.58 - 1)} = \frac{0.2 \times 0.1}{50 \times 0.58} = \mathbf{6.89 \times 10^{-4}\ cm.}$$

Ex. 12 *A thin piece of glass of refractive index* 1.52 *is placed in one of the interfering beams of a bi-prism. The central bright fringe is observed to be displaced to the position previously occupied by the* 5th *bright band. Find the thickness of the glass sheet.* [Given: $\lambda = 5890$ Å] **[CUH 2000]**

Soln. If x and $x + \Delta$ be the optical path lengths of the fifth bright fringe, then the optical path difference

$$\Delta = (x + \Delta) - x = n\lambda \quad \text{or,} \quad \Delta = 5\lambda \quad [\because n = 5 \text{ in this case}].$$

The glass plate of thickness t (say) introduces a path difference $(\mu - 1)t$, which is equal to Δ.

$$\therefore \Delta = (\mu - 1)t = 5\lambda \quad \text{or,} \quad t = \frac{5\lambda}{\mu - 1} = \frac{5 \times 5890 \times 10^{-8}}{1.52 - 1} = \mathbf{5.6634 \times 10^{-4}\ cm.}$$

Ex. 13 *A parallel beam of light of wavelength* $\lambda = 5890$ Å *is incident on a glass plate* ($\mu = 1.5$) *such that the angle of refraction into the plate is* 60°. *What should be the minimum thickness of the glass plate, which would make the plate dark in reflected light?*

Soln. For minimum thickness (t) of the plate the condition that the plate will appear dark in reflected light is

$$2\mu t \cos r = 2n\frac{\lambda}{2} = \lambda \quad \text{or,} \quad t = \frac{\lambda}{2\mu \cos r} = \frac{5890 \times 10^{-8}}{2 \times 1.5 \times \cos 60°} = \frac{5890 \times 10^{-8}}{2 \times 1.5 \times 0.5} = \mathbf{3927\ Å.}$$

Ex. 14 *White light is incident on a soap film of* $\mu = 1.33$ *and thickness* $t = 5000$ Å. *What wavelengths in the visible region are reflected by the film?*

Soln. The condition of constructive interference is $2\mu t \cos r = (2n + 1)\frac{\lambda}{2}$, where $n = 0, 1, 2,$ etc.
Here $t = 5000$ Å $= 5000 \times 10^{-8}$ cm, $\mu = 1.33$.
If the incidence be normal, then $r = 0°$ and $\cos r = 1$.

$$\therefore \lambda = \frac{4\mu t}{(2n+1)} = \frac{4 \times 1.33 \times 5000 \times 10^{-8}}{(2n+1)} = \frac{26600 \times 10^{-8}}{(2n+1)} \text{ cm.}$$

Now, if $n = 0$, $\lambda_1 = 26600 \times 10^{-8}$ cm (infrared)

$$n = 1,\ \lambda_2 = \frac{26600 \times 10^{-8}}{3} = 8866.7 \times 10^{-8} \text{ cm (infrared)}$$

$$n = 2,\ \lambda_3 = \frac{26600 \times 10^{-8}}{5} = 5320 \times 10^{-8} \text{ cm (visible)}$$

$$n = 3,\ \lambda_4 = \frac{26600 \times 10^{-8}}{7} = 3800 \times 10^{-8} \text{ cm (ultraviolet).}$$

So the only wavelength $\lambda_3 = \mathbf{5320 \times 10^{-8}}$ cm in the visible region will be reflected by the film.

Ex. 15 *The fringes of equal thickness are formed when two glass plates are kept over each other with a small gap in between. If a parallel beam of light of wavelength* 6000 Å *is used and the angle of the wedge is* 20.62″, *find the fringe width.*

Soln. Here the fringe width $\beta = \frac{\lambda}{2\mu\alpha}$, where the wavelength index $\mu = 1$ (for air),

and the angle of the wedge is $\alpha = 20.62'' = \frac{20.62°}{3600} = \frac{20.62}{3600} \times \frac{\pi}{180}$ radian.

$$\therefore \beta = \frac{6000 \times 10^{-8} \times 180 \times 3600}{2 \times 1 \times 20.62 \times \pi} = \mathbf{3\ mm.}$$

Ex. 16 *Find the minimum thickness of coating with* MgF_2 *required for no reflection at the centre of visible spectrum* ($\lambda = 5.5 \times 10^{-5}$ cm). [Given: μ for $MgF_2 = 1.38$]

Soln. The condition for destructive interference for normal incidence on the film is $2\mu t = (2n + 1)\frac{\lambda}{2}$, where $n = 0, 1, 2,$ etc.
For $n = 0$,

$$t = \frac{\lambda}{4\mu} = \frac{5.5 \times 10^{-5}}{4 \times 1.38} = \mathbf{0.996 \times 10^{-5}\ cm.}$$

Ex. 17 *In a Newton ring experiment, the diameter of the* 5th *dark ring is* 0.336 cm *and the diameter of the* 15th *dark ring is* 0.590 cm. *Find the radius of the plano-convex lens, if the wavelength of light used is* 5890 Å. **[WBUT 2004]**

Soln. The radius of the plano-convex lens is given by

$$R = \frac{D_{n+m}^2 - D_n^2}{4m\lambda},$$

where $D_{n+m} = D_{5+10} = 0.590$ cm, $D_n = D_5 = 0.336$ cm, $m = 10$, and $\lambda = 5890\ \text{Å} = 5890 \times 10^{-8}$ cm.

$$\therefore R = \frac{(0.590)^2 - (0.336)^2}{4 \times 10 \times 5890 \times 10^{-8}} = \frac{(0.590 + 0.336)(0.590 - 0.336)}{4 \times 589 \times 10^{-6}} = \frac{0.926 \times 0.254}{4 \times 589} \times 10^6$$

$$= 0.00009983 \times 10^6 = \mathbf{99.83\ cm.}$$

Ex. 18 *Newton's rings are observed in the reflected light of wavelength* 5900 Å. *The diameter of the* 10th *dark ring is* 0.5 cm. *Find the radius of curvature of the lens used.*

Soln. If R be the radius of curvature of the lens and λ the wavelength of light used, then the diameter of the nth dark ring is given by

$$D_n = \sqrt{4nR\lambda}, \quad n = 1, 2, 3, \cdots$$

$$\text{or,} \quad D_n^2 = 4nR\lambda \quad \text{or,} \quad R = \frac{D_n^2}{4n\lambda}.$$

Here $n = 10$, $D_{10} = 0.5$ cm, $\lambda = 5900 \times 10^8$ cm.

$$\therefore R = \frac{(0.5)^2}{4 \times 10 \times 5900 \times 10^{-8}} = \frac{0.25}{236000 \times 10^{-8}} = \mathbf{105.9\ cm.}$$

Ex. 19 *In a Newton's ring experiment, the diameter of the* 5th *and* 15th *dark rings are* 0.336 cm, *and* 0.590 cm *respectively. If the radius of curvature of the curved surface of the plano-convex lens used be* 100 cm, *find the wavelength of the light used.* **[CU 1971]**

Soln. The wavelength of the light used is

$$\lambda = \frac{D_{n+m}^2 - D_n^2}{4mR} = \frac{(0.590)^2 - (0.336)^2}{4 \times 10 \times 100} = \frac{0.926 \times 0.254}{4 \times 10 \times 100} = \mathbf{5879 \times 10^{-8}\ cm.}$$

Ex. 20 *Light containing two wavelengths* λ_1 *and* λ_2 *falls normally on a plano-convex lens of radius of curvature* R *resting on a glass plate. If the* n*th dark ring due to* λ_1 *coincides with the* $(n+1)$*th dark ring due to* λ_2*, prove that the radius of the* n*th dark ring of* λ_1 *is* $= \sqrt{\dfrac{\lambda_1\lambda_2 R}{\lambda_1 - \lambda_2}}$.

Soln. We know that the radius of the nth dark ring due to λ_1 is

$$r_n = \sqrt{n\lambda_1 R}.$$

Similarly, the radius of the $(n+1)$th dark ring due to λ_2 is

$$r_{n+1} = \sqrt{(n+1)\lambda_2 R}.$$

By the problem, $r_1 = r_{n+1} = r$ (say).

$$\therefore r = \sqrt{n\lambda_1 R} = \sqrt{(n+1)\lambda_2 R} \quad \text{or,} \quad n\lambda_1 R = (n+1)\lambda_2 R \quad \text{or,} \quad n = \frac{\lambda_2}{\lambda_1 - \lambda_2}.$$

$$\therefore r = \sqrt{n\lambda_1 R} = \sqrt{\frac{\lambda_1\lambda_2 R}{\lambda_1 - \lambda_2}} \quad \left[\because n = \frac{\lambda_2}{\lambda_1 - \lambda_2}\right].$$

Ex. 21 *The diameter of the nth Newton's ring changes from* 1.2 *to* 1 cm, *when the air space between the lens and the plate is replaced by a transparent liquid. Find the refractive index of the liquid.*

Soln. $(D_n^2)_{\text{liq}} = \dfrac{4nR\lambda}{\mu}$; $(D_n^2)_{\text{air}} = 4nR\lambda$.

$\therefore$ refractive index of the liquid is

$$\mu = \frac{(D_n^2)_{\text{liq}}}{(D_n^2)_{\text{air}}} = \frac{(1.2)^2}{1^2} = \mathbf{1.44}.$$

Ex. 22 *Two plano-convex lenses each of radius of curvature* 100 cm *are placed with their curved surfaces in contact with each other. Newton's rings are formed by using a light of wavelength* 6×10^{-5} cm. *Find the distance between* 10th *and* 20th *rings.*

Soln. From Fig. 4.30 the thickness of the air film at any point

$$t = t_1 + t_2 = \left(\frac{D^2}{8R_1} + \frac{D^2}{8R_2}\right).$$

If the incidence is normal, then for the nth dark ring,

$$2t = n\lambda \quad \text{or,} \quad 2\left[\frac{D_n^2}{8R_1} + \frac{D_n^2}{8R_2}\right] = n\lambda \quad \text{or,} \quad \frac{D_n^2}{4}\left[\frac{1}{R_1} + \frac{1}{R_2}\right] = n\lambda.$$

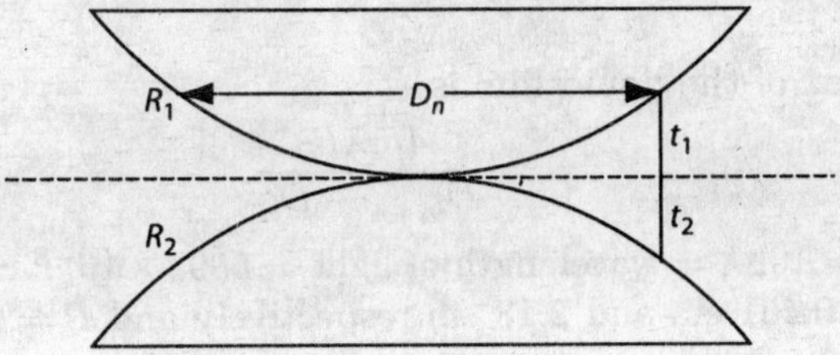

Fig. 4.30

But $D_n^2 = \dfrac{4n\lambda R_1 R_2}{R_1 + R_2}$ and $r_n^2 = \dfrac{n\lambda R_1 R_2}{R_1 + R_2}$

$$\therefore r_n = \sqrt{\frac{4n\lambda R_1 R_2}{R_1 + R_2}},$$

the radius of the nth dark ring.

Here $\lambda = 6 \times 10^{-5}$ cm, $R_1 = R_2 = 100$ cm, $n_1 = 10$, $n_2 = 20$.

$$\therefore r_{20} - r_{10} = \left[\frac{20 \times 6 \times 10^{-5} \times 100^2}{200}\right]^{1/2} - \left[\frac{10 \times 6 \times 10^{-5} \times 100^2}{200}\right]^{1/2}$$

$$= \left(\sqrt{20} - \sqrt{10}\right)\left[\frac{6 \times 10^{-5} \times 10^4}{200}\right] = \mathbf{0.751\ cm.}$$

Ex. 23 *A bi-prism is kept at a distance of* 5 cm *from a narrow slit illuminated by sodium light. Distance between two virtual sources formed by the bi-prism is* 0.05 cm. *If the distance of the screen from bi-prism be* 75 cm, *calculate the fringe width.* [*Wavelength of light* $\lambda = 5890$ Å.]

[BU 2002]

Soln. The fringe width

$$\beta = \frac{D}{2d}\lambda,$$

where D = the distance between the slit and the screen $= 75 + 5 = 80$ cm
$2d$ = the distance between the virtual sources $= 0.05$ cm
and λ = wavelength of light used $= 5890 \times 10^{-8}$ cm.

$$\therefore \beta = \frac{80 \times 5890 \times 10^{-8}}{0.05} = \mathbf{0.0942\ cm.}$$

Ex. 24 *In a Newton's ring apparatus, the radius of curvature of the lower surface of the convex lens is* 10 m. *The* k*th and* $(k+6)$*th dark rings are found to have radii* 3 mm *and* 7 mm *respectively. Find the wavelength of light used.* **[CU 1982]**

Soln. If r_k and r_{k+6} be the radii of the kth and $(k+6)$th dark rings respectively, then

$$r_k^2 = k\lambda R \quad \text{and} \quad r_{k+6}^2 = (k+6)\lambda R,$$

where R is the radius of curvature of the lens.

$$\therefore r_{k+6}^2 - r_k^2 = 6\lambda R.$$

So

$$\lambda = \frac{r_{k+6}^2 - r_k^2}{6R} = \frac{(0.7)^2 - (0.3)^2}{6 \times 10 \times 100} = \mathbf{0.66 \times 10^{-4}\ cm.}$$

Ex. 25 *Newton's rings are seen with reflected light using a plane glass plate and a plano-convex lens, the radius of curvature of the curved wavelength of light used is* 5893 Å *and if the diameters of the* n*th and* $(n+1)$*th bright rings are* 2.18 mm *and* 4.51 mm, *find the refractive index of the liquid.* **[NBU 1983]**

Soln. The refractive index of the liquid film is

$$\mu = \frac{4m\lambda R}{D_{m+n}^2 - D_n^2},$$

where $m = (n+10) - n = 10$, λ = wavelength of light $= 5893 \times 10^{-8}$ cm. Diameters of $(m+n)$th and nth bright rings are 0.451 cm and 2.18 cm respectively and $R = 90$ cm.

$$\therefore \mu = \frac{4 \times 10 \times 5893 \times 10^{-8} \times 90}{(0.451)^2 - (0.218)^2} = \frac{36 \times 5893}{669 \times 233} = \mathbf{1.36.}$$

Ex. 26 *A film of r.i.* 1.70 *is placed between a plate glass plane and an equi-convex lens. The focal length of the lens is* 1 m. *Determine the radius of the* 10th *dark ring when light of wavelength is* 6000 Å. **[WBUT 2008; CUH 1996]**

Soln. The focal length of the equi-convex lens is given by

$$\frac{1}{f} = (\mu - 1)\frac{2}{R},$$

where $\mu = 1.70$, the r.i. of the film and
$f = 100$, the focal length of the equi-convex lens
$R = ?$

Now, $\frac{1}{100} = (1.7 - 1)\frac{2}{R}$. $\therefore R = 140$ cm.
Again, the radius of the nth dark ring is

$$r_n = \sqrt{\frac{n\lambda R}{\mu}}, \quad n = 10, \text{ the order number of the dark ring}$$

$$= \sqrt{\frac{10 \times 6000 \times 10^{-8} \times 140}{1.7}} = \sqrt{\frac{84}{1.7} \times 10^{-3}}$$

or, $r_{10} = 0.22$ cm $=$ **0.022 m.**

Ex. 27 *In a Newton's ring experiment, the diameter of a dark ring is* 0.32 cm, *when the wavelength of monochromatic light is* 6000 Å. *What would be the diameter of that ring when the wavelength of light changes to* 5000 Å? **[WBUT 2010]**

Soln. If R be the radius of curvature of the lens and λ, the wavelength of light used, then the diameter of the nth order dark ring is given by

$$D_n = \sqrt{4nR\lambda}, \quad n = 1, 2, 3, \cdots.$$

If D_1 and D_2 be the diameters of nth order dark rings for wavelength 6000 Å and 5000 Å respectively, then

$$D_1 = \sqrt{4nR \times 6000}$$

$$\text{and} \quad D_2 = \sqrt{4nR \times 5000}.$$

$$\therefore \quad \frac{D_2}{D_1} = \sqrt{\frac{5}{2}} \quad \text{or,} \quad D_2 = 0.32 \times \sqrt{\frac{5}{6}} = \mathbf{0.29\ cm.}$$

Ex. 28 *In Young's double slit experiment the distance between two slits is* 0.5 mm. *The wavelength of light is* 5000 Å *and the separation between the sources and the screen is* 50 cm. *Calculate the Fringe width in this case.* **[WBUT 2010]**

Soln. If $2d$ be the separation between the two slits and D that between the sources and screen, then the fringe width for light of wavelength λ is

$$\beta = \frac{D}{2d}\lambda.$$

Here, $D = 50$ cm, $2d = 0.5$ mm $= 0.05$ cm and $\lambda = 5000$ Å $= 5000 \times 10^{-8}$ cm $= 5 \times 10^{-5}$ cm.

$$\therefore \ \beta = \frac{50}{0.05} \times 5 \times 10^{-5} = \mathbf{0.05\ cm.}$$

EXERCISES

4(A)–Objective Type Questions

Select the correct answer:

4.1. That light is a wave is evidenced by (a) black body radiation (b) photoelectric effect (c) radioactive emission (d) interference phenomenon.

4.2. A pair of sources may be coherent if they emit light waves of (a) same amplitude (b) constant phase difference (c) same wavelength (d) same frequency.

4.3. When interference of light takes place at some region the light energy is (a) created (b) destroyed (c) redistributed. **[WBUT 2004]**

4.4. Two waves having the intensities in the ratio of 9 : 1 produce interference. The ratio of maximum to minimum intensity is equal to (a) 10 : 8 (b) 9 : 1 (c) 4 : 1 (d) 2 : 1. **[WBUT 2006]**

4.5. In the phenomenon of interference, the law of conservation of energy is (a) never violated (b) always violated (c) sometime violated (d) violated depending on frequency.

4.6. Two light waves interfere constructively at a point if their path difference is (a) odd multiple of $\lambda/2$ (b) even multiple of $\lambda/2$ (c) fraction of $\lambda/2$ (d) none of these.

4.7. In Young's double slit experiment, the separation between the slits is halved and the distance between the slits and screen is doubled. The fringe width is (a) unchanged (b) halved (c) quadrupled (d) doubled.

4.8. If Young's double slit experiment be carried out in water, then the fringe width compared to that in air (a) disappears (b) decreases (c) increases (d) remains unchanged.

4.9. Two light waves having zero phase difference and common wavelength λ interfere destructively at a point if the path difference is (a) 5λ (b) $\frac{3\lambda}{4}$ (c) 10λ (d) $12\frac{\lambda}{2}$.

4.10. In Young's double slit experiment, if the distance between the coherence sources is d, the distance between the source and screen is D and wavelength of light used is λ, then the fringe width β is (a) $\frac{d}{\lambda D}$ (b) $\frac{Dd}{\lambda}$ (c) $\frac{\lambda D}{d}$ (d) $\frac{\lambda d}{D}$.

4.11. The fringe width β with light of wavelength $\lambda = 400$ nm is 0.4 mm. If $\lambda = 600$ nm, then β will be (a) 0.66 mm (b) 0.6 mm (c) 0.4 mm (d) 0.12 mm.

4.12. When a soap bubble is illuminated with white light, multiple colours are observed due to (a) refraction (b) diffraction (c) interference (d) polarization.

4.13. A change of phase occurs when light wave is reflected from water-glass interface, which is (a) $\pi/2$ (b) 0 (c) π (d) 2π.

4.14. When light wave is refracted from glass-air interface the change of phase of the reflected wave is (a) π (b) 0 (c) 2π (d) $\pi/2$.

4.15. If a thin glass plate be illuminated with light of wavelength λ, then in reflected light it appears (a) bright (b) dark (c) coloured (d) none of these.

4.16. Newton's rings appear due to a phenomenon of light called (a) polarization (b) diffraction (c) scattering (d) interference.

4.17. Newton's rings are the fringes of (a) equal thickness (b) equal inclination (c) equal chromatic order (d) equal width.

4.18. In reflected light the centre of Newton's ring system is (a) bright (b) dark (c) coloured (d) none of these.

4.19. The radii of dark Newton's rings are proportional to square root of (a) even natural numbers (b) odd natural numbers (c) all natural numbers (d) none of these.

4.20. Newton's rings are produced by division of (a) wavefront (b) amplitude (c) phase (d) frequency.

4.21. Fill in the blanks:

(a) The fringe width of Newton's rings—with increase of order number.

(b) If μ be the refractive index of a film for anti-reflection coating for light of wavelength λ, then the required thickness of the film $t =$ —.

(c) With white light the centre of Newton's ring system looks —.

(d) With white light only first few Newton's rings are coloured but the rest are —.

4.22. Newton's rings are example of interference fringes of (a) equal inclination, coherent is obtained by amplitude division (b) equal thickness, coherence is obtained by amplitude division (c) equal thickness, coherence is obtained by wavefront division (d) equal inclination, coherence is obtained by wavefront division.

4.23. Double slit interference pattern is the limiting case of double slit diffraction pattern when the (a) distance between the slits tends to zero (b) distance between the source and the slits tends to infinity (c) slit widths tend to zero (d) distance between the slits tends to infinity. [**WBUT 2008**]

4.24. A source of light emits light of frequencies between ν to $\nu + \Delta\nu$. Coherence time of the emergent light beam is τ_c, then (a) $\tau_c \propto \Delta\nu$ (b) $\tau_c \propto \frac{1}{\Delta\nu}$ (c) $\tau_c \propto \nu$ (d) $\tau_c \propto \frac{1}{\nu}$. [**WBUT 2008**]

4.25 Two sources are said to be coherent when the waves produced by them have (a) same wavelength (b) same wavelength and same phase (b) same wavelength and constant phase difference (d) same amplitude and constant phase difference. [**WBUT 2010**]

4.26 In an arrangement of viewing Newton's ring if the lens, which rests on a glass plate were moved upwards by one wavelength (of viewing light), which of the following will be observed? (a) The central spot becomes bright (d) No change of fringe pattern is observed (c) The rings shift towards the centre (d) The rings move out from the centre. **[WBUT 2010]**

ANSWERS

4.1. (d) **4.2.** (b) **4.3.** (c) **4.4.** (c) **4.5.** (a) **4.6.** (b) **4.7.** (c) **4.8.** (b) **4.9.** (d) **4.10.** (c) **4.11.** (b) **4.12.** (c) **4.13.** (c) **4.14.** (b) **4.15.** (b) **4.16.** (d) **4.17.** (a) **4.18.** (b) **4.19.** (c) **4.20.** (b) **4.21.** (a) decrease (b) $\frac{\lambda}{4\mu}$ (c) dark (d) white **4.22.** (b) **4.23.** (c) **4.24.** (b) **4.25.** (c) **4.26.** (c).

4(B)–Short Answer Type Questions

4.1. What do you mean by interference of light?

4.2. What are the essential conditions for sustained interference?

4.3. What are coherent and non-coherent sources of light? **[WBUT 2001, '03; VH '02]**

4.4. Show that energy is conserved in interference phenomena. **[WBUT 2002; CU '04; NBU '01]**

4.5. What are the basic requirements for obtaining permanent interference pattern? **[WBUT 2003, '04]**

4.6. Can you observe interference with wide sources? Explain why?

4.7. Can two real sources of light act as coherent sources? **[PU 1994]**

4.8. If in Young's double slit experiment one slit is covered with blue transparent paper and the other with red transparent paper, can you expect interference phenomenon?

Ans. No, because the two slits in this case will emit light of two different wavelengths and the two sources will not be coherent. So interference phenomenon cannot be expected.

4.9. Can you observe interference pattern if two coherent sources are separated by a distance less than a wavelength of light? **[PU 1993]**

Ans. No. We know that the fringe width $\beta = \frac{D\lambda}{2d}$, where $2d$ is the separation of two coherent sources and λ is the wavelength of light used.

If $2d < \lambda$, then β will be very large and uniform illumination on the screen will be produced but no interference phenomenon.

4.10. Can you obtain interference with wide slits? Why? **[PU 1995]**

Ans. No. A wide slit may be thought as an integration of a large number of narrow sources placed side by side. Each pair of narrow sources will produce interference pattern on the screen. The superposition of such interference pattern will produce general illumination on the screen. So no interference will be obtained.

4.11. Can light from two different candles produce interference? Why?

4.12. Why a broad source of light is necessary to observe colours in thin films? **[KU 1995; PU 1993]**

4.13. Describe what are non-reflection films? **[PU 1994, '92]**

4.14. How will you estimate the difference of film thickness between two points of a wedge shaped film? **[KU 1995]**

Soln. Let AB and AC be two transparent glass plates enclosing a wedge-shaped film of refractive index μ. Let the angle of the wedge be α [Fig. 4.31].

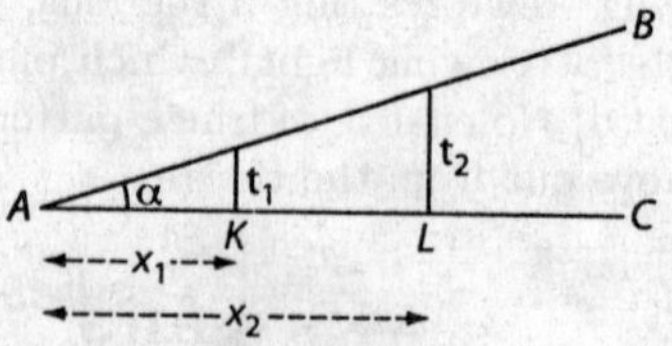

Fig. 4.31

If t_1 and t_2 be the thickness of the film at K and L such that $AK = x_1$ and $AL = x_2$, then

$$x_1 = \frac{t_1}{\alpha} \quad \text{and} \quad x_2 = \frac{t_2}{\alpha}, \quad \therefore x_2 - x_1 = \frac{t_2 - t_1}{\alpha}.$$

If N be the number of interference fringes between x_1 and x_2, then the fringe width

$$\beta = \frac{x_2 - x_1}{m} = \frac{t_2 - t_1}{N\alpha}.$$

But the fringe width in wedge shaped film is

$$\beta = \frac{\lambda}{2\mu\alpha},$$

where λ is the wavelength of light and μ is the refractive index of the film. Then

$$\frac{t_2 - t_1}{N\alpha} = \frac{\lambda}{2\mu\alpha} \quad \text{or,} \quad t_2 - t_1 = \frac{N\lambda}{2\mu}.$$

4.15. How does the interference pattern by reflection in thin films differ from that of refraction? Why? **[HPU 1995]**

4.16. How will you test the flatness of a surface by interference of light?

Ans. The flatness of a surface can be tested by the application of interference by wedge shaped film. If the two plates enclosing a wedge be illuminated with monochromatic light, straight parallel fringes will be obtained. Since these fringes are the fringes of equal thickness, the fringes will be perfectly straight lines if the surface is perfectly plane. If the surface is of varying thickness, then the fringes will be irregular in shape.

4.17. What happens if instead of monochromatic light white light be used in Newton's ring experiment?

Ans. We know that the diameter of the nth order bright ring $D_n \propto \sqrt{\lambda}$, where λ is the wavelength of the light used. If white light be used, then for a particular order number n there will be superposition of rings of different colours (i.e., different wavelengths). The superposition effect will be large for higher order fringes having smaller widths and general illumination will be observed. For lower value of n, the fringe width is larger and the superposition effect is less pronounced and the fringes will be highly coloured. The central fringe, however, will be dark in reflected light.

4.18. Show that the diameter of the dark Newton rings are proportional to the square root of the natural numbers.

4.19. Show that the diameters of the bright Newton rings are proportional to the square root of the odd natural numbers.

4.20. Show that the width of Newton's rings decreases with order number of rings. **[PU 1991]**

Soln. The diameter of the nth dark ring, $D_n = 2\sqrt{nR\lambda}$, the symbols have their usual meanings.

$$\therefore \frac{dD_n}{dn} = 2\frac{d}{dn}(nR\lambda)^{1/2} = 2\sqrt{R\lambda}\,\frac{n^{-\frac{1}{2}}}{2} \quad \text{or,} \quad \frac{dD_n}{dn} = \sqrt{\frac{R\lambda}{n}}, \quad \text{i.e.,} \quad \frac{dD_n}{dn} \propto \frac{1}{\sqrt{n}}.$$

This shows that the rate of change of diameters of Newton's rings is inversely proportional to the square root of order number. In other words, the width of Newton's rings decreases with order number.

4.21. In Newton's ring experiment, the fringes are circular and the central fringes is dark. Why? **[KU 1982]**

Ans. The reflected rays from the upper surface of the air film and from the glass-plate has a path difference $\lambda/2$ as glass is denser than air. The thickness of the air film at the centre, where the plano-convex lens touches the glass plate is zero. So the path difference between the two reflected rays from the upper and lower surfaces of the air film is $\lambda/2$, which produces destructive interference. Hence, the central fringe is dark.

The thickness of air film increases as one moves from centre to outwards. Since these fringes are fringes of equal thickness and along the circumference of a circle, the thickness is constant, we observe concentric circular fringes with the central fringe dark.

4.22. Explain why an excessively thin film appears black in reflected light? **[KU 1988]**

4.23. Why do we see colours when white light falls on a thin film of transparent medium?

4.24. Do you expect colours if the thickness of the film be large? Why? **[cf. PU 1994, '93]**

4.25. Explain what will happen in Newton's ring experiment when air in the interspace is replaced by transparent liquid (water). **[KU 1995]**

[*Hints.* $D^2_{\text{air}} = 4Rn\lambda$, $D^2_{\text{liq}} = \dfrac{4Rn\lambda}{\mu}$

$$\therefore \left(\frac{D_{\text{liq}}}{D_{\text{air}}}\right)^2 = \frac{1}{\mu}, \quad \therefore \frac{D_{\text{liq}}}{D_{\text{air}}} = \frac{1}{\sqrt{\mu}} = \frac{1}{\sqrt{1.33}} = 0.867 \quad [\because \mu_{\text{water}} = 1.33].$$

$\therefore D_{\text{liq}} = 0.867 \times D_{\text{air}}$, $\because D_{\text{liq}} < D_{\text{air}}$, the fringe system will be found to shrink in size.]

4(C)–Numerical Problems

4.1. Two straight and narrow parallel slits 3 mm apart are illuminated by monochromatic light of wavelength 5.9×10^{-5} cm. Fringes are obtained on a 30 cm distant screen from the slits. Find the fringe width. [*Ans.* 0.0059 cm]

4.2. A bi-prism of obtuse angle 176° is made of glass of refractive index 1.5. A slit illuminated by monochromatic light is placed 20 cm behind the bi-prism and the width of the interference fringes, formed on a screen 80 cm in front of the bi-prism is found to be 8.25×10^{-3} cm. Find the wavelength of light used. [*Ans.* 5762 Å]

4.3. In Fresnel's bi-prism experiment, on inserting a thin glass plate of thickness 6×10^{-4} cm in the path of one of the interfering beams, it is found that the central fringe shifts to the position previously occupied by the 5th bright fringe. If the wavelength of the light used is 6000 Å, find the refractive index of glass. [*Ans.* 1.5] **[CU 2001]**

4.4. Fringes are produced by Fresnel's bi-prism in the focal plane of a telescope, which is 100 cm from the slit. A lens inserted between the prism and the telescope gives two images of the slit in two positions. In one case, the two images of the slit are 4.05 mm and in the other case, 2.9 mm apart. If sodium light ($\lambda = 5893 \times 10^{-8}$ cm) is used, find the fringe width.

[*Ans.* 0.172 mm] **[CU 1985; BU 1994]**

4.5. In a double slit experiment a light of $\lambda = 5460$ Å is exposed to slits, which are 0.1 mm apart. The screen is placed 2 m away from the slits. What is the angular position of the 10th maximum and first minimum? [*Ans.* $\theta = 3.127°, \theta = 0.156°$]

4.6. In a Young's double slit experiment, the slits are separated by 0.2 cm and the screen is placed 1 m away. The slits are illuminated by yellow light ($\lambda = 5896$ Å). Calculate the fringe width.

[*Ans.* 29.48×10^{-5} cm] **[CU 2005]**

4.7. Find the thickness of soap film that gives constructive second order interference of reflected red light (λ = 700 nm). The index of refraction of the film is 1.33 (Assume that parallel beam of incident light is directed at 30° to the normal.) **[BUH 2002]**

4.8. A uniform soap film ($\mu = 1.83$) has thickness 0.3 μm. What does it show when seen in reflected light along the normal? **[BUH 2000]**

4.9. When a wedge shaped air film is viewed by a monochromatic source of light incident normally, the interference fringes 4 mm apart are observed. If the air space is filled with water ($\mu = 1.33$), how far apart will the fringes be observed? [*Ans.* 3×10^{-3} m] **[KU 1991]**

4.10. In a Newton's ring experiment, the diameter of the 10th ring changes from 1.40 cm to 1.27 cm, when a liquid is introduced between the lens and the plate. Calculate the refractive index of the plate. [*Ans.* 1.215]

4.11. In Newton's ring experiment, the diameters of the 4th and 12th dark rings are 0.400 cm and 0.700 cm respectively. Find the diameter of the 20th dark ring. [*Ans.* 0.0906 cm]

4.12. In a Newton's ring experiment if the diameters of 3rd and 23rd rings are 0.181 cm and 0.501 cm, respectively and if the radius of curvature of the curved surface of the plano-convex lens be 50 cm, calculate the wavelength of the light. [*Ans.* 5456 Å] **[CU 1986, 2002]**

4.13. Newton's rings are formed with reflected light of wavelength 5895×10^{-8} cm with a liquid between the plane surface of the glass plate and curved surface of a convex lens. The diameter of the 5th dark ring is 0.3 cm and the radius of curvature of the curved surface of the lens is 100 cm. Calculate the refractive index of the liquid. [*Ans.* 1.31] **[BU 1999, NBU 2001]**

4.14. A soap film is illuminated by white light at an angle of 30°. The reflected light is examined by a spectrometer in which dark band corresponding to $\lambda = 6 \times 10^{-7}$ m is found. Find the thickness of the film, given μ for film = 1.33.

[*Hints.* If t be the minimum thickness, then

$$t = \frac{\lambda}{2\mu \cos r} = \frac{\lambda}{2\mu\sqrt{1 - \sin^2 r}}.$$

But

$$\mu = \frac{\sin i}{\sin r} \quad \text{or,} \quad \sin r = \frac{\sin i}{\mu} = \frac{\sin 30°}{1.33} = \frac{0.5}{1.33} = 0.3759.$$

$$\therefore t = \frac{6 \times 10^{-7}}{2 \times 1.33\sqrt{(1 - (0.3759)^2}} = \mathbf{2.434 \times 10^{-7}\ m.}]$$

4.15. Find the thickness of a non-reflecting layer to be deposited on glass surface corresponding to wavelength $\lambda = 6000$ Å. [$\mu = 1.35$] [*Ans.* 1.11×10^{-7} m]

4.16. In a Young's double slit interference experiment, distance between the coherent sources is 1.15 cm. Calculate the fringe width that would be observed on a screen placed at a distance of 85 cm from the source. The wavelength of light used is 5893 Å. **[WBUT 2008]**

[*Hints.* Fringe width $\beta = \frac{D\lambda}{2d} = \frac{85 \times 5893 \times 10^{-8}}{0.115} = 4355.69 \times 10^{-5}$ cm $= \mathbf{4355.69 \times 10^{-7}}$ **m.**]

4(D)–Broad Answer Type Questions

4.1. Deduce an expression for the intensity of light at a point due to superposition of waves coming from two light sources. Hence, find the conditions of destructive and constructive interference. **[CU 2005]**

4.2. (a) What are the conditions to get sustained interference pattern?

(b) In an interference experiment d is the distance between two coherent sources of light with wavelength λ, and D is the source-screen distance. Show that the separation between the two consecutive dark bands is given by $\beta = \frac{D\lambda}{d}$. **[WBUT 2004]**

4.3. (a) Establish the conditions of constructive and destructive interference of light coming from two coherent sources.

(b) Show that energy is conserved in interference phenomena.

(c) If the distance between the two coherent sources of light wth wavelength λ is d and D be the source-screen distance, then show that fringe width separation, $X = \frac{D\lambda}{d}$. **[WBUT 2002]**

4.4. (a) What are coherent and non-coherent sources of light? Outline the theory for the determination of wavelength of light by Fresnel's bi-prism.

(b) A mica-sheet of refractive index 1.58 is introduced in one of the interfering beams and the central fringe gets shifted by 0.2 cm. Distance between the coherent sources is 0.1 cm and the screen is placed at a distance of 50 cm from the sources. Determine the thickness of the mica sheet. **[WBUT 2001]**

4.5. Explain the phenomenon of interference of light. Describe and explain Young's experiment demonstrating interference of light.

4.6. (a) Give Stokes' treatment to explain the change of phase when reflection takes place at a denser medium.

(b) Explain what do you understand by (i) temporal coherence and (b) spatial coherence.

4.7. How can the wavelength of monochromatic light be measured with the help of a Fresnel's bi-prism? Give the theory of the method and the arrangement of the apparatus. **[NBU 1982, CU 1985]**

4.8. Describe how the wavelength of monochromatic beam of light can be measured by Newton's ring experiment. Deduce the working formula. **[CU 1982, 2004, NBU 1983, BU 1999]**

4.9. Give the theory of Newton's rings to determine the wavelength of monochromatic light and show how from their study the refractive index of a liquid can be determined.

[CUH 2000, cf. BUH 2002]

4.10. (a) Give the theory of Newton's rings and how from their study the wavelength of monochromatic light can be determined.

(b) Why the central fringe is dark? How can it be made bright?

(c) If the monochromatic source be replaced by a white source, then what change in the ring system do you expect?

4.11. Describe how Newton's ring apparatus can be employed to measure (a) the radius of curvature of a long focal length spectacle glass (b) refractive index of a liquid.

Chapter 5 | Module 2.2

Diffraction of Light

5.1 Introduction

Very much like sound wave, light wave also suffers bending round the sharp edge of an obstacle if the dimension of the obstacle is comparable with the wavelength of the wave and spreads into the geometrical shadow. The amount of bending, however, depends upon (i) *the size of the obstacle* and (ii) *the wavelength of the wave, the bending being more for light of greater wavelength.*

This phenomenon is known as *diffraction of light* and was first accurately described by an Italian scientist *F. M. Grimaldi in* 1665. The word *diffraction* is derived from the Latin word *diffract* meaning '*break in pieces*'. The *wave nature of light* is further confirmed by the phenomenon of diffraction of light.

Definition: *The phenomenon of bending of light waves around obstacles or apertures of sizes comparable with the wavelengths of light and resulting thereby in their spreading into the geometrical shadow of the object is called diffraction.*

The appropriate interpretation of diffraction of light wave first proved by *Fresnel.* He showed that diffraction is due to the *mutual interference of secondary waves* originating from different points of the same wavefront. On the basis of *the principle of secondary wavelets* and *that of superposition, Fresnel* could calculate the position of diffraction bands, which agree with experiment.

5.2 Distinction between Interference and Diffraction

Though diffraction patterns are formed due to interference of waves still the phenomenon of diffraction has some basic difference from interference. The differences are listed below:

Table 5.1

Interference	*Diffraction*
1. In interference, *two separate wavefronts originating from two coherent sources* superpose at a point on the screen.	1. In diffraction, secondary wavelets originating from the *different parts of the same wavefront* superpose.
2. The width of interference fringes may or may not be the same.	2. The diffraction bands are never of equal width.

Interference	*Diffraction*
3. All bright fringes have the same intensity.	3. The intensity of the bright band usually decreases with increase of order number.
4. All dark fringes have zero intensity.	4. The intensity of dark bands is not zero.
5. Intensity distribution of interference pattern is uniform.	5. Intensity distribution of diffraction pattern is non-uniform.

5.3 Classification of Diffraction

The diffraction phenomenon can be divided into two main classes, viz., (i) *Fresnel class of diffraction* (ii) *Fraunhofer class of diffraction.*

In *Fresnel class of diffraction*, either the source of light or the screen or both are at *finite distances* from the diffraction aperture or obstacle. The wavefront in Fresnel class of diffraction is either *spherical* or *cylindrical.*

In Fraunhofer class of diffraction, the source of light and the screen are effectively at *infinite distance* from the diffracting aperture and the wavefront is *plane.*

The above two types of diffraction are diagrammatically represented in Figs 5(a) and (b) respectively.

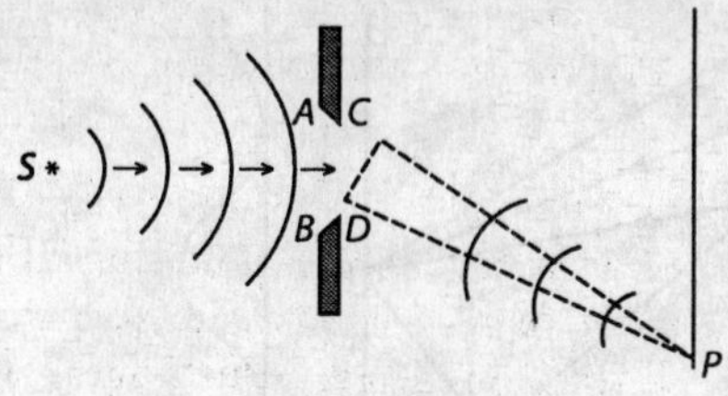

Fig. 5.1(a): Fresnel's diffraction.

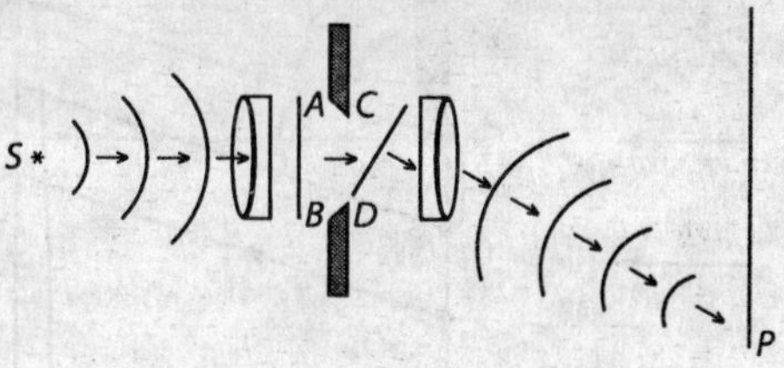

Fig. 5.1(b): Fraunhofer diffraction.

Table 5.2

Fresnel diffraction	*Fraunhofer diffraction*
1. The distance of the source, or the screen or both from the diffracting elements are *finite.*	1. The distance of the source and the screen from the diffracting elements are effectively *infinite.*
2. The wavefronts are not plane but either spherical or cylindrical.	2. The wavefronts are plane.
3. No mirror or lenses are used for study.	3. Converging lens is used to focus parallel rays.
4. In the plane of the aperture, the phase of the secondary wavelets is not the same at all points.	4. The secondary wavelets are in same phase at every point in the plane of the aperture.

Fresnel diffraction	*Fraunhofer diffraction*
5. The centre of the diffraction pattern may be bright or dark depending upon the number of *Fresnel zones.*	5. The centre of the diffraction pattern is always bright for all paths parallel to the axis of the lens.
6. The resultant effect at any point on the screen is the combined effect of all the secondary waves originating from various *zones.*	6. The interference of parallel rays brought to focus produces the resultant diffraction pattern on the screen.

5.4 Fraunhofer Diffraction at a Single-Slit

Let a plane wavefront of *monochromatic light of wavelength* λ be incident normally on a narrow slit AB of width b placed perpendicular to the plane of the paper [Fig. 5.2]. We have to calculate the intensity distribution due to diffraction on the screen XY placed at the focal point of the lens L. We assume that (i) *each point on the slit is a source of secondary wavelets. Interference takes place between the wavelets originating from two such secondary sources.* (ii) *the slit consists of a large number of secondary point sources, which are equi-spaced.*

If Δ be the separation between any two point sources $A_1, A_2, A_3, \cdots$, then

$$b = (n-1)\Delta, \tag{5.1}$$

where n is the number of point sources.

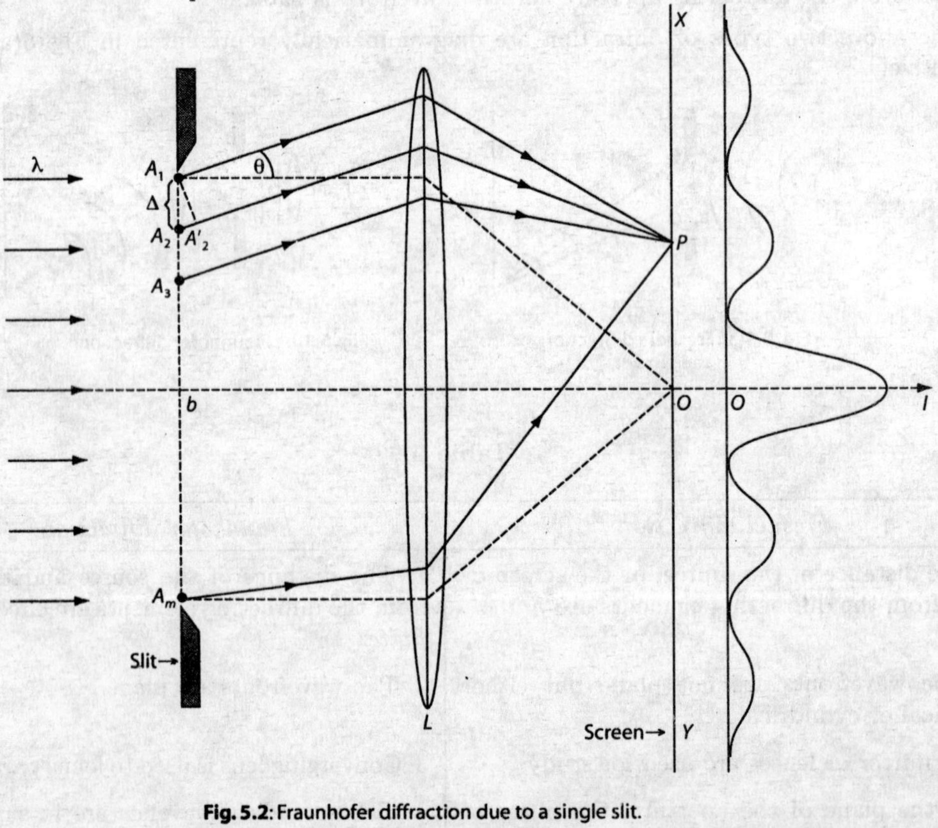

Fig. 5.2: Fraunhofer diffraction due to a single slit.

As the point P on the screen is at a large distance from the slit, the amplitudes of the secondary waves from A_1 and A_2 will be nearly the same. If the diffracted rays from A_1 and A_2

make an angle θ with the normal to the plane of the slit, then the path difference between them is $A_2A_2' = \Delta \sin\theta$, where $A_1A_2 = \Delta$.

The corresponding phase difference is

$$\phi = \frac{2\pi}{\lambda}(\Delta \sin\theta). \tag{5.2}$$

So if the field at P due to disturbance coming from A_1 be represented by

$$E_1 = a \cos \omega t,$$

where a is the *amplitude* and ω the *circular frequency*, then the field at P due to the disturbance from A_2 is

$$E_2 = a \cos(\omega t - \phi).$$

Similarly, the fields at P due to the disturbances from $A_3, A_4, \cdots, A_n$ are

$$E_3 = a \cos(\omega t - 2\phi)$$
$$E_4 = a \cos(\omega t - 3\phi)$$
$$\ldots, \ldots\ldots\ldots\ldots\ldots\ldots\ldots$$
$$E_n = a \cos[\omega t - (n-1)\phi].$$

Thus, the resultant field at P is

$$\begin{aligned} E &= E_1 + E_2 + E_3 + \cdots + E_n \\ &= a[\cos \omega t + \cos(\omega t - \phi) + \cos(\omega t - 2\phi) + \cdots + \cos[\omega t - (n-1)\phi]. \end{aligned} \tag{5.3}$$

Now,

$$\cos \omega t + \cos(\omega t - \phi) + \cdots + \cos[\omega t - (n-1)\phi] = \frac{\sin \frac{n\phi}{2}}{\sin \frac{\phi}{2}} \cos\left[\omega t - \tfrac{1}{2}(n-1)\phi\right]. \tag{5.4}$$

$$\therefore E = a\frac{\sin \frac{n\phi}{2}}{\sin \frac{\phi}{2}} \cos\left[\omega t - \tfrac{1}{2}(n-1)\phi\right] = E_\theta \cos\left[\omega t - \tfrac{1}{2}(n-1)\phi\right], \tag{5.5}$$

where

$$E_\theta = a\frac{\sin \frac{n\phi}{2}}{\sin \frac{\phi}{2}}. \tag{5.6}$$

Now, if $n \to \infty$, $\Delta \to 0$, then $n\Delta \to b$.

$$\therefore \frac{n\phi}{2} = \frac{n}{2} \times \frac{2\pi}{\lambda}\Delta \sin\theta = \frac{n\pi\Delta}{\lambda}\sin\theta = \frac{\pi b \sin\theta}{\lambda} \text{ and } \frac{\phi}{2} = \frac{1}{2} \times \frac{2\pi}{\lambda}\Delta \sin\theta = \frac{\pi b \sin\theta}{\lambda n} \to 0.$$

$$\therefore E_\theta \simeq \frac{a \sin \frac{n\phi}{2}}{\frac{\phi}{2}} = \frac{na \sin\left(\frac{\pi b \sin\theta}{\lambda}\right)}{\frac{\pi b \sin\theta}{\lambda}} = A\frac{\sin\beta}{\beta}, \tag{5.7}$$

where

$$A = na \text{ and } \beta = \frac{\pi b \sin\theta}{\lambda}. \tag{5.8}$$

$$\therefore E = A\frac{\sin\beta}{\beta}\cos(\omega t - \beta) \quad \left[\because \frac{1}{2}(n-1)\phi \simeq \frac{n\phi}{2} = \frac{\pi b \sin\theta}{\lambda} = \beta\right]. \tag{5.9}$$

The intensity distribution on the screen due to the diffraction at the single-slit is

$$I = I_0\frac{\sin^2\beta}{\beta^2}, \tag{5.10}$$

where $I_0 = A^2 = n^2a^2$ is the intensity at $\theta \to 0$.

Eqn (5.10) shows that the intensity at any point on the screen XY is a function of β and hence, θ ($\because \beta = \frac{\pi b \sin\theta}{\lambda}$). *This shows that a series of alternate maxima and minima of intensity will be obtained on the screen.*

Intensity distribution:

(1) **Position of the principal maxima**: When $\theta \to 0$,

$$\beta = \frac{\pi b \sin\theta}{\lambda} \to 0 \text{ and } \lim_{\beta\to 0} \frac{\sin\beta}{\beta} \to 1.$$

$\therefore$ from eqn (5.10),

$$I = I_0 \left(\lim_{\beta\to 0} \frac{\sin\beta}{\beta} \right)^2 = I_0. \tag{5.11}$$

This corresponds to the *principal maximum*, which occurs in the region of the screen, where $\beta \to 0$, i.e., $\theta \to 0$, i.e., around the point O of Fig. 5.2.

(2) **Positions of minima**: From eqn (5.10) it is clear that $I = 0$, when $\sin\beta = 0$, i.e., when $\beta = \pm m\pi$, where $m = 1, 2, 3, \cdots$

$$\text{or, } \frac{\pi b \sin\theta}{\lambda} = \pm\pi \text{ or, } b\sin\theta = \pm m\lambda. \tag{5.12}$$

Eqn (5.12) gives the *condition of diffraction minima.* The value of $\theta = 0$ is not admissible because this corresponds to the position of the principal maximum.

The first minimum occurs at $\theta = \pm\sin^{-1}(\lambda/b)$, the second minimum at $\theta = \pm\sin^{-1}(2\lambda/b)$, etc.

Since $\sin\theta$ cannot exceed unity, *the maximum value of the order number of the minimum* $m \le \frac{b}{\lambda}$.

[$\because \sin\theta \le 1$, $b\sin\theta = m\lambda$ gives $\sin\theta = \frac{m\lambda}{b}$ or, $\frac{m\lambda}{b} \le 1$ or, $m \le \frac{b}{\lambda}$.]

The minima are situated on either side of the principal maximum.

Since $\sin\theta = \pm m\frac{\lambda}{b}$, the positions of first, second, third, etc. minima are given by $\theta_1 = \pm\frac{\lambda}{b}$, $\theta_2 = \pm\frac{2\lambda}{b}$, $\theta_3 = \pm\frac{3\lambda}{b}$, etc., which shows that *the minima are of equal width* (i.e., $\frac{\lambda}{b}$).

(3) **Positions of secondary maxima**: We have from eqn (5.10),

$$I = I_0\frac{\sin^2\beta}{\beta^2}. \quad \therefore \frac{dI}{d\beta} = I_0\left[\frac{2\sin\beta\cos\beta}{\beta^2} - \frac{2\sin^2\beta}{\beta^3}\right].$$

For I to be minimum,

$$\frac{dI}{d\beta} = 0: \quad \therefore 2\sin\beta\left[\frac{\cos\beta}{\beta^2} - \frac{\sin\beta}{\beta^3}\right] = 0,$$

i.e., either $\sin\beta = 0$ or, $\beta = \pm m\pi$ corresponds to the positions of the minima

$$\text{or, } \frac{\cos\beta}{\beta^2} - \frac{\sin\beta}{\beta^3} = 0 \text{ or, } \beta - \tan\beta = 0 \text{ or, } \tan\beta = \beta. \tag{5.13}$$

The root $\sin\beta = 0$, i.e., $\theta = 0$ corresponds to the positions of the *central maximum.* The other roots of eqn (5.13) can be found by plotting $y = \beta$ and $y = \tan\beta$ on the same graph paper and then finding the coordinates of the point of intersections [Fig. 5.3(b)]. The intersections occur at $\beta = 1.43\pi$, 2.46π, etc. Hence, the secondary maxima occur at $\beta = 1.43\pi, 2.46\pi$, etc.

Taking $\beta \simeq 0, \frac{3\pi}{2}, \frac{5\pi}{2}$ for all practical purposes, we can find out the magnitudes of the intensity of the principal maximum (I_0), secondary first maximum (I_1), second maximum and so on as follows:

(i) *For the principal maximum* $\theta = 0$. $\therefore \beta = 0$ and $I = I_0$.

(ii) *For the first secondary maximum,* $\beta = \frac{3\pi}{2}$.

$$\therefore I_1 = I_0\left(\frac{\sin\frac{3\pi}{2}}{\frac{3\pi}{2}}\right)^2 = I_0\left(\frac{-1}{3\pi/2}\right)^2 = \frac{4I_0}{9\pi^2} = \frac{I_0}{22}.$$

(iii) *For the second secondary maximum,* $\beta = \frac{5\pi}{2}$.

$$\therefore I_2 = I_0\left(\frac{\sin\frac{5\pi}{2}}{\frac{5\pi}{2}}\right)^2 = I_0\left(\frac{1}{\frac{5\pi}{2}}\right)^2 = \frac{4I_0}{25\pi^2} = \frac{I_0}{61} \text{ and so on.}$$

$$\therefore I_0 : I_1 : I_2 : I_3 = 1 : \frac{1}{22} : \frac{1}{61} : \frac{1}{121}. \quad (5.14)$$

This indicates that the intensity of the principal maximum is the highest and the intensity of the secondary maxima decreases on either side of I_0 with the increase of the order number m.

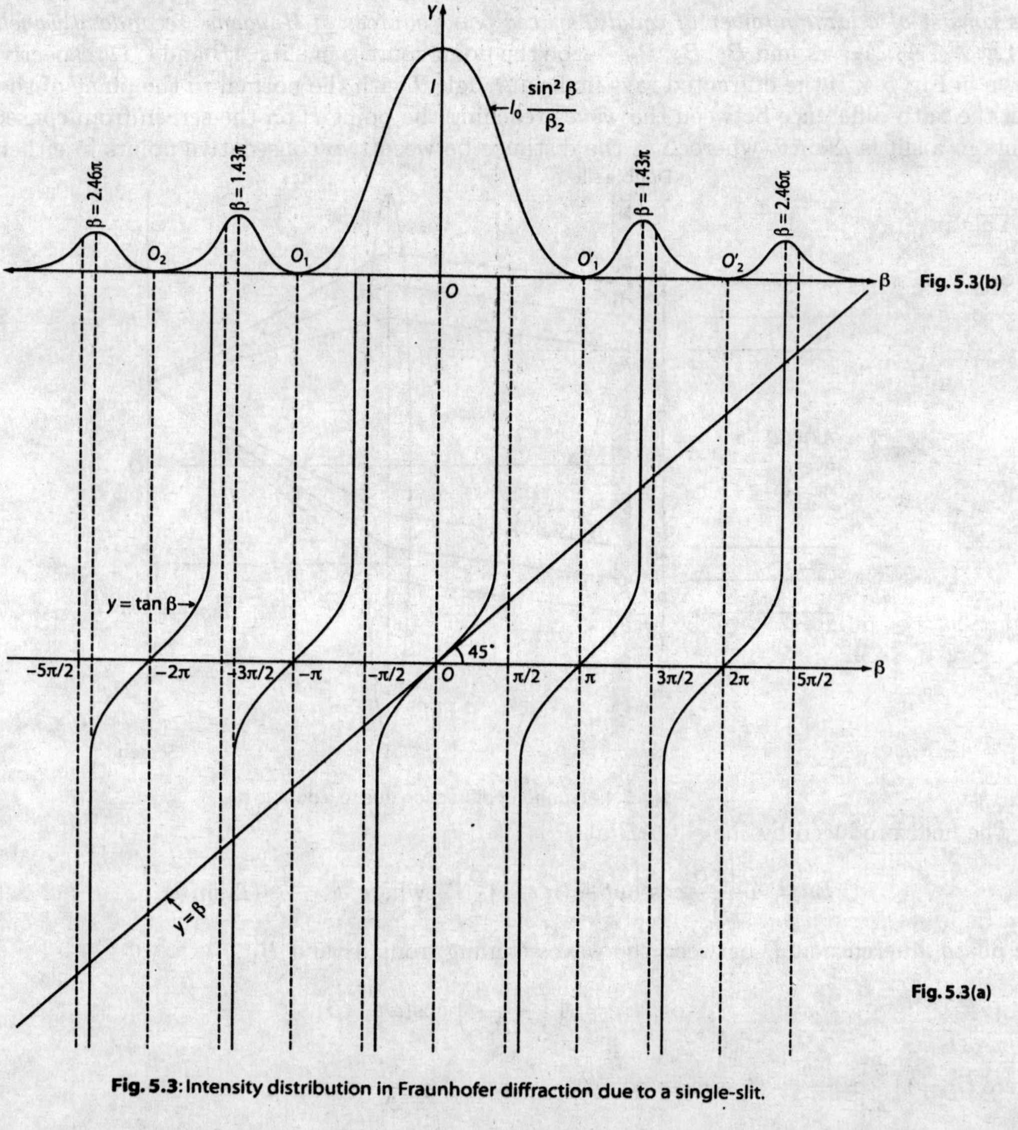

Fig. 5.3: Intensity distribution in Fraunhofer diffraction due to a single-slit.

As the secondary maxima occur at $\beta = 0, 1.43\pi, 2.46\pi$, etc. it is evident that the secondary maxima *are not equi-spaced. That is why in single-slit experiment we take readings for the positions of the minima and not of the maxima.*

Effect of white light: If instead of monochromatic light, white light be used, then the secondary maxima will be coloured but the central maximum will be white. For $b\sin\theta = m\lambda$ gives $\theta = \frac{\lambda}{b}$ for the first order dark band ($\because\ \sin\theta \to \theta$ as $\theta \to 0$). Again, $\lambda_r > \lambda_v$. $\therefore\ \theta_r > \theta_v$. This indicates that the central maximum for which $\theta \to 0$ for all colours (i.e., for all λ) will be white. The other maxima will be coloured with red being farther apart from the blue.

The intensity distribution due to single-slit diffraction is shown in Fig. 5.3(b).

5.5 Fraunhofer Diffraction Due to a Double-Slit

Let AB and CD be two slits of small width b and a plane wavefront of λ be incident normally on the slits. We assume that (i) *each point on the slit is a source of secondary wavelets* and (ii) *the slits consist of a large number of equally spaced point sources of Huygens' secondary wavelets.*

Let $A_1, A_2, A_3, \cdots$ and $B_1, B_2, B_3, \cdots$ be the point sources in slits AB and CD respectively as shown in Fig. 5.4. If the diffracted rays make an angle θ with the normal to the plane of the slits, then the path difference between the waves reaching the point P on the screen from consecutive points in a slit is $\Delta\sin\theta$, where Δ is the distance between two consecutive points in either slit.

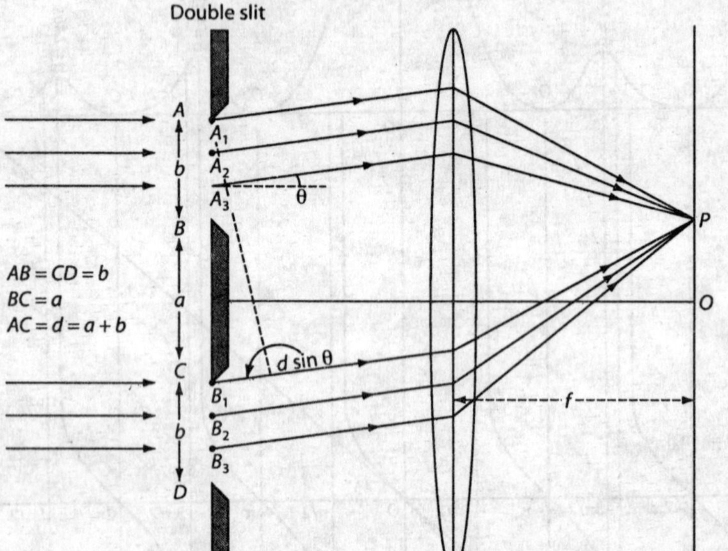

Fig. 5.4: Fraunhofer diffraction due to a double slit.

The field produced by the slit AB at P is

$$E_1 = A\frac{\sin\beta}{\beta}\cos\left[\omega t - (n-1)\frac{\phi}{2}\right], \text{ where } \phi = \frac{2\pi}{\lambda}(\Delta\sin\theta), \tag{5.15}$$

the phase difference at P between the waves coming from A_1 and B_1

$$\text{or, } E_1 = A\left(\frac{\sin\beta}{\beta}\right)\cos(\omega t - \beta),$$

where $(n-1)\frac{\phi}{2} \simeq \frac{n\phi}{2} = \beta$ as $n \to \infty, n\Delta \to b$ and $\beta = \frac{\pi b}{\lambda}\sin\theta$.

Similarly, the field at P due to the slit CD is

$$E_2 = A\frac{\sin\beta}{\beta}\cos(\omega t - \beta - \phi_1), \tag{5.16}$$

$$\text{where } \phi_1 = \frac{2\pi}{\lambda}(d\sin\theta) \tag{5.17}$$

represents the phase difference between secondary waves from *two corresponding points* reaching P. d is the distance between two corresponding points in the double-slit (i.e., $A_1B_1 = A_2B_2 = \cdots = d$).

Therefore, the resultant field at P is

$$E = E_1 + E_2 = A\frac{\sin\beta}{\beta}[\cos(\omega t - \beta) + \cos(\omega t - \beta - \phi_1)]$$

$$= 2A\frac{\sin\beta}{\beta}\cos\gamma\cos\left(\omega t - \beta - \frac{\phi_1}{2}\right), \tag{5.18}$$

where

$$\gamma = \frac{\phi_1}{2} = \frac{\pi d\sin\theta}{\lambda} \tag{5.19}$$

$$\text{or,}\quad E = \mathcal{A}\cos\left(\omega t - \beta - \frac{\phi_1}{2}\right), \tag{5.20}$$

where $\mathcal{A} = 2A = \dfrac{\sin\beta}{\beta}\cos\gamma$ is the resultant amplitude at P.

Now, the intensity at P is

$$I = \mathcal{A}^2 = 4A^2\frac{\sin^2\beta}{\beta^2}\cos^2\gamma = \left(4I_0\frac{\sin^2\beta}{\beta^2}\right)\times\cos^2\gamma, \text{ (where } A^2 = I_0)$$

$$\text{or, } I = I_1 \times I_2, \tag{5.21}$$

where $I_1 = I_0\dfrac{\sin^2\beta}{\beta^2}$ represents the intensity distribution produced by one slit.

$I_2 = \cos^2\gamma$ represents the intensity distribution of *diffraction pattern* in the *interference pattern* produced by two point sources separated by d.

It is interesting to note that if $b \to 0$, then $\beta = \frac{\pi b\sin\theta}{\lambda} \to 0$ and $\lim\limits_{\beta\to 0}\frac{\sin^2\beta}{\beta^2} \to 1$, then there is no variation of $\frac{\sin^2\beta}{\beta^2}$ with θ.

Diffraction maxima and minima

(a) *Principal maximum*: The factor $I_1 = I_0\frac{\sin^2\beta}{\beta^2}$ gives the principal maximum in the direction $\beta \to 0$, i.e., $\theta \to 0$ corresponding to the point O on the screen [Fig. 5.4]. *This principal maximum has on either side alternate minima and secondary maxima of gradually diminishing intensity.*

(b) *Minima*: The minima are in the direction given by

$$\sin\beta = 0 \text{ or, } \beta = \frac{\pi b\sin\theta}{\lambda} = \pm m\lambda \text{ or, } b\sin\theta = m\lambda, \tag{5.22}$$

where $m = 1, 2, 3$, etc. excluding 0 (why?). *The $\pm$ sign before $m\lambda$ represents that the minima are situated symmetrically on either side of the principal maximum.*

(c) *The positions of the secondary maxima*: The positions of the secondary maxima are obtained at

$$\beta = \pm\frac{3\pi}{2}, \pm\frac{5\pi}{2}, \pm\frac{7\pi}{2}, \text{ etc.} = \pm(2m \pm 1)\frac{\pi}{2}, \tag{5.23}$$

where $m = 1, 2, 3, \cdots, m = 0$ is excluded (why?)

(d) *Interference maxima and minima*: From the term $I_2 = \cos^2\gamma$, the maxima occur for the values of θ for which $\cos^2\gamma = 1$ or, $\gamma = \pm n\pi$, where $n = 0, 1, 2, 3, \cdots$

$$\text{or, } \frac{\pi d \sin\theta}{\lambda} = \frac{\pi(a+b)\sin\theta}{\lambda} = \pm n\pi,$$

where $d = a + b$ is the *distance between two corresponding points of the double-slit*

$$\text{or, } (a+b)\sin\theta = \pm n\lambda \tag{5.24}$$

$$\text{or, } (a+b)\theta = \pm n\lambda \quad [\because \text{ when } \theta \to 0, \sin\theta \to \theta] \tag{5.25}$$

$$\text{or, } \theta = \pm\frac{n\lambda}{a+b} = \frac{\lambda}{a+b}, \frac{2\lambda}{a+b}, \frac{3\lambda}{a+b}, \cdots. \tag{5.26}$$

So the *angular separation between two consecutive interference maxima is*

$$\Delta\theta = \frac{\lambda}{a+b}. \tag{5.27}$$

Again, from $I_2 = \cos^2\gamma$, we note that $I_2 = 0$, when $\cos^2\gamma = 0$ or, $\gamma = \pm(2n+1)\frac{\pi}{2}$ $(n = 0, 1, 2, 3, \cdots)$

$$\text{or, } \frac{\pi(a+b)\sin\theta}{\lambda} = \pm(2n+1)\frac{\pi}{2} \text{ or, } (a+b)\sin\theta = \pm(2n+1)\frac{\lambda}{2}, \tag{5.28}$$

is the condition of interference minima

$$\text{or, } \sin\theta = \frac{(2n+1)\lambda}{2(a+b)}.$$

Putting $\sin\theta \simeq \theta$ as $\theta \to 0$, we get

$$\theta = \frac{(2n+1)\lambda}{2(a+b)} \text{ or, } \theta = \frac{\lambda}{2(a+b)}, \frac{3\lambda}{2(a+b)}, \frac{5\lambda}{2(a+b)}, \cdots.$$

So the angular separation between any two consecutive interference minima is

$$\Delta\theta = \frac{3\lambda}{2(a+b)} - \frac{\lambda}{2(a+b)} = \frac{\lambda}{a+b}. \tag{5.29}$$

It is to note that if the wavelength λ is constant, i.e., for monochromatic light, the angular width $(\Delta\theta)$ of interference minima is inversely proportional to the value of $(a+b)$. Furthermore, since both for interference maxima as well as minima, $\Delta\theta = \frac{\lambda}{a+b}$, we conclude that the term $I = \cos^2\gamma$ *gives a set of equidistant dark and bright interference fringes on the screen.*

Missing orders of interference maxima: If the slit width b is kept constant and the width of the opaque space a is gradually altered, then according to the condition of diffraction minima

$$b\sin\theta = \pm m\lambda,$$

the diffraction pattern remains unaltered but according to the condition of interference maxima

$$d\sin\theta = (a+b)\sin\theta = \pm n\lambda,$$

the spacing between the interference maxima changes. *It may so happen that for a particular value of the angle of diffraction θ, the conditions of interference maximum and diffraction minimum are simultaneously satisfied, then the corresponding interference maximum will be missed from the resultant spectrum.*

Let the conditions

$$(a+b)\sin\theta = n\lambda \quad \text{[Intensity maxima of interference pattern]}$$

$$b\sin\theta = m\lambda \quad \text{[Intensity minima of diffraction pattern]}$$

be simultaneously satisfied for a particular value of θ, then nth order interference maximum is formed, where the mth order diffraction minima would be formed. As a result the nth order interference maxima will be missing.

From the above conditions,

$$\frac{a+b}{b} = \frac{n}{m}. \tag{5.30}$$

(i) If $b = a$, then $n = 2m = 2, 4, 6, \cdots$ [$\because m = 1, 2, 3, \cdots$]. Thus, the $2, 4, 6, \cdots$ interference maxima will be missing from the resultant pattern. Because these will be formed at the positions of first, second, third diffraction minima. The central maxima, then will have three interference maxima [Fig. 5.5].

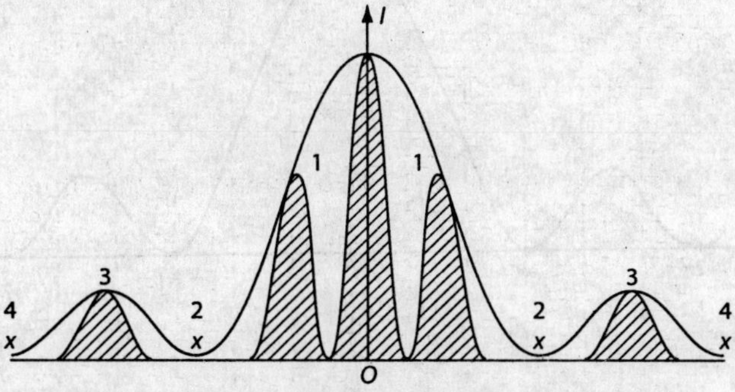

Fig. 5.5: Missing orders for 2*b* = *a*.

(ii) If $2b = a$, then from relation $n = 3m = 3, 6, 9$, etc. [$\because m = 1, 2, 3, \cdots$]. Thus, the third, sixth, ninth interference maxima will be absent. The central diffraction maximum will contain 5 interference maxima [Fig. 5.6].

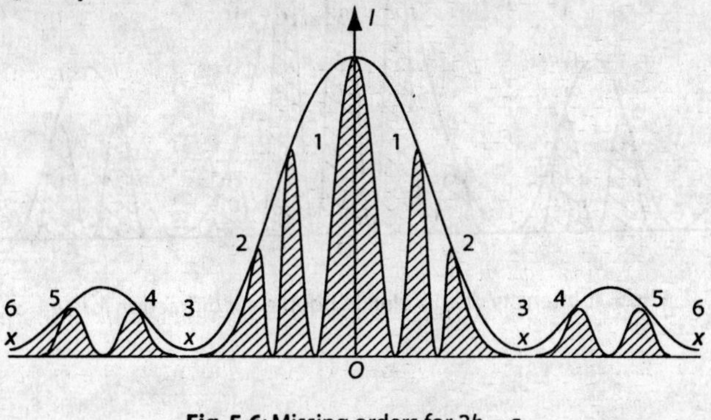

Fig. 5.6: Missing orders for 2*b* = *a*.

(iii) If $a = 0$, i.e., $a + b = b$, then $n = m = 1, 2, 3, \cdots$ all the interference maxima will be absent. The diffraction pattern in this case is similar to that due to a single-slit of width $2b$.

Intensity distribution curve: Fig. 5.7(a) represents the intensity (I_2) distribution due to interference between two secondary wavelets coming from two corresponding points, Fig. 5.7(b) represents the intensity (I_1) distribution of diffraction due to a single-slit and the resultant intensity ($I = I_1 \times I_2$) distribution due to a double-slit for $2b = b$ is shown by Fig. 5.7(c).

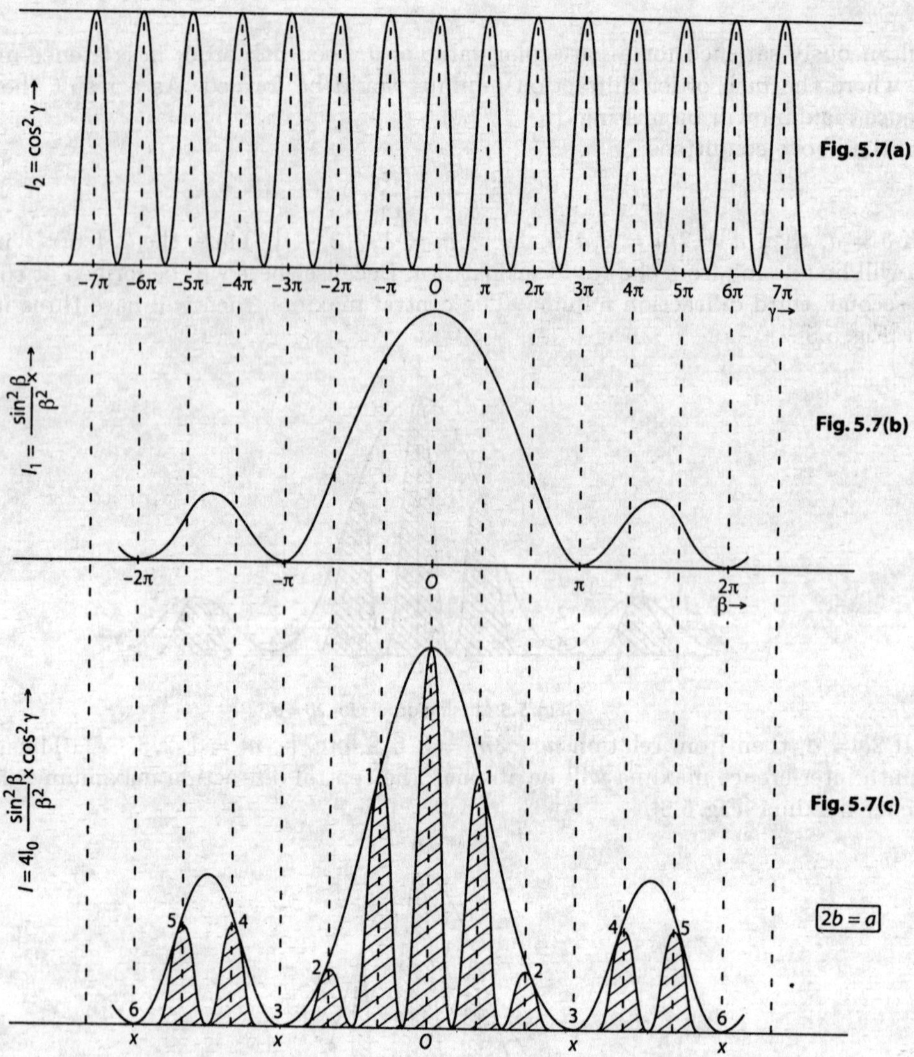

Fig. 5.7: Intensity distribution in double-slit diffraction for $2b = a$.

5.6 Fraunhofer Diffraction Due to N-slits (Theory of a Diffraction Grating)

A *diffraction grating* is an arrangement of a large number of equidistant parallel rectangular slits of equal width separated by equal opaque partitions. A *plane transmission grating* may be formed by ruling equidistant parallel lines with a fine diamond point on an optically plane glass-plate. The ruled lines are opaque to light while the space in between two lines is transparent.

Let $ABCDEFGH$ represents the section of a plane transmission grating supposed perpendicular to the plane of the paper [Fig. 5.8]. Let the points A and C, C and E, E and G, etc. be separated by a distance $d = a + b$, where $a =$ the width of an opaque space and $b =$ that of the transparent space. The points separated by d are called *pair of corresponding points.*

Let a plane wavefront of monochromatic light be incident normally on the surface of the grating. The diffracted light would consist of a large number of secondary wavelets issuing from the clear spaces of the grating. From the different points of the incident wavefront, large number of rays are diffracted in various directions. The lens L focuses the diffracted rays at an angle of diffraction θ at P on the screen. The point P will be either bright or dark according as the rays produce constructive or destructive interference with each other [Fig. 5.8].

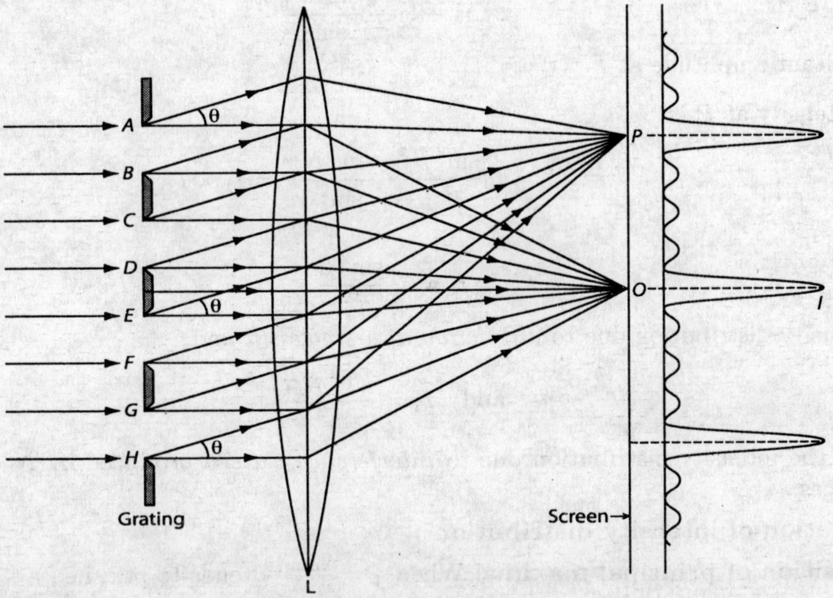

Fig. 5.8: Fraunhofer diffraction due to N-slits.

Theory: If each slit of the grating consists of N equally spaced point sources separated by a distance Δ, then the resultant field at P will be due to the superposition of the waves $E_1, E_2, E_3, \cdots, E_N$ from the slits $AB, CD, EF, \cdots$, where

$$E_1 = A\frac{\sin\beta}{\beta}\cos(\omega t - \beta)$$

$$E_2 = A\frac{\sin\beta}{\beta}\cos[\omega t - \beta - \phi_1]$$

$$E_3 = A\frac{\sin\beta}{\beta}\cos[\omega t - \beta - 2\phi_1]$$

....

$$E_n = A\frac{\sin\beta}{\beta}\cos[\omega t - \beta - (N-1)\phi_1],$$

where $\beta = \dfrac{\pi b \sin\theta}{\lambda}$ and $\phi_1 = \dfrac{2\pi d \sin\theta}{\lambda}$.

$\therefore$ the resultant field at P is

$$\begin{aligned} E &= E_1 + E_2 + E_3 + \cdots + E_N \\ &= A\frac{\sin\beta}{\beta}[\cos(\omega t - \beta) + \cos(\omega t - \beta - \phi_1) + \cdots + \cos\{\omega t - \beta - (N-1)\phi_1\}] \\ &= A\frac{\sin\beta}{\beta}\frac{\sin N\gamma}{\sin\gamma}\cos\left[\omega t - \beta - \frac{1}{2}(N-1)\phi_1\right], \end{aligned} \tag{5.31}$$

where

$$\gamma = \frac{\phi_1}{2} = \frac{\pi d \sin\theta}{\lambda}. \tag{5.32}$$

$$\therefore E = \mathcal{A}\cos\left[\omega t - \beta - \frac{1}{2}(N-1)\phi_1\right], \tag{5.33}$$

where

$$\mathcal{A} = \frac{A\sin\beta}{\beta}\frac{\sin N\gamma}{\sin\gamma} \tag{5.34}$$

is the resultant amplitude at P.

The intensity at P is

$$I = \mathcal{A}^2 = I_0\frac{\sin^2\beta}{\beta^2} \times \frac{\sin^2 N\gamma}{\sin^2\gamma} = I_1 \times I_2, \tag{5.35}$$

where

$$I_1 = I_0\frac{\sin^2\beta}{\beta^2} \tag{5.36}$$

is the intensity distribution due to *diffraction at a single-slit* and

$$\text{and} \quad I_2 = \frac{\sin^2 N\gamma}{\sin^2\gamma} \tag{5.37}$$

represents the intensity distribution due to *interference pattern* produced by N equally spaced point sources.

Interpretation of intensity distribution

(a) **Position of principal maxima**: When $\gamma = \frac{\pi d \sin\theta}{\lambda}$ tends to $m\pi$, i.e.,

$$\lim_{\gamma\to m\pi}\frac{\sin N\gamma}{\sin\gamma} = \lim_{\gamma\to m\pi}\frac{N\cos N\gamma}{\cos\gamma} = \pm N. \tag{5.38}$$

Then the resultant amplitude is

$$E = N\frac{A\sin\beta}{\beta} \tag{5.39}$$

and intensity

$$I = N^2 I_0\frac{\sin^2\beta}{\beta^2}, \text{ [where } I_0 = A^2 \text{ and } \beta = \frac{\pi b\sin\theta}{\lambda}\text{]}. \tag{5.40}$$

Again, for $\gamma = \frac{\pi d\sin\theta}{\lambda} = m\pi$, we have

$$d\sin\theta = m\lambda \quad [m = 1, 2, 3, \cdots]. \tag{5.41}$$

$$\therefore \sin\theta = \frac{m\lambda}{d}.$$

So

$$\beta = \frac{\pi b}{\lambda} \times \frac{m\lambda}{d} = \frac{\pi b m}{d}. \tag{5.42}$$

Eqn (5.40) represents the maximum intensity and hence such maxima are called the *principal maxima* with intensity

$$I_{pm} = N^2 I_0\frac{\sin^2\beta}{\beta}, \tag{5.43}$$

the condition for such maxima being

$$d\sin\theta = m\lambda. \tag{5.44}$$

Actually at the position of these principal maxima the fields $E_1, E_2, \cdots, E_N$ are all in phase and so they add up to give the resultant field N times the field due to a single-slit. That is why the intensity I_{pm} is of large value, if $\frac{\sin^2\beta}{\beta^2}$ is not very small.

The number of principal maxima: From the condition of occurrence of the principal maxima we have $d\sin\theta = m\lambda$ or, $\sin\theta = \frac{m\lambda}{d}$.

But $|\sin\theta| \leq 1$,

$$\therefore \frac{m\lambda}{d} \leq 1 \text{ or, } m \leq \frac{d}{\lambda}. \tag{5.45}$$

Thus, the *order number of the principal maxima is finite and is given by* $m \leq \frac{d}{\lambda}$, *the maximum value being*

$$m = \frac{d}{\lambda}. \tag{5.46}$$

(b) **Diffraction minima**: At diffraction minima,

$$I = 0 \text{ or, } I_0\frac{\sin^2\beta}{\beta^2} \times \frac{\sin^2 N\gamma}{\sin^2\gamma} = 0.$$

If either

$$\beta = \frac{\pi b\sin\theta}{\lambda} = n\pi, \text{ i.e., } b\sin\theta = n\lambda\ (n = 1, 2, 3, \cdots) \tag{5.47}$$

$$\text{or, } N\gamma = s\pi,\ s \neq N, 2N, \cdots \tag{5.48}$$

Eqn (5.46) gives the condition for minima due to single-slit diffraction pattern. These minima are called the *diffraction minima of N-slit diffraction pattern.* The direction of these minima are given by

$$N\gamma = s\pi \quad (s \pm N, 2N, \cdots)$$

$$\text{or, } \frac{N\pi d\sin\theta}{\lambda} s\pi$$

$$\text{or, } d\sin\theta = \frac{s}{N}\lambda. \tag{5.49}$$

If we allow $s = N, 2N$, etc., then $d\sin\theta = \lambda, 2\lambda$, etc., which represents the *condition of the principal maxima.*

Finally, the condition of minima is given by

$$d\sin\theta = \frac{\lambda}{N}, \frac{2\lambda}{N}, \cdots, \frac{(N-1)\lambda}{N}, \cdots, \frac{(N+1)\lambda}{N}, \cdots, \frac{(2N-1)\lambda}{N}, \frac{(2N+1)\lambda}{N}, \frac{(2N+2)\lambda}{N}, \cdots. \tag{5.50}$$

(c) **Secondary maxima**: In the diffraction pattern produced by a grating having N number of slits, there are at least $(N-1)$ minima between two principal maxima. Between two such consecutive minima, there are $(N-2)$ maxima.

Since

$$I_2 = \frac{\sin^2 N\gamma}{\sin^2\gamma}.$$

$$\therefore \frac{dI_2}{d\gamma} = 2\left(\frac{\sin N\gamma}{\sin\gamma}\right)[N\cot N\gamma - \cot\gamma]$$

$$\text{and } \frac{d^2I_2}{d\gamma^2} = 2I_2[N\cot N\gamma - \cot\gamma] + 2I_2\left(1 - N^2 + \cot^2\gamma - N^2\cot^2 N\gamma\right).$$

Thus, if $\frac{dI_2}{d\gamma} = 0$, then $\frac{d^2I_2}{d\gamma^2}$ = negative, i.e., I_2 is maximum.

The condition of this maximum is $\frac{dI_2}{d\gamma} = 0$

$$\text{or,} \quad N \cot N\gamma - \cot \gamma = 0 \quad \text{or,} \quad N \cot N\gamma = \cot \gamma$$
$$\text{or,} \quad N \tan \gamma = \tan N\gamma. \tag{5.51}$$

Eqn (5.51) also represents the condition for secondary maxima. The value of γ satisfying equation $N \tan \gamma = \tan N\gamma$ will give the positions of the secondary maxima except $\gamma = m\pi$, where the principal maxima have already been formed.

Now, $\because N \tan \gamma = \tan N\gamma$

$$\text{or,} \quad \frac{N^2 \cos^2 N\gamma}{\cos^2 \gamma} = \frac{\sin^2 N\gamma}{\sin^2 \gamma}$$

$$\text{or,} \quad \frac{N^2(1 - \sin^2 N\gamma)}{1 - \sin^2 \gamma} = \frac{N^2 \sin^2 N\gamma}{N^2 \sin^2 \gamma} = \frac{N^2}{1 + (N^2 - 1)\sin^2 \gamma} \quad \left[\because \frac{a}{b} = \frac{c}{d} = \frac{a+c}{b+d}\right].$$

Hence, the intensity of the secondary maxima is

$$I_{sm} = I_1 \times \frac{N^2}{1 + (N^2 - 1)\sin^2 \gamma} = \frac{I_{pm}}{1 + (N^2 - 1)\sin^2 \gamma} \quad [\because I_{pm} = I_1 \times N^2].$$

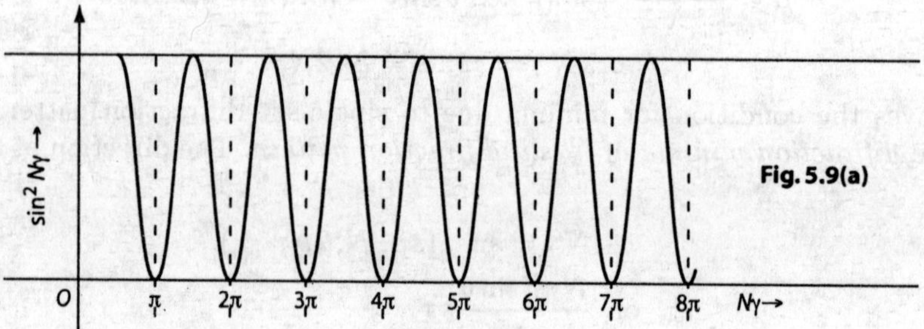

Fig. 5.9: Intensity distribution in the diffraction pattern of a plane transmission grating.

$\therefore$ the ratio of the intensities of the principal maxima to the secondary maxima is

$$\frac{I_{pm}}{I_{sm}} = 1 + (N^2 - 1)\sin^2 \gamma. \tag{5.52}$$

Thus, for $N \to \infty, I_{pm} \to \infty$ and in that case, $I_{sm} \to 0$, i.e., the secondary maxima will be invisible.

Fig. 5.9 shows the intensity distribution in the diffraction pattern of a plane transmission grating with $N = 6$.

5.6.1 Spectrum Produced by a Grating

We have the condition of the principal maxima of the diffraction spectrum produced by a grating of grating element $d = a + b$ is $d \sin\theta = m\lambda$, where λ is the wavelength of light, θ the angle of diffraction and m is the order number of the spectrum.

For the first order spectrum, $m = 1$ and if θ_1 be the angle of diffraction, then $d\sin\theta_1 = \lambda$.

Hence, the first order spectrum is obtained in a direction inclined at an angle θ_1 with the direction of incidence.

Similarly, for the second order spectrum $d\sin\theta_2 = 2\lambda$, where $m = 2$ and the angle of diffraction $\theta = \theta_2$.

In this way different order of spectra are obtained on both sides of the direct ray as shown in Fig. 5.10.

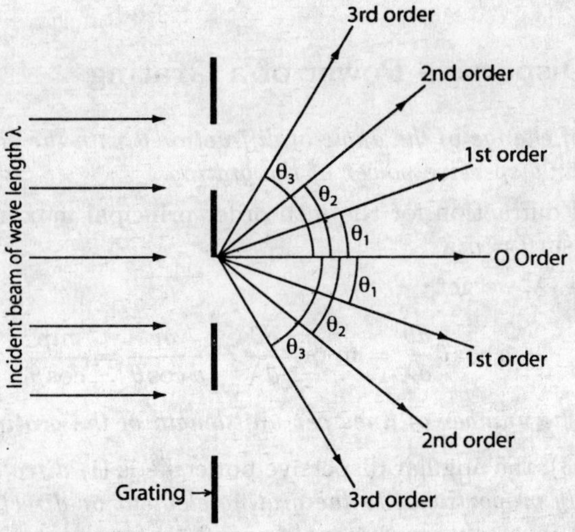

Fig. 5.10: Grating spectrum for monochromatic source.

Fig. 5.11 shows the diffraction pattern if the source of light be *white*. Here in each order of diffraction the different colours corresponding to different wavelengths are diffracted at different angles. The conditions for the principal maxima for the first order ($n = 1$) are

$$d\sin\theta_v = \lambda_v \text{ (for violet light)}$$
$$d\sin\theta_r = \lambda_r \text{ (for red light).}$$

Since $\lambda_r > \lambda_v$, $\theta_r > \theta_v$, there is no overlapping of the different colours in the grating spectra, which does occur in a prismatic spectrum.

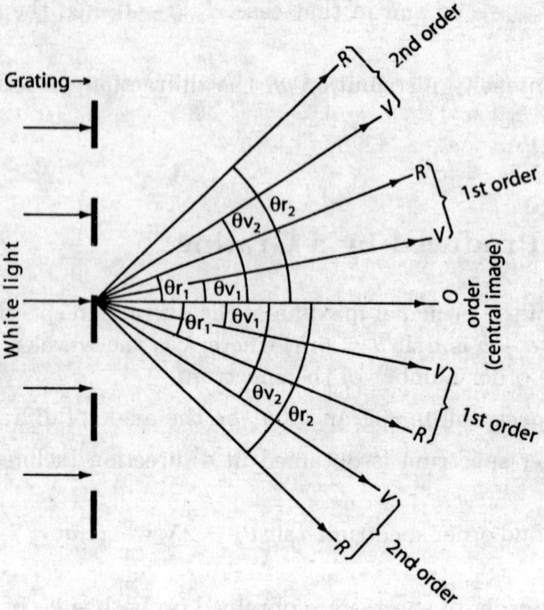

Fig. 5.11: Grating spectrum for white light.

5.6.2 Angular Dispersive Power of a Grating

Definition: *The rate of change of the angle of diffraction θ with the change in wavelength λ (i.e., $\frac{d\theta}{d\lambda}$) is called the angular dispersive power of the grating.*

If θ be the angle of diffraction for the mth order principal maximum of wavelength λ, then for normal incidence $d\sin\theta = m\lambda$.

Differentiating w.r.t. λ, we get

$$d\cos\theta\frac{d\theta}{d\lambda} = m \text{ or, } \frac{d\theta}{d\lambda} = \frac{m}{d\cos\theta} = \frac{mp}{\cos\theta}, \tag{5.53}$$

where $p = \frac{1}{d} = \frac{1}{a+b}$ *is the number of lines per unit length of the grating.*

Thus, from eqn (5.53), the angular dispersive power $\frac{d\theta}{d\lambda}$ is (i) *directly proportional to the order number m*, (ii) *inversely proportional to the grating element or directly proportional to the lines per unit length* (p), (iii) *inversely proportional to* $\cos\theta$. So for a large value of $\theta, \cos\theta$ is small and $\frac{d\theta}{d\lambda}$ is large.

Normal spectrum: If λ and $\lambda + d\lambda$ be the wavelengths of two spectral lines separated by dx in the focal plane of the telescope objective, then

$$dx = f\,d\theta \text{ [} f \text{ is the focal length of the telescope objective].}$$

$$\therefore \frac{dx}{d\lambda} = f\frac{d\theta}{d\lambda} = f\frac{mp}{\cos\theta} \quad \text{or,} \quad dx = f\frac{mp}{\cos\theta}d\lambda. \tag{5.54}$$

Thus, for a particular value of f, p and λ, $dx \propto m$, i.e., dx is larger for higher order number.

5.6.3 Determination of Wavelength of Monochromatic Light by Grating

Theory: If λ be the wavelength of the light and θ be the angle of diffraction for the mth order principal maximum, then $(a+b)\sin\theta = m\lambda$, where $(a+b)$ is the grating element and $\frac{1}{a+b} = p$ is the number of lines per unit length. Then

$$\lambda = \frac{\sin\theta}{mp}. \tag{5.55}$$

Measuring θ for the order number m of the diffraction maxima with the help of a spectrometer we can calculate the unknown wavelength of the monochromatic light by eqn (5.55).

Adjustment of spectrometer: Before performing the experiment, the following adjustment are required: (i) The spectrometer is first levelled and focussed for parallel rays. (ii) Then the grating is placed on the prism table with the surface normal to the parallel rays emerging from the collimator, the collimator-slit being illuminated with the given monochromatic source of light. The rulings of the grating are also made parallel to the vertical collimator-slit [Fig. 5.12].

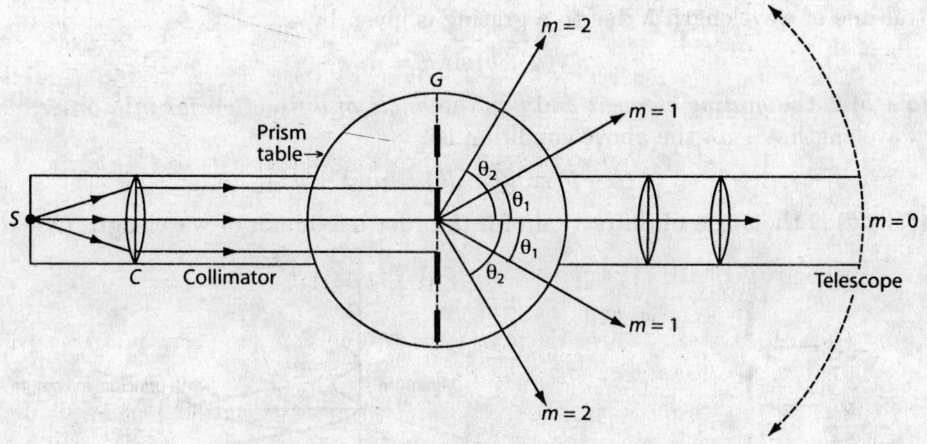

Fig. 5.12: Diffraction spectrum produced by a grating.

Measurement of θ: The telescope of the spectrometer is rotated to receive a given order spectrum. The vertical cross-wire of the eye-piece is made to coincide with the line whose λ is to be determined. The spectrometer reading is carefully noted. Next, the telescope is rotated to the other side of the direct beam and the spectrum of the same order is received. The vertical cross-wire is again made to coincide with the line corresponding to λ and reading of the spectrometer is noted. If θ_m and θ'_m be the readings of the spectrometer in the two positions of the telescope and θ, the angle of diffraction corresponding to the mth order spectrum, then

$$\theta_m - \theta'_m = 2\theta \text{ and } \theta = \frac{\theta_m - \theta'_m}{2}.$$

Putting this value of θ and knowing m and p, the wavelength λ of the given monochromatic light can be determined using eqn (5.55).

5.6.4 Rayleigh Criterion of Resolution

According to Rayleigh, *two point objects lying close to each other are said to be just resolved by an optical instrument if the central maximum in the diffraction pattern of one falls on the first minimum in the pattern of the other.* This is *Rayleigh's criterion resolution.*

In case of two close spectral lines of wavelength λ and $\lambda + d\lambda$, where $d\lambda$ is very small, *the two spectral lines will be just resolved when the central maximum of* λ *falls on the first minimum due to* $\lambda + d\lambda$.

In other words, *the least angular separation between the principal maxima of the diffraction pattern of the two spectral lines in a given order should be equal to half the angular width of either principal maximum.*

5.6.5 Resolving Power of a Grating

Resolving power: The ability of a grating to distinguish two very close spectral lines of wavelengths λ and $\lambda + d\lambda$ is called its *resolving power.* The resolving power of a grating is given by

$$\text{RP} = \frac{\lambda}{d\lambda},$$

where $d\lambda$ is the smallest wavelength difference between λ and $\lambda + d\lambda$.

Expression for resolving power: For normal incidence, the mth order principal maximum of spectral line of wavelength λ due to a grating is given by

$$(a+b)\sin\theta = m\lambda, \tag{5.56}$$

where $(a+b)$ is the *grating element* and θ is the *angle of diffraction for* m*th order.*

For wavelength $\lambda + d\lambda$ the above condition is

$$(a+b)\sin(\theta + d\theta) = m(\lambda + d\lambda), \tag{5.57}$$

where $(\theta + d\theta)$ is the angle of diffraction of mth order maximum of wavelength $(\lambda + d\lambda)$.

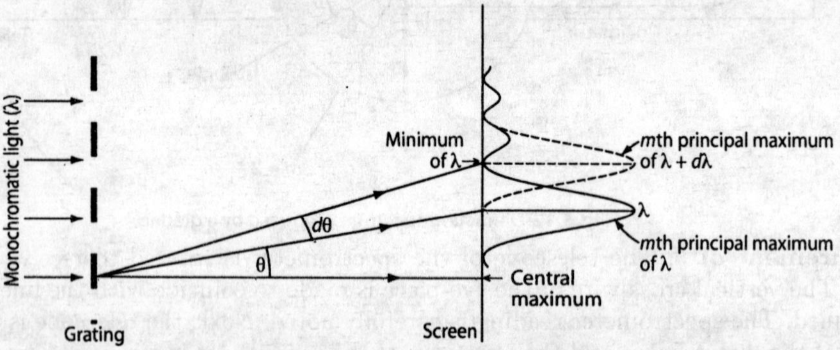

Fig. 5.13: Rayleigh criterion of resolution.

By Rayleigh's criterion of resolution the spectral line corresponding to λ and $\lambda + d\lambda$ will be just resolved if the position of mth order principal maximum of $\lambda + d\lambda$ coincides with the first minimum adjacent to mth order maximum of λ. Thus, the mth order principal maximum of $\lambda + d\lambda$ and the first minimum for λ should have the same angle of diffraction.

Again, in a grating, the condition of minima is

$$N(a+b)\sin\theta = s\lambda, \tag{5.58}$$

where N = total number of slits in the grating and $s \neq 0, N, 2N, \cdots$ because other integral values of s satisfy the condition of maxima.

Thus, the first minimum due to λ adjacent to mth principal maximum in the direction $(\theta + d\theta)$ is obtained by putting $s = (mN + 1)\lambda$.

The first minimum in the direction of $(\theta + d\theta)$ satisfies the condition

$$N(a+b)\sin(\theta + d\theta) = (mN+1)\lambda. \tag{5.59}$$

Again, from eqn (5.57),

$$N(a+b)\sin(\theta+d\theta) = mN(\lambda+d\lambda). \tag{5.60}$$

So for just resolution of the spectral lines of λ and $\lambda + d\lambda$, we have

$$(mN+1)\lambda = mN(\lambda+d\lambda) \text{ or, } \lambda = mNd\lambda \text{ or, RP} = \frac{\lambda}{d\lambda} = mN. \tag{5.61}$$

Thus, RP of a grating is (i) *proportional to the order number* m, (ii) *proportional to* N, (iii) *independent of* $(a+b)$.

Furthermore, from eqn (5.56), we have

$$m = \frac{(a+b)\sin\theta}{\lambda}.$$

$$\therefore\ \text{RP} = mN = \frac{N(a+b)\sin\theta}{\lambda} \text{ or, RP} = \frac{\lambda}{d\lambda} = W\cdot\frac{\sin\theta}{\lambda}, \tag{5.62}$$

where $W = N(a+b)$ represents the width of the grating.

Thus, RP of a grating is proportional to the width W. For normal incidence, $\theta = 90°$.

$$\therefore\ (\text{RP})_{\text{max}} = \left(\frac{\lambda}{d\lambda}\right)_{\text{max}} = \frac{W}{\lambda} = \frac{N(a+b)}{\lambda}. \tag{5.63}$$

5.6.6 Resolving Power of a Microscope

The ability of a microscope to form distinctly separate images of two neighbouring small objects is known as its *resolving power.*

The resolving power of a microscope is measured by the reciprocal of the least angular separation between the two close objects so that their images just to be separated if viewed through the microscope. So the resolving power of a microscope is given by

$$\text{RP} = \frac{1}{d\theta},$$

where *$d\theta$ is the least angular separation for the images* of two close objects to be separated and is called the *limit of resolution. Thus, smaller the limit of resolution, greater is the resolving power.*

Expression for resolving power: Let AB be the aperture of the microscope objective and P and Q be two self-luminous point objects separated by a distance d. P and Q are very close to the objective and d is normal to the axis as shown in Fig. 5.14.

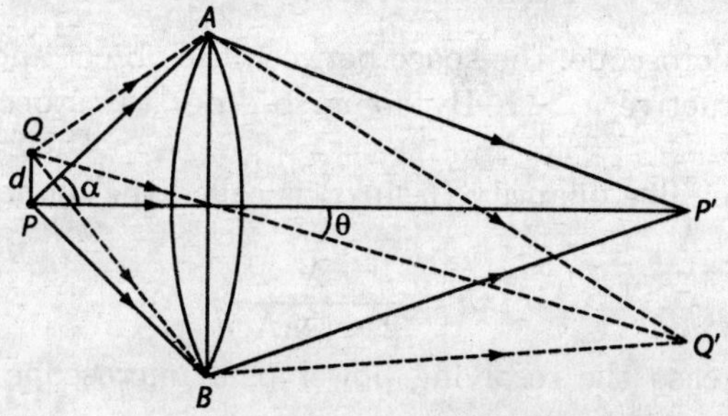

Fig. 5.14: Resolving power of a telescope.

The images of P and Q are the Fraunhofer diffraction pattern due to circular aperture of the objective. The images P' and Q' of P and Q consist of central bright discs surrounded by a number of alternate dark and bright rings.

According to Rayleigh's criterion of resolution, *the objects P and Q will be just resolved if the first dark ring of Q' passes through the centre of the principal bright disc of P'*. For this to occur the path difference between the extreme rays QBP' and PAP' should be 1.22λ [Fig. 5.14].

Calculation of path difference: Path difference between the rays QBP' and PAP' is given by

$$\begin{aligned}\Delta &= (QB + BP') - (QA + AP') \quad \text{[Fig. 5.14]}\\ &= QB - QA \ (\because BP = AP)\\ &= (QD + DB) - (CA - CQ) \quad \text{[Fig. 5.15]}\\ &= (QD + PB) - (PA - CQ) \quad [\because DB = PB = PA = CA]\\ &= QD + CQ,\end{aligned}$$

where $PC \perp CA$ and $PD \perp QB$. [Fig. 5.15]

$$\therefore \Delta = QD + CQ = d \sin\alpha + d\sin\alpha = 2d\sin\alpha \quad [\because PQ = d],$$

where α is the semi-vertical angle of the cone of rays incident on AB.

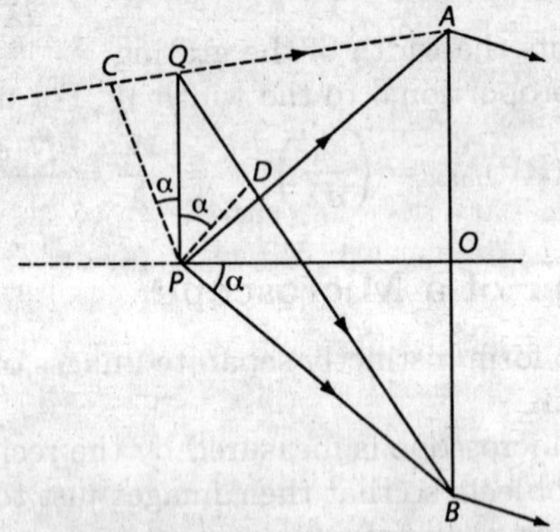

Fig. 5.15: Calculation of path difference.

So the images P' and Q' will be just resolved, if

$$2d\sin\alpha = 1.22\lambda \text{ or, } d = \frac{1.22\lambda}{2\sin\alpha},$$

where d is the limit of resolution.

In case the objects are illuminated by external source of light, then *Abbe* showed that

$$2d\sin\alpha = \lambda \text{ or, } d = \frac{\lambda}{2\sin\alpha}.$$

In a high resolution microscope, the space between the object and the objective is filled with *cedar wood oil* whose refractive $n > 1$. By doing so d decreases because, then $d = \frac{\lambda}{2n\sin\alpha}$ and resolving power increases.

Numerical aperture: The quantity ($n\sin\alpha$) is called the *numerical aperture* (N.A.) of the microscope. In terms of N.A.,

$$d = \frac{\lambda}{2 \times (\text{N.A.})}.$$

In conclusion, to increase the resolving power of a microscope (i) λ *should be small* and (ii) *N.A. should be high.*

N.B. (i) The wavelength λ for ultraviolet light being small, in *ultramicroscope* this radiation is used for greater resolving power. (ii) Resolving power of electron microscope (where electron beam is used instead of white light) is higher as compared to optical microscopes.

5.7 Answers of Some Important Questions

5.1. *What do you mean by diffraction of light?*

Ans. The phenomenon of bending of light waves around obstacles or apertures of sizes comparable with the wavelengths of light and resulting thereby in their spreading into the geometrical shadow of the objects is called *diffraction of light.*

There are two classes of diffraction, viz., (i) *Fresnel class* and (ii) *Fraunhofer class of diffraction.*

5.2. *State the difference between interference and diffraction.*

Ans. See Section 5.2 (p. 122).

5.3. *How Fraunhofer class of diffraction differs from Fresnels class?*

Ans. See Section 5.3 (p. 123).

5.4. *Sound wave can undergo diffraction at the corner of a building but not light wave. Why?*

Ans. The condition that a wave can undergo diffraction is that *its wavelength must be comparable with the dimension of the diffracting obstacle.* Now, the wavelength of sound wave is ~ 320 m/s whereas that light wave is $\sim 5893 \times 10^{-8}$ cm. So the sharpness of the corner of a building satisfies the condition of diffraction for the sound wave but not for the light wave. Hence, though the corner of a building can diffract sound wave, it fails to diffract light wave.

5.5. *What happens to the single-slit diffraction pattern if the slit width is gradually increased?*

Ans. The single-slit diffraction pattern consists of alternate bright and dark bands with monochromatic light of wavelength λ. In case of white light, the central maximum is white and on either side of it, the diffraction bands are coloured. At a large distance from the central band, general illumination occurs due to overlapping of different colours.

For the 1st order minima, $\sin\theta = \frac{\lambda}{a}$, where θ = the angle of diffraction and a = the slit width. If the width a of the slit is large, then $\sin\theta$ and hence θ will be small; as a result the maxima and minima will be formed very close to the central maximum. On the other hand, if the slit width a is small, then the maxima and minima will be quite wide on either side of the central maximum. Hence, if the slit width of the single-slit be gradually increased, the diffraction maxima and minima will come closer and closer.

5.6. *In single-slit diffraction experiment, readings for the position of minima are taken. Why?*

Ans. Secondary maxima, in single-slit diffraction pattern, are *unequally spaced* on either side of the central maximum, whereas the minima are *equally spaced.* That is why in single-slit experiment reading for the positions of minima are taken.

5.7. *Show that double-slit interference pattern is the limiting case of double-slit diffraction pattern when the slit width tends to zero.*

Ans. The intensity distribution in the double-slit diffraction pattern is given by

$$I = 4I_0 \frac{\sin^2\beta}{\beta^2}\cos^2\gamma,$$

where

$$\beta = \frac{\pi b}{\lambda}\sin\theta \text{ and } \gamma = \frac{\pi}{\lambda}\sin\theta,$$

where b = width of the single-slit and d = the distance between two corresponding points in the double-slit, θ = the angle of diffraction.

Now, $I_0\frac{\sin^2\beta}{\beta^2}$ represents the intensity distribution due to a single-slit and $\cos^2\gamma$ represents the intensity due to interference pattern.

If the slit width b is very small so that

$$\lim_{\beta\to 0}\frac{\sin^{2\beta}}{\beta^2} \to 1,$$

then $I = 4I_0\cos^2\gamma$, which is the intensity distribution in the *double-slit interference pattern when* $b \to 0$.

5.8. *Show that if the width of the opaque space (a) of a double-slit be zero, then the double-slit diffraction pattern will be reduced to a single-slit diffraction pattern.*

Ans. If m and n be respectively the *order of diffraction minima* and *interference maxima* in the diffraction pattern due to a double-slit of width b, then nth order interference maxima will be missed if

$$\frac{b+a}{b} = \frac{n}{m}.$$

If the width of the opaque space $a = 0$, then $n = m$, $m = 1, 2, 3, \cdots$.

This means that all the orders of interference maxima will be missing from the double-slit diffraction pattern. Only the diffraction pattern of a single-slit of width $2a$ will be observed.

5.9. *What is the maximum value of the order number of principal maxima that can be obtained using a grating of N number of rulings?*

Ans. Vide Section 5.6, (p. 132).

5.10. *If the grating element $(a+b)$ of a grating becomes of the order of wavelength λ, what effect would be observed?* **[HPU 1994]**

Ans. If θ_m be the angle of diffraction for the mth order principal maximum due to a grating of element $(a+b)$, then $(a+b)\sin\theta_m = m\lambda$.

For the first order $m = 1$, $\therefore \sin\theta_1 = \frac{\lambda}{(a+b)}$.

If $(a+b) = \lambda$, then $\sin\theta_1 = 1$ or, $\theta_1 = 90°$.

This is the theoretical limit and the first order maximum is not possible to be observed.

Again, if $(a+b) < \lambda$, then $\sin\theta_1 > 1$, which is impossible. Such a grating is, therefore, unable to produce spectrum and hence is of no use.

5.11. *Give the difference between double-slit diffraction pattern and double-slit interference pattern.*

Ans. The following table gives the difference between the above two patterns.

Table 5.3

Double-slit diffraction pattern	*Double-slit interference pattern*
1. It consists of a central maximum with equally spaced interference maxima and minima on either side.	1. It consists of a number of interference maxima and minima.
2. The intensity of secondary maxima on either side of the principal maximum decreases gradually.	2. The intensity of the interference maxima is almost constant.
3. The intensity of the central maximum is four times the intensity of the central maximum due to a single-slit.	3. There is no such central maximum.
4. The minima on either side of the central maximum are equally spaced.	4. The interference maxima and minima are equally spaced.
5. The spacing of diffraction maxima and minima depends on the slit with b.	5. The spacing of interference maxima and minima depends on $(a+b)$, where a is the slit separation.
6. The intensity of interference maxima and minima is not constant due to the overlapping of diffraction maxima and minima and reduces to zero on either side of the central maximum.	6. There is no such overlapping of interference maxima and minima and diffraction maxima and minima.
7. The double-slit diffraction pattern reduces to single-slit diffraction pattern when $a \to 0$.	7. The double-slit diffraction pattern reduces to Young's double-slit interference pattern when $b \to 0$.

5.12. *On which factors, the resolving power of a plane transmission grating depends?*
Ans. The resolving power (RP) of a plane transmission grating is given by

$$\text{RP} = \frac{\lambda}{d\lambda} = nN,$$

where n is the order number of the spectrum and N is the total number of lines in the grating.

From the above relation it is evident that (i) RP *increases with the order* (n) *of the spectrum* (ii) RP *increases with the total number of lines in the effective part of the grating.*

RP of a grating does not depend on the grating element $(a + b)$.

5.13. *What are limit of resolution and numerical aperture of a microscope? What factors govern the resolving power of a microscope?*
Ans. *Limit of resolution*: The minimum distance by which two points in the object are separated from each other so that their images produced by the microscope and just seen as separate is called its *limit of resolution.*

In *oil immersion type objective* the limit of resolution x is given by

$$x = \frac{\lambda_0}{2n \sin \theta},$$

where λ_0 is the wavelength of light in vacuum, n is the r.i. of the *cedar wood oil* and θ is the semi-vertical angle of the cone of light rays received by the objective.

The quantity $n \sin \theta$ is called the *numerical aperture* (NA) of the microscope. It is seen that the limit of resolution (x) is (i) proportional to λ_0 and (ii) inversely proportional to the NA of the microscope.

The resolving power of the microscope is given by

$$\text{RP} = \frac{1}{x} = \frac{2(\text{NA})}{\lambda_0}.$$

Hence, the resolving power (RP) of the microscopic is (i) directly proportional to NA and (ii) inversely proportional to λ_0.

5.14. *What is Rayleigh criterion of resolution?*
Ans. See Section 5.6.4 (p. 139).

5.8 Solved Problems

Ex. 1 *In an arrangement of Fraunhofer diffraction we use a slit of width* 0.2 mm *and first minimum is at* 5 mm *on either side of the central maximum. If the distance between the lens and the screen is* 2 m, *calculate the wavelength of light.* **[HPU 1997]**

Soln. The distance of the first secondary minimum from the centre of the principal maximum is

$$x = \lambda \frac{D}{a}, \quad x = 5 \text{ mm} = 0.5 \text{ cm},$$

where D = distance between the slit and the screen = $2m$ = 200 cm: a = width of the slit = 0.2 mm = 0.02 cm.

∴ wavelength of light

$$\lambda = \frac{ax}{D} = \frac{2 \times 10^{-2} \times 5 \times 10^{-1}}{200} = 5 \times 10^{-5} \text{ cm} = \mathbf{5000\ A}.$$

Ex. 2 *A screen is placed* 2 m *away from a narrow slit. Find the slit width if the first minimum lies* 5 mm *on either side of the central maximum when plane waves of* 5000 Å *are incident on the screen.*

Soln. The slit width a is given by

$$a = \frac{D\lambda}{x},$$

where D = distance between the screen and the slit = 2m = 200 cm; x = distance of the first minimum from the centre of the central maximum = 5 mm = 0.5 cm and λ = the wavelength of light = 5000×10^{-8} cm.

$$\therefore a = \frac{200 \times 5000 \times 10^{-8}}{0.5} = \mathbf{0.02\ cm.}$$

Ex. 3 *A convex lens of focal length of* 20 cm *is placed after a slit of width* 0.6 mm. *If a plane wave of wavelength* 6000 Å *falls on the slit normally, calculate the separation between the second minima on either side of the central maximum.* **[PU 1996]**

Soln. The condition of nth minimum on the screen is $a \sin\theta_n = n\lambda$.

Here λ = the wavelength of light = 6000 Å = 6000×10^{-8} cm = 6×10^{-5} cm; $n = 2$ and a = the width of the slit = 0.6 mm = 0.06 cm.

$$\therefore \sin\theta_2 = \frac{2\lambda}{a} = \frac{2 \times 6 \times 10^{-5}}{0.06} = 0.002.$$

The screen is placed on the focal plane of the lens [Fig. 5.16].

$$\therefore \sin\theta_2 = \tan\theta_2 = \frac{x}{f} = \frac{2\lambda}{a},$$

where x is the distance between the centre of the principal maximum and the second minimum.

$$\therefore x = \frac{2\lambda}{a} \times f = 0.002 \times 20 = 0.04\ \text{cm}.$$

The separation between the second minima = $2x = 0.04 \times 2 =$ **0.08 cm.**

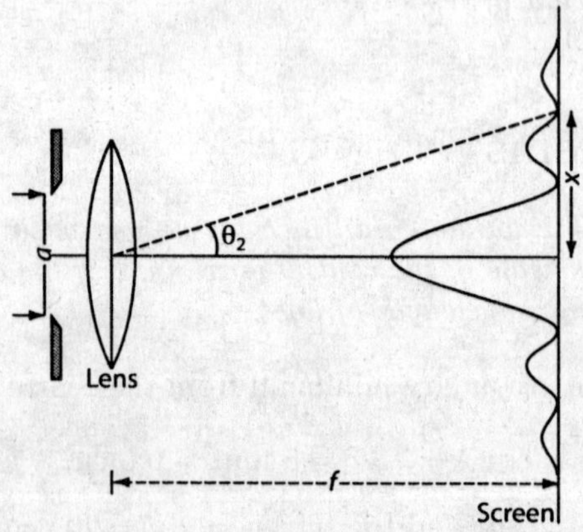

Fig. 5.16

Ex. 4 *A lens whose focal length is* 40 cm *forms a Fraunhofer diffraction pattern of a slit* 0.3 mm *width. Calculate the distances of the first dark band and of the next bright band from the axis [wavelength of the light used* = 5890 Å].

Soln. In a single-slit the position of nth minimum is given by $a \sin\theta = n\lambda$, $n = 1, 2, 3, \cdots$.

For the first minima, the angle of diffraction is given by

$$\sin\theta = \frac{n\lambda}{a} = \frac{\lambda}{a} \quad (\because n = 1)$$
$$= \frac{5890\times10^{-8}}{0.03} \quad [\because a = 0.3 \text{ mm} = 0.03 \text{ cm}]$$
$$= 0.00196.$$

Since θ is very small, $\sin\theta \simeq \theta = 0.00196$ radian. If x be the linear distance of the first dark band from the axis, then $\theta = \frac{x}{f}$ or, $x = f\theta$, where f = the focal length of the lens = 40 cm.

$\therefore x = 40 \times 0.001916 =$ **0.0766 cm.**

If θ_1 be the angle of diffraction of the first bright band on either side of the central maximum, then

$$a\sin\theta_1 = \frac{3\lambda}{2} \text{ or, } \sin\theta_1 = \frac{3\lambda}{2a} = \frac{3}{2}\times 0.00196 = 0.00294.$$
$$\therefore \theta_1 \simeq \sin\theta_1 = 0.00294 \text{ radian.}$$

The linear distance of the first bright band from the central maximum is $y_1 = f\theta_1 = 40 \times 0.00294 = 0.1176$ cm.

$\therefore$ the distance between the *first dark band* and the *next bright band* is $dy = y_1 - y = 0.1176 - 0.0766 =$ **0.041 cm.**

Ex. 5 *A slit of width* 12×10^{-5} cm *is illuminated with light of wavelength* 6000 Å. *Find the half-angular width of the central bright band.*

Soln. The condition for the first minimum from the central maximum is $b\sin\theta = \lambda$.

$$\therefore \sin\theta = \frac{\lambda}{b} \text{ or, } \theta = \sin^{-1}\left(\frac{\lambda}{b}\right).$$

Here wavelength of light λ = 6000 Å = 6000×10^{-8} cm and the width of the slit $b = 12\times10^{-5}$ cm.

$$\therefore \theta = \sin^{-1}\left(\frac{6000\times10^{-8}}{12\times10^{-5}}\right) = \sin^{-1}(0.5) = 30°.$$

So the half-angular width of the central maximum is **30°**.

Ex. 6 *Monochromatic light of wavelength λ falls normally on a single-slit of width b. Show that intensities of the first and second secondary maxima are approximately* 4.5% *of I_0 and* 1.62% *of I_0 respectively.*

Soln. The intensity of a secondary maximum

$$I = I_0\frac{\sin^2\beta}{\beta^2}, \quad I = I_{\max} \text{ for } \beta = (2n+1)\frac{\pi}{2},$$

where I_0 is the intensity of the *central maximum.*

Now, the first and second secondary maxima occur at $\beta = 3\pi/2$ and $\beta = 5\pi/2$ respectively.

$\therefore$ the intensity of the first secondary maximum is

$$I_1 = I_0\frac{\sin^2(3\pi/2)}{(3\pi/2)^2} = \frac{4I_0}{9\pi^2} \simeq 0.045I_0 \text{ or, } \frac{I_1}{I_0} \simeq 0.045.$$
$$\therefore \mathbf{I_1 = 4.5\% \text{ of } I_0}.$$

Similarly,

$$I_2 = I_0\frac{\sin^2(5\pi/2)}{(5\pi/2)^2} = I_0\frac{4}{25\pi^2} \text{ or, } I_2/I_0 = \frac{4\times49}{25\times22\times22} \simeq 0.0162.$$
$$\therefore \boldsymbol{I_2 = 1.62\% \text{ of } I_0}.$$

Ex. 7 *In the Fraunhofer diffraction due to a double-slit, a monochromatic light of wavelength* 5000 Å *is made incident normally on the plane of the double-slit of width* 0.02 mm *and the separation between the two slits is* 0.10 mm. *If the distance of the screen from the slits is* 100 cm, *then calculate* (i) *the distance between the central maximum and the first minimum of the fringe envelope and* (ii) *the distance between any two consecutive double-slit dark fringes.*

Soln. Here slit width $b = 0.02$ mm $= 0.002$ cm, slit separation $a = 0.1$ mm $= 0.01$ cm.
The wavelength of light $\lambda = 5000$ Å $= 5000 \times 10^{-8}$ cm $= 5 \times 10^{-5}$ cm.
The distance between the screen and the slit $D = 100$ cm.
(i) The angle of diffraction of the first maximum is given by

$$\sin\theta \simeq \theta = \frac{\lambda}{(a+b)}.$$

The angle of diffraction of the first minimum is given by

$$\sin\theta_1 \simeq \theta_1 = \frac{\lambda}{2(a+b)}.$$

If the distance of the first minimum from the central maximum on the screen is x_1, then

$$\theta_1 = \frac{x_1}{D} = \frac{\lambda}{2(a+b)} \text{ or, } x_1 = \frac{\lambda D}{2(a+b)} = \frac{5 \times 10^{-5} \times 100}{2(0.002 + 0.01)} = \frac{5 \times 10^{-3}}{2 \times 0.012} = \mathbf{0.2083\ cm.}$$

(ii) If θ_2 be the angle of diffraction for the second dark band, then

$$\sin\theta_2 \simeq \theta_2 = \frac{3\lambda}{2(a+b)}.$$

If x_2 be the linear distance of the second dark band from the central maximum, then

$$\theta_2 = \frac{x_2}{D} = \frac{3\lambda}{2(a+b)} \text{ or, } x_2 = \frac{3D\lambda}{2(a+b)} = \frac{3 \times 100 \times 5 \times 10^{-5}}{2 \times 0.012} = 0.6249 \text{ cm.}$$

$\therefore$ the distance between two consecutive double-slit dark bands is, $\Delta x = x_2 - x_1$
$= 0.6249 - 0.2083 =$ **0.4166 cm.**

Ex. 8 *The width of each slit of a double-slit is* 0.15 mm *and they are separated by a distance of* 0.45 mm. *If the double-slit produces Fraunhofer diffraction, find the missing orders.*

Soln. Here width of each slit $b = 0.15$ mm and the width of the opaque space $a = 0.45$ mm.
When the nth order interference maximum coincides with mth order diffraction minima, then

$$\frac{n}{m} = \frac{a+b}{b}.$$

Here $\frac{n}{m} = \frac{0.45+0.15}{0.15} = 4$ or, $n = 4m$.
For $m = 1, 2, 3$, etc. $n = 4, 8, 12, \cdots$, etc. This indicates that 4th, 8th, 12th, etc. orders of interference maxima will be missed from the diffraction spectra.

Ex. 9 *Fraunhofer double-slit diffraction pattern is observed in the focal plane of a lens of focal length* 0.5 m. *The wavelength of the incident light is* 500 nm. *The distance between two maxima adjacent to zero order maximum is* 5 mm *and the fourth order maximum is missing. Find the width of each slit and the distance between their centres.*

Soln. Since the 4th order maximum is missing, we have

$$\frac{a+b}{b} = 4 \text{ or, } a = 3b.$$

For the first order maximum, the condition is $(a+b)\sin\theta_1 = \lambda$.

$$\therefore \theta_1 \simeq \frac{\lambda}{a+b} = \frac{\lambda}{4b}.$$

Now, the distance between two maxima adjacent to zero order maximum is

$$x = f \times 2\theta_1 = f \times \frac{2\lambda}{4b} = f \times \frac{\lambda}{2b}.$$

$$\therefore b = \frac{f\lambda}{2x} = \frac{0.5 \times 500 \times 10^{-9}}{2 \times 5 \times 10^{-3}}\ \text{m} = \mathbf{0.025\ mm.}$$

$$\therefore a = 3b = 3 \times 0.025\ \text{mm} = \mathbf{0.075\ mm.}$$

Ex. 10 *Consider a diffraction grating of width* 5 cm *with slit width* 0.0001 cm *separated by a distance of* 0.0002 cm. *What is the corresponding grating element? How many orders would be observable, if* $\lambda = 5.5 \times 10^{-5}$ cm? **[WBUT 2001]**

Soln. The width of each slit $b = 0.0002$ cm.
Width of the opaque space $a = 0.0001$ cm.
$\therefore$ grating element $= (a+b) = (0.0001 + 0.0002)$ cm $=$ **0.0003 cm.**
Now, the condition of the mth order principal maximum is $(a+b)\sin\theta = m\lambda$.
For maximum order number, $\theta = 90°$ and $\sin 90° = 1$.
$\therefore$ maximum order number $m = \frac{a+b}{\lambda} = \frac{3\times10^{-4}}{5.5\times10^{-5}}$ or, $m = 5.45$.
So the maximum observable order number $=$ **5.**

Ex. 11 *A parallel beam of light is incident normally on a plane grating having* 4250 *lines per* cm *and the second order spectrum is formed at an angle* 30°. *Calculate the wavelength of the monochromatic light.* **[WBUT (Question Bank)]**

Soln. No. of lines/cm, $p = \frac{1}{a+b}$, where $a+b =$ grating element.

$$\therefore a+b = \frac{1}{p} = \frac{1}{4250} = 2.35 \times 10^{-4}\ \text{cm}.$$

Now, the condition for mth order principal maximum in a grating is $(a+b)\sin\theta = m\lambda$.
Here $m = 2, \theta = 30°$ and $(a+b) = 2.35 \times 10^{-4}$ cm.

$$\therefore \lambda = \frac{(a+b)\sin\theta}{m} = \frac{2.35 \times 10^{-4} \times \sin 30°}{2} = \frac{2.35 \times 10^{-4} \times 0.5}{2} = 5875 \times 10^{-8}\ \text{cm} = \mathbf{5875\ Å}.$$

Ex. 12 *Show that in a diffraction grating with grating element* 1.5×10^{-6} m *and light of wavelength* 550 nm, *third and higher order principal maxima are not visible.* **[GNDU 1996]**

Soln. Here the grating element $(a+b) = 1.5 \times 10^{-6}$ m $= 1.5 \times 10^{-4}$ cm.
Wavelength of light $\lambda = 550$ nm $= 550 \times 10^{-7}$ cm.
The maximum order number observable is

$$m = \frac{a+b}{\lambda} = \frac{1.5 \times 10^{-4}}{550 \times 10^{-7}} = 2.72 < 3.$$

So only second order is observable, third and higher order not possible.

Ex. 13 *A diffraction grating used at normal incidence gives a line* (5400 Å) *in certain order superimposed on another line* (4050 Å) *of the next higher order. If the angle of diffraction is* 30°, *how many lines/cm are there in the grating?* **[CU 1990; PU 1993, '99]**

Soln. Here $\lambda_1 = 5400\,\text{Å} = 5400 \times 10^{-8}$ cm; $\lambda_2 = 4050\,\text{Å} = 4050 \times 10^{-8}$ cm and $\theta_1 = \theta_2 = 30°$, order number $m = ?$

$\therefore$ the two lines corresponding to mth and $(m+1)$th *order superimposed*, we have $(a+b)\sin\theta = m\lambda_1$ and $(a+b)\sin\theta = (m+1)\lambda_2$.

$$\therefore \frac{m+1}{m} = \frac{\lambda_1}{\lambda_2} = \frac{5400 \times 10^{-8}}{4050 \times 10^{-8}} = \frac{540}{405} \text{ or, } \frac{1}{m} = \frac{540}{405} - 1 = \frac{135}{405} = \frac{1}{3}.$$

$$\therefore m = 3.$$

Again, $(a+b)\sin 30° = 3 \times 5400 \times 10^{-8}$

$$\therefore a + b = \frac{3 \times 5400 \times 10^{-8}}{0.5} = 324 \times 10^{-6}.$$

$\therefore$ the number of rulings/cm, $p = \frac{1}{a+b} = \frac{10^6}{324} = 3086.42 \simeq$ **3086.**

Ex. 14 *Show that only the first order spectrum is possible in a grating if the grating element is less than twice the wavelength of light used.*

Soln. For the mth order principal maximum, we have $(a+b)\sin\theta = m\lambda$, where $a + b =$ the grating element and $\theta =$ the angle of diffraction

$$\because \sin\theta = \frac{m\lambda}{a+b} \text{ or, } |\sin\theta| = \frac{m\lambda}{< 2\lambda} = \frac{m}{< 2} \quad [\because (a+b) < 2\lambda]$$

$$\therefore m < 2 \; [\because |\sin\theta| \not> 1]$$

So, only $m = 1$ is possible. Hence, the result follows.

Ex. 15 *Light of wavelengths* 5890 *and* 5890 Å *falls normally on a plane transmission grating of* 2 cm *width and having* 10000 *lines. Find the angular separation of two lines of the above wavelengths in the first order spectrum.*

Soln. The number of lines/cm, $p = \frac{10000}{2} = 5000$ lines/cm.

We know

$$(a+b)\sin\theta = m\lambda \text{ or, } \sin\theta = \frac{m\lambda}{a+b} = pm\lambda \left[\because p = \frac{1}{a+b}\right].$$

Here order number $m = 1$,

$$\lambda_1 = 5890\,\text{Å} = 5890 \times 10^{-8} \text{ cm}$$

$$\lambda_2 = 5896\,\text{Å} = 5896 \times 10^{-8} \text{ cm}.$$

Now, for the line of $\lambda_1, \sin\theta_1 = 5890 \times 10^{-8} \times 5000 = 0.2945.$

$\therefore \theta_1 = 17.1275°.$

For $\lambda_2, \sin\theta_2 = 5000 \times 5896 \times 10^{-8} = 0.2948.$

$\therefore \theta_2 = 17.1455°.$

$\therefore \theta_2 - \theta_1 = 17.1455 - 17.1275 = 1.08' = 1'5''.$

Thus, the angular separation in the first order = **1′5″.**

Ex. 16 *What is the maximum number of lines of a grating, which will resolve third order spectrum of two lines having wavelengths* 5890 Å *and* 5896 Å. **[WBUT 2004]**

Soln. The resolving power of a grating RP $= \frac{\lambda}{d\lambda} = mN$. Here $m = 3, N = ?$

Here $\lambda = \frac{5890+5896}{2} = 5893\,\text{Å} = 5893 \times 10^{-8}$ cm and $d\lambda = 5896 - 5890 = 6\,\text{Å}$.

$$\therefore \frac{5893}{6} = 3N \text{ or, } N = \frac{5893}{3 \times 6} = \frac{5893}{18} \simeq 327.39.$$

So the required maximum number of lines of the grating = **327.**

Ex. 17 *A parallel beam of light of wavelength* 5890 Å *falls normally on a plane transmission grating having* 4250 lines/cm. *Find the angle of diffraction for maximum intensity in the first order.* **[WBUT 2006]**

Soln. If θ be the angle of diffraction for the principal maximum of order $m = 1$, then

$$(a+b)\sin\theta = m\lambda \text{ or, } \sin\theta = \frac{1\times\lambda}{a+b} = \lambda p,\ p = \frac{1}{a+b}.$$

The lines/cm = 4250

$$\therefore \sin\theta = 5890\times10^{-8}\times4250 = 589\times425\times10^{-6} = 0.2503.$$

$$\therefore \theta = \mathbf{14°30'}.$$

Ex. 18 *Two lines in the second order spectrum formed by a plane transmission grating are just resolved. If the lines are due to wavelengths* 5890 Å *and* 5890 Å, *find the number of lines in the grating.*

Soln. We know that the resolving power of the grating is RP $= \frac{\lambda}{d\lambda} = Nm$.

Here $\lambda = \frac{5890+5896}{2} = 5893$ Å, $d\lambda = 5896 - 5890 = 6$ Å and $m = 2$.

$$\therefore \frac{5893}{6} = N\times2 \text{ or, } N = \frac{5893}{6\times2} = \frac{5893}{12} = 491.08, \text{ i.e., } N = \mathbf{491\ (nearly)}.$$

Ex. 19 *A plane transmission grating gives an angle of diffraction of a line to be* 30° *in the first order for a wavelength* 6×10^{-5} cm. *Find the number of lines per* cm *of the grating.* **[CU 2003]**

Soln. The number of lines per cm

$$N = \frac{\sin\theta_m}{\lambda m} = \frac{\sin 30°}{6\times10^{-5}\times1}\quad [\because \theta_m = 30°, m = 1 \text{ and } \lambda = 6\times10^{-5} \text{ cm}]$$

$$= \frac{1}{2\times6\times10^{-5}} = \mathbf{8333\ lines/cm.}$$

Ex. 20 *Convex lens of focal length* 40 cm *is employed to focus the Fraunhofer diffraction pattern of a single-slit of* 0.3 mm *width. Calculate the linear distance of the first dark band from the central band. The wavelength of light is* 589 nm. **[WBUT 2008]**

Soln. The distance of the first dark band from the central band is given by $x = f\frac{\lambda}{a}$, where f = the focal length of the lens = 40 cm; λ = the wavelength of light = 589 nm = 5.89×10^{-5} cm and a = width of the slit = 0.3 mm = 0.03 cm.

$$\therefore x = \frac{40\times5.89\times10^{-5}}{0.03} = 7853.33\times10^{-5} = \mathbf{0.0785\ cm.}$$

Ex. 21 *Calculate the least width that a grating must have to resolve two components of the sodium D-lines in the second order, the grating having* 800 *lines*/cm. *The wavelength for* D_1 *and* D_2 *lines of sodium are* 5890 Å *and* 5896 Å *respectively.* **[WBUT 2008]**

Soln. The average value of wavelength

$$\lambda = \frac{5890+5896}{2} = 5893 \text{ Å} = 5893\times10^{-10} \text{ m}.$$

Order number of diffraction band $m = 2$ and $\Delta\lambda = 5896 - 5890 = 6$ Å $= 6\times10^{-10}$ m.

Now, the resolving power of the grating is

$$\frac{\lambda}{d\lambda} = Nm \text{ or, } N = \frac{1}{m}\times\frac{\lambda}{d\lambda} = \frac{5893\times10^{-10}}{2\times6\times10^{-10}} = 491.1 \text{ lines}.$$

$\therefore$ minimum number of lines required for resolution = 492.

So the least width of the grating = $\frac{492}{800}$ = **0.615 cm.**

Ex. 22 *Deduce the missing orders for a double-slit diffraction pattern if the slit widths are* 0.16 mm *and they are* 0.8 mm *apart.* **[WBUT 2008]**

Soln. If the slit width b is gradually increased and the width of the opaque space a is kept constant, then according to the condition of diffraction minima, $b\sin\theta = \pm m\lambda$, the diffraction pattern remains unaltered. But according to the condition of interference maxima,

$$d\sin\theta = (a+b)\sin\theta = \pm n\lambda,$$

the spacing between the interference maxima changes. It may so happen that, for a particular value of the angle of diffraction θ, the condition for interference maximum and diffraction minimum are simultaneously satisfied, then the corresponding interference maximum will be missed from the resultant spectrum.

If the above condition is satisfied, then nth order interference maximum is formed, where the mth order diffraction would be formed. Then the nth order interference maximum would be missed.

Now,

$$\frac{(a+b)\sin\theta}{b\sin\theta} = \frac{n\lambda}{m\lambda} \text{ or, } \frac{a+b}{b} = \frac{n}{m}.$$

In the problem, $b = 0.16$ mm and $a = 0.8$ mm.

$$\therefore \frac{n}{m} = \frac{0.80 + 0.16}{0.16} = \frac{0.96}{0.16} = 6.$$

$\therefore n = 6\text{m} = 6, 12, 18,$ etc. for $m = 1, 2, 3, \cdots$ respectively.

Thus, the 6th, 12th, 18th, etc. order interference maxima will be missed.

Ex. 23 *Two stars situated at a distance of* 9.5×10^{12} km *from a telescope of diameter* 20 cm *are sending light of wavelength* 600 nm. *Find the distance of the separation of the stars for which they are just resolved.* **[WBUT 2008]**

Soln. Limit of angular separation of the stars for which they are just to be resolved is

$$d\theta = \frac{1.22\lambda}{D}, \text{ where } \lambda = 6.0 \times 10^{-5} \text{ cm and } D = 20 \text{ cm}$$

$$= \frac{1.22 \times 6.0 \times 10^{-5}}{20} = 0.366 \times 10^{-5} \text{ rad.}$$

If x be the distance of separation and d the distance of the stars from the telescope, then

$$\frac{x}{d} = d\theta = 0.366 \times 10^{-5}.$$

$$\therefore x = (0.366 \times 10^{-5}) \times (9.5 \times 10^{12}) = \mathbf{3.447 \times 10^7 \text{ km}}.$$

EXERCISES

5(A) Objective Type Questions

Select the correct answer:

5.1. In Fraunhofer diffraction the incident wavefront is (a) plane (b) spherical (c) cylindrical. **[WBUT 2004, '05]**

5.2. The phenomenon in which light waves penetrate into the geometrical shadow is called (a) interference (b) polarization (c) diffraction (d) dispersion.

5.3. If D be the distance between the screen and the slit, b the width of the slit, then the width of the central maximum in the diffraction pattern with light of wavelength λ is (a) $\frac{D\lambda}{2b}$ (b) $\frac{2D\lambda}{b}$ (c) $\frac{b}{2D\lambda}$ (d) $\frac{\lambda b}{2D}$.

5.4. In a single-slit experiment if the width is reduced (a) the fringes become wider (b) the fringes become brighter (c) the fringes become narrower (d) none of these.

5.5. In a single-slit diffraction pattern the minima are (a) unequally spaced (b) equally spaced (c) irregularly spaced (d) none of these.

5.6. In a single-slit diffraction pattern the ratio of intensity of the first secondary maximum to the central maximum is (a) $\frac{I_1}{I_0} = \frac{1}{28}$ (b) $\frac{I_1}{I_0} = \frac{1}{22}$ (c) $\frac{I_1}{I_0} = \frac{1}{121}$ (d) $\frac{I_1}{I_0} = \frac{1}{5}$.

5.7. If a and b be the width of the opaque space and that of the slit of a double-slit, then the diffraction pattern reduces to that of a single-slit, if (a) $a = 2b$ (b) $a = 0$ (c) $a = \frac{b}{2}$ (d) $a = 4b$.

5.8. If a be the width of the opaque space and b that of the slit, then if $2a = b$, the number of interference maxima enclosed by central maximum is (a) 3 (b) 6 (c) 5 (d) 7.

5.9. If b and a be the slit width and the separation of the slits respectively, then the number of lines/length N is given by (a) $N = (a + b)$ (b) $N = \frac{1}{a+b}$ (c) $N = \frac{1}{ab}$ (d) $N = \frac{a}{b}$.

5.10. If $N = 5000$ lines/cm and $\lambda = 4200$ Å, then the maximum number of orders (m) possible is (a) $m = 4$ (b) $m = 5$ (c) $m = 6$ (d) $m = 3$.

5.11. If the slit separation a in a double-slit is increased keeping the slit width b constant, then the envelope of the fringe pattern becomes (a) broader (b) narrower (c) constant (d) none of these.

5.12. If the slit separation a is increased keeping the slit width b constant, then the number of interference maxima enclosed by the central diffraction maxima will (a) decrease (b) remain the same (c) increase (d) none of these.

5.13. In a grating the resolving power is more if the order number m is (a) less (b) more (c) infinity (d) moderate.

5.14. If W be the width of a grating, λ the wavelength of the incident light and θ the angle of diffraction, then the resolving power $(\lambda/d\lambda)$ is (a) $\frac{W\lambda}{\sin\theta}$ (b) $\frac{\sin\theta}{W\lambda}$ (c) $\frac{W\sin\theta}{\lambda}$ (d) $W\lambda\sin\theta$.

5.15. The resolving power of a grating (a) decreases with order of the spectrum (b) decreases with the total number of lines (c) is independent of the grating element $(a + b)$ (d) is independent on the total width of the grating (W).

5.16. The angle of diffraction of the first order maximum in a diffraction grating (a) increases, if the separation between the rulings is increased (b) decreases, if the separation between the rulings is increased (c) does not change, if the separation between the rulings is increased (d) change depends on the wavelength of the incident light.

5.17. A high power microscope objective is immersed in (a) water (b) mustard oil (c) ceder wood oil, (d) kerosine oil.

5.18. Ultramicroscope uses (a) visible light (b) red light (c) γ-rays (d) ultraviolet light.

5.19. The intensity of principal maxima in the spectrum of grating with N number of lines is proportional to (a) $\frac{1}{N}$ (b) N (c) N^2 (d) $\frac{1}{N^2}$. **[WBUT 2008, '10]**

ANSWERS

5.1. (a) **5.2.** (c) **5.3.** (b) **5.4.** (a) **5.5.** (b) **5.6.** (b) **5.7.** (b) **5.8.** (c) **5.9.** (b) **5.10.** (a) **5.11.** (a) **5.12.** (c) **5.13.** (b) **5.14.** (c) **5.15.** (c) **5.16.** (b) **5.17.** (c) **5.18.** (b) **5.19.** (d).

5(B)–Short Answer Type Questions

5.1. Define the term *Diffraction.* **[WBUT 2003]**

5.2. Classify diffraction of light wave. **[WBUT 2003]**

5.3. How do diffraction spectra differ from interference spectra? **[WBUT 2002]**

5.4. Distinguish between Fresnel and Fraunhofer diffraction. **[WBUT 2001]**

5.5. Discuss the nature of grating spectra. **[WBUT 2001]**

5.6. What is Fresnel's diffraction of light? **[WBUT 2005]**

5.7. Explain the difference between interference and diffraction. **[WBUT 2006]**

5.8. What happens in the diffraction pattern due to a single-slit when the slit width is gradually increased?

5.9. Distinguish between a single-slit and a double-slit diffraction patterns.

Table 5.4

Single-slit diffraction pattern	*Double-slit diffraction pattern*
1. It consists of a central brightest maximum along with minima and secondary maxima of rapidly decreasing intensity on either side.	1. It consists of equally spaced interference maxima and minima within the central maximum. The intensity of the central maximum due to a double-slit is four times that due to a single-slit.
2. The spacing between diffraction maxima and minima depends on the slit width b.	2. The spacing of interference maxima and minima depends on a and b, where a is the separation of slits.
3. There is no missing order of interference pattern as the phenomenon of interference does not arise	3. Under certain conditions missing orders of interference maxima arise.

5.10. Find the missing order in a double-slit diffraction pattern.

5.11. Derive an expression for the angular dispersion of a plane grating.

5.12. What will be effect on the diffraction pattern of a grating if the grating element is of the order of λ or less.

5.13. If X-rays are allowed to fall on an ordinary diffraction grating, will the X-rays be diffracted? **[BUH 2003]**

5.14. Distinguish between Fraunhofer and Fresnel diffraction. **[BUH 1999, 2002, '05, CUH 1991, '94, '99, 2002]**

5.15. Within a given order of diffraction pattern of a plane transmission grating, which colour (red or violet) is located closest to the next higher order? **[BUH 2001]**

5.16. How is the resolving power of a plane diffraction grating defined? **[CUH 2002]**

5.17. How large a diffraction grating is needed to separate two wavelengths present? **[BUH 2000]**

5.18. How will you increase the resolving power of a grating?

5.19. For a plane diffraction grating, calculate the maximum number of order possible.

5.20. What do you mean by absent spectra with a diffraction grating?

5.21. What is dispersive power of a grating? Show that the dispersive power of a grating is $\frac{d\theta}{d\lambda} = \frac{mp}{\cos\theta}$, where the different letters have their usual meaning.

5.22. Write down the expression for the intensity of light due to Fraunhofer diffraction in a transmission grating and hence find the condition for secondary minima in the interference pattern. **[WBUT 2010]**

5(C)–Numerical Problems

5.1. A single-slit of width 1 mm is illuminated by light of wavelength 589 nm. Find the angular spread of the central maximum of the diffraction pattern observed. **[WBUT 2005]**

[*Hint.* $b\sin\theta = \lambda$, $\therefore \sin\theta = \frac{\lambda}{b}$. $\because \theta \to 0$, for the central maximum, $\sin\theta \sim \theta$. $\therefore \theta = \frac{\lambda}{b}$. The angular spread is $2\theta = \frac{2\lambda}{b}$. Use $\lambda = 589 \times 10^{-9}$ m $= 589 \times 10^{-7}$ cm and $b = 0.1$ cm in $2\theta = \frac{2\lambda}{b}$ to get the angular spread of the central maximum.]

5.2. A light wave of $\lambda = 6 \times 10^{-5}$ cm falls normally on a slit of width 0.2 mm. Calculate (a) the total angular width of the central maximum (b) the linear width of the central maximum on a screen placed 2 m away. [*Ans.* (a) 6×10^{-3} radians (b) 1.2 cm]

5.3. A lens of focal length 40 cm forms a Fraunhofer diffraction pattern of a slit 0.3 mm wide. Calculate the distance of the first dark band from the axis. [$\lambda = 5890$ Å]. [*Ans.* 0.0785 cm]

5.4. A slit of width 12×10^{-5} cm is illuminated by light of wavelength 6000 Å. Find the half-angular width of the central bright maximum in the Fraunhofer diffraction pattern. [*Ans.* 30°]

5.5. A screen is placed at a distance of 40 cm from a narrow slit illuminated by a parallel beam of light of wavelength 5890 Å. Calculate the width of the slit, if the first dark line lies 2 mm on either side of the central maximum. [*Ans.* $\sim$ 0.012 cm]

5.6. The distance of the screen from a double-slit of slit width 0.02 mm and spacing between the two slits 0.10 mm is 100 cm. If the wavelength of the light used is 5000 Å, calculate (a) the distance between the central maximum and the first minimum of the fringe envelope and (b) the distance between any two consecutive double-slit dark fringes. [*Ans.* (a) $\sim$ 2.083 mm, (b) 4.166 mm]

5.7. A double-slit has each slit width $a = 8.8 \times 10^{3}$ cm and a separation between the sits $b = 7.0 \times 10^{-2}$ cm illuminated with light of wavelength $\lambda = 6.328 \times 10^{-5}$ cm. How many interference maxima will occur between the two diffraction minima on either side of the central maximum? **[VU 2002]**

5.8. A plane beam of sodium light is allowed to be incident normally on a plane transmission grating having 4250 lines/cm and a second order spectral line is observed to be deviated through 30°. Calculate the wavelength of the spectral line. [*Ans.* $\lambda = 5883 \times 10^{-8}$ cm]

5.9. The separation of the sodium line (mean $\lambda = 5893 \times 10^{-8}$ cm) in the second order spectrum of a transmission grating containing 5000 lines per cm is 2.5 minutes for normal incidence. What is the difference in wavelength between the two yellow lines? [*Ans.* $\Delta\lambda = 5.81$ Å]

5.10. Light of wavelength 5×10^{-5} cm is incident normally on a plane transmission grating of width 3 cm and 15000 lines. Find the angle of diffraction in the first order. [*Ans.* 14°29′]

5.11. A parallel beam of light is incident normally upon a plane diffraction grating having 14500 lines per inch. The first order spectrum of two bright lines are at a deviation of 19°39′ and 19°40′30″. Calculate the difference in their wavelengths. [*Ans.* 7.2 Å]

5.12. In a diffraction grating you notice that the 3rd, 6th and 9th order maxima are missing. What conclusion can you draw about the nature of the ruling? **[BUH 1996]**

5.13. How many orders will be visible if the wavelength of the incident radiation is 5000 Å and the number of lines on the grating is 2620 in one inch? [*Ans.* 19]

5.14. A grating has grating element 5×10^{-5} m. What is its resolving power in the second order if the width of grating is 0.02 m? [*Ans.* 800]

5.15. Two spectral lines have wavelengths λ and $\lambda + d\lambda$ respectively ($d\lambda \ll \lambda$). Show that the angular separation $d\theta$ in the nth order of the spectrum by a grating is given by $d\theta = \frac{d\lambda}{[(a+\frac{b}{n})^2 - \lambda^2]^{1/2}}$.

5.16. Calculate the RP of a microscope, if N.A. is 0.12 and the wavelength of light used $= 6 \times 10^{-8}$ cm. [*Ans.* 4000]

5(D)–Broad Answer Type Questions

5.1. (a) Distinguish between Fresnel and Fraunhofer diffraction. Discuss the nature of grating spectra. (b) Deduce an expression for the resolving power of a grating. **[WBUT 2001]**

5.2. (a) How does diffraction spectra differ from interference spectra? (b) Explain Fraunhofer diffraction by a single-slit with necessary theory. Point out also the graphical representation of intensity distribution. **[WBUT 2002]**

5.3. (a) Define the term *Diffraction.* (b) Classify diffraction of light waves. (c) Discuss Fraunhofer double-slit diffraction. **[WBUT 2003]**

5.4. (a) Show that the intensity of the first secondary maximum formed by a single-slit Fraunhofer diffraction process is nearly 4.5% of that of the principal maximum. (b) What do you mean by resolving power of an optical instrument? **[WBUT 2004]**

5.5. (a) What is Fresnel's diffraction of light? (b) Derive an expression for the intensity distribution due to Fraunhofer diffraction at a single-slit. **[WBUT 2005]**

5.6. The intensity distribution for single-slit diffraction is $I = I_0 \frac{\sin^2 \beta}{\beta^2}$, where $\beta = \frac{\pi b \sin \theta}{\lambda}$, b is the width of the slit and λ is the wavelength of light. Show that secondary maxima are given by the equation $\tan \beta = \beta$. **[WBUT 2006]**

5.7. Describe with necessary theory the diffraction pattern due to diffraction by a narrow slit illuminated by parallel beam of monochromatic light incident normally on the slit. **[CU 1999]**

5.8. Deduce an expression for the intensity pattern due to diffraction in a single experiment. Also show your result graphically. **[NBU 2001]**

5.9. Explain the formation of diffraction pattern when a monochromatic beam of light passes through a narrow rectangular aperture. **[BU 1983]**

5.10. (a) Show that the resultant intensity distribution in the Fraunhofer diffraction pattern produced by two parallel slits each of width b and separated by a distance d is a product of the single-slit (width b) diffraction pattern and the interference pattern produced by two point sources separated by a distance d. (b) Discuss the positions of maxima and minima. What do you mean by missing order? **[VU 2002]**

5.11. Explain the action of a diffraction grating and describe how to use it to measure the wavelength of light. **[NBU 1983, 2000; CU 2000]**

5.12. Explain what is meant by the resolving power of a grating. Show that the resolving power is given by the product of the total number of rulings and the order number of the spectrum.

[BU 1999; CU 2002]

5.13. Show that Fraunhofer diffraction by a slit on the first minimum on either side of the central maximum is given by $\theta = \lambda/W$, where W is the width of the slit and $W \gg \lambda$. **[BUH 2000]**

Calculate the position of the first maximum and first minimum as a function of angle in the region far from the slit. **[BUH 2001]**

5.14. A monochromatic beam of light of wavelength λ is incident on a slit having width $W = 5\lambda$. Sketch the intensity pattern as a function of angle in the region far from the slit. Calculate the positions of the first maximum and first minimum. **[BUH 2001]**

5.15. Find out an expression for the intensity at a point due to Fraunhofer diffraction through a plane transmission grating. **[BUH 2006, '03]**

5.16. (a) State and explain Rayleigh's criterion of resolution. (b) Find out an expression for the resolving power of microscope. How can the resolving power of a microscope be increased?

5.17 (a) Derive the intensity distribution of Fraunhofer class of diffraction due to a single slit. Sketch the intensity distribution. (b) A single slit forms diffraction pattern of Fraunhofer class with white light. The second maximum in the pattern for red light of wavelength 7000 Å coincides with the third maximum of an unknown wavelength. Calculate the unknown wavelength. **[WBUT 2010]**

Chapter 6 | Module 3.1

Polarization of Light

6.1 Introduction

The phenomena of *interference* and *diffraction* of light proved beyond any doubt that light is some form of *wave motion*. The above two properties could not decide whether light wave is longitudinal or transverse or whether the vibrations are linear or circular. Maxwell's electromagnetic theory requires that light waves are transverse. But long before the discovery of electromagnetic theory Arago (1786–1853) and Fresnel (1788–1827) jointly discovered through their experiments on polarization that *light wave is transverse* in nature. Thus, the phenomenon of polarization conclusively proves that *light wave is transverse in character.*

The concept of polarization is essentially needed in the wave propagation through waveguides and optical fibres, biological activities, etc.

Applications of polarized light. Polarized light finds important applications in industry and engineering, e.g., (i) Liquid crystal display (LCD) used widely in wrist-watches, (ii) calculators, (iii) TV screens, (iv) video displays, etc.

6.2 General Concept of Polarization

To have a clear idea regarding polarization we consider two slits S_1 and S_2 parallel to each other and a string AB passing through them [Fig. 6.1]. The end A is attached to a fixed point and the end B is moved up and down perpendicular to AB in any arbitrary manner. Transverse waves will be generated in the string, which are parallel to S_1 or have some components parallel to S_1, which will emerge along CD. Since S_1 and S_2 are parallel, these transverse vibrations pass through S_2 and reach B with undiminished amplitude [Fig. 6.1].

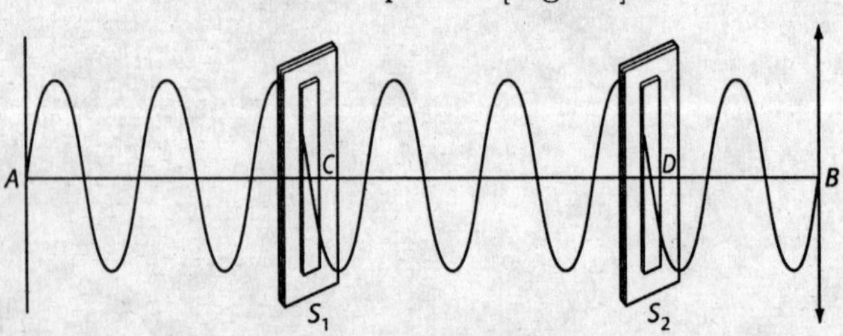

Fig. 6.1: Wave emerges along *AB*, when slits S_1 and S_2 are parallel.

If now S_2 is rotated through 90° so that the slit S_2 is perpendicular to slit S_1, then the vibrations passing through S_1 are unable to pass through S_2 and there is no vibrations between S_2 and B [Fig. 6.2]. If S_2 makes an angle θ such that $0° < \theta < 90°$ with S_1, then the vibrations pass through S_2 but with diminished amplitude. So the amplitude of vibrations between S_2 and B varies as the S_2 is rotated w.r.t. S_1. This is possible as the vibrations are *transverse.*

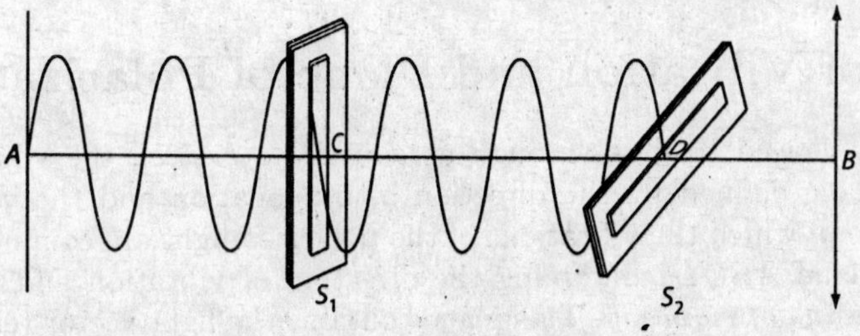

Fig. 6.2: Wave cut-off when the slits S_1 and S_2 are crossed.

If the vibrations were *longitudinal,* then the amplitude of vibrations between S_2 and B would remain unchanged with rotation of S_2. We thus conclude that *the vibrations generated in the string must be transverse in nature.*

6.3 Polarization of Light Waves

Light from a source S falls on a tourmaline crystal P cut parallel to its *crystallographic axis* [Fig. 6.3]. If a second crystal A cut similarly be placed in the path of light such that the axes of A and P are parallel, then the intensity of the emergent light from A is the maximum.

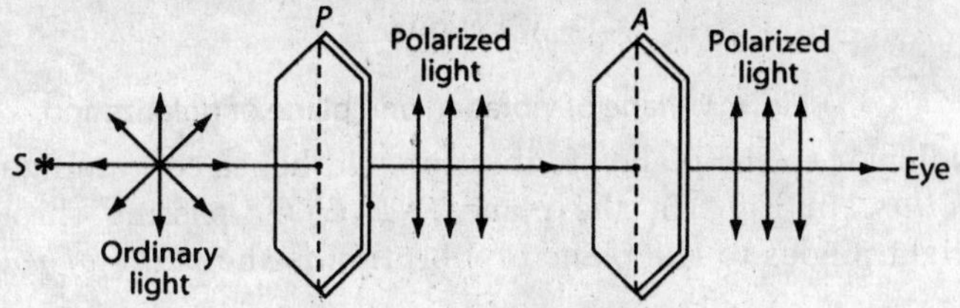

Fig. 6.3: Slits of P and A are parallel.

If now P is kept fixed and the crystal A is rotated about the beam as axis, then the intensity of the emergent light from A will decrease and ultimately reduce to zero when the axis of A is perpendicular to P [Fig. 6.4]. If A is kept rotating, then the intensity of the emergent light will again increase and becomes maximum when the axis of A is parallel to P. *This result leads to the conclusion that light wave is transverse.* If light were longitudinal vibrations, then no extinction of the emergent light from A would occur.

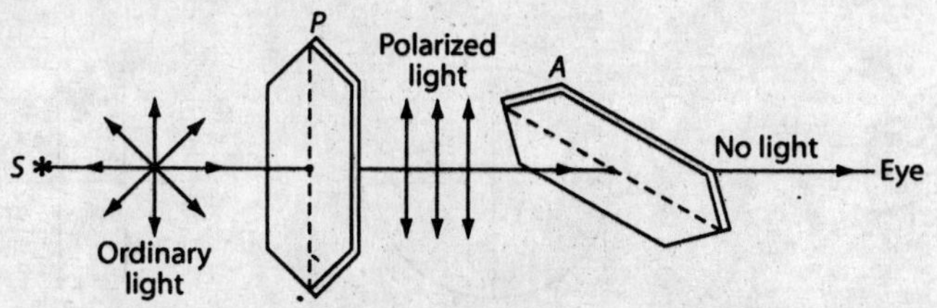

Fig. 6.4: Slits of P and A are at 90° to each other.

The above experiment shows that ordinary light after passing through P is not symmetrical about the direction of propagation. Instead the vibrations are confined in a single line in a plane perpendicular to the propagation direction. So light after passing through P acquires a property, which prevent it from passing through A, when the axis of A is at 90° to P. The property of one *sidedness* acquired by light after passing through P is called *polarization* and the corresponding light is called *polarized light.* The experiment also shows that *light waves are transverse in nature.*

6.4 Plane of Vibration and Plane of Polarization

If ordinary light is allowed to pass through a *tourmaline crystal*, the vibrations are found to be confined in a plane defined by the direction of propagation and the vibrating light vector [Fig. 6.5]. This plane in which the vibrations of the polarized light are confined is called the *plane of vibration.* The plane $ABCD$ containing the direction of vibration and also the direction of propagation is the *plane of vibration.* This plane contains the light vector (electric vector $\vec{E}$) and the wave propagation vector ($\vec{k}$).

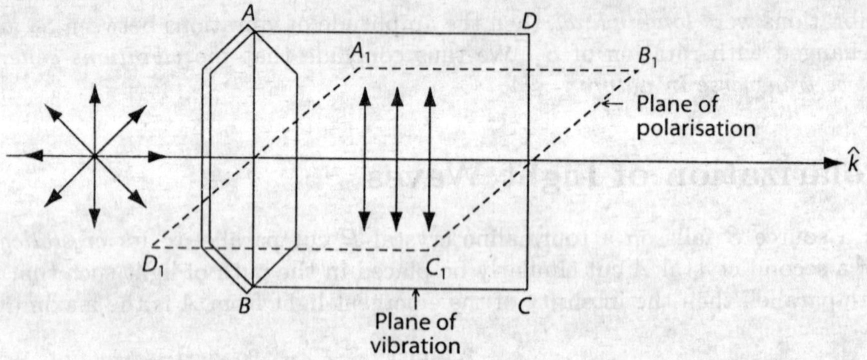

Fig. 6.5: Plane of vibration and plane of polarization.

The plane perpendicular to the plane of vibration and hence containing no vibration is called the *plane of polarization.* In Fig. 6.5, the plane $A_1B_1C_1D_1$ passing through the direction of propagation and at right angles to the plane of vibration is the *plane of polarization.*

6.5 Pictorial Representation of Light Vibration: Unpolarized Light

In a light wave, the light vectors vibrate in a direction perpendicular to the direction of propagation. In a source, the light emitting atoms oscillates independently and as such the light waves emitted have their own planes of vibrations. So the waves have vibrations in all possible directions with equal probability.

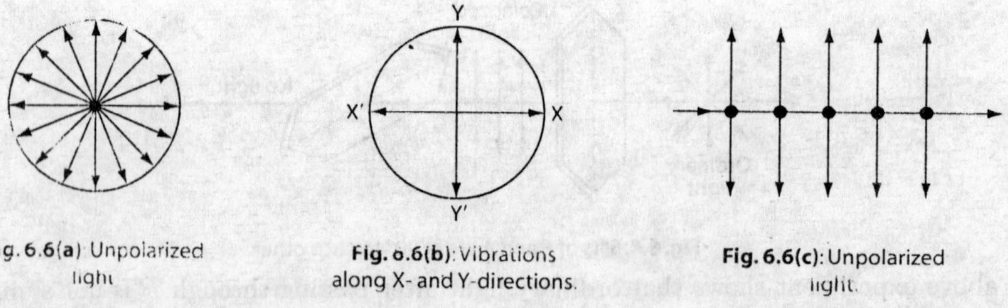

Fig. 6.6(a): Unpolarized light **Fig. 6.6(b)**: Vibrations along X- and Y-directions. **Fig. 6.6(c)**: Unpolarized light.

Now, a wave propagating in a direction perpendicular to the plane of the paper, it will have vibrations in the plane of the paper in all possible directions with same amplitude [Fig. 6.6(a)]. These vibrations may be considered as two mutually perpendicular vibrations along XX' and YY' [Fig. 6.6(b)]. If light propagates along the plane of the paper, then this unpolarized light will consist of vibrations in the plane of the paper (but perpendicular to the direction of propagation) and is represented by *arrows* and vibrations perpendicular to the plane of the paper represented by *dots* [Fig. 6.6(c)]. Fig. 6.6(c) thus pictorially represents *unpolarized light.*

Plane polarized light. If the vibrations of the light vector are confined to a single plane, then it is a *plane polarized light.* The vibrations in this case may be parallel to the plane and are represented by *arrows* [Fig. 6.7(a)] or perpendicular to the plane of the paper and are represented by *dots* [Fig. 6.7(b)].

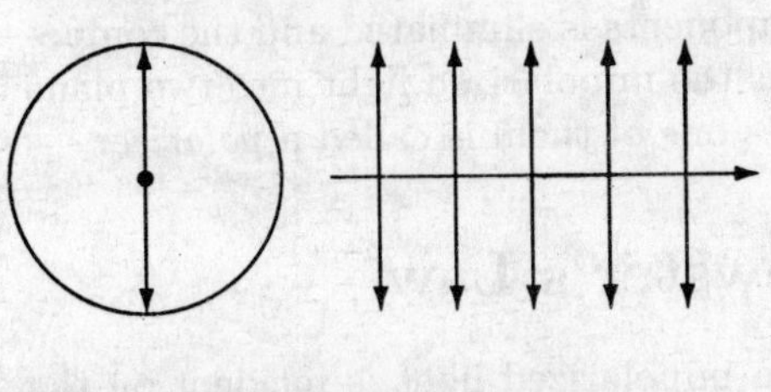

Fig. 6.7(a): Vibrations parallel to the plane of the paper.

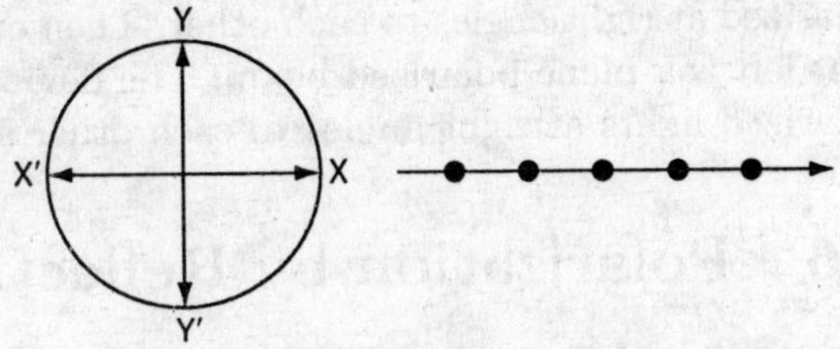

Fig. 6.7(b): Vibrations perpendicular to the plane of the paper.

Circularly polarized light. In circularly polarized light the resultant electric vector of constant amplitude rotates while proceeding in the form of a helix. On a plane perpendicular to the direction of propagation, the projection of the wave gives a circle with constant radius [Fig. 6.8].

If the vector rotates in the clockwise direction w.r.t. the direction of propagation, then it is called a *right-circularly polarized light.* If, on the other hand, the vector rotates in the anti-clockwise direction w.r.t. the direction of propagation, then it is a *left-circularly polarized light.*

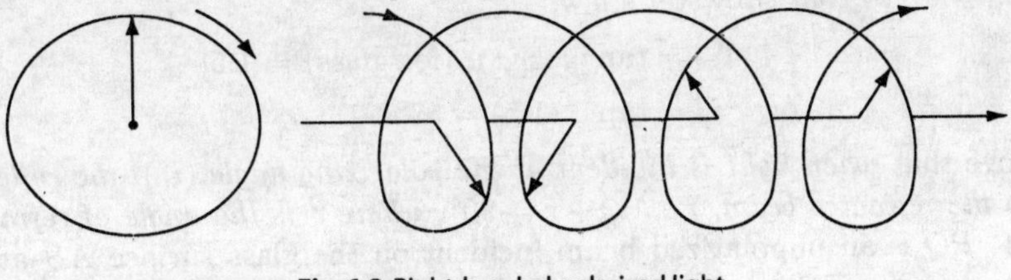

Fig. 6.8: Right-handed polarized light.

Elliptically polarized light. If the resultant electric vector of the polarized light varies periodically between a maximum and a minimum as it rotates, then its tip traces an ellipse. Such a polarized light is called *elliptically polarized light* [Fig. 6.9]. Elliptically polarized light can also be *right-handed* or *left-handed.*

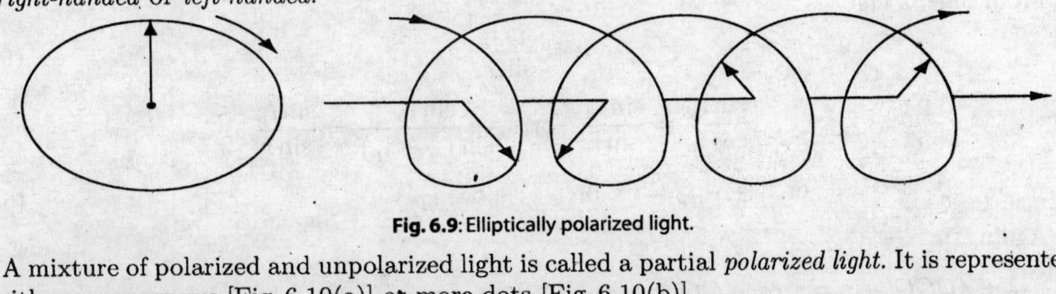

Fig. 6.9: Elliptically polarized light.

A mixture of polarized and unpolarized light is called a partial *polarized light.* It is represented by either more arrows [Fig. 6.10(a)] or more dots [Fig. 6.10(b)].

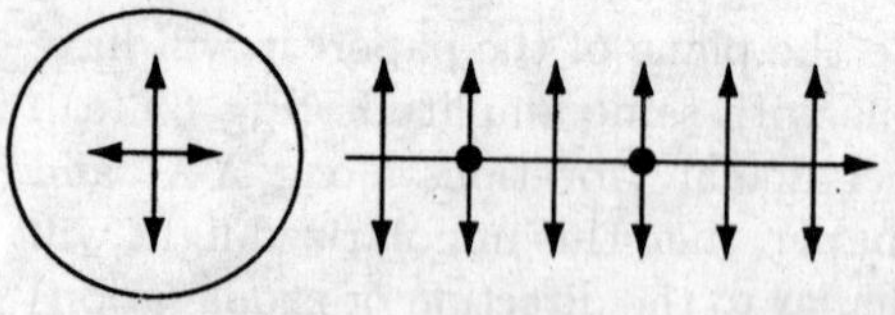

Fig. 6.10(a): Partially polarized light with more arrows.

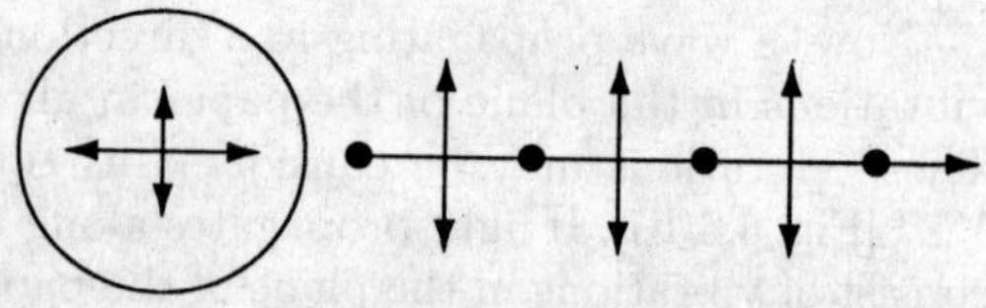

Fig. 6.10(b): Partially polarized light with more dots.

- **Production of plane polarized light.** Plane polarized light can be experimentally produced by any of the following three methods: (1) by simple reflection (2) by double reflection (3) by dichroism (or selective absorption).

An unpolarized light consists of optical vibrations in all possible directions with equal probabilities. To produce plane polarized light the unpolarized light is divided into two components polarized at right angles to each other. Then one of the components is eliminated and the component left is a plane polarized beam. The device that divides the unpolarized light into two plane polarized lights at right angles to each other and eliminates one of them is called a *polarizer*.

6.6 Polarization by Reflection: Brewster's Law

In 1815 *David Brewster* (1781–1868) discovered that when unpolarized light is incident on the surface of a transparent material, both the reflected and refracted beams are partially plane polarized. *For a particular angle of incidence (the polarizing angle) the reflected and refracted beams of light are completely polarized, the plane of vibration being at right angles to the plane of incidence.*

Brewster's law. *The tangent of the angle of polarization is numerically equal to the refractive index of the reflector relative to the surrounding medium.*

If i_p be the polarizing angle and μ the refractive index of the material of the reflector, then

$$\mu = \tan i_p. \tag{6.1}$$

For glass, $i_p = 57°$. From Brewster's law,

$$1.5 = \tan i_p \quad [\because \mu \text{ (for glass)} = 1.5]$$

$$\text{or,} \quad i_p = \tan^{-1}(1.5) = 57°.$$

We can prove that *when light is incident at the polarizing angle (i_p), the reflected beam is at right angles to the refracted beam, i.e., $i_p + r = 90°$, where r is the angle of refraction.*

In Fig. 6.11, PQ is an unpolarized beam incident on the glass surface AB at the polarizing angle i_p. OQ and OO' are the corresponding reflected and refracted rays.

If r be the angle of refraction, then from Brewster's law,

$$\mu = \tan i_p = \frac{\sin i_p}{\cos i_p}. \tag{6.2}$$

From *Snell's law*,

$$\mu = \frac{\sin i_p}{\sin r}. \tag{6.3}$$

$$\therefore \quad \frac{\sin i_p}{\cos i_p} = \frac{\sin i_p}{\sin r} \quad \text{or,} \quad \frac{\sin i_p}{\sin\left(\frac{\pi}{2} - i_p\right)} = \frac{\sin i_p}{\sin r}$$

$$\text{or,} \quad \frac{\pi}{2} - i_p = r \quad \text{or,} \quad i_p + r = \frac{\pi}{2}. \tag{6.4}$$

Again,

$$i_p + \angle QOO' + r = \pi \text{ or, } \angle QOO' = \pi - (i_p + r) = \pi - \frac{\pi}{2} = \frac{\pi}{2} \quad \left(\because i_p + r = \frac{\pi}{2}\right). \tag{6.5}$$

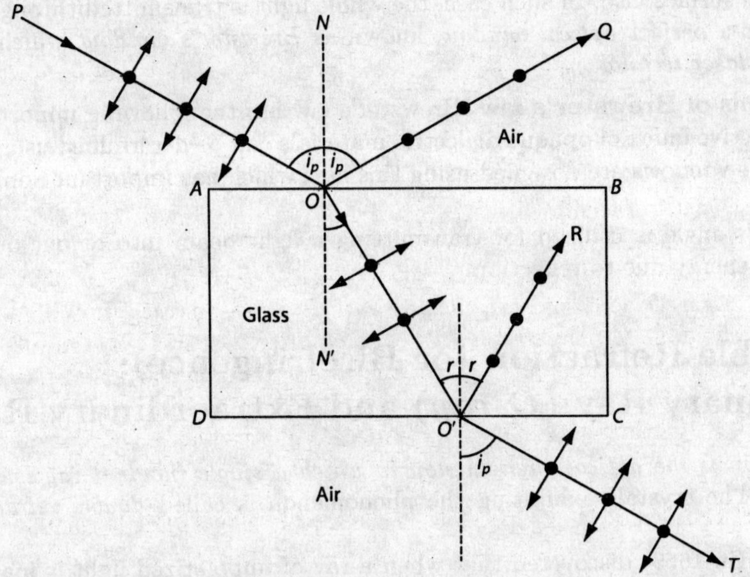

Fig. 6.11: Complete polarization by reflection when the angle of incidence is the polarizing angle i_p.

Eqn (6.5) shows that *at the polarizing angle of incidence the reflected ray is at right angles to the refracted ray.*

Brewster's law is also true even when light is reflected at the surface of a *rarer medium.* If light is incident from the glass medium at polarizing angle r along OO' [Fig. 6.11], then the refracted ray $O'T$ will make an angle i_p with the normal to DC at O'.

Then the refractive index of the medium is given by

$$\tan r = \frac{\sin r}{\cos r} = \frac{\sin r}{\sin i_p} = \frac{1}{\mu}. \tag{6.6}$$

The ray $O'T$ is then plane polarized in the plane of incidence.

Brewster's window. If a beam of *plane polarized* light is incident at Brewster's angle (i_p) on a glass plate having parallel faces, then no light is reflected from the first face [Fig. 6.12]. Since the critical angle (θ_c) at the second surface is given by

$$\theta_c = \sin^{-1}\left(\frac{1}{\mu}\right) = \sin^{-1}\left(\frac{1}{1.5}\right) = 41°.$$

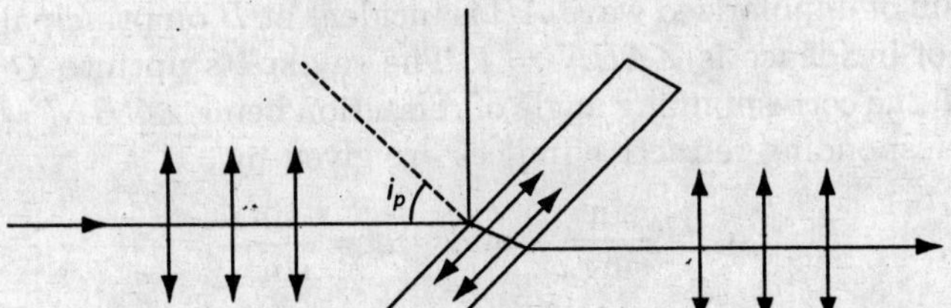

Fig. 6.12: Brewster's window.

The angle of incidence on the second surface is $r = 33.70°$ ($\because i_p = 56.30°$), which is less than $\theta_c(= 41°)$. Total internal reflection will not take place, as a result of which no light is reflected from the second surface also. In such case, the whole light is transmitted through the glass plate. In effect we get a perfect *optical window,* known as *Brewster's window,* which finds important applications in *laser technology.*

Applications of Brewster's law. Brewster's law has the following important applications:

1. The refractive index of opaque dielectric materials can be determined using Brewster's law.

2. Brewster's windows are designed using this law, which has important application in He-Ne gas laser tube.

3. Brewster's angle is utilized for transmitting a light beam into or out of an optical fibre without loss of energy due to reflection.

6.7 Double Refraction (or Birefringence): Ordinary Ray (*O-ray*) and Extraordinary Ray (E-ray)

Double refraction is the optical phenomenon in which a single incident ray is split up into two refracted rays. The crystals exhibiting the phenomenon is called *double refracting crystals* or *birefringent.*

Bartholinus (in 1669) discovered that when a ray of unpolarized light is made incident on a crystal such as calcite ($CaCO_3$), it breaks up into two refracted rays as shown in Fig. 6.14. One of these rays has vibrations perpendicular to the *principal section* of the crystal and obey the laws of refraction. This ray is called the *ordinary ray* or *O-ray.* The other ray has vibrations in the plane of the principal section and does not obey the laws of refraction. This ray is called the *extraordinary ray* or *E-ray. Thus, O-ray and E-ray are plane polarized having vibrations at right angles to each other.*

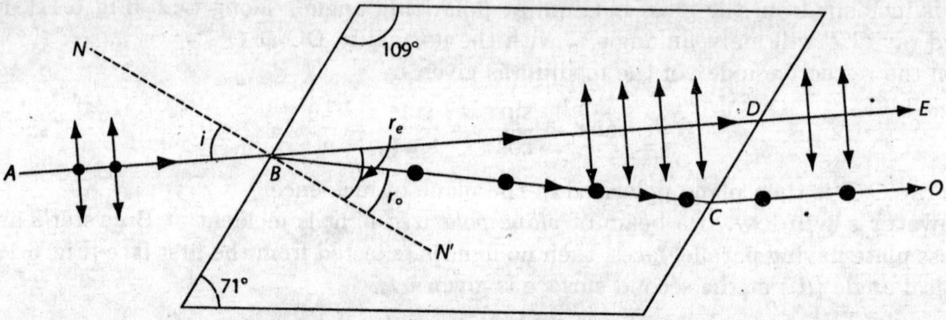

Fig. 6.13: Double refraction.

In Fig. 6.13, a beam of unpolarized wave AB is incident at B on principal section of the crystal such that the angle of incidence is $\angle ABN = i$. The ray splits up into *O-ray* (BC) and *E-ray* (BD) after refraction, the corresponding angle of refraction being $\angle CBN' = r_o$ and $\angle DBN' = r_e$ respectively. The corresponding refractive indices are given by

$$\mu_o = \frac{\sin i}{\sin r_o} \quad \text{and} \quad \mu_e = \frac{\sin i}{\sin r_e}.$$

In *calcite crystal,* $r_o < r_e$ and so $\mu_o > \mu_e$.

But $\mu_o = c/v_o$ and $\mu_e = c/v_e$, where c is the velocity of light in vacuum and v_O and v_e are the velocities of the *O-ray* and *E-ray* respectively. Therefore, it follows that $v_e > v_o$, i.e., in calcite, the *E-ray* travels faster than the *O-ray.*

The following experiment demonstrates the phenomenon of double refraction [Fig. 6.14]. A dot mark is put with ink on a piece of paper and the image of the ink dot is received on a screen S through a double refracting crystal. Two images corresponding to *O-ray* and *E-ray* will be observed on S. If the crystal be rotated in the clockwise direction with the incident ray as axis, then it will be observed that the O-image is fixed but the E-image rotates in the clockwise direction [Fig. 6.14]. This simple experiment proves that (i) *the single incident ray AB breaks up into O-ray (BO) and E-ray (BE)* and (ii) *the velocity of O-ray is the same in all directions but that for the E-ray is different in different directions.*

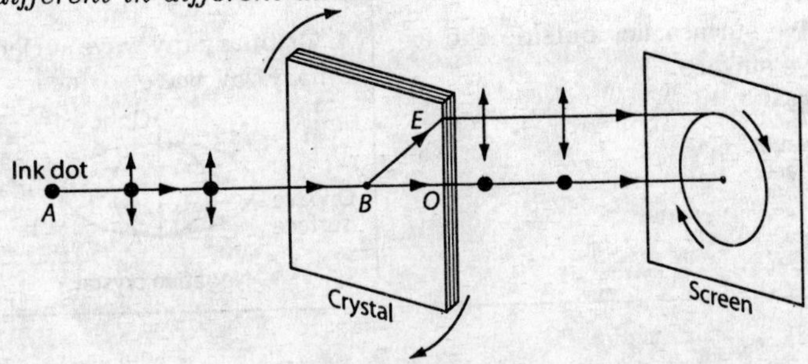

Fig. 6.14: Demonstration of double refraction.

Isotropic refracting medium has the same physical properties in all directions but in *anisotropic* crystals (e.g., calcite, quartz, tourmaline, etc.) the physical properties are *direction dependent, i.e., different in different directions.* That is why *anisotropic crystals only can exhibit the phenomenon of double refraction.*

Negative crystals and positive crystals. We have seen that the velocities of the *O-ray* (v_o) and that of the *E-ray* (v_e) are not equal in an anisotropic crystal except along the optic axis. Also according to the *Huygens's concept* of the wavefront, the wave surface of the *O-ray* is a sphere (as v_o is the same in all directions) and the wave surface of the *E-ray* is a spheroid (as v_e is direction dependent).

So a crystal in which $v_e > v_o$ and as such the ordinary wave surface lies within the extraordinary wave surface is called a *negative crystal.*

Example. Calcite Fig. 6.15(a).

A crystal in which the extraordinary wave surface lies within the ordinary wave surface (as $v_e < v_o$) is called a *positive crystal.*

Example. Quartz Fig. 6.15(b).

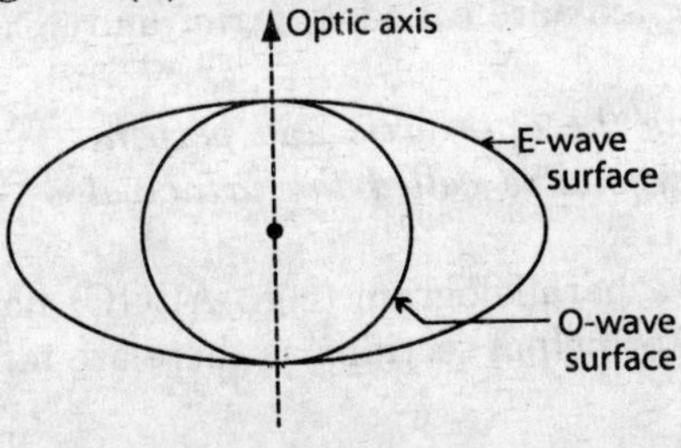

Fig. 6.15(a): Negative crystal.

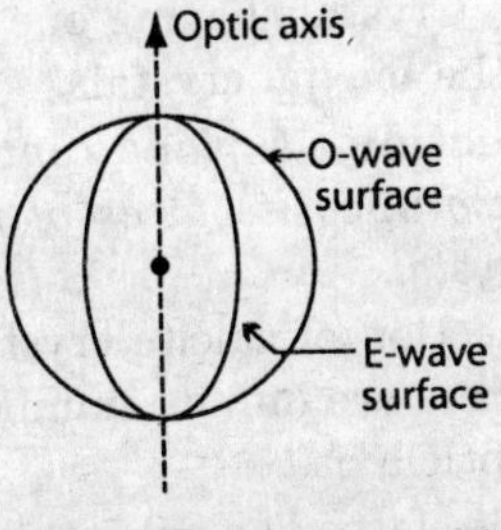

Fig. 6.15(b): Positive.crystal.

Difference between positive and negative crystals

Positive crystals	*Negative crystals*
1. The velocity of *E-ray* (v_e) is direction dependent but the velocity of the *O-ray* (v_o) is constant in all directions of the crystal. Along the optic axis, $v_e = v_o =$ maximum.	1. v_e changes with direction but v_o is constant in all directions of the crystal. Along the optic axis, $v_e = v_o$.

Positive crystals	*Negative crystals*
2. Along the optic axis, $v_e = v_o$, but in any other direction, $v_o > v_e$. v_e is the minimum along the perpendicular direction to optic axis.	2. $v_e = v_o$, along the optic axis but along any other direction, $v_e > v_o$. v_o is the maximum along the perpendicular direction of optic axis.
3. The refractive index of *E-ray* (μ_e) is greater than that of *O-ray* (μ_o), i.e., $\mu_e > \mu_o$.	3. $\mu_e < \mu_o$.
4. Ordinary ray wave surface lies outside the extraordinary ray wave surface.	4. Ordinary ray wave surface lies inside the extraordinary ray wave surface.
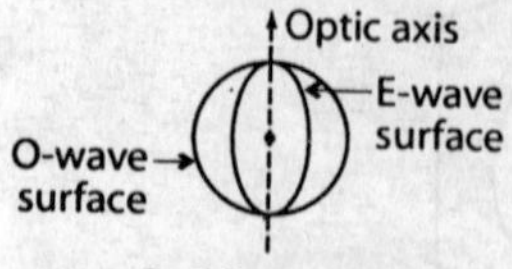 Positive crystal	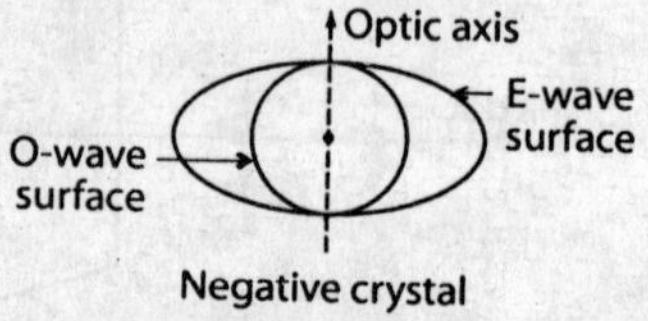 Negative crystal

6.8 Geometry of Calcite: Optic Axis and Principal Section

Calcite is *hydrated calcium carbonate* ($CaCO_3$) and is known as *iceland spar.* It is *transparent* and *colourless.* It belongs to *rhombo-hedral class of hexagonal system of crystals.*

The crystal has six faces, which are parallelograms with angles 101°55′ and 78°5′ each as shown in Fig. 6.16. The three obtuse angles meet at diagonally opposite corners. Out of the remaining six corners two are *acute* and one is *obtuse angle* as is evident from Fig. 6.16.

Optic axis. *A line passing through any one of the blunt corners* (A *or* G) *and equally inclined to the three faces meeting at the corners is the direction of the optic axis.* Optic axis is a direction and not any particular line. So *any line parallel to the line described above is also the optic axis.*

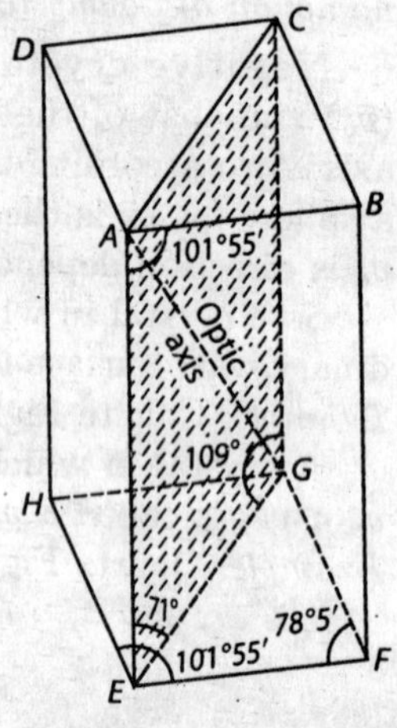

Fig. 6.16: Geometry of calcite crystal.

Optic axis has the following characteristics: (i) *Optic axis in a crystal is a direction and not any particular line.* (ii) *The crystal is symmetrical about the optic axes.* (iii) *The O-ray and E-ray travel with equal velocity along optic axes.* Crystals having one optic axis are called *uniaxial* and two optic axes, the biaxial crystals.

Principal section. *A plane containing the optic axis and perpendicular to the two opposite faces of the crystal is called the principal section of the crystal.*

A principal section of calcite crystal is a parallelogram (e.g., AEGC) having adjacent angles 71° and 109°. There are infinite number of principal sections as there are infinite number of lines parallel to the optic axis.

As the opposite faces of a calcite crystal are always parallel, the *O-ray* and *E-ray* emerges out parallel to the incident ray. The vibrations of *O-ray* are perpendicular to the principal section, i.e., *the O-ray is polarized in the principal section.* But the vibrations of *E-ray* are in the principal section and hence the *E-ray is polarized in the plane perpendicular to the principal section.*

Principal plane of *O-ray*. The plane in the crystal drawn through the optic axis and the *O-ray* is known as the *principal plane of the O-ray.*

Principal plane of *E-ray*. The plane in the crystal drawn through the optic axis and the *E-ray* is known as the *principal plane of the E-ray.*

6.9 Nicol Prism

Using the property of double refraction of calcite, Nicol in 1832 constructed an optical device (the Nicol prism) for the production and analysis of polarized light.

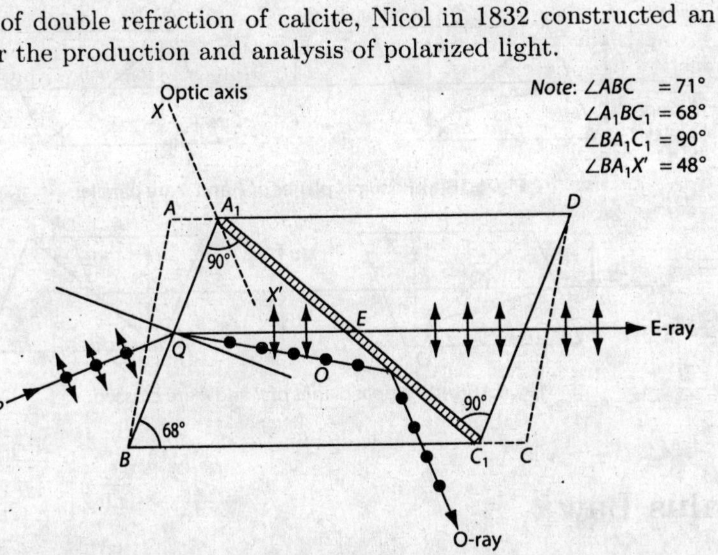

Fig. 6.17: The Nicol prism.

Construction. It is constructed out of a *calcite crystal* $ABCD$ whose length is *three times its breadth.* The two end faces AB and CD are grounded to reduce $\angle ABC$ from 71° to 68° and $\angle BCD$ from 109° to 112°. The crystal is then cut into two halves along the plane A_1C_1 passing through the blunt corners and perpendicular to both the principal section and end faces. A_1C_1 makes an angle 90° with C_1D and A_1B. The surfaces are polished optically flat and then cemented together by *Canada balsam.* Canada balsam is a transparent liquid having $\mu = 1.55$ for $\lambda = 5893$ Å. Then the crystal is enclosed in a tube blackened inside [Fig. 6.17].

Action as polarizer. A beam of unpolarized light PQ is incident on the face A_1B. The beam splits into two refracted rays QO and QE called *O-ray* and *E-ray* respectively. *O-ray* has vibrations perpendicular to the principal section and *E-ray* has vibrations in the principal section of the crystal.

The refractive index of the *Canada balsam layer* (μ_c) is such that $\mu_e < \mu_c < \mu_o$. Due to the large length of the crystal the *O-ray* meets the Canada balsam layer at an angle greater than the critical angle (69°) and hence suffers *total internal reflection* and is finally absorbed by the blackened surface of the enclosure. On the other hand, the *E-ray* is transmitted through the Canada balsam layer and emerges out of the crystal. Thus, the emergent light from the Canada balsam layer is *plane polarized in the plane of the principal section of the crystal.*

Action as analyzer. The polarizer P renders the incident *unpolarized light* as plane polarized with vibration in the principal section of P [Fig. 6.18(a)]. If this light falls on a second Nicol A whose principal section is parallel to that of P, then vibrations of the light wave will be in the principal plane of A and as such the ray will be the *E-ray.* It will be completely transmitted and the intensity of the emergent light will be the *maximum* [Fig. 6.18(a)].

If the analyzer A be rotated about the line of propagation of light so that the principal sections of P and A are at right angles to each other, the *E-ray* from P will enter the analyzer A as *O-ray.* This ray will be totally reflected by the Canada balsam layer towards the edge of A and will be absorbed by the black point. So no light will be emerging out of A. The Nicol P and A are said be *crossed* at this position [Fig. 6.18(b)].

Here Nicol P and A are called the *polarizer* and *analyzer* respectively.

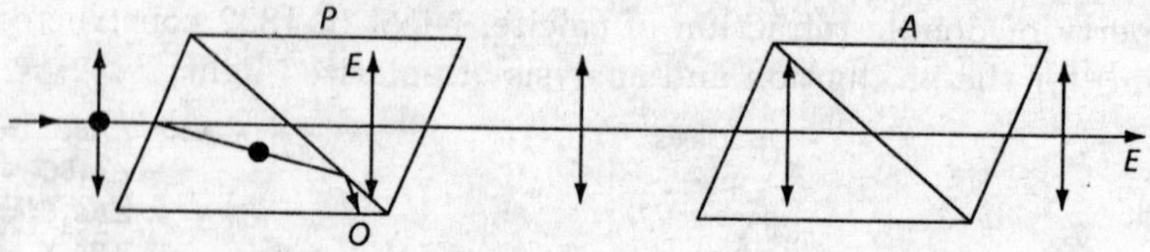

Fig. 6.18(a): Principal planes of *P* and *A* are parallel.

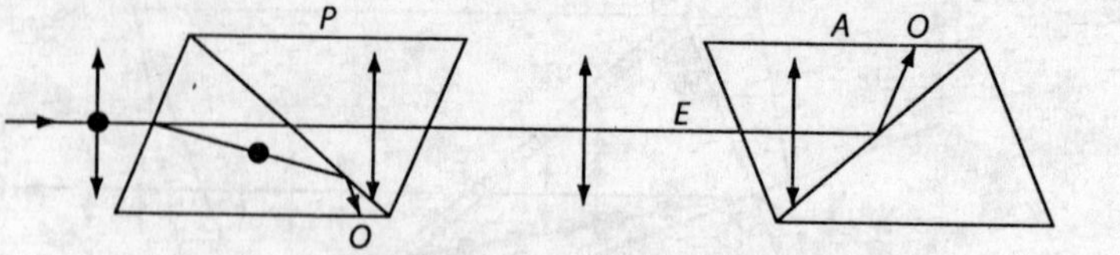

Fig. 6.18(b): Principal planes of *P* and *A* are crossed.

6.10 Malus Law

Malus law relates the intensity of the transmitted light from the analyzer with the angle between the transmission planes of the polarizer and the analyzer. The law states that, *the intensity of light transmitted from the analyzer is proportional to the square of the cosine of the angle between the planes of transmission of analyzer and polarizer.*

Proof. Let the angle between the planes of polarizer (OP) and the rotating analyzer (OA) at any instant be θ. The amplitude a of vibration of the polarized light emerging from the polarizer can be resolved into two perpendicular components, viz., $a\cos\theta$ along OA and $a\sin\theta$ along OB as shown in Fig. 6.19. The parallel component $a\cos\theta$ is transmitted through the plane of transmission (OA) of the analyzer and the perpendicular component $a\sin\theta$ is eliminated.

Thus, the intensity of the emergent light from the analyzer is

$$I_\theta = a^2\cos^2\theta = I\cos^2\theta, \tag{6.7}$$

where $I = a^2$ is the intensity of the polarized light incident on the analyzer. Since I is constant, we have $I_\theta \propto \cos^2\theta$. Thus, eqn (6.7) represents the *Malus law*.

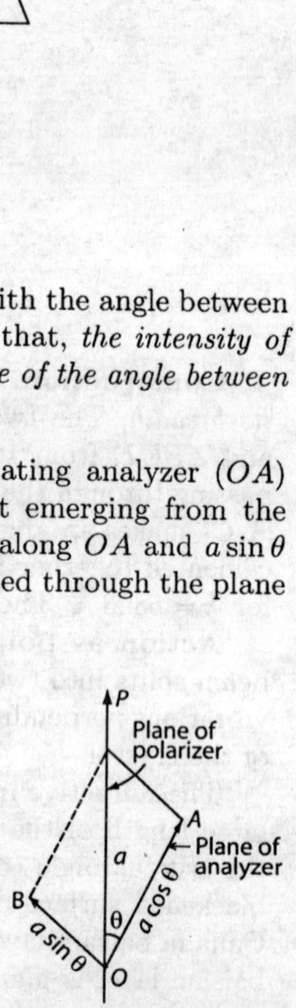

Fig. 6.19: Representation of Malus law.

Special Cases. (i) If the polarizer and analyzer are parallel to each other, then $\theta = 0°$ or, $180°$. Then $\cos^2\theta = 1$ and $I_\theta = I$. (ii) If the polarizer and the analyzer are perpendicular to each other, then $\theta = 90°$ and $\cos\theta = 0$. $\therefore I_\theta = 0$.

Verification of Malus law. A beam of unpolarized light is passed through a polarizer (P) and an analyzer (A). The emergent beam from the analyzer is received by a photocell to measure the intensity of light [Fig. 6.21]. The analyzer is rotated with the beam as axis and the orientation of the analyzer w.r.t. the polarizer for the maximum intensity is marked as $\theta = 0$.

Next θ is varied by rotating the analyzer (A) and for each value of θ the corresponding intensity I_θ is recorded photo-electrically by the milliammeter (mA).

Then I_θ is plotted against the value of $\cos^2\theta$ and straight line graph is obtained passing through the origin [Fig. 6.21]. This verifies *Malus law* experimentally.

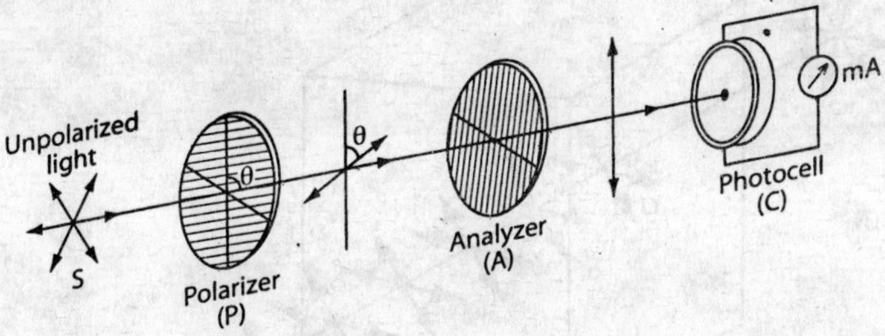

Fig. 6.20: Verification of Malus law.

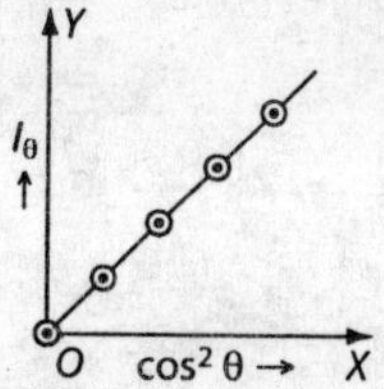

Fig. 6.21: Variation of I_θ with $\cos^2\theta$.

6.11 Polarization by Scattering

If a narrow beam of natural light passes through a medium consisting of suspension of ultramicroscopic particles, then the scattered light is partially polarized. If the dimension of the particles be smaller than the wavelength of light, then the intensity of the scattered light (follows the *Rayleigh scattering formula*) is inversely proportional to the wavelength λ, i.e., $I \propto 1/\lambda^4$ and the scattered light is plane polarized. The *degree of polarization* depends on the *angle of scattering.*

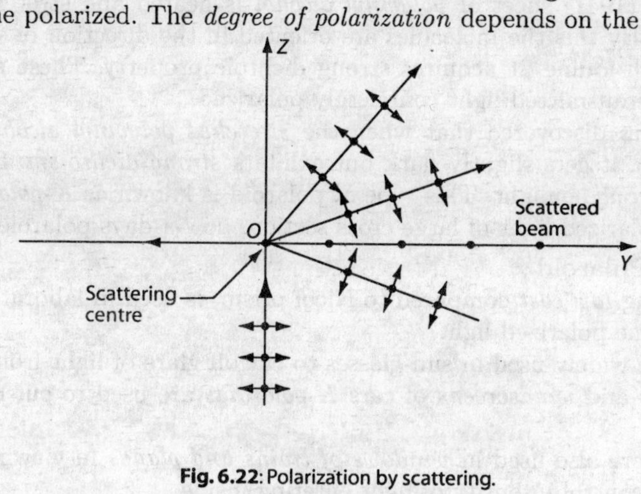

Fig. 6.22: Polarization by scattering.

The beam scattered at 90° w.r.t. the direction of incidence is completely *plane polarized.* The direction of vibration of the scattered light will be perpendicular to the plane containing direction of propagation and direction of vibration [Fig. 6.22].

The *blue colour of the sky* is due to the scattered sunlight by the small particles present in the atmosphere. Brewster showed that *the scattered blue light of the sky is polarized.*

6.12 Polarization by Selective Absorption

Dichroism. Doubly refracting crystals like *tourmaline* [an aluminosilicate containing Al_2O_3, B_2O_3 and SiO_2] absorb one of the doubly refracting rays strongly while the other is allowed to pass through with no loss. *This phenomenon of selective absorption is called dichroism and the crystals possessing this property are known as dichroic crystals.*

If an unpolarized light passes through tourmaline of proper thickness the *O-ray* is completely absorbed and the *E-ray* is transmitted without any loss. Thus, the transmitted light is linearly polarized [Fig. 6.23]. Thus, a tourmaline crystal of thickness ~ 1 mm may be used as a polarizer.

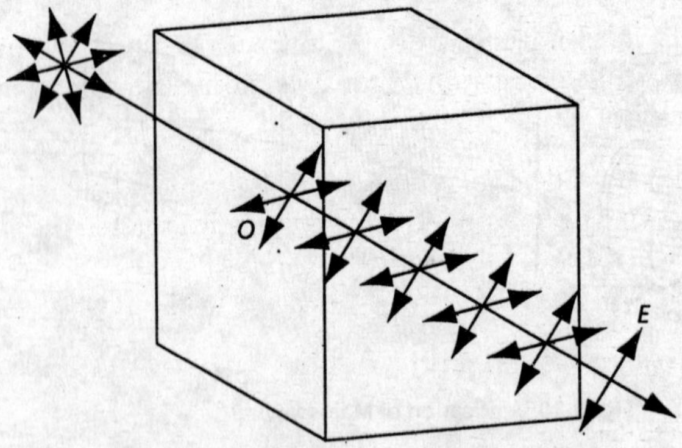

Fig. 6.23: Polarization by selective absorption.

Polaroid. It is a thin and large transparent film of crystalline polarizing material, which can produce plane polarized light when ordinary light passes through it. *Quinine sulphate peroxides* called *herapathite* are aligned in a solid matrix with their optic axis all parallel when they behave as a single crystal of large dimensions. It is highly dichroic and produces plane polarized light by selective absorption.

In recent years (1941) sheet of *polyvinyl alcohol* is heated and stretched from 3 to 8 times its original length. By this the molecules are oriented in the direction of stress. When the sheet is impregnated with iodine, it acquires strong dichroic property. These are called *H-polaroids*, which renders the transmitted light completely polarized.

Land and Rogers discovered that when the *stretched polyvinyl alcohol film* is heated with catalyst (e.g., HCl), it gets slightly dark but exhibits *strong dichroism.* It is very stable and is not bleached by strong sunlight. This type of polaroid is known as *K-polaroid.*

To get plane polarized light of large cross section, now-a-days polaroids are widely used.

Applications of Polaroid.

1. Polaroid being *low cost* compared to Nicol prism, is used in laboratory for the production and analysis of plane polarized light.
2. Polaroids are widely used in sun-glasses to cut off glare of light from horizontal surfaces.
3. In *headlights* and *windscreens* of cars *K-polaroids* are used to cut off dazzling light of an approaching car.
4. *K-polaroids* are also used in *windows of trains and planes* in viewing stereoscopic motion of pictures and to control intensity of light entering inside.

6.13 Huygens Theory of Double Refraction

Huygens wave theory in its original form is unable to explain the phenomenon of double refraction in uniaxial crystals. He, therefore, extended his theory of *secondary wavelets* and explained the phenomenon of *double refraction* as below:

1. Due to the incidence of a wavefront on a doubly refracting crystal, every point on the crystal becomes the origin of two wavefronts, viz., *ordinary and extraordinary wavefronts to account for the O-ray and E-ray.*
2. The *O-ray* always obeys the laws of refraction and has the same velocity in all possible directions. That is why the *ordinary wavefront is spherical.*
3. As the *E-ray* does not obey the laws of refraction and has different velocity in different directions, the E-wave surface is an *ellipsoid of revolution* with the optic axis as the axis of revolution.

4. A *uniaxial crystal* does not show double refraction along the optic axis as the *E-ray* and *O-ray* travel with the same velocity along that direction. *The sphere and the spheroid touch each other along the optic axis.*

5. If the ellipsoid lies outside the sphere, then it implies that the velocity of *E-ray* is more than that of the *O-ray* in the crystal. Such crystals are called *negative crystals*, e.g., calcite, tourmaline, etc. [Fig. 6.15(a)].

6.14 Theory of Production of Different Types of Polarized Light

Plane polarized light. If the light vectors vibrate simple harmonically along a fixed straight line perpendicular to the direction of propagation (i.e., the orientation of the light vector does not change but its magnitude undergoes variation), then the polarized light is called *plane polarized* [Fig. 6.24].

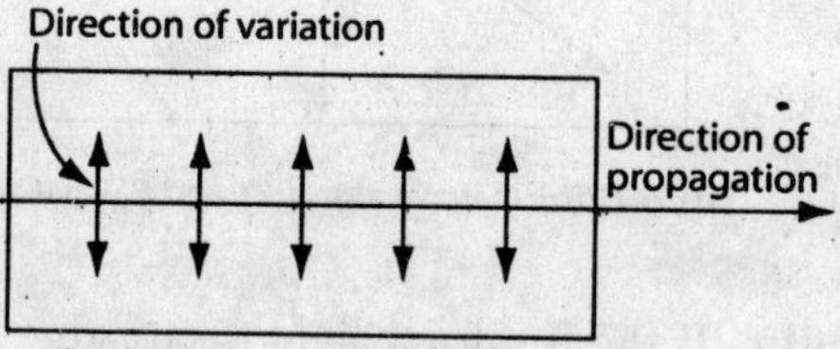

Fig. 6.24: Plane polarized light.

Circularly polarized light. If two plane polarized waves superpose, then under certain conditions, the resultant light vector may undergo rotation with constant period in a plane perpendicular to the direction of propagation. If the resultant light vector remains constant in magnitude during rotation but its orientation varies regularly so that the tip of the resultant vector traces a circle, then light is called *circularly polarized* [Fig. 6.25].

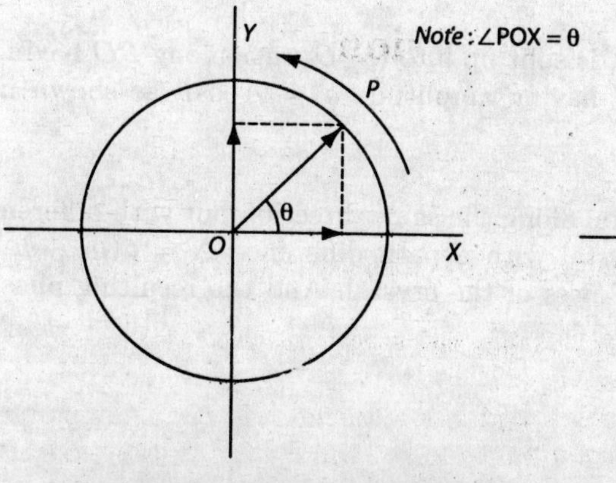

Fig. 6.25: Circularly polarized light.

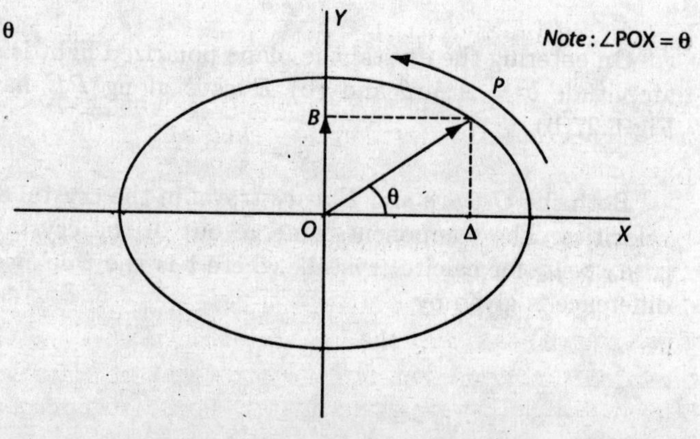

Fig. 6.26: Elliptically polarized light.

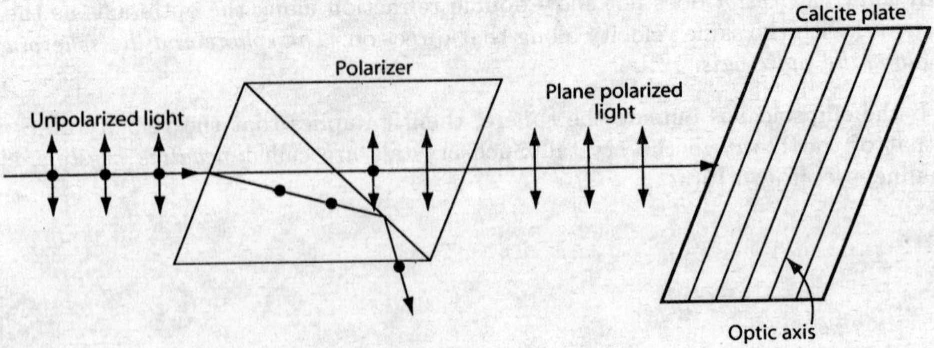

Fig. 6.27(a): Plane polarized light incident on crystal plate.

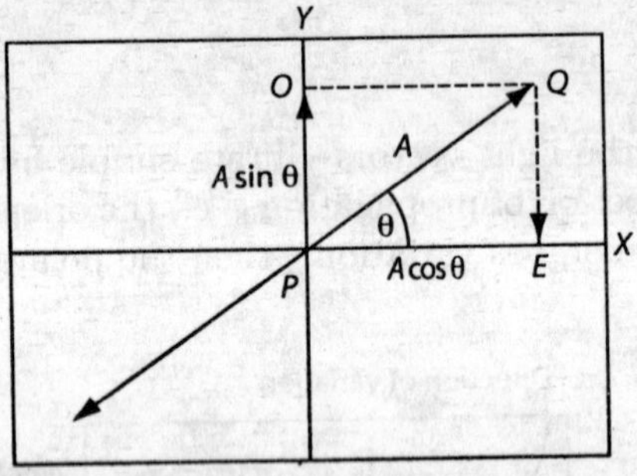

Fig. 6.27(b): *O*-wave along *PY* and *E*-wave along *PX*.

Elliptically polarized light. If the resultant light vector due to the superposition of two plane polarized light waves undergoes rotation with constant period in a plane perpendicular to the direction of propagation of light, then under certain conditions, both the magnitude and the orientation of the vector may vary, so that the tip of the vector traces an *ellipse*, and the light is called *elliptically polarized light.*

Theory. We consider a beam of plane polarized light obtained from a Nicol prism falling normally on a calcite crystal, cut with optic axis parallel to its faces [Fig. 6.27(a)]. Let θ be the angle, which the plane of vibration of the polarized light makes with the optic axis of the crystal.

On entering the crystal the plane polarized light is split up into (a) *O-wave* along PO having amplitude $b = A\sin\theta$ and (b) *E-wave* along PE having amplitude $a = A\cos\theta$ as shown in Fig. 6.27(b).

Both the *O-wave* and *E-wave* travel in the crystal along the same direction but with different velocities. The components emerge out of the crystal with a path difference $\Delta = t\,(\mu_o - \mu_e)$ [$\because \mu_o > \mu_e$ for calcite crystal], where t is the thickness of the crystal. And the resulting phase difference is given by

$$\text{Phase difference } \phi = \frac{2\pi}{\lambda}\Delta = \frac{2\pi}{\lambda}t\,(\mu_o - \mu_e).$$

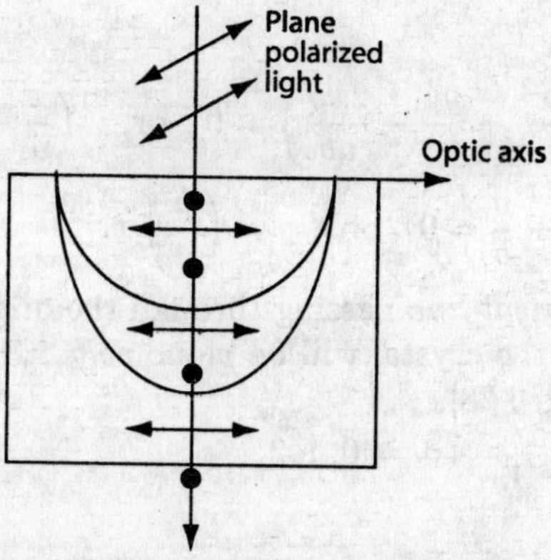

Fig. 6.27(c): *O*-wave and *E*-wave travel along the same direction with different velocities.

Now, the equations of *E-ray* and *O-ray* can be written as

$$x = A\cos\theta\sin(\omega t + \phi) \quad \text{or,} \quad x = a\sin(\omega t + \phi) \tag{6.8}$$

$$\text{and} \quad y = A\sin\theta\sin\omega t \quad \text{or,} \quad y = b\sin\omega t, \tag{6.9}$$

where ω is the angular frequency of vibration.

From eqn (6.8),

$$\frac{x}{a} = \sin(\omega t + \phi) = \sin\omega t\cos\phi + \cos\omega t\sin\phi.$$

But from eqn (6.9),

$$\frac{y}{b} = \sin\omega t \text{ and } \sqrt{1 - \frac{y^2}{b^2}} = \cos\omega t.$$

$$\therefore \quad \frac{x}{a} = \frac{y}{b}\cos\phi + \sqrt{1 - \frac{y^2}{b^2}}\sin\phi \quad \text{or,} \quad \left(\frac{x}{a} - \frac{y}{b}\cos\phi\right)^2 = \left(1 - \frac{y^2}{b^2}\right)\sin^2\phi$$

$$\text{or,} \quad \frac{x^2}{a^2} + \frac{y^2}{b^2} - \frac{2xy}{ab}\cos\phi = \sin^2\phi. \tag{6.10}$$

Eqn (6.10) represents an *ellipse* confined within a rectangle of sides $2a$ and $2b$. The actual nature of the resultant emergent light depends on the value of ϕ.

Special cases.

Case (i) When $\phi = 2m\pi \ (m = 0, 1, 2, \cdots)$
$= 0, 2\pi, 4\pi, \cdots$.

$\sin\phi = 0$ and $\cos\phi = 1$.

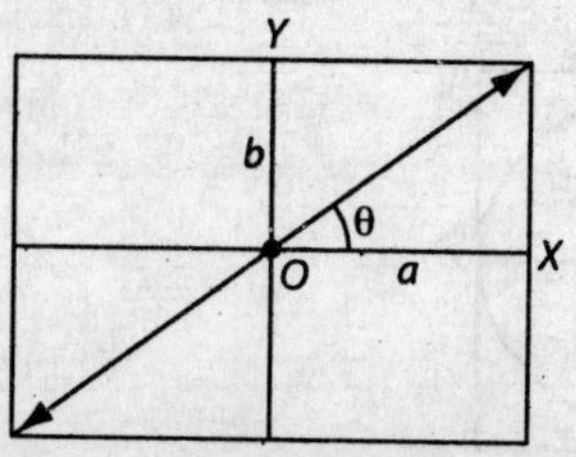

Fig. 6.28: Emergent light is plane polarized.

Then eqn (6.10) reduces to

$$\left(\frac{x^2}{a^2}+\frac{y^2}{b^2}-\frac{2xy}{ab}\right)=0 \quad \text{or,} \quad \left(\frac{x}{a}-\frac{y}{b}\right)^2=0$$

$$\text{or,} \quad \frac{x}{a}-\frac{y}{b}=0 \quad \text{or,} \quad y=\left(\frac{b}{a}\right)x. \tag{6.11}$$

Eqn (6.10) represents *a straight line* passing through the origin with a positive slope of (b/a). Thus, the emergent light from the crystal will be *plane polarized* with the vibrations in the same plane as the incident light [Fig. 6.28].

Case (ii) When $\phi = (2m+1)\pi \qquad (m = 0, 1, 2, \cdots)$
$= \pi, 3\pi, 5\pi, \cdots$.

From eqn (6.10), we get

$$\frac{x^2}{a^2}+\frac{y^2}{b^2}+\frac{2xy}{ab}=0 \quad \text{or,} \quad \left(\frac{x}{a}+\frac{y}{b}\right)^2=0$$

$$\text{or,} \quad \frac{x}{a}+\frac{y}{b}=0 \quad \text{or,} \quad y=-\frac{b}{a}x. \tag{6.12}$$

Eqn (6.12) represents *a straight line* passing through the origin having negative slope $(-\frac{b}{a})$. The emergent light in this case will be *plane polarized, with the direction of vibration making an angle* $(-\theta)$ *with the X-axis* [Fig. 6.29].

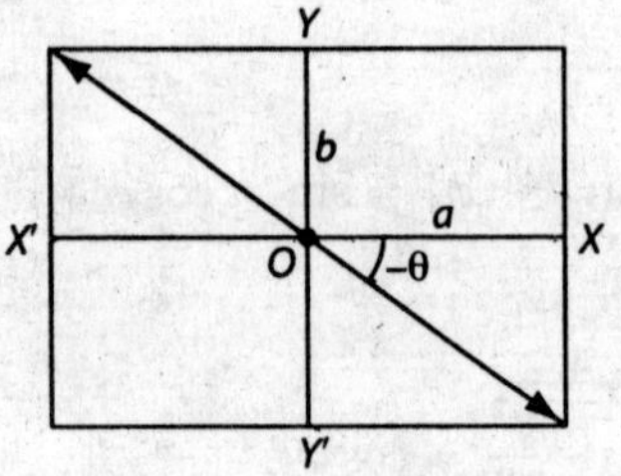

Fig. 6.29: Emergent light is plane polarized with negative slope.

Case (iii) When $\phi = (2m+1)\frac{\pi}{2} = \frac{\pi}{2}, \frac{3\pi}{2}, \frac{5\pi}{2}, \cdots$ with $m = 0, 1, 2, \cdots$
$\sin\phi = 1$ and $\cos\phi = 0$.

The eqn (6.10), then reduces to

$$\frac{x^2}{a^2}+\frac{y^2}{b^2}=1. \tag{6.13}$$

Eqn (6.13) represents *a symmetrical ellipse.* The emergent light from the crystal, then is *elliptically polarized* as shown in Fig. 6.30 with the axes of the ellipse parallel to X- and Y-axes.

Case (iv) When $\delta = \frac{\pi}{2}, \frac{3\pi}{2}, \frac{5\pi}{2}, \cdots$ and $a = b$, then the eqn (6.13) reduces to

$$x^2+y^2=a^2. \tag{6.14}$$

Thus, the emergent light from the crystal will be *circularly* polarized [Fig. 6.31].
Now, if $A\cos\theta = A\sin\theta$, then $a = b$, $\theta = 45°$.

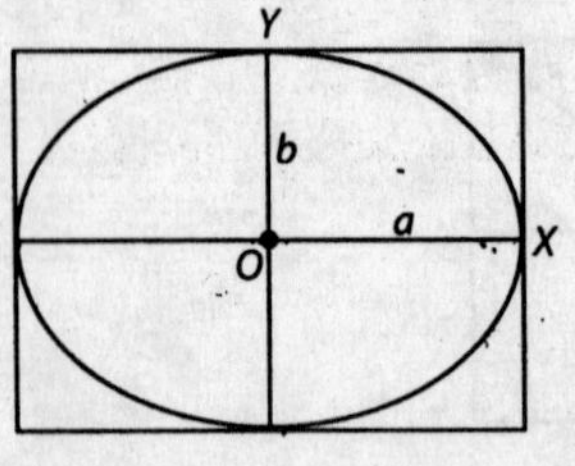

Fig. 6.30: Emergent light is elliptically polarized.

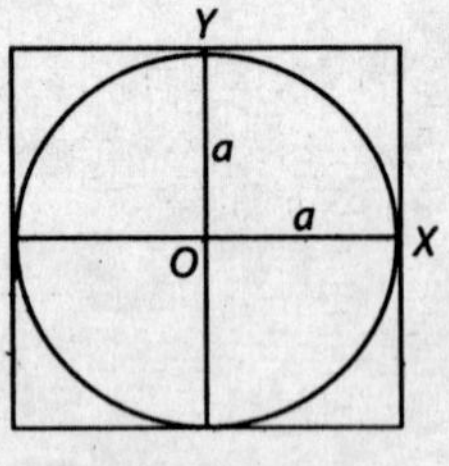

Fig. 6.31: Emergent light is circularly polarized.

Hence, the incident plane polarized light will make an angle 45° with the optic axis of the crystal.

6.15 Left-handed and Right-handed Polarized Light

Elliptically polarized light as well as circularly polarized light both may either be *left-handed* or *right-handed.*

If $\phi = \frac{\pi}{2}, \frac{5\pi}{2}, \frac{9\pi}{2}, \cdots$ the equation of motion of *E-wave* and *O-wave* from eqns (6.8) and (6.9) are

$$\left.\begin{aligned} x &= a\cos\omega t \\ y &= b\sin\omega t \end{aligned}\right\}. \tag{6.15}$$

$$\therefore \quad \frac{x^2}{a^2} + \frac{y^2}{b^2} = 1 \quad \text{and} \quad \tan\omega t = \left(\frac{y}{x}\frac{a}{b}\right)$$

is positive, i.e., ωt *is positive.* So the tip of the resultant light vector $\vec{r} = \hat{i}x + \hat{j}y$ traces out an ellipse in the *anti-clockwise direction.* The corresponding polarized light is then said to be *left-handed elliptically polarized light* [Fig. 6.32].

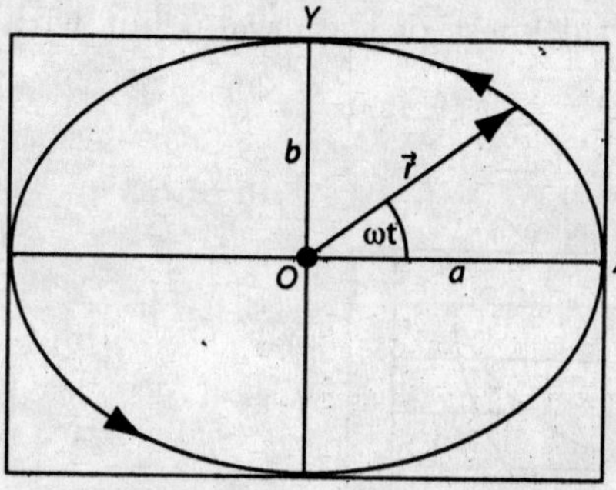

Fig. 6.32: Left-handed elliptically polarized light.

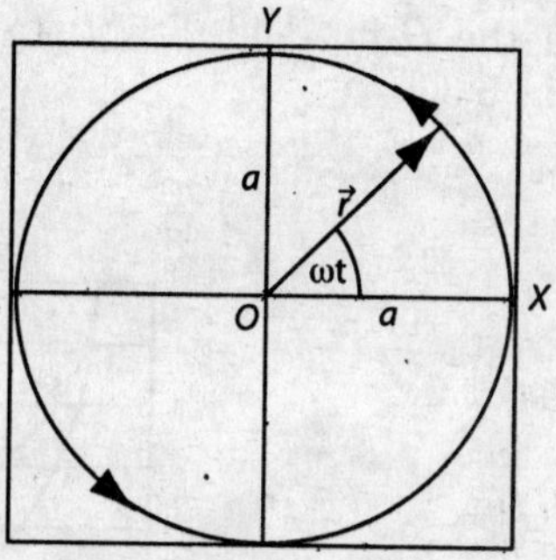

Fig. 6.33: Left-handed circularly polarized light.

In addition, if $a = b$, then $x^2 + y^2 = a^2$ and $\tan\omega t = \frac{y}{x}$, i.e., ωt is *positive.* So the resultant light is *left-handed circularly polarized light* [Fig. 6.33].

If, however, $\delta = \frac{3\pi}{2}, \frac{7\pi}{2}, \frac{11\pi}{2}, \cdots$ the equations for *E-* and *O-waves* are

$$\left.\begin{aligned} x &= -a\cos\omega t \\ y &= b\sin\omega t \end{aligned}\right\}. \tag{6.16}$$

$$\therefore \quad \frac{x^2}{a^2} + \frac{y^2}{b^2} = 1 \quad \text{and} \quad \tan\omega t = -\left(\frac{y}{x}\frac{a}{b}\right),$$

i.e., ωt is negative, then the tip of resultant light vector $\vec{r} = -\hat{i}x + \hat{j}y$ describes an ellipse in the *clockwise direction* and the emergent light is *right-handed elliptically polarized light* [Fig. 6.34].

If, in addition, $a = b$, then

$$\begin{aligned} x &= -a\cos\omega t \\ y &= a\sin\omega t; \end{aligned}$$

so $x^2 + y^2 = a$.

Hence, the emergent light is *right-handed circularly polarized light* [Fig. 6.35].

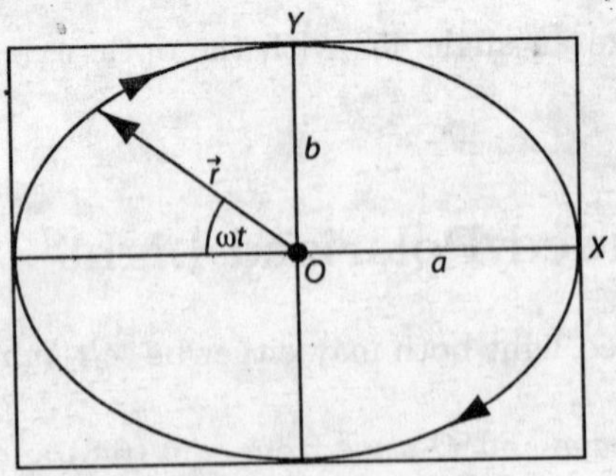

Fig. 6.34: Right-handed elliptically polarized light.

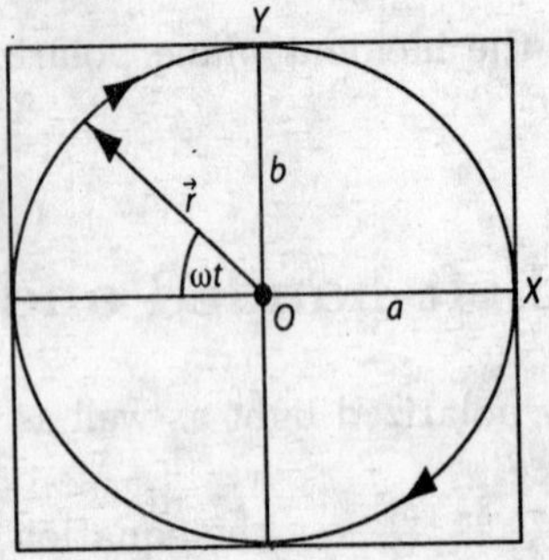

Fig. 6.35: Right-handed circularly polarized light.

6.16 Retardation Plates

Quarter-wave plate: *It is a plate of doubly refracting crystal of such a thickness as to introduce a path difference of* $\lambda/4$ *(or a phase retardation of* $\pi/2$*) between the O- and E-waves after emergence.*

Let a plane parallel plate be obtained by cutting a doubly refracting crystal with faces parallel to optic axis. If now a beam of light of wavelength λ be incident normally on the face of the crystal, then the *O-ray* and *E-ray* travel the same thickness of the crystal but with *different velocities* [Fig. 6.36].

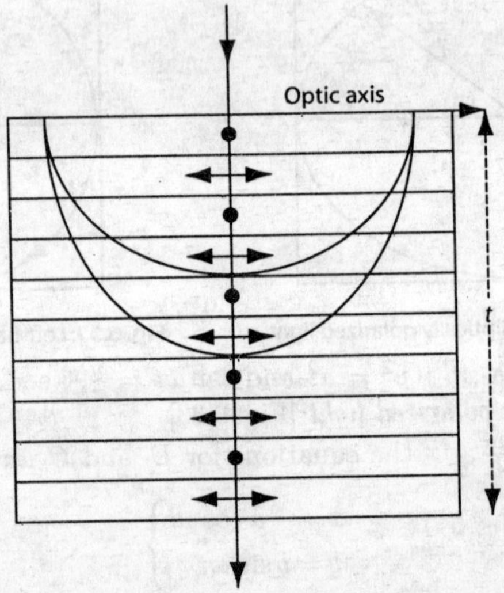

Fig. 6.36: *O*-ray and *E*-ray travel with different velocities.

For *negative crystal* $v_e > v_o$ and hence $\mu_o > \mu_e$, where v_e, μ_e and v_o, μ_o are the velocity and refractive index of *E-ray* and *O-ray* respectively. If t be the thickness of the crystal, then the path difference between the *E-ray* and *O-ray* after emergence is $\Delta = (\mu_o - \mu_e)t$.

For a *quarter wave plate,* $\Delta = \lambda/4$.

$$\therefore \quad (\mu_o - \mu_e)t = \frac{\lambda}{4} \quad \text{or,} \quad t = \frac{\lambda}{4(\mu_o - \mu_e)}. \tag{6.17}$$

For a *positive crystal,* $\mu_e > \mu_o$.

$$\therefore \quad t = \frac{\lambda}{4(\mu_e - \mu_o)}. \tag{6.18}$$

Eqns (6.17) and (6.18) give respectively *the minimum thickness of a quarter-wave plate for negative and positive crystals.*

Quarter-wave plate is used for the production of *circularly and elliptically polarized light.* Further a quarter-wave plate in conjunction with a Nicol is used to analyse all kinds of polarized light.

Half-wave plate: *It is a doubly refracting crystal plate of such a thickness as to introduce a path difference of $\lambda/2$ (or a phase retardation of π) between the O- and E-waves after emergence.*

If t be the thickness of the crystal, then for *half-wave plate of a negative crystal* ($\mu_o > \mu_e$)

$$(\mu_o - \mu_e)t = \frac{\lambda}{2} \quad \text{or,} \quad t = \frac{\lambda}{2(\mu_o - \mu_e)}. \tag{6.19}$$

For a *positive crystal,* $\mu_e > \mu_o$ and the *thickness of a half-wave plate for such a crystal* is

$$t = \frac{\lambda}{2(\mu_e - \mu_o)}. \tag{6.20}$$

Eqns (6.19) and (6.20) give the *minimum thickness of a half-wave plate with negative and positive crystals respectively.*

A half-wave plate has important applications in the construction of *half-shade plate* in polarimeters.

6.17 Production of Plane, Circularly and Elliptically Polarized Light

6.17.1 Plane Polarized Light

If a beam of unpolarized monochromatic light is passed through a Nicol prism the beam splits into *O-ray* and *E-ray* [Fig. 6.37]. The *O-ray* suffers total internal reflection at the *Canada balsam* layer. The *E-ray,* which is plane polarized emerges out of the Nicol prism.

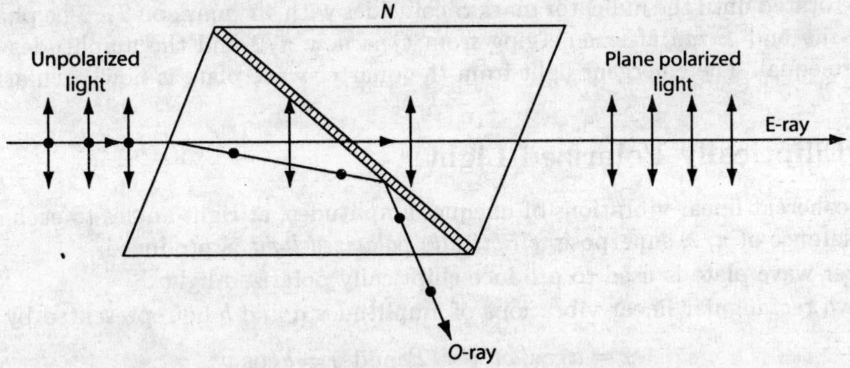

Fig. 6.37: Production of plane polarized light.

Instead of Nicol prism, now-a-days polaroids are used to produce plane polarized light.

6.17.2 Circularly Polarized Light

We know that, if two coherent linear vibrations of equal amplitudes and at right angles to each other having phase difference $\pi/2$ superpose, circularly polarized light is produced. The *circularly polarized light* can be produced by using a quarter-wave plate.

Now, if $x = a\cos(\omega t + \frac{\pi}{2})$ and $y = a\sin\omega t$ be two rectangular vibrations of equal amplitude and having phase difference $\pi/2$, then due to their superposition the resultant vibration is given by $x^2 + y^2 = a^2$, which is *circularly polarized.*

The experimental set-up for producing circularly polarized light is shown in Fig. 6.38. The following steps are followed:

(i) A parallel beam of *unpolarized monochromatic beam of light* is passed throgh a polarizer P. The transmitted light from P is *plane polarized.*

(ii) The analyzer A is kept at a certain distance in the line of incidence of unpolarized light and rotated with the line as axis until the field of view is dark. Now, the polarizer P and the analyzer A are in the *crossed* position [Fig. 6.38].

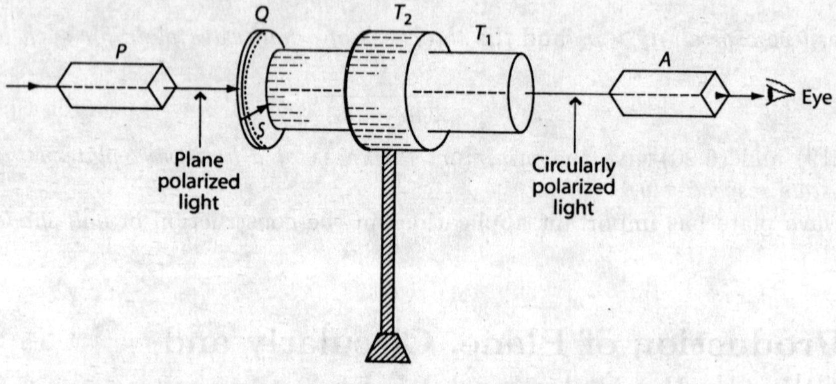

Fig. 6.38: Experimental set-up to produce circularly and elliptically polarized light.

(iii) Now, a quarter-wave plate Q mounted on a tube T_1 is placed in between P and A. The tube T_1 is capable of rotation about another fixed tube T_2. Now, the field of view of A may be bright. Q is rotated so that the field of view of A again becomes dark.

(iv) Keeping Q fixed, T_1 is rotated until zero mark on T_1 coincides with the indicator mark S on Q.

(v) Q is rotated until the indicator mark S coincides with 45° mark on T_1. The phase difference between *O-ray* and *E-ray* after emerging from Q is now $\pi/2$ and the amplitudes of these two variables are equal. The emergent light from the quarter-wave plate is now circularly polarized.

6.17.3 Elliptically Polarized Light

When two coherent linear vibrations of unequal amplitudes, at right angles to each other having a phase difference of $\pi/2$, superpose, *elliptically polarized light* is produced.

A quarter-wave plate is used to produce elliptically polarized light.

If the two rectangular linear vibrations of amplitudes a and b be represented by

$$x = a\cos(\omega t + \pi/2) \text{ and } y = b\cos\omega t,$$

then the resultant vibration due to superposition is given by $x^2/a^2 + y^2/b^2 = 1$, which represents an *elliptically polarized light.*

The experimental arrangement for producing elliptically polarized light is the same as shown in Fig. 6.38.

The steps followed are:

(i) A parallel *monochromatic beam of unpolarized light* is allowed to fall on the polarizer P. The emergent light from P is plane polarized and then falls on the analyzer A. A is set in the *crossed* position when the field of view of A is dark.

(ii) Now, the *quarter-wave plate* ($\lambda/4$-plate) Q is inserted between P and A. The field of view of A may now be bright.

(iii) The $\lambda/4$-plate is rotated till the field of view of A is dark again. Keeping Q fixed, T_1 is rotated until zero mark on T coincides with the indicator marks on Q.

(iv) Finally Q is rotated to an angle θ ($\theta \neq 45°$). Then the amplitudes of O- and E-wave from Q are unequal and we get elliptically polarized light.

6.18 Analysis of Polarized Light

Polarized light may be of the following types: 1. Unpolarized (or natural) light 2. Plane (or linearly) polarized light 3. Elliptically polarized light 4. Circularly polarized light.

There are other three types, which are mixtures of unpolarized and any of the other three. Using a polarizer and a quarter-wave plate the actual state of polarization of a light beam can be determined as discussed below:

1. A polarizer P is placed in the path of light of unknown state of polarization. If the transmitted light intensity from the polarizer is *twice extinguished* during one complete rotation of the polarizer, then the incident light is *plane polarized.*

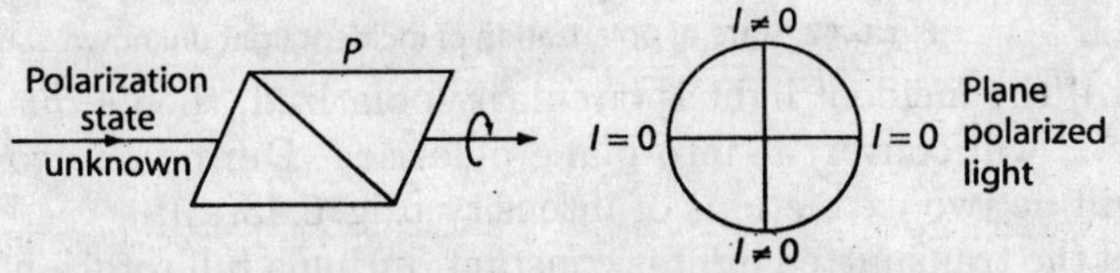

Fig. 6.39: Incident light is plane polarized.

2. If during one complete rotation of the polarizer the intensity of the transmitted light varies between a maximum (I_{max}) and a minimum (I_{min}) twice, then the incident light is either partially polarized or elliptically polarized.

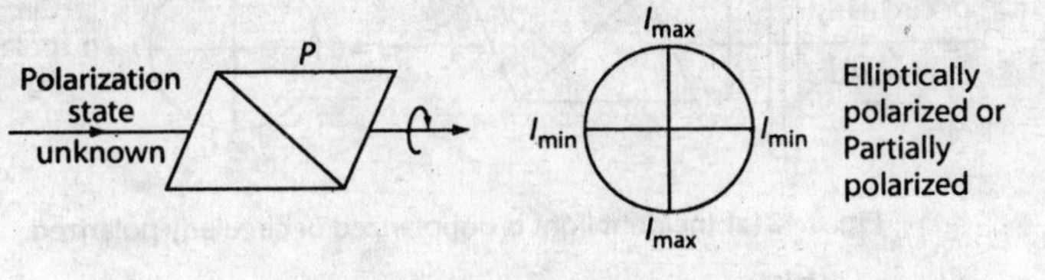

Fig. 6.40: Incident light is partially polarized.

3. To ascertain whether the given light is elliptically polarized or partially polarized we place a quarter-wave plate before the polarizer P. If now the incident light is *elliptically polarized*, then the quarter-wave plate converts it into a *plane polarized light.* Then the light intensity will be twice extinguished during a complete rotation of P.

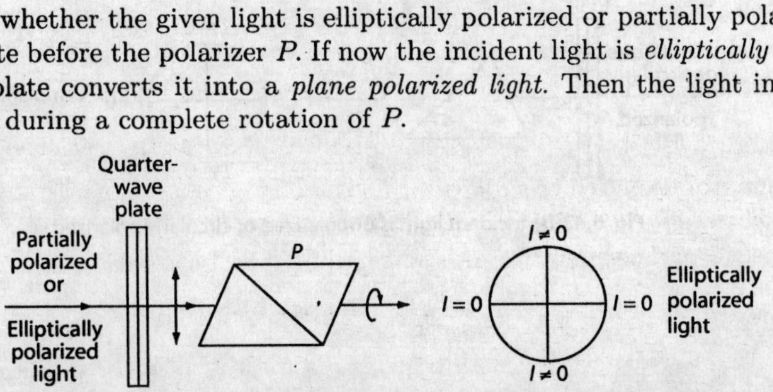

Fig. 6.41(a): Incident light is partially polarized or elliptically polarized.

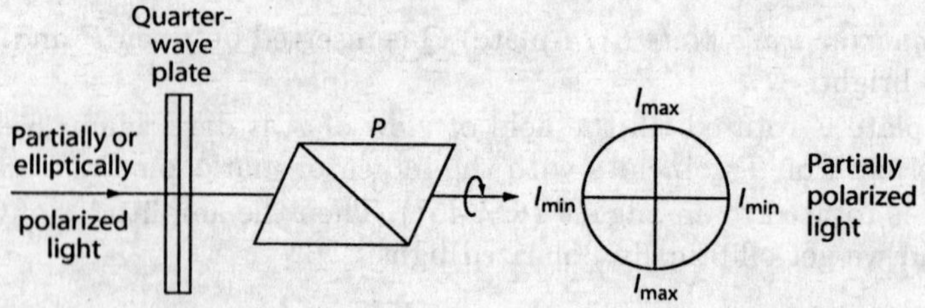

Fig. 6.41(b): Incident light is partially or elliptically polarized.

If, on the other hand, the light intensity varies between a maximum and a minimum twice during a complete rotation of P, then the intensity is partially polarized.

4. If, on rotation of the polarizer through 360°, the intensity of the transmitted light remains constant, then the incident light is either circularly polarized or unpolarized.

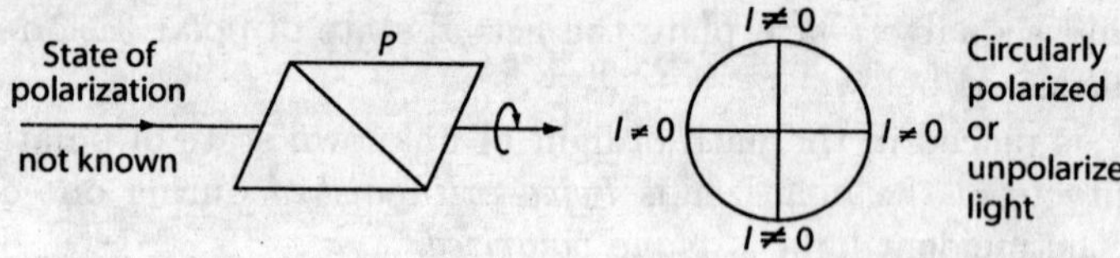

Fig. 6.42: State of polarization of incident light unknown.

5. In case of (4), if the incident light is circularly polarized, then a quarter-wave plate placed before the polarizer P will convert it into plane polarized. During one complete rotation of the polarizer P there will be two extinctions of intensity [Fig. 6.43(a)].

If the intensity of the transmitted light is constant during a full rotation of P, then the incident light is unpolarized [Fig. 6.43(b)].

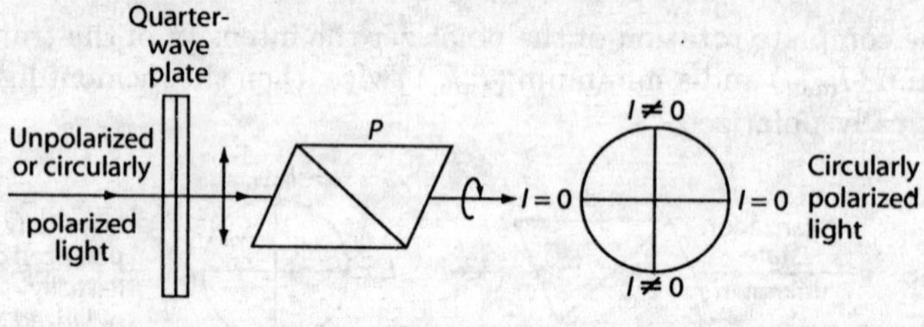

Fig. 6.43(a): Incident light is unpolarized or circularly polarized.

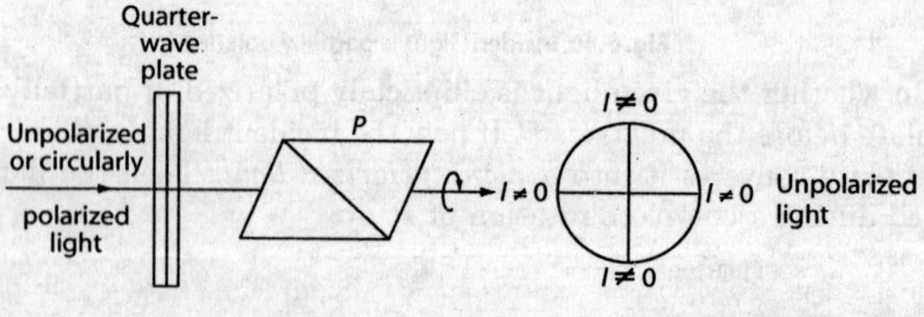

Fig. 6.43(b): Incident light is unpolarized or circularly polarized.

SUMMARY OF THE ANALYSIS

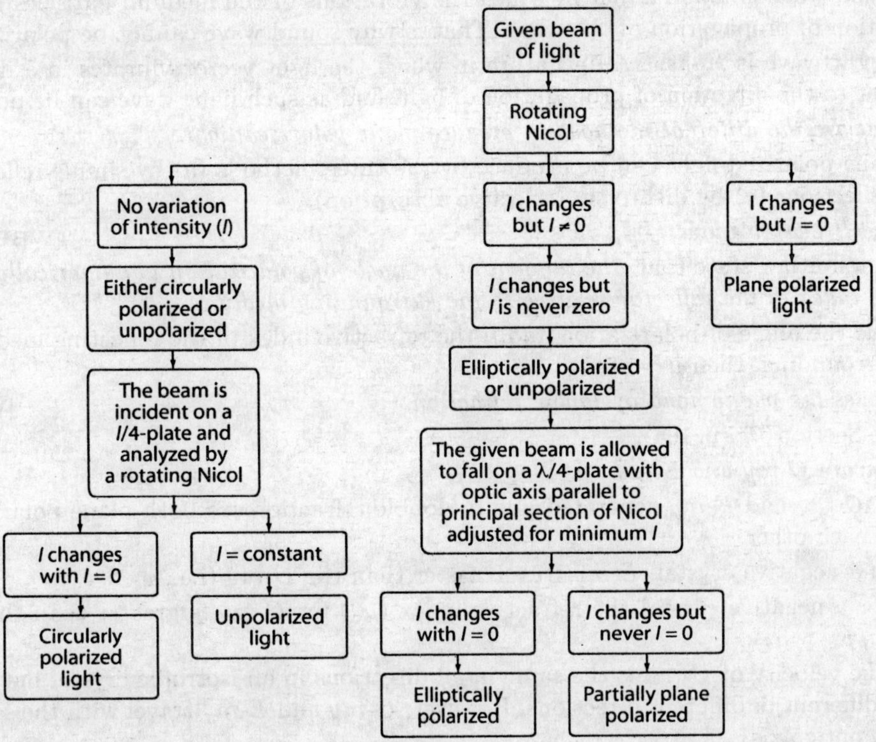

6.19 Answers of Some Important Questions

6.1. *What is polarization of light?* **[WBUT 2002, '04]**

Ans. J.C. Maxwell first theoretically predicted that light is a transverse electromagnetic wave. The transverse nature of light has been established by polarization experiments. In transverse wave, the particles of the medium execute periodic oscillations in a direction perpendicular to the direction of propagation of the light wave. In electromagnetic wave, the light emitting atoms oscillate at random and emit individual wave trains in all possible directions. The corresponding light wave is called *natural light*, which is *unpolarized.*

If, however, the oscillations are confined to only one direction, then we get a *plane polarized light.* The electric vector of transverse electromagnetic wave is identified as the *light vector.* Hence, in plane polarized light the light vectors (electric vector) vibrate in a single plane called the *plane of vibration* and the plane perpendicular to it is the *plane of polarization.*

Polarized light is of three types, viz., (a) plane polarized light (b) elliptically polarized light and (c) circularly polarized light.

6.2. *What do you mean by plane of vibration and plane of polarization of a plane polarized light?* **[WBUT 2004]**

Ans. See Section 6.4 (p. 160).

6.3. *What information regarding light do we get from the study of the phenomenon of polarization?*

Ans. From the study of polarization phenomenon we are confirmed that (a) light has a wave nature (b) light is transverse in characteristic.

6.4. *Though light and sound both are waves, the former can be polarized but the latter not—why?*
Ans. Sound wave is *longitudinal* in which the vibrations of the medium particles take place in the direction of propagation of the wave. That is why sound wave cannot be polarized.

But light wave is *transverse* in nature in which the light vector vibrates in a direction perpendicular to the direction of propagation of light and as such light wave can be polarized.
6.5. *What are the different methods to produce plane polarized light?*
Ans. Plane polarized light can be produce by the three methods (a) by simple reflection (b) by double reflections (c) by dichroism (selective absorption).
6.6. *State Brewster's law.* **[WBUT 2005, '06]**
Ans. Brewster law state that, *the tangent of the angle of polarization is numerically equal to the refractive index of the reflector relative to the surrounding medium.*

If i_p be the angle of polarization and μ the refractive index of the reflecting medium relative to the surrounding, then $\mu = \tan i_p$.
6.7. *Discuss the phenomena of double refraction.* **[WBUT 2005]**
Ans. See Section 6.7 (p. 164).
6.8. *Compare O-ray and E-ray.*
Ans. (a) *O-ray* and *E-ray* generated due to double refraction are both plane polarized at right angles to each other.

(b) In a negative crystal, *E-ray* travels faster than the *O-ray* (i.e., $v_e > v_o$).

(c) For a negative crystal the refractive index (μ_o) for *O-ray* is greater than that for *E-ray* (μ_e), (i.e., $\mu_o > \mu_e$).

(d) The velocity of *O-ray* is the same in all directions in an isotropic crystal but that for the *E-ray* is different in different directions. However, *O-ray* and *E-ray* travel with the same velocity along the optic axis.
6.9. *What are the differences between positive and negative crystals?*
Ans. See Section 6.7 (p. 164).
6.10. *State and explain Malus law.* **[CUH 2001]**
Ans. See Section 6.10 (p. 168).
6.11. *What is optic axis? What are its characteristics?*
Ans. See Section 6.8 (p. 166).
6.12. *Describe the construction and use of the Nicol prism as a polarizer and analyzer.*
[BUH 2000, CUH 2002; WBUT 2004]
6.13. *What do you mean by uniaxial and biaxial crystals?*
Ans. The crystals in which there is a single direction along, which *E-rays* and *O-rays* are transmitted with the same and uniform velocity are called *uniaxial crystals.*

Examples. Calcite, quartz.

There is another class of crystals called *bi-axial crystals* in which there are two directions of one uniform velocity for *E-rays* and *O-rays*.

Examples. Borax, mica, Aragonite, etc.
6.14. *Explain briefly the action of a dichroic polarized.* **[WBUT 2008]**
Ans. See Section 6.12 (p. 169).
6.15. *What are* (a) *Quarter-wave plate and* (b) *Half-wave plate?*
Ans. (a) Quarter-wave plate ($\lambda/4$-plate): It is a plate of doubly refracting crystal of such a thickness as to introduce a path difference of $\lambda/4$ (or a phase difference of $\pi/2$) between *O-* and *E-rays* after emergence.

If λ be the wavelength of light and μ_o and μ_e, the refractive indices for the *O-* and *E-rays* respectively, then the thickness of the quarter-wave plate is given by

$$t = \frac{\lambda}{4(\mu_o - \mu_e)} \quad \text{(if the crystal is negative)}$$

$$\text{and} \quad t = \frac{\lambda}{4(\mu_e - \mu_o)} \quad \text{(if the crystal is positive).}$$

(b) Half-wave plate ($\lambda/2$-plate): If a plate of doubly refracting crystal introduces a path difference of $\lambda/2$ (or a phase difference of π) between the *O*- and *E-rays* after emergence, then it is called a half-wave plate.

$$t = \frac{\lambda}{2(\mu_o - \mu_e)} \quad \text{for a negative crystal and}$$

$$t = \frac{\lambda}{2(\mu_e - \mu_o)} \quad \text{for a positive crystal.}$$

6.20 Solved Problems

Ex. 1 *Find the state of polarization when x- and y-components of the electric field are*

(i) $E_x = E_0 \sin(\omega t + kz)$ $\quad$ (ii) $E_x = E_0 \cos(\omega t + kz)$

$E_y = E_0 \cos(\omega t + kz)$ $\quad$ $E_y = (E_0/\sqrt{2}) \cos(\omega t + kz + \pi)$.

Soln. (i) Here $E_x^2 + E_y^2 = E_0^2[\sin^2(\omega t + kz) + \cos^2(\omega t + kz)] = E_0^2$.

This represents a circular vibration.

Also at $z = 0$, $\vec{E} = \hat{i}E_x + \hat{j}E_y = \hat{i}E_0 \sin\omega t + \hat{j}E_0 \cos\omega t$.

This shows that the resultant vibration rotates in the clockwise direction. Hence, the light is right-handed circularly polarized.

(ii) In this case,

$$E_y = -\frac{E_0}{\sqrt{2}} \cos(\omega t + kz) = -\frac{E_x}{\sqrt{2}}.$$

So the light is linearly polarized making an angle $\theta = \tan^{-1}(-1/\sqrt{2})$ with the *X*-axis.

Ex. 2 *A ray of light is incident on the surface of a glass plate of refractive index* 1.732 *at the polarizing angle. Calculate the angle of refraction of the ray.* **[WBUT 2004]**

Soln. According to Brewster's law, $\mu = \tan i_p$.

$\therefore i_p = \tan^{-1}\mu = \tan^{-1}(1.732) = 60°$.

If r be the angle of refraction, then since $i_p + r = 90°$,

$\therefore r = 90° - i_p = 90° - 60° = \mathbf{30°}$.

Ex. 3 *A polarizer and an analyzer are oriented so that the maximum intensity is achieved. To what fraction of its maximum value is the intensity reduced when the analyzer is rotated through* 60°?

Soln. If θ be the angle between the polarizer and the analyzer and I_0 be the maximum original intensity, then according to *Malus law* the transmitted intensity through the analyzer is $I = I_0 \cos^2\theta$ or, $I/I_0 = \cos^2\theta$.

The reduction in intensity is $(I_0 - I)$ and the fraction of the intensity reduced is

$$\left(\frac{I_0 - I}{I_0}\right) = 1 - \frac{I}{I_0} = 1 - \cos^2\theta = 1 - \cos^2 60° = 1 - \left(\frac{1}{2}\right)^2 = \frac{3}{4}.$$

$\therefore$ % of reduction intensity is $= \frac{3}{4} \times 100 = \mathbf{75\%}$.

Ex. 4 *A glass plate is to bc used as a polarizer. Find the angle of polarization for it. Also find the angle of refraction.* [μ *for glass* = 1.54]

Soln. According to Brewster's law, $\mu = \tan i_p$ or, $\tan i_p = 1.54$.

$\therefore i_p = \tan^{-1}(1.54) = \mathbf{57°}$.

If r be the angle of refraction, then $i_p + r = 90°$ or, $r = 90° - i_p = 90° - 57° = \mathbf{33°}$. [The Snell's law, $\mu = \frac{\sin i_p}{\sin r}$, may also be used to find r.]

Ex. 5 *What is the refractive index of a piece of glass for the light of wavelength* 5460 Å *is plane polarized when reflected at an angle* 60°? **[CUH 2001]**

Soln. According to *Brewster's law,* the refractive index of glass is $\mu = \tan i_p$, where i_p is the angle of polarization = 60°.

$\therefore \mu = \tan 60° = \mathbf{\sqrt{3}}$.

Ex. 6 *At what angle* β *above the horizon is the sun when a person observing its rays reflected in water* ($\mu = 1.33$) *finds them linearly polarized along the horizontal?* **[NBUH 2003]**

Soln. If i_p be the angle of polarization and μ the refractive index of water, then $\mu = \tan i_p$ or, $1.33 = \tan i_p$ or, $i_p = \tan^{-1}(1.33) = 57°$.

$\therefore \beta = 90° - 57° = \mathbf{33°}$.

Ex. 7 *Refractive index of glass is 1.5. Calculate the Brewster's angle for it. Also calculate the angle of refraction.* **[BHU 1990, HPU 1997]**

Soln. If i_p be the polarizing angle and μ the refractive index, then $\mu = \tan i_p$ or, $\tan i_p = \mu = 1.5$ or, $i_p = \tan^{-1}(1.5) = \mathbf{56.30°}$.

Again, $i_p + r = 90°$. $\therefore r = 90° - 56.3° = \mathbf{33.7°}$.

Ex. 8 *If the critical angle for refraction of glass to air is* 40°, *calculate the polarizing angle for glass.* **[HPU 1996]**

Soln. If θ_c be the critical angle, then

$$\mu = \frac{1}{\sin\theta_c} = \frac{1}{\sin 40°} = \frac{1}{0.6428} = 1.56.$$

$\therefore \tan i_p = 1.56$ or, $i_p = \tan^{-1}(1.56) = 57.3°$.

$\therefore$ polarizing angle $i_p = \mathbf{57.3°}$.

Ex. 9 *Two polarizing sheets have their polarizing directions parallel so that the intensity of the transmitted light is a maximum. Through what angle must either sheet be turned so that intensity becomes one half of the initial value?*

Soln. According to *Malus law* the transmitted intensity $I = I_0 \cos^2\theta$.

If $I = I_0/2$, then $I_0/2 = I_0 \cos^2\theta$ or, $\cos^2\theta = 1/2$ or, $\cos\theta = \pm 1/\sqrt{2}$.

$\therefore \theta = \cos^{-1}\left(\pm 1/\sqrt{2}\right) = \pm 45°, \pm 135°$.

Ex. 10 *Unpolarized light falls on two polarizing sheets one placed top of the other. What must be the angle between their characteristic directions, if the intensity of transmitted light is one-third of intensity of the incident beam?*

Soln. The intensity of the incident beam transmitted by the first polarizer is $I_1 = \frac{1}{2}I_0$, where I_0 = intensity of the incident beam.

If θ be the angle between the characteristic directions of two polarizers, the intensity transmitted by the second polarizer is

$$I_2 = I_1 \cos^2\theta \text{ [by Malus law]}$$
$$= \tfrac{1}{2} I_0 \cos^2\theta.$$

By the condition of the problem, $I_2 = \frac{1}{3}I_0$.

$\therefore \frac{1}{2}I_0 \cos^2\theta = \frac{1}{3}I_0$ or, $\cos^2\theta = \frac{2}{3}$ or, $\cos\theta = \sqrt{\frac{2}{3}}$.

$\therefore \theta = \cos^{-1}\left(\sqrt{\frac{2}{3}}\right) = \mathbf{35.26^\circ}$.

Ex. 11 *Calculate the thickness of quarter-wave plate with* $\lambda = 5.8 \times 10^{-7}$ m, $\mu_e = 1.553$ *and* $\mu_o = 1.544$. **[CU 1985]**

Soln. For a quarter-wave plate of quartz the required thickness of the plate is given by

$$t = \frac{\lambda}{4(\mu_e - \mu_o)} = \frac{5.8 \times 10^{-7}}{4(1.553 - 1.544)} = \frac{5.8 \times 10^{-7}}{4 \times 0.009}$$
$$= 1.61 \times 10^{-5}\,\text{m} = \mathbf{1.61 \times 10^{-3}\,cm}.$$

Ex. 12 *Calculate the thickness of a quarter-wave plate of quartz for light of wavelength* 5893 Å. *The refractive indices of quartz for E-ray and O-ray are* 1.5533 *and* 1.5442 *respectively.* **[cf. BUH 1999, CUH 1999]**

Soln. If t be the required thickness of the quartz plate, then

$$t = \frac{\lambda}{4(\mu_e - \mu_o)} = \frac{5893 \times 10^{-8}}{4(1.5533 - 1.5422)} = \frac{5893 \times 10^{-8}}{4 \times 0.0111} = \mathbf{0.0013\,cm}.$$

Ex. 13 *A plane polarized light of wavelength* 5893 Å *is incident on a thin quartz plate cut with faces parallel to the optic axis. Calculate* (a) *the minimum thickness of the plate, which introduces a phase difference* 60° *between the ordinary and extraordinary rays* (b) *the minimum thickness of the plate for which the O-ray and E-ray waves will combine to produce plane polarized light.* [$\mu_o = 1.544$ *and* $\mu_e = 1.533$]

Soln. (a) A phase difference 60° corresponds to a path difference $\frac{\lambda}{360^\circ} \times 60^\circ = \frac{\lambda}{6}$.

Now, the path difference between *E-ray* and *O-ray* in traversing a distance t through the crystal is $t(\mu_e - \mu_o)$.

By the condition of the problem $t(\mu_e - \mu_o) = \lambda/6$

$$\text{or, } t = \frac{\lambda}{6(\mu_e - \mu_o)} = \frac{5893 \times 10^{-8}}{6 \times (1.553 - 1.544)} = \frac{5893 \times 10^{-8}}{6 \times 0.009} = \mathbf{0.00109\,cm}.$$

(b) If the plate introduces a phase difference π or path difference $\lambda/2$, the *E-ray* and *O-ray* after emergence will combine to produce plane polarized light.

So $t(\mu_e - \mu_o) = \frac{\lambda}{2}$

$$\text{or, } t = \frac{\lambda}{2(\mu_o - \mu_o)} = \frac{5893 \times 10^{-8}}{2 \times 0.009} = \mathbf{0.003274\,cm}.$$

Ex. 14 *Calculate the thickness of a quartz half-wave plate for the line* 6563 Å *for which the extraordinary and ordinary refractive indices are* $\mu_e = 1.55085$ *and* $\mu_o = 1.54184$. **[CU 1997]**

Soln. If t be required thickness of the quartz crystal, then

$$t = \frac{\lambda}{2(\mu_e - \mu_o)} = \frac{6563 \times 10^{-8}}{2(1.55085 - 1.54184)} = \frac{6563 \times 10^{-8}}{2 \times 0.00901} = \mathbf{3.64 \times 10^{-3}\ cm.}$$

Ex. 15 *A given calcite plate behaves as a half-wave plate for a particular wavelength* λ. *Assuming variation of refractive index with* λ *to be negligible, how would the above plate behave for another light of wavelength* 2λ.

Soln. For calcite, $\mu_o > \mu_e$, the thickness of calcite half-wave plate is given by

$$t = \frac{\lambda}{2(\mu_o - \mu_e)} = \frac{2\lambda}{4(\mu_o - \mu_e)} = \frac{\lambda'}{4(\mu_o - \mu_e)}.$$

This represents the thickness of a quarter-wave plate for wavelength λ'. So, a half-wave plate for λ will behave as a quarter-wave plate for $\lambda' = 2\lambda$, if μ does not vary with λ.

Ex. 16 *A thin quartz crystal of thickness t has been so cut that the optic axis is parallel to the surface of the crystal. If two beams of light* (λ = 589 nm) *polarized parallel and perpendicular to the optic axis are incident normally on the crystal in phase, how thick must the crystal be in order that the beams emerge* 90° *out of phase? Given: index of refraction of the crystal for light polarized parallel to the optic axis is* 1.55 *and that for light perpendicular to optic axis is* 1.54.

[BUH 1998]

Soln. If μ_e and μ_o be the refractive incidences of the crystal for *E-ray* and *O-ray* respectively and t the thickness, then the path difference after emergence is $t(\mu_e - \mu_o)$.

Now, 90° phase difference corresponds to $\frac{\lambda}{360°} \times 90° = \frac{\lambda}{4}$ path difference.

$\therefore t(\mu_e - \mu_o) = \frac{\lambda}{4}$

$$\text{or, } t = \frac{\lambda}{4(\mu_e - \mu_o)} = \frac{5890 \times 10^{-8}}{4(1.55 - 1.54)} = \frac{5890 \times 10^{-8}}{4 \times 0.01} = \mathbf{0.00147\,cm.}$$

Ex. 17 *A beam of plane polarized light is changed into circularly polarized light by passing it through a slice of crystal* 0.003 cm *thick. Calculate the birefringence of the crystal assuming this to be the minimum thickness that will produce the effect* [$\lambda = 6 \times 10^{-5}$ cm].

Soln. If t be the thickness of the crystal, then since the plane polarized light is converted into circularly polarized light,

$$t = \frac{\lambda}{4(\mu_e - \mu_o)} = \frac{\lambda}{4\Delta\mu} \quad \text{or,} \quad \Delta\mu = \frac{\lambda}{4t} = \frac{6 \times 10^{-5}}{4 \times 0.003} = \mathbf{0.005}$$

is the birefringence of the crystal.

Ex. 18 *Two linearly polarized light waves are in phase but have different amplitudes. They are represented by*

$$\vec{E}_1(z,t) = \hat{i}A_1\cos(kz - \omega t) + \hat{j}B_1\cos(kz - \omega t)$$
$$\vec{E}_2(z,t) = \hat{i}A_2\cos(kz - \omega t) + \hat{j}B_2\cos(kz - \omega t).$$

Show that, $\vec{E}_1 + \vec{E}_2 = \vec{E}$ *is also linearly polarized. Find the polarization direction.* **[CUH 2001]**

Soln. $\vec{E}(z,t) = \vec{E}_1(z,t) + \vec{E}_2(z,t) = \left[\hat{i}(A_1 + A_2) + \hat{j}(B_1 + B_2)\right]\cos(kz - \omega t).$

Here, the amplitude $\hat{i}(A_1 + A_2) + \hat{j}(B_1 + B_2)$ is a constant. So the resultant wave is linearly polarized.

Again, let $\vec{E} = \hat{\imath}E_x + \hat{\jmath}E_y$, then

$$E_x = (A_1 + A_2)\cos(kz - \omega t)$$

$$E_y = (B_1 + B_2)\cos(kz - \omega t).$$

$$\therefore \quad \frac{E_y}{E_x} = \frac{B_1 + B_2}{A_1 + A_2} \quad \text{or,} \quad E_y = \frac{B_1 + B_2}{A_1 + A_2}E_x = (\tan\theta)E_x.$$

So, it represents a linearly polarized vibration in the XY-plane making an angle

$$\theta = \tan^{-1}\left(\frac{B_1 + B_2}{A_1 + A_2}\right) \text{ with } X\text{-axis.}$$

EXERCISES

6(A)–Objective Type Questions

Select the correct answer:

6.1. Polarization conclusively proved that light wave is (a) longitudinal (b) progressive (c) stationary (d) transverse.

6.2. Which of the following phenomena proves the transverse characteristics of light? (a) interference (b) dispersion (c) polarization. **[WBUT 2004]**

6.3. The angle between the planes of vibration and polarization of a beam of plane polarized light is (a) 90° (b) 45° (c) 180°. **[WBUT 2005]**

6.4. The plane of vibration and plane of polarization of a beam of plane polarized light (a) are identical to each other (b) are orthogonal to each other (c) make an angle, which depends on the colour of the light (d) rotate w.r.t. each other along the path of the beam. **[WBUT 2006]**

6.5. Plane polarized light can be produced by (a) simple reflection (b) Nicol prism (c) pile of plates (d) all of the above.

6.6. The phase difference ϕ between *E-ray* and *O-ray* on traversing a thickness t of calcite crystal is (a) $\frac{2\pi t}{\lambda}(\mu_o - \mu_e)$ (b) $\frac{2\pi t}{\lambda}(\mu_e - \mu_o)$ (c) $\frac{2\pi t}{\lambda}\mu_e\mu_o$ (d) $\frac{2\pi t}{\lambda}(\mu_e/\mu_o)$.

6.7. Nicol prism works on the principle of (a) scattering (b) refraction (c) reflection (d) double refraction.

6.8. If θ be the angle of polarization, the refractive index μ of the material is given by (a) $\sin\theta$ (b) $\cos^{-1}\theta$ (c) $\tan\theta$ (d) $\cos\theta$.

6.9. If the refractive index of water is 1.33, the angle of polarization of light reflected from water is (a) $\cos^{-1}(1.33)$ (b) 53.1° (c) 36.9° (d) none of these.

6.10. If θ_p and r be the angle of polarization and refraction of a medium respectively, then $(\theta_p + r)$ is (a) $> 90°$ (b) $< 90°$ (c) $= 90°$ (d) none of these.

6.11. When light is incident at Brewster's angle on the surface of a transparent medium, the reflected light is (a) plane polarized (b) partially plane polarized (c) circularly polarized (d) elliptically polarized.

6.12. If I_0 be the intensity of the incident of an unpolarized light on a polarizer, then the intensity transmitted through an analyzer making an angle θ with the polarizer is (a) $I_0\cos^2\theta$ (b) I_0 (c) $\frac{1}{2}I_0\cos^2\theta$ (d) zero.

6.13. An unpolarized light of intensity I_0 passes through a polaroid. The intensity of the transmitted light is (a) I_0 (b) $\frac{I_0}{2}$ (c) $\frac{I_0}{4}$ (d) $\frac{3I_0}{4}$.

6.14. A light beam is incident on a polarizer. On rotating the polarizer the transmitted intensity is found to vary between maximum and zero. The incident light on the polarizer is (a) unpolarized (b) partially plane polarized (c) plane polarized (d) circularly polarized.

6.15. If on rotating the analyzer, the emergent light does not change in intensity, then it is (a) either plane polarized or partially polarized (b) either unpolarized or circularly polarized (c) either partially polarized or elliptically polarized (d) only circularly polarized.

6.16. If the emergent light from a rotating polarizer shows maximum and minimum (not zero) intensities, then the incident light is (a) partially plane polarized (b) unpolarized (c) circularly polarized (d) plane polarized.

6.17. An ink dot on a white paper is viewed through a calcite crystal by rotating about an axis perpendicular to the plane of the paper. It will be observed that (a) two dots are rotating (b) one dot is rotating about the other (c) two dots are stationary (d) a single dot is fixed.

6.18. When the *O-ray* and *E-ray* travel along the optic axis of an uniaxial crystal (a) $\mu_e > \mu_o$ (b) $\mu_e < \mu_o$ (c) $\mu_e = \mu_o$ (d) $\mu_e = \frac{1}{\mu_o}$.

6.19. If two vibrations given by

$$E_x = E_0 \sin(\omega t - kz)$$
$$E_y = E_0 \cos(\omega t - kz),$$

superpose, then the resultant vibration is (a) plane polarized (b) circularly polarized (c) elliptically polarized (d) none of these.

6.20. If a $\frac{\lambda}{4}$-plate of quartz is placed between two crossed polaroid such that $\frac{\lambda}{2}$-plate is oriented at 45° to the axis of the polaroid, the combination will transmit (a) no light (b) $\frac{1}{4}$ of the intensity of incident light (c) $\frac{1}{2}$ of the intensity of incident light (d) none of these.

6.21. A half-wave plate for λ will behave for 2λ as a (a) quarter-wave plate (b) half-wave plate (c) glass plate (d) none of these.

6.22. A light ray is passing through a calcite crystal. If the plane of vibration of the light ray is perpendicular to the optic axis, then the ray (a) is an *O-ray* (b) is an *E-ray* (c) has both non-zero O- and E-components (d) cannot propagate through the crystal. **[WBUT 2008]**

6.23. In the propagation of light wave, the angle between the plane of the polarization and plane of vibration is (a) 0° (b) $\pi/2$ (c) $\pi/4$ (d) π. **[WBUT 2010]**

6.24. In Nicol prism, the *O-ray* is totally internally reflected and the *E-ray* is transmitted. This statement is (a) true (b) false (c) partly true (d) partly false. **[WBUT 2010]**

ANSWERS

6.1. (d) **6.2.** (c) **6.3.** (a) **6.4.** (b) **6.5.** (d) **6.6.** (a) **6.7.** (d) **6.8.** (c) **6.9.** (c) **6.10.** (c) **6.11.** (a) **6.12.** (a) **6.13.** (b) **6.14.** (c) **6.15.** (b) **6.16.** (a) **6.17.** (b) **6.18.** (c) **6.19.** (b) **6.20.** (c) **6.21.** (a) **6.22.** (a) **6.23.** (b) **6.24.** (a).

6(B)–Short Answer Type Questions

6.1. Define polarization. **[WBUT 2002]**

6.2. State Brewster's law in polarization. **[WBUT 2004, '05]**

6.3. A polarizer and an analyzer are oriented so that the maximum intensity is achieved. To what fraction of its maximum value is the intensity reduced when the analyzer is rotated through 60°? **[WBUT 2006]**

6.4. What type of phenomenon provides direct experimental proof of the transverse nature of light? **[BUH 1993]**

6.5. For maximum polarization by reflection what is the angle between the reflected and refracted rays? **[BUH 1993]**

6.6. Distinguish between polarized and unpolarized lights. **[NBU 2001]**

6.7. Define positive and negative crystals in connection with double refraction of light. **[BUH 1994]**

6.8. State Brewster's law and Malus law.

6.9. Prove that the angle between the reflected and refracted rays for complete polarization of the reflected ray is 90°.

6.10. What do you mean by double refraction?

6.11. Light waves can be polarized but sound waves cannot. Why?

6.12. Explain the terms (a) optic axis (b) principal plane.

6.13. Distinguish between positive and negative crystals.

6.14. How would you detect plane polarized light from partially polarized light?

6.15. What do you mean by plane polarized light, circularly polarized light and elliptically polarized light.

6.16. What do you mean by dichroism? What are polaroid?

6.17. What are quarter-wave plate and half-wave plate?

6(C)–Numerical Problems

6.1. Find the state of polarization when the x- and y-components of the electric field are

(a) $E_x = E_0 \sin(\omega t + kz)$, $E_y = E_0 \cos(\omega t + kz)$ (b) $E_x = E_0 \cos(\omega t + kz)$, $E_y = \frac{E_0}{\sqrt{2}} \cos(\omega t + kz + \pi)$

[*Ans.* (a) Circularly polarized, (b) Linearly polarized]

6.2. Calculate the polarizing angle for diamond surface, if the angle of refraction of a beam of light through it is 12°, when the incident beam makes an angle 60°. [*Ans.* 76.5°]

6.3. The angle of refraction of an unpolarized beam of light incident at a polarizing angle on a glass block is 32.5°. Calculate the refractive index of glass. [*Ans.* 1.57] **[KU 2003]**

6.4. Refractive index of glass is 1.5, calculate the Brewster's angle for it. Also calculate the angle of refraction. [*Ans.* 56°, 33°30′] **[NBU 2000]**

6.5. At what angle the light should be incident on a glass plate ($\mu = 1.569$) to get a plane polarized light by reflection. [*Ans.* 57°30′]

6.6. A polarizer and an analyzer are oriented so that maximum light is transmitted. To what fraction of its maximum value is the intensity of the transmitted light reduced when the analyzer is rotated through 60°? [*Ans.* 0.25]

6.7. Two Nicol prisms are crossed to each other. Now, one of them is rotated through 60°. What percentage of incident unpolarized light will pass through the system? [*Ans.* 37.5%]

6.8. When light is incident at an angle of 60° to the normal, the reflected light is plane polarized. What is the refractive index of the transparent reflecting medium? What is the angle of refraction corresponding to the angle of incidence of 60°? What is the angle between the reflected and refracted components? [*Ans.* 1.732, 30°, 90°]

6.9. Calculate the thickness of (a) a quarter-wave plate (b) half-wave plate; given that $\mu_e = 1.533$, $\mu_o = 1.544$ and $\lambda = 5000$ Å. [*Ans.* 1.136×10^{-3} cm, 2.272×10^{-3} cm]

6.10. Light reflected from a smooth ice surface is found to be completely polarized. Find the angle of the incidence of light, if the refractive index of ice is 1.739. [*Ans.* ~ 52°36′] **[CU 2005]**

6.11. A right-circularly polarized beam ($\lambda = 525$ nm) is incident normally on a doubly refracting crystal with optic axis parallel to the surface. The thickness of the crystal is 0.003 mm. Given: $n_O - n_e = 0.175$. Find the state of polarization of the emergent beam. [*Ans.* right-circularly polarized]

6.12. Consider the waves: (a) $E_x = E_1 \cos(\omega t - kz)$, $E_y = 0$ (b) $E_x = E_2 \cos(\omega t - kz)$, $E_y = E_2 \cos(\omega t - kz - \pi/2)$ (c) $E_x = E_3 \cos(\omega t + kz)$, $E_y = E_3 \cos(\omega t + kz + \pi/2)$ (d) $E_x = E_4 \cos(\omega t + kz)$, $E_y = E_4 \cos(\omega t - kz)$. Identify the state of polarization of each of the waves.

[*Ans.* (a) linear (b) right-circular (c) left-circular (d) linear]

[BUH 1999, 2003]

6.13. Four perfect polarizing plates are stacked so that the axis of each is turned 30° clockwise w.r.t. the preceding plate. How much unpolarized beam of light is transmitted by the stack? **[BUH 2001]**

[*Hints.* Use Malus law.]

6.14. A left-circularly polarized beam ($\lambda_0 = 5893$ Å) is incident normally on a calcite crystal with its optic axis cut parallel to the surface of thickness 0.005141 mm. What will be the state of polarization of the emergent beam? [*Ans.* right-circularly polarized]

6.15. Calculate the thickness of a mica sheet required for making quarter-wave plate for $\lambda = 5460$ Å. The indices of refraction for *O-ray* and *E-ray* are 1.586 and 1.592 respectively. [*Ans.* 0.002275 cm]

[WBUT 2002]

6.16. Two polarizers are at crossed position (angle between the polarizing planes are 90°), a third polarizer with angle θ with the first one is placed between them. An unpolarized light of intensity I is incident on the first one and passes through all three polarizers. Find the intensity of light that comes out. **[WBUT 2008]**

[**Soln.** Let the polarizers A and B are at crossed position and a third polarizer C is placed between A and B such that C makes an angle θ with A. If I be the intensity of the unpolarized light incident, A, then by Malus law, the average intensity transmitted through A is given by

$$I_A = I\langle\cos^2\alpha\rangle = \tfrac{1}{2}I,$$

where α is the angle between the polarizing direction of the polarizer and that of vibration of the light vector of incident light.

The intensity of light transmitted through C is $I_C = I_A\cos^2\theta = \frac{1}{2}I\cos^2\theta$.

The intensity of light transmitted through B is $I_B = I_C\cos^2(90° - \theta) = \frac{1}{2}I\cos^2\sin^2\theta$.]

6.17. Calculate the polarizing angle for light travelling from water of refractive index 1.33 to glass of refracting index 1.53. **[WBUT 2008]**

[**Soln.** If i_p be the polarizing angle, then according to Brewster's law,

$$\tan i_p = {}_w\mu_g = \frac{\mu_g}{\mu_w},$$

where μ_g and μ_w are respectively the refractive indices of glass and water.

Here $\mu_g = 1.53$

$\mu_w = 1.33.$

$$\therefore i_p = \tan^{-1}\left(\frac{1.53}{1.33}\right) = \tan^{-1}1.15 = 49°.]$$

6.18. Two polarizers are placed at crossed position. Angle between the polarizing planes are 90°. A third polarizer with angle θ with the first one is placed between them. An unpolarized light of intensity I is incident on the first one and passes through all three polarizers. Find the intensity of the light that comes out. **[WBUT 2010]**

6(D)–Broad Answer Type Questions

6.1. (a) What is polarization of light? (b) What do you mean by positive crystal and negative crystal? (c) How can you distinguish between an unpolarized light, a circular polarized light and a plane polarized light? **[WBUT 2004]**

6.2. (a) Define polarization of light. (b) Discuss the phenomenon of double refraction. (c) Show that when a ray is incident at the Brewster's angle, the reflected ray is perpendicular to the refracted ray.

6.3. (a) Discuss the phenomenon of double refraction. What do you mean by positive crystal and negative crystal? (b) A right circularly polarized beam is incident on a calcite half-wave plate. Show that the beam will be left-circularly polarized. **[WBUT 2001]**

6.4. (a) Define polarization. (b) How elliptically and circularly polarized light are distinguished? **[WBUT 2002]**

6.5. State and explain Brewster's law of polarization indicating clearly the nature of polarization of the reflected and refracted rays. **[CUH 2005]**

Show that the reflected and refracted components are mutually perpendicular to each other.

6.6. Describe how polarized light can be obtained with the help of Nicol prism. **[BUH 2000; CUH 1991, '94, '96, '99, 2002]**

6.7. How can you distinguish between circularly polarized and unpolarized lights? **[BUH 1999]**

6.8. What are the differences between plane polarized light and circularly polarized light? Discuss the method of producing plane polarized lights by reflection. **[WBUH 1999]**

6.9. (a) How are unpolarized, plane polarized, circularly polarized and elliptically polarized light distinguished?

(b) Define birefringence. What do you mean by positive and negative crystals? Give examples of each. **[BUH 1997, 2006]**

(c) Discuss how circularly polarized light can be produced from (i) unpolarized light (ii) plane polarized light. **[BUH 1997]**

6.10. (a) Explain the action of a Nicol prism.

(b) How are unpolarized, plane polarized, circularly polarized and elliptically polarized light distinguished? **[BUH 2000, '02]**

(c) Define optic axis, positive and negative crystals in connection with double refraction of light.

(d) What is birefringence? Why are the two rays into which a ray of light breaks up in a birefringent crystal called ordinary ray and extraordinary ray? **[BUH 2000]**

6.11. State and explain Malus law. How will you prove the law experimentally?

6.12. What is dichroism? How does it produce plane polarized light? Use a dichroic crystal to produce plane polarized light. **[PU 1992, '95; HPU 1995]**

6.13. (a) Explain the polarization of light by scattering. **[HPU 1993; GNDU 1996]**

(b) Explain (i) blue colour of the sky (ii) red colour at sunrise and sunset. **[HPU 1997, PU 1990]**

6.14. Explain how the state of polarization of a given beam of light can be determined with the help of the Nicol prism and a quarter-wave plate.

6.15. (a) How is circularly polarized light distinguished from the unpolarized light? (b) How would you distinguish between elliptically polarized light and mixture of plane polarized and unpolarized lights?

6.16. (a) How can you get an elliptically polarized light from a linearly polarized light by using an optical device? (b) A plane polarized light is incident on a piece of quartz cut parallel to the axis. Find the thickness for which the ordinary and extraordinary rays combine to form plane polarized light, given that $\mu_o = 1.5442$ and $\mu_e = 1.5533$, $\lambda = 5 \times 10^{-5}$ cm. **[WBUT 2010]**

Chapter 7 | Module 3.2

Laser

7.1 Introduction

The word LASER is an acronym for *Light Amplification by Stimulated Emission of Radiation* and is derived by taking the first letter of different words as follows:

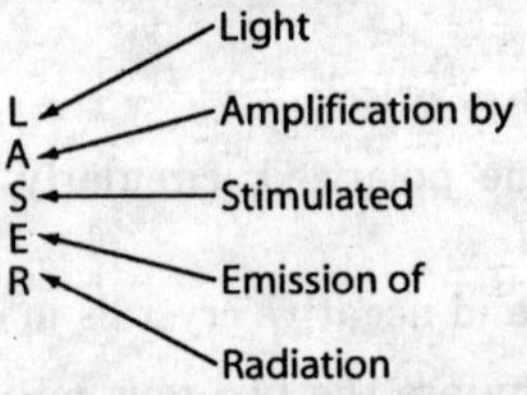

Einstein in 1917 first theoretically predicted that *the process of stimulated emission must exist along with spontaneous emission. Charles H. Townes and others* of Columbia University operated a microwave device (in 1960) for *Microwave Amplification by Stimulated Emission of Radiation* (MASER). In the same year (1960) *T. H. Maiman* of the Hughes Research Laboratories first achieved laser action at optical frequency ($\sim 10^{15}$ Hz) in Ruby crystal (Al_2O_3). *Ali Javan and others* in 1961 invented the first gas laser (viz., He : Ne laser), the wavelength of the laser beam being $\sim 0.633\,\mu$m. The first fibre laser (barium crown glas dopd with Nd^{3+} ions) was designed by *Elias Snitzer. CKN Patil* in 1963 invented CO_2 laser with wavelength $\lambda \simeq 10.6\,\mu$m. Later on Ar-ion laser ($\lambda \sim 0.515\,\mu$m), Nd: VAG laser ($\lambda \sim 0.515\,\mu$m) were intended by *W. Bridges* (1963) and *J.E. Geusic and others* (1964) respectively.

7.2 Properties of Laser Light

A laser is fundamentally a light source having some remarkable properties not available from any other light source. These remarkable properties are as follows:

1. **Directionality.** The divergence of a laser beam is usually limited to less than 10^{-5} radians by diffraction. This property finds applications in surveying, remote sensing, lidar, etc.

2. **High Power.** The power levels of continuous wave laser is $\sim 10^5$ watt, whereas for pulsed laser the total energy is ~ 50000 J. Due to this large power content laser beam is used in welding, cutting, laser fusion, star wars, etc.

3. **Tight focussing.** Due to high degree of directionality laser beam can be focussed within a small area $\sim$ a few $(\mu\text{m})^2$ leading to large amount of energy ($\sim 12 \times 10^5$ watt/m^2). This property of laser beam is used in material processing, surgery, compact disc (CD), etc.

4. **Spectral purity.** The spectral width ($\Delta\lambda$) of laser beam is very small, i.e., $\Delta\lambda \sim 10^{-6}$ Å, which can further be reduced by adopting suitable technique. This property is successfully applied in spectroscopy, holography, optical communication, etc.

5. **High degree of coherence.** A predictable correlated motion of several waves having a fixed amplitude and phase relation is known as coherence.

Coherence is of two types. (i) spatial coherence and (ii) temporal coherence. The property spatial coherence is used in performing different types of interference experiment. The temporal coherence is used in Michelson-Morley experiment.

6. **Continuous power output.** Laser has the capability of providing continuous power output from microwatt to kilowatt.

7.3 The Basic Principle of Laser

According to *N. Bohr*, the atoms are characterized by *discrete energy levels. Einstein* proposed that such atoms can interact with incident electromagnetic waves in the following three ways, viz., 1. *Stimulated absorption* 2. *Spontaneous emission* 3. *Stimulated emission.*

The above three processes form the *basic principle of laser.*

1. **Stimulated absorption.** Let an atom have two energy states E_1 and E_2 ($E_2 > E_1$) as shown in Fig. 7.1. When a photon of frequency ν be incident on such an atom so that $\nu = \frac{E_2 - E_1}{h}$, then the atom is *pumped* to the excited state E_2. The rate of *stimulated absorption* (or, simply *absorption*) depends on the two factors: (i) *the intensity of the incident radiation* and (ii) *the number of atoms in state* E_1.

By this process the density of the photons in the incident beam decreases. The atom in state E_2 can exist for a small interval time $\tau \simeq 10^{-8}$ s, the life-time of the excited state.

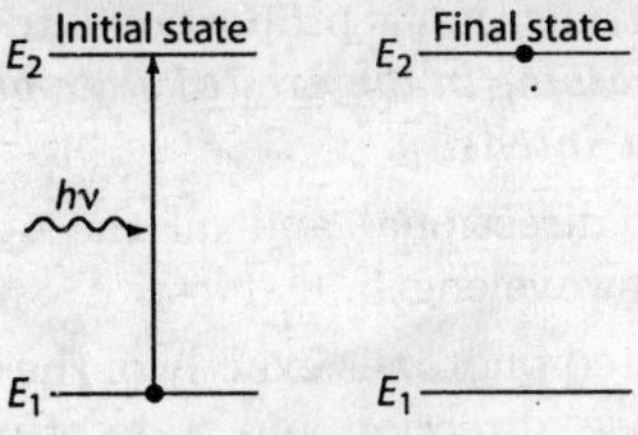

Fig. 7.1: Stimulated absorption.

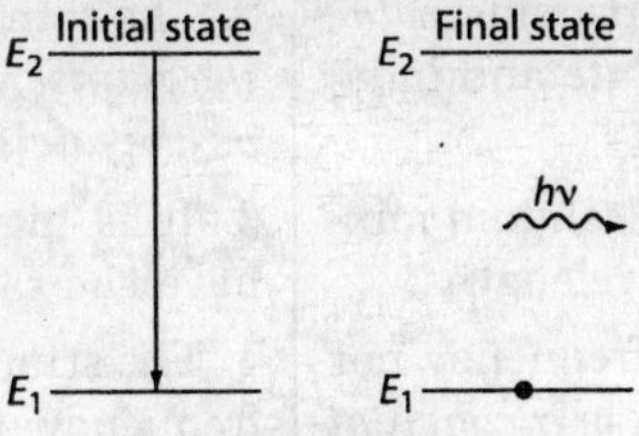

Fig. 7.2: Spontaneous emission.

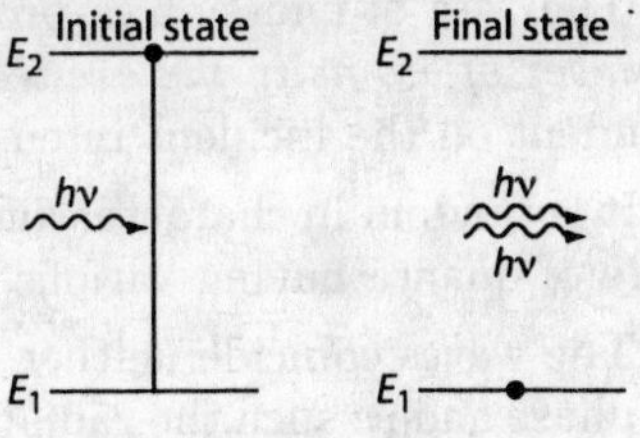

Fig. 7.3: Stimulated emission.

2. **Spontaneous Emission.** The life-time of the excited state being very small ($\sim 10^{-8}$ s); an atom in energy state E_2 transits to the state E_1 giving rise to the emission of a photon of frequency

$$\nu = \frac{E_2 - E_1}{h}, \tag{7.1}$$

where $h = 6.626 \times 10^{-34}$ J-s is the Planck's constant [Fig. 7.2]. Since this process can take place in absence of any radiation, it is called *spontaneous emission.*

It is to be remembered that: (i) *the rate of spontaneous transition is proportional to the number of atoms in the excited state* E_2 (ii) *independent on the intensity of exciting radiation.*

The *spontaneous emission* is random in character having *no definite phase relationship among the emitted waves of various wavelength.* The radiation is thus *incoherent having broad spectrum.*

3. **Stimulated emission.** *Einstein* proposed that an atom in the excited state E_2 can transit to the ground state E_1 being triggered by an incident photon of *appropriate frequency* [Fig. 7.3] and the emitted radiation is known as *stimulated emission.* This leads to the *amplification* of the incident radiation.

The rate of stimulated emission depends on (i) *the number of atoms in the excited state* and (ii) *the intensity of the external radiation.*

As the *direction of propagation, the frequency, phase* and *state of polarization of stimulated emission* are the same as the incident photon, it is highly *coherent.*

These two photons emitted in the above process can interact with two other atoms in the excited state E_2 and can again induce stimulated emission giving four photons. In this way a *chain process* is continuously increasing the *density of photon* and the light amplification (laser action) occurs. For this chain process to sustain the essential condition is that *the number of atoms in the excited state must be greater than that in the ground state, that is,* $N_2 > N_1$. The highly populated energy state is called the *population inversion.* So *population inversion is the prerequisite for laser action.*

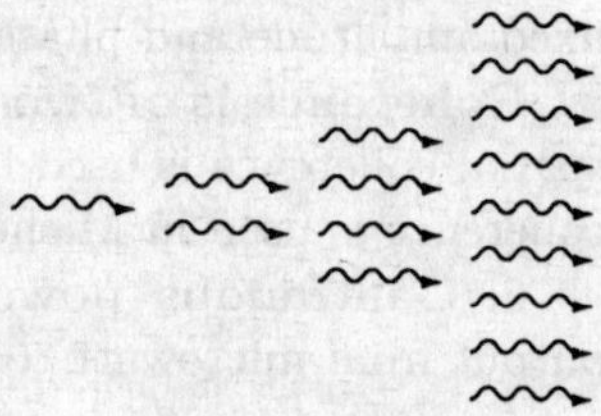

Fig. 7.4: Generation of photons by chain process.

7.4 Spontaneous Emission Versus Stimulated Emission of Radiation

Table 7.1 shows a comparative study between spontaneous and stimulated emission of radiation.

Table 7.1

Spontaneous emission	*Stimulated emission*
1. This process takes place in absence of any radiation.	1. This process is triggered by an incident radiation of proper frequency.
2. The rate of emission is proportional to *the number of atoms in the excited state* and independent on the incident intensity.	2. The rate of transition is proportional to both *the number of atoms in the excited state as well as the incident intensity.*
3. It is random in character and a random mixture of quanta having various wavelengths.	3. It is highly directional and all the quanta have the same wavelength.
4. The waves coincide neither in frequency nor in phase and as such the radiation is *incoherent* and has a broad spectrum.	4. The stimulated photon is exactly of the same frequency, phase, direction and state of polarization as the incident photon. So it is *coherent.*
5. It has no relevance with light amplification.	5. It can develop a chain reaction leading to light amplification and high intensity coherent radiation.

7.5 Comparison between Laser and Ordinary Light

Table 7.2 shows a comparative study between laser and ordinary light.

Table 7.2

Laser	*Ordinary light*
1. It is highly directional, the divergence being limited to $\sim 10^5$ radians.	1. It is random in character, the divergence being larger than laser beam.
2. High peak power and large energy per pulse in pulsed output laser.	2. The power of ordinary light is much greater than laser.

Laser	*Ordinary light*
3. Laser is highly monochromatic.	3. Ordinary monochromatic light has a wider range of frequency in comparison to laser.
4. It can be focussed in an area $\sim \lambda^2$.	4. It cannot be focussed in an area $\sim \lambda^2$.
5. Laser is perfectly coherent, since the phase and frequency of all the emitted waves are the same.	5. It is incoherent, there being no constant phase and frequency relationship.

7.6 Relation between Spontaneous and Stimulated Emission Probabilities—Einstein's Theory

To find out a relation between *spontaneous* and *stimulated* emission probabilities, Einstein assumed that,

(i) *Since the system consists of a very large number of atoms, to calculate the rate of relative transition between two energy states, the theory of probability is to be applied.*

(ii) *The atomic system is in equilibrium with the electromagnetic radiations.*

Let us consider a system having two energy states E_1 and E_2 ($E_2 > E_1$) having number of atoms N_1 and N_2 per unit volume respectively at temperature T [Fig. 7.5]. Then according to *Maxwell-Boltzmann distribution law* the numbers of atoms in states E_1 and E_2 are $N_1 = N_0 e^{-E_1/kT}$ and $N_2 = N_0 e^{-E_2/kT}$.

$$\therefore \frac{N_2}{N_1} = e^{-(E_2-E_1)/kT}, \tag{7.2}$$

where k is the *Boltzmann constant* and N_0 is the number of atoms in the ground state E_0. Eqn (7.2) shows that $N_2 < N_1$ as $E_2 > E_1$.

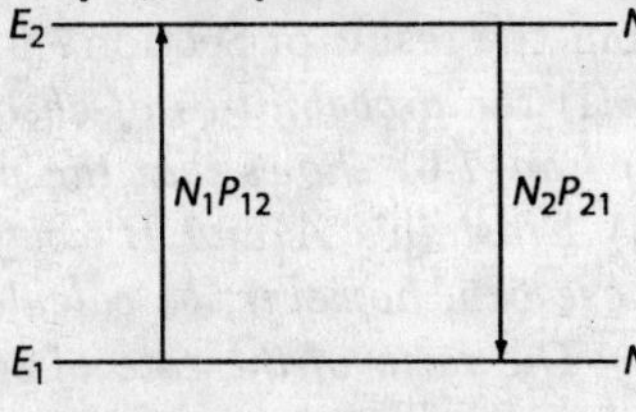

Fig. 7.5: A systems considering of two energy levels E_1 and E_2.

The *probability* that an atom will absorb photon of frequency ν and make a transition from E_1 to E_2 per unit time is

$$P_{12} \propto E(\nu).$$

So the number of atoms that will make a transition from E_1 to E_2 per unit time is

$$N_1 P_{12} = N_1 B_{12} E(\nu), \tag{7.3}$$

where B_{12} is called the *Einstein's coefficient of absorption* and $E(\nu)$ is the *energy density of frequency* ν.

Now, the probability of *downward transition* from E_2 to E_1 per unit time consists of two parts, viz., (a) *the rate of spontaneous emission* $N_2 A_{21}$ and (b) *the rate of stimulated emission* $N_2 B_{21} E(\nu)$.

Therefore, the total probability of transition from E_2 to E_1 per unit time is

$$N_2 P_{21} = N_2 A_{21} + N_2 B_{21} E(\nu)$$

$$\text{or,} \quad N_2 P_{21} = N_2 [A_{21} + B_{21} E(\nu)].$$

In *thermal equilibrium* the emission and absorption must balance, i.e.,

$$N_1 B_{12} E(\nu) = N_2 [A_{21} + B_{21} E(\nu)]$$

$$\text{or,} \quad E(\nu)[N_1 B_{12} - N_2 B_{21}] = N_2 A_{21}$$

$$\text{or,} \quad E(\nu) = \frac{N_2 A_{21}}{[N_1 B_{12} - N_2 B_{21}]} = \frac{A_{21}/B_{21}}{\left(\frac{N_1}{N_2}\right)\left(\frac{B_{12}}{B_{21}}\right) - 1}.$$

Einstein proved that *under equilibrium condition* $B_{12} = B_{21}$.

$$\therefore \quad E(\nu) = \frac{A_{21}/B_{21}}{\frac{N_1}{N_2} - 1} = \frac{A_{21}/B_{21}}{e^{(E_2-E_1)/kT} - 1}$$

$$\text{or,} \quad E(\nu) = \frac{A_{21}/B_{21}}{e^{h\nu/kT} - 1} \quad (\because E_2 - E_1 = h\nu). \tag{7.4}$$

Eqn (7.4) relates *energy density of photons having frequency* ν *under equilibrium with the atoms in states* E_1 *and* E_2 *at temperature* T.

The above relation must be in accord with *Planck's radiation formula*,

$$E(\nu) = \frac{8\pi h\nu^3}{c^3}\frac{1}{e^{\frac{h\nu}{kT}} - 1}. \tag{7.5}$$

Comparing the above two equations, we get

$$A_{21}/B_{21} = \frac{8\pi h\nu^3}{c^3}, \tag{7.6}$$

where c is the velocity of light.

7.7 Important Discussion

From the result of Section 7.6, the following points need to be discussed:

(i) *the probabilities of absorption and stimulated emission are equal* ($\because B_{12} = B_{21}$).

(ii) *Eqn (7.6) shows that the probability of spontaneous emission increases directly as* ν^3.

(iii) *Einstein's A and B coefficients cannot be determined by classical electromagnetic theory. These can, however, be calculated using perturbation theory due to Dirac.*

(iv) *The ratio of the rate of stimulated emission to the rate of spontaneous emission is given by*

$$\rho = \frac{B_{21}N_2E(\nu)}{A_{21}N_2} = \frac{B_{21}}{A_{21}} \times \frac{8\pi h\nu^3}{c^3}\frac{1}{e^{h\nu/kT} - 1} \quad \text{[from eqn (7.5)]} \tag{7.7}$$

$$\text{or,} \quad \rho = \frac{1}{e^{h\nu/kT} - 1} \quad \text{[from eqn (7.6)]}. \tag{7.8}$$

We conclude from eqn (7.7) that:

(a) **Condition for higher stimulated emission:** In *microwave* region $h\nu \ll kT$ and hence the probability of *stimulated emission is greater than that of spontaneous emission if the incident radiation is in the microwave region.* So to have higher probability of stimulated emission compared to that of spontaneous emission the incident radiation must be in the *microwave region.*

(b) In the **visible region**: If $\lambda \simeq 6500$ Å and $T \simeq 1000$ K, then $\rho \sim 10^{-58}$. Hence, the *spontaneous emission forms the major component of emission* and the stimulated emission is *almost absent.*

(c) **Condition for higher stimulated emission than absorption:**

We have rate of stimulated emission $= N_2B_{21}E(\nu)$ and the rate of stimulated absorption $= N_1B_{12}E(\nu)$.

$$\therefore \quad \text{at thermal equilibrium,} \quad \frac{\text{Stimulated emission rate}}{\text{Stimulated absorption rate}} = \frac{N_2B_{21}E(\nu)}{N_1B_{12}E(\nu)}$$

$$\text{or,} \quad \rho_1 = \frac{N_2}{N_1} \quad (\because B_{12} = B_{21}). \tag{7.9}$$

Since at thermal equilibrium $N_1 \gg N_2$, $\rho_1 \ll 1$, i.e., the stimulated absorption rate dominates the stimulated emission rate. This, therefore, cannot lead to *light amplification* (Laser action).

To have light amplification, $N_2 \gg N_1$, i.e., the state E_2 must be more populated than state E_1. Under this condition stimulated emission rate will be greater than the corresponding stimulated absorption rate when the photon density is increasing and *light amplification by stimulated emission of radiation (Laser) occurs.*

The state for which $N_2 \gg N_1$ is called the *population inversion* when the number of atoms in the higher energy state (E_2) is very much greater than that of lower energy state (E_1). Hence, *population inversion is an essential condition for laser action to take place.*

(d) **The population inversion** ($N_2 \gg N_1$) : can be realised if there be a *metastable state* (E_0) such that $E_1 < E_0 < E_2$, having *life-time* considerably large ($\sim 10^{-3}$ s).

7.8 Theory of Amplification of Light

Let a beam of light having intensity I_0 passes through a medium of length x. The intensity of the beam after leaving the medium by *Beer's law* is

$$I(x) = I_0 e^{-\alpha x}, \tag{7.10}$$

where α is the *absorption coefficient of the medium.* Due to *stimulated absorption* by the medium, the intensity of the light beam will decrease. If the medium is *active*, i.e., inversely populated, then the *stimulated emission* will predominate over *stimulated absorption.* As a result the light beam undergoes *amplification* instead of attenuation during passing through the medium.

$$\therefore \quad \boxed{I(x) = I_0 e^{\beta x}}, \tag{7.11}$$

where β is the *gain (i.e., amplification) coefficient.* Hence, *if the medium is inversely populated, the density of photon (i.e., intensity) increases when the beam of light passes through it.*

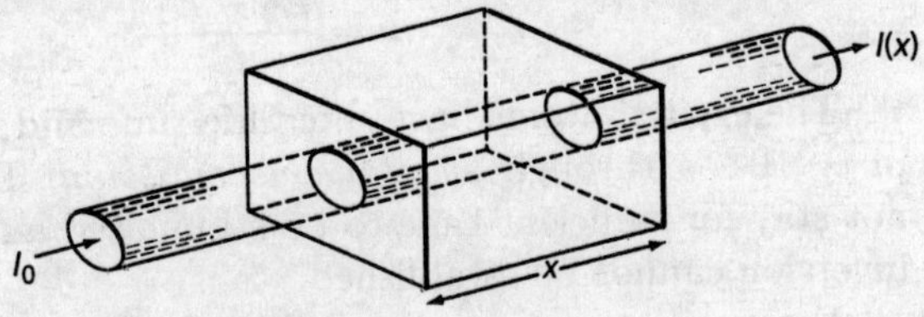

Fig. 7.6: A beam of light passes through a medium in thermal equilibrium.

7.8.1 Gain Coefficient (β)

If n be the number of photons in the medium at any instant t, then

$$\frac{dn}{dt} = \text{rate of increase of photon density}$$

$$= \text{(stimulated emission rate)} - \text{(stimulated absorption rate)}$$
if spontaneous emission rate is negligible.

$$\text{or,} \quad \frac{dn}{dt} = N_2 B_{21} E(\nu) - N_1 B_{12} E(\nu)$$

$$= (N_2 - N_1) E(\nu) B_{21} \quad [\because B_{12} = B_{21}]. \tag{7.12}$$

But, intensity of light is $I = nh\nu c$ [c = velocity of light in the medium]

$$\therefore \quad \frac{dI}{dx} = \frac{dn}{dx} h\nu c$$

$$\text{or,} \quad dn = \left(\frac{dI}{dx}\right) dx \frac{1}{h\nu c}$$

$$\text{or,} \quad \frac{dn}{dt} = \left(\frac{dI}{dx}\right)\left(\frac{dx}{dt}\right)\frac{1}{h\nu c} = \left(\frac{dI}{dx}\right)\frac{1}{h\nu} \quad \left[\because \frac{dx}{dt} = c\right]. \tag{7.13}$$

But since $I = I_0 e^{\beta x}$,

$$\therefore \quad \frac{dI}{dx} = \beta I_0 e^{\beta x} = \beta I(x). \tag{7.14}$$

Using the value of $\frac{dI}{dx}$ from eqn (7.14) in eqn (7.13), we get

$$\frac{dn}{dt} = \beta I(x)\frac{1}{h\nu} = \beta n h\nu c\frac{1}{h\nu} = \beta nc. \tag{7.15}$$

Comparing eqns (7.12) and (7.15), we get

$$\beta = (N_2 - N_1)B_{21}\frac{E(\nu)}{nc} \quad \text{or,} \quad \boxed{\beta = (N_2 - N_1)B_{21}\frac{h\nu}{c}}. \tag{7.16}$$

Light amplification will occur if β is *positive*, i.e., $N_2 > N_1$, i.e., when *population inversion* is achieved. Under such condition the light intensity (density) of photon increases while passing through the medium.

7.9 Metastable State

An atom can be excited to higher energy level (E_2) from a lower energy level (E_1) by supplying energy of frequency ν such that

$$\nu = \frac{E_2 - E_1}{h}.$$

The excited atoms have short life-time and release their energy in $\sim 10^{-8}$ s through *spontaneous emission.* Thus, the atoms do not stay for sufficient time to be stimulated and hence population inversion cannot be established.

For population inversion to be established the life-time of the excited state must be $\sim 10^{-6}$ to 10^{-3} s, which is 10^3 to 10^6 times the life-time of the ordinary energy levels. Such an energy state is called a *metastable state* (E_3).

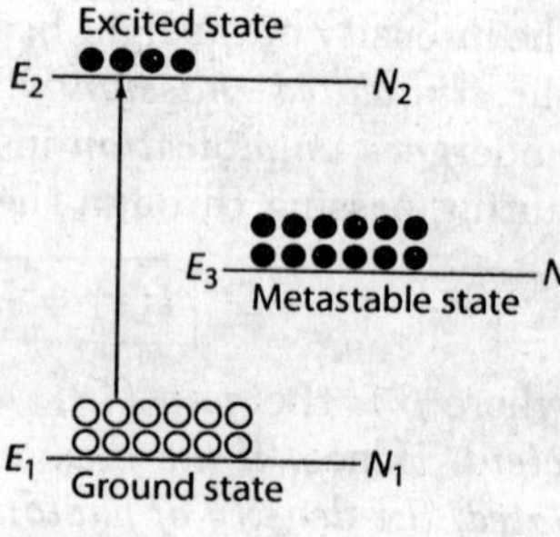

Fig. 7.7: Metastable state E_3 accumulating larger number of atoms.

Thus, *a metastable state allows accumulation of a large number of excited atoms.* The population of the metastable state is larger than the population of the lower energy state (E_1), i.e., $N_3 > N_1$, and establishes the *condition of population inversion.*

It is to be noted that (a) *population inversion and laser action is impossible without metastable state,* (b) *metastable state in a crystal can be obtained by doping impurity atoms having energy levels in the forbidden gap of the crystal,* (c) *population inversion readily takes place as the life-time of these states is large* [Fig. 7.7].

7.10 Active Medium

An inversely populated medium in which light gets amplified is called an *active medium.* An active medium may be solid, liquid or gas. In such a medium only a small fraction of the atoms are responsible for light amplification and are called the *active centres.* The rest part of the medium behaves as host and supports the active centre in the bulk. An active medium possesses the following characteristics:

The medium should (i) *possess good mechanical and thermal properties* (ii) *possess good optical properties* (iii) *have transparency to stimulated radiation and laser output* (iv) *maintain the desired properties of the active species in it.*

7.11 Pumping

The mechanism by which population inversion can be achieved is called pumping.

There are different mechanisms of pumping.

(a) **Optical pumping**: Optical pumping is the mechanism by which atoms of a medium are raised to higher energy state (E_2) from lower energy state (E_1), when exposed to electromagnetic radiation of frequency $\nu = \frac{E_2 - E_1}{h}$.

The materials (like Ruby) have very broad absorption band, sufficient amount of energy is absorbed from the emission band of flash lamp, and population inversion is created. Nowadays the system is made more efficient and reliable by *using laser diodes instead of flash lamp.*

(b) **Electrical discharge**: Gas laser (like He : Ne laser) has very narrow absorption band, so optical pumping is *ineffective.* Therefore, for most of the gas lasers *electrical discharge* is used for creating population inversion. The accelerated electrons in the discharge tube collide with neutral atoms and a fraction of them are raised to higher energy state according to the following scheme $A + e_{\text{fast}} \to A^* + e_{\text{slow}}$, where A and A^* are the atoms in the *normal state* and *excited state* respectively and e is the electron making collision with A.

(c) **Inelastic collision between atoms**: In this method, an atom X is excited to X^* state by inelastic collision with a fast electron (e_{fast}) in a mixture of two gases. X^*, then collides with Y atom and raise it to its excited state Y^*, and X transit to its normal state as follows:

(d) **Chemical reaction**: Excitation may also occur by the energy released in a chemical reaction creating population inversion. For example, in the chemical reaction of H_2 and F_2 the product HF molecule is raised to excited state. The number of such excited molecules is much more larger than those in the ground state and hence population inversion occurs. $H_2 + F_2 \longrightarrow 2HF^*$.

The energy is sufficient to pump CO_2 laser.

(e) **Injection current**: The injection current through the junction of a semiconductor laser creates population inversion among the charge carriers. This is known as *pumping by injection current.*

7.12 Population Inversion

If the number of atoms (N_2) in higher energy state (E_2) is greater than the number of atoms (N_1) in the lower energy state (E_1), then the situation is called *population inversion.* Population inversion is a prerequisite for light amplification in a *non-equilibrium distribution of atoms* among the various energy levels of the system.

According to *Boltzmann's law* the number of atoms N_i in a given energy state E_i at equilibrium temperature is

$$N_i = N_0 e^{-E_i/kT}, \tag{7.17}$$

where N_0 is the number of atoms in the ground state and $k = 1.38 \times 10^{-23}$ J/K, the Boltzmann's constant.

If $kT \gg E_i$, $e^{-E_i/kT} \to 1$. Therefore, $N_i = N_0$ in the extreme case. In no case N_i can be greater than N_o. In other words, higher energy level (E_i) can never be more populated than ground level E_0 only by means of optical pumping. So *by direct pumping, population inversion is impossible in a two-level system.*

A natural question, then arises, how is population inversion achieved?

Three level scheme: Fig. 7.8 shows the Boltzmann distribution of energy states with no pumping. If the atoms in E_0 are *intensely pumped,* a good number of them are excited to E_2 through *stimulated absorption.* If, the energy level E_2 is short-lived, then the atoms decay fast to level E_1. If E_1 is a metastable state, then the atoms are accumulated at E_1 [Fig. 7.9]. Thus, with intense pumping from E_0 to E_2 and then rapid decay to E_1, it is possible to achieve non-equilibrium distribution of atoms. Then E_1 is more populated than E_0, when laser transition takes place between E_1 and E_0.

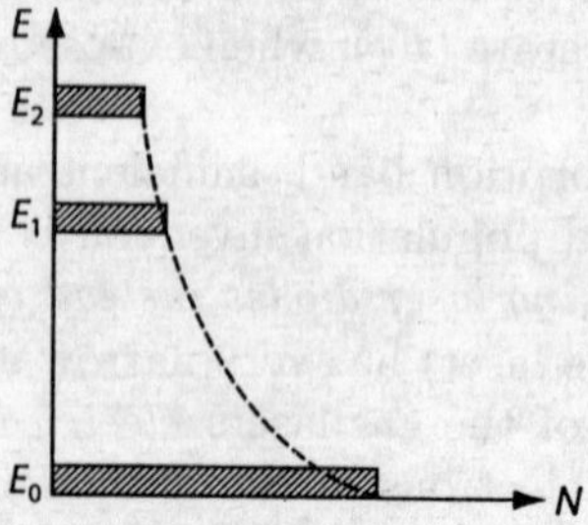

Fig. 7.8: Boltzmann distribution without pumping.

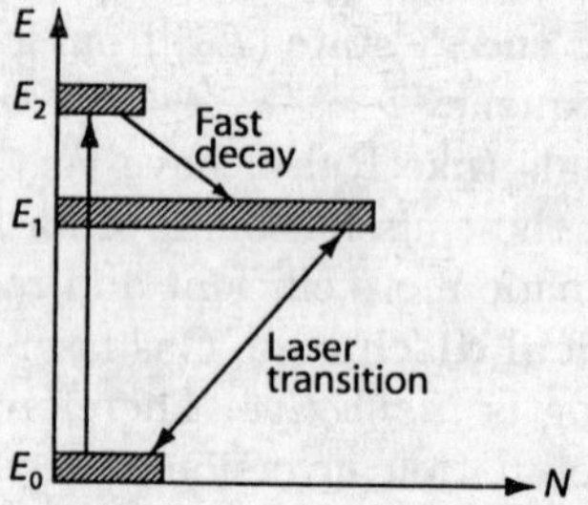

Fig. 7.9: Population inversion created by intense pumping.

It is not possible to continuously maintain E_1 more populated than E_0 due to depopulation of E_1. Such a system (e.g., the Ruby laser) works in pulsed mode only.

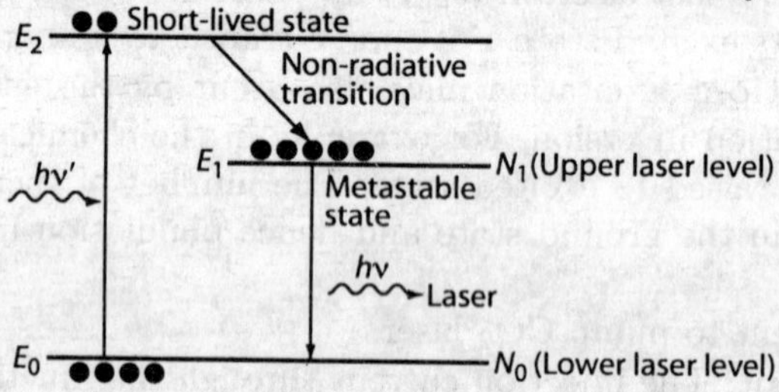

Fig. 7.10: Mechanism of population inversion ($N_1 > N_0$) in three level scheme.

N.B. *From Boltzmann relation (7.17) we note that $N_i > N_0$, if T is negative. That is why the state of population inversion is sometimes referred to as negative temperature state. It does not mean temperature below* 0 K *is attainable. It simply means that population inversion is a non-equilibrium state. Practically population inversion is obtained at ordinary temperature.*

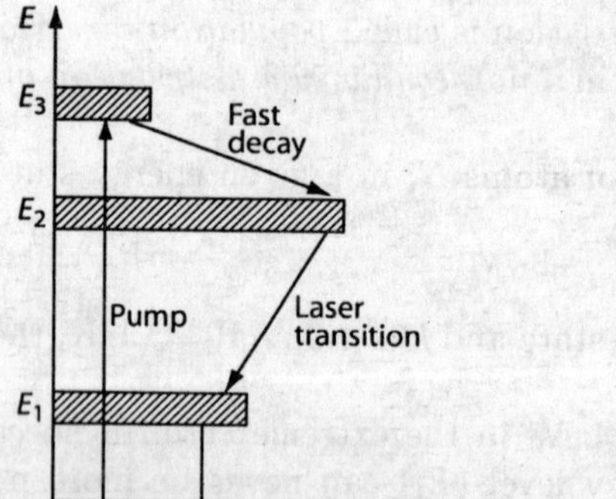

Fig. 7.11: Population of energy levels in a four-level scheme.

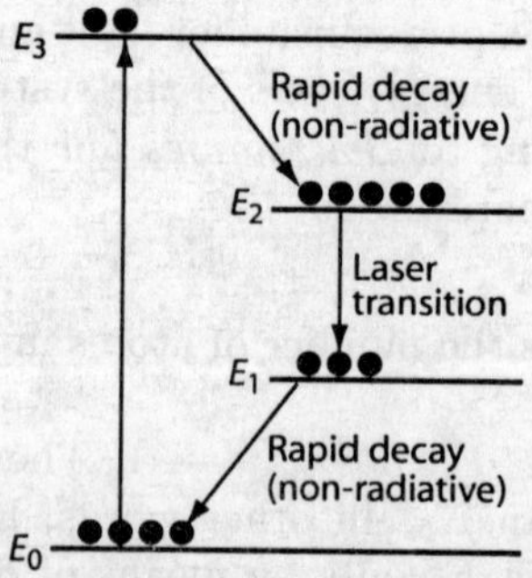

Fig. 7.12: Scheme of different transitions in a four-level system.

Four-level Scheme. In four-level scheme, atoms are lifted from ground state (E_0) to the highest state E_3 by pumping [Fig. 7.11]. From E_3 state atoms decay to E_2 state (metastable state) and population in this rapidly increases. A population inversion of $E_2 \rightarrow E_1$ transition can be achieved if the life-time of $E_3 \rightarrow E_2$ transition is shorter compared to $E_2 \rightarrow E_1$ transition with moderate pumping. Since here E_1 is the *lower laser level,* it is comparatively easier to maintain population inversion between E_2 and E_1 continuously with moderate pumping and get continuous laser beam. For this, $E_1 \rightarrow E_0$ transition must be very fast. If this transition is slower, then the laser source will give *pulsed beam* only. He – Ne laser is a good example of four-level laser. The mechanism of creation of population inversion is schematically represented in Fig. 7.12.

7.13 Threshold Condition for Light Amplification

According to theory once population inversion is created light amplification should take place continuously with time. But the practical situation is different due to the following reasons: (a) *during light amplification some photons are lost due to absorption and scattering* (b) *due to photons taken out as an output of the system.*

If the population density of the medium is *sufficiently large to compensate the loss,* then only lasing action will occur. The population inversion density for which the gain is just sufficient to compensate for the loss is termed as *threshold inversion density.* The gain corresponding to the inversion density under threshold condition is called *threshold gain* (β_{th}).

Calculation of β_{th} : Let two plane mirrors M_1, M_2 with reflectivity R_1, R_2 be placed at the two ends of the active medium of length l as shown in Fig. 7.13.

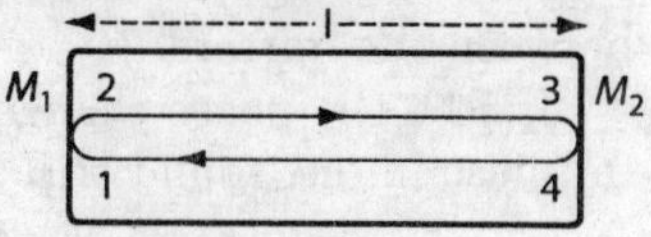

Fig. 7.13: A round trip of photons.

Let I_0 = intensity of the incident beam (1) on M_1, then the intensity of the reflected beam (2) at $M_1 = I_0 R_1$.

If β = gain of the medium, then the intensity of beam (3) is $I_0 R_1 e^{\beta l}$ and that of the beam (4) reflected at M_2 is $= I_0 R_1 R_2 e^{2\beta l}$.

$\therefore$ gain of the medium (G)

$$= \frac{\text{Intensity of the beam at the end of one round-trip}}{\text{Intensity of the beam at the start of round-trip}} = \frac{I_0 R_1 R_2 e^{2\beta l}}{I_0}$$

or, $G = R_1 R_2 e^{2\beta l}$.

At threshold, $G = I_0$, as the intensity of the beam does not change

$$\text{or,} \quad I_0 R_1 R_2 e^{2\beta_{th} l} = I_0, \quad [\text{where } \beta = \beta_{th} \text{ at threshold}]$$

$$\text{or,} \quad e^{2\beta_{th} l} = \frac{1}{R_1 R_2} \quad \text{or,} \quad 2\beta_{th} l = \ln\left(\frac{1}{R_1 R_2}\right)$$

$$\text{or,} \quad \boxed{\beta_{th} = \frac{1}{2l} \ln\left(\frac{1}{R_1 R_2}\right)}. \tag{7.18}$$

If the value of β_{th} satisfies the above relation, then continuous light amplification will occur.

7.14 Main Components of the Laser

Following are the main components of laser.

1. **The active medium**: The active medium (A) may be either a *solid, liquid, gas* or *semiconductor junction* having the capacity to sustain stimulated emission.

2. **An external source of excitation**: The pumping source (E) supplies pumping energy to create *population inversion.*

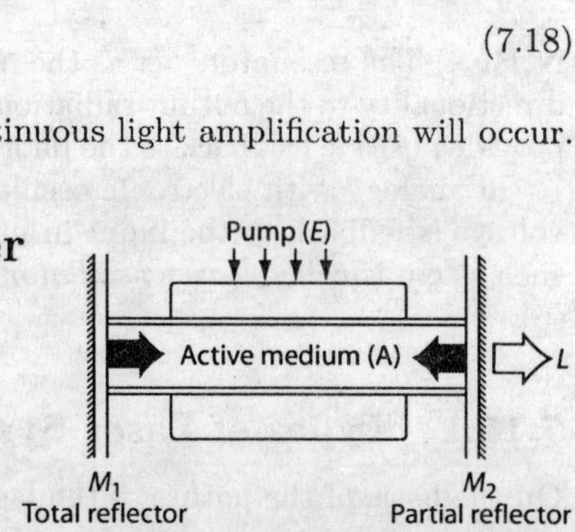

Fig. 7.14: Basic laser system.

3. **The optical resonator.** The optical resonator may be a pair of plane parallel mirrors M_1 and M_2 as shown in Fig. 7.14. The mirror M_1 is a total reflector and M_2 a partial one. M_2 allows a part of the laser beam L to pass out.

Action: The medium is irradiated with pump energy E. The excited atoms release this energy by stimulated emission. A photon released from one excited atom encounters a second excited atom in its path and stimulates it to release a second photon. These two coherent photons make the intensity of the beam twice. The intensity of the light beam increases rapidly by more stimulated photons as it passes through the medium. The beam being reflected by more stimulated photons as it passes through the medium. The beam being reflected by M_1 undergoes another passage through the excited medium and gets further amplified. Reaching M_2, a part L of the amplified beam passes out as a laser beam.

7.15 Cavity Resonance

Stationary waves of definite wavelengths are formed in the active medium A [Fig. 7.15] inside the closed vessel, which behaves as a *cavity resonator*. Integral number of half-waves are excited between M_1 and M_2, which act as two nodes [Fig. 7.15]. As the distance d between M_1 and M_2 is much greater than the wavelength (λ) of light, a large number of half-waves are generated between the mirrors.

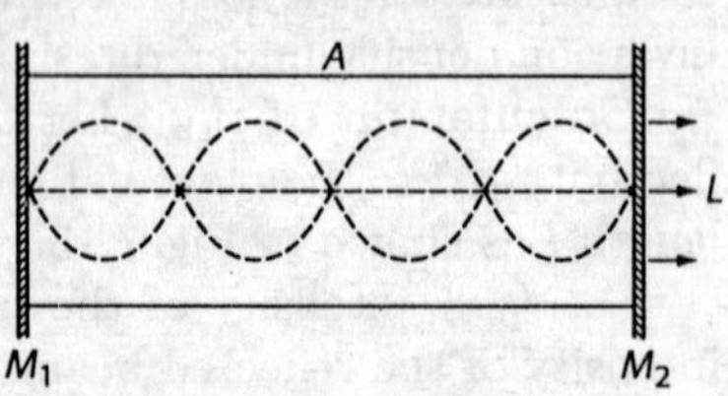

Fig. 7.15: Optical resonator.

If δ be the phase change at each reflection, then the change in phase in one round trip between M_1 and M_2 is

$$\phi = \frac{2\pi}{\lambda} \times 2d + 2\delta.$$

For resonance, $\phi = 2\pi n$ (where n is an integral)

$$\text{or,}\quad \frac{2\pi}{\lambda} \times 2d + 2\delta = 2\pi n \quad \text{or,}\quad \lambda = \frac{4\pi d}{2n\pi - 2\delta} \quad \text{or,}\quad \frac{1}{\lambda} = \frac{2n\pi}{4\pi d} - \frac{2\delta}{4\pi d}$$

$$\text{or,}\quad \frac{c}{\lambda} = \frac{nc}{2d} - \frac{\delta c}{2\pi d} \quad \text{or,}\quad \nu = \frac{nc}{2d} - \frac{\delta c}{2\pi d}, \tag{7.19}$$

where $c = \nu\lambda$ is the velocity of light.

Light waves of frequency (ν) given by eqn (7.19) only resonate and excite standing waves in the cavity.

The frequency difference between two consecutive modes of vibrations is

$$\Delta\nu = \nu_{n+1} - \nu_n = \left[\frac{(n+1)c}{2d} - \frac{\delta c}{2\pi d}\right] - \left[\frac{nc}{2d} - \frac{\delta c}{2\pi d}\right] \quad \text{or,}\quad \Delta\nu = \frac{c}{2d}. \tag{7.20}$$

N.B. (i) The resonators act as the frequency selector. (ii) They provide directionality to the output radiation. (iii) The mirrors of the resonators provide positive feedback to the photons amplified in the active medium.

In analogy with electronic oscillator (in which a part of the output voltage is fedback to the input in proper phase, i.e., positive feedback) the system is called *Laser oscillator* [Fig. 7.16].

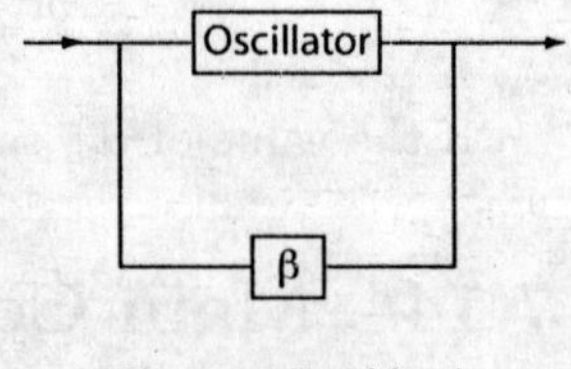

Fig. 7.16: Feed-back oscillator.

7.15.1 Types of Laser Systems

On the basis of the nature of the lasing medium, lasers are of the following four types (i) *Solid-state laser* (ii) *Semiconductor laser* (iii) *Gas laser* (iv) *Liquid laser*.

N.B. Optical resonant cavity is a pair of optically plane parallel mirrors and is called Fabry-Perot optical resonator.

Table 7.3 shows a list of commonly used laser and the lasing wavelengths.

Table 7.3

Sl. No.	*Types of laser*	*Example*	*Laser wavelength*
(i)	Solid-state lasers	(a) Ruby (b) Nd-YAG and glass (c) Alexandrite	(a) 694.30 nm (b) 1.064 μm (c) 700–800 nm
(ii)	Semi conductor laser	(a) GaAlAs diode (b) InGaAsP diode	(a) 750–900 nm (b) 1100–1600 nm
(iii)	Gas laser	(a) He-Ne (b) Argon ion (c) Nitrogen (d) CO_2 (e) Excimer (i) XeCl (ii) XeF	(a) 632.80 nm (b) 450–520 nm (c) 337 nm (d) 10.6 μm (e) (i) 308 nm (ii) 351 nm
(iv)	Liquid Laser	Organic dye	300–1000 nm 400–900 nm

7.16 Pulsed Ruby Laser

Laser action was first observed in Ruby by *Maiman* in 1960. Synthetic Ruby crystal (coloured red or pink) are grown from sapphire (Al_2O_3) doped with 0.01 to 0.05% chromium by weight. For 0.05% doping the density of chromium ion (Cr^{+++}) is $\sim 1.6 \times 10^{25}$ per cubic metre. Cr^{+++} having a set of energy levels plays the role of active centres.

Construction of the set-up. The experimental set-up of a Ruby laser is shown in Fig. 7.17.

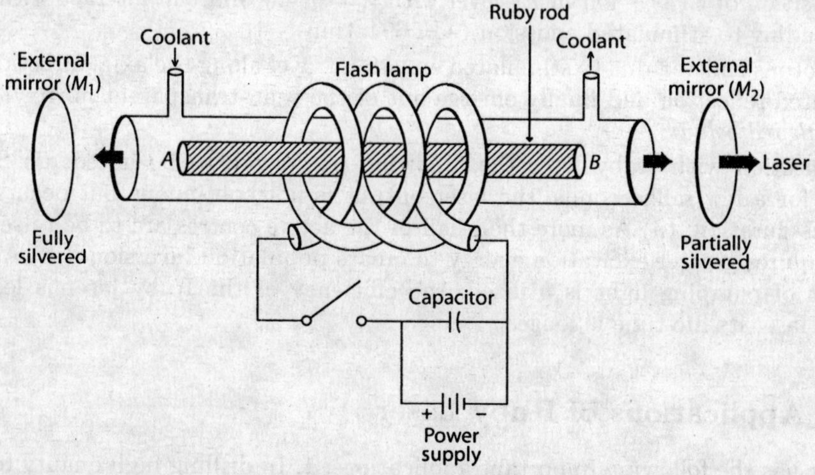

Fig. 7.17: Set-up for Ruby laser.

The Ruby rod is cylindrical of 10 cm long and 0.5 cm in diameter. The ends A and B of the rod are polished optically flat and parallel and also perpendicular to the axis of the rod. One end A of the rod is fully silvered and the other end B is partially silvered. For external mirror configuration the mirrors M_1 and M_2 are placed as shown in Fig. 7.17.

The helical flash lamp surrounding the Ruby rod is excited by the discharge of the capacitor using a secondary high voltage ($\sim 20\,\text{kV}$) trigger pulse.

To protect the crystal from the heat dissipated in the flash tube, a coolant (water) is circulated around the Ruby rod.

7.16.1 Working

The energy levels of Cr^{+++} ions in the crystal lattice are shown in Fig. 7.18, which form basically a *three-level system.*

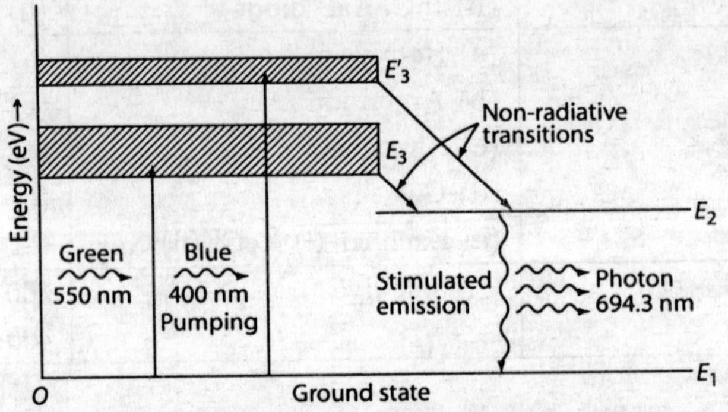

Fig. 7.18: Energy levels of Cr^{+++} in Ruby laser.

When the xenon discharge starts, a large number of Cr^{+++}-ions are excited from the ground state E_1 to the band of levels E_3. The excited levels being *highly unstable* (life-time $\tau \sim 10^{-8}$ s), the Cr^{+++}-ions undergo a *non-radiative transition* from E_3 to E_2. E_2 is a metasatble state ($\tau \sim 3 \times 10^{-3}$ s) and Cr^{+++}-ions accumulate there. The *population inversion gets established* in a short time in E_2 if the pumping occurs at a faster rate. The *spontaneous photon* emitted due to the transition of Cr^{+++}-ion in E_2-level with $\lambda = 694.33$ nm can interact with excited ions in E_2 state leading to stimulated emission ($\lambda = 694.3$ nm).

The photons emitted due to stimulated emission travel along the axial direction and amplified after repeated reflection and finally emerge out of the semi-transparent mirror as a *strong beam of laser with red colour.*

In connection with Ruby laser the following points are to be noted: (i) Since the xenon flash lasts for a few milliseconds, the laser output is not continuous but occurs in the form of pulses of μs-duration. (ii) As more than half of the active centres are to be raised to the excited state, it requires greater excitation energy to create population inversion. (iii) As only the green component of pumping light is utilized, the efficiency of the Ruby laser is less. (iv) Being a solid-state laser its life-time is longer.

7.16.2 Applications of Ruby Laser

Ruby laser has the following important applications: 1. In drilling high quality holes 2. In holography 3. Trimming resistors and integrated circuit masks 4. In distance measurement using pulse echo technique 5. In defense, as target designators and range finders 6. In the filed of scientific research, used in plasma production and fluorescent spectroscopy 7. In atmosphere ranging, scattering and lidar measurements.

7.17 Continuous He-Ne Visible Laser

He-Ne laser is the first successful gas laser developed by *Ali Javan, W. Bennet* and *D. Herriot* in 1961.

Construction: The schematic diagram of a typical He-Ne laser is shown in Fig. 7.19.

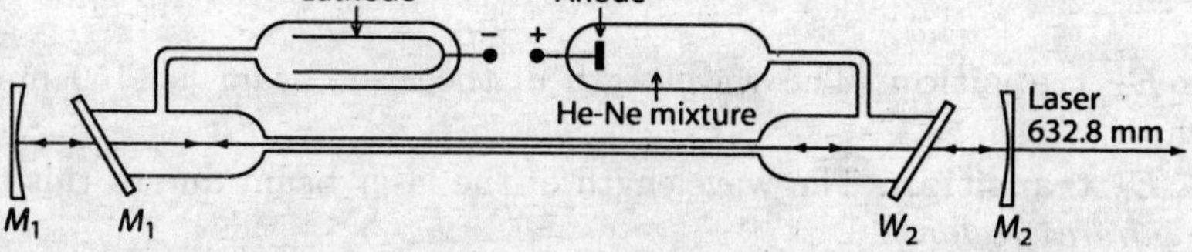

Fig. 7.19: He-Ne laser.

Pure He and Ne gas in the ratio 10 : 1 at a total pressure of *one torr* is enclosed in a discharge tube of 50 cm long and 1 cm diameter. The gain of the system varies directly with the tube length and inversely with the diameter. Electric discharge is passed through the tube by two electrodes connected to a high voltage source. Two spherical mirrors M_1 and M_2 (99.5% reflectivity) are fixed at the two ends of the tube, which form the *Fabry-Perot resonator*. Two optically flat windows W_1 and W_2 (the *Brewster windows*) cemented at the ends at Brewster's angle $\theta = \tan^{-1} n$ ($n =$ the r.i. of the reflectors) to get a *plane polarized beam.*

7.17.1 Working

In He-Ne laser there is a *four-level pumping scheme.* As the electric discharge passes through the tube, some He atoms are excited by collision with the accelerated electrons to metastable states F_2 and F_3 [Fig. 7.20]. The number of such excited He atoms increases as the optical transition to lower state is forbidden.

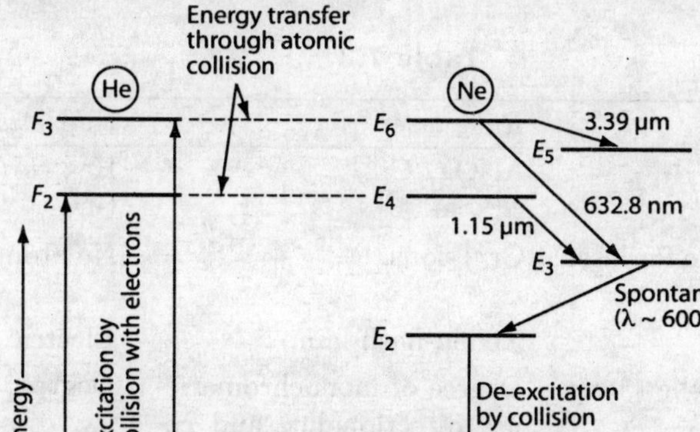

Fig. 7.20: Energy levels of He and Ne and transitions between the levels.

Some of the excited states of Ne are lower but very close to the metastable states F_2, F_3 of He. When He atoms in states F_2, F_3 collide with Ne atoms in the ground state E_1, exchange of energy from He atoms to Ne atoms occurs as a result of which Ne atoms are excited to E_4 and E_6-levels. He atoms after collision are de-excited to F_1 state. The process may be represented as below:

$$e^* + \text{He} \longrightarrow \text{He}^* + e$$

$$\text{He}^* + \text{Ne} \longrightarrow \text{He} + \text{Ne}^*.$$

He star (He^*) represents excited state and e, the electron.

The discharge continuously populates E_4- and E_6-levels of Ne creating population inversion between E_4- (or, E_6-) and E_5- and E_3-levels. Consequently, three laser transitions occur. These are:

(i) **$E_6 \rightarrow E_3$ transition.** The wavelength of the generated laser beam is 632.8 nm and the *colour is red.*

(ii) **$E_4 \rightarrow E_3$ transition.** The wavelength of the laser beam is 1150 nm and lies in the *infrared region.*

(iii) **$E_6 \rightarrow E_5$ transition.** The wavelength of the laser beam due to this transition is 3.39 μm in the *far infrared region.*

Lasing transitions, $E_5 \rightarrow E_2$ and $E_3 \rightarrow E_2$ can make downward transition to E_2-level emitting *incoherent light due to spontaneous transitions.* The state of population inversion between lasing levels depopulated at a faster rate.

The accumulated Ne atoms in the metastable state E_2 drift towards the discharge tube wall and collide. By this they lose energy and are de-excited to ground state E_1.

7.17.2 Applications of He-Ne Laser

On account of high degree of monochromaticity, directionality and coherency He-Ne laser has found following important applications: 1. It is used in holography, in laser printing. 2. Widely used for interferometric experiments in the laboratories as monochromatic light source. 3. It is used to read bar codes on the products, special characteristics or other symbols. 4. It is used to test smoothness of surfaces.

Table 7.3 shows Ruby Laser versus He-Ne laser (*A comparative study*).

Table 7.3

Sl. No.	*Particulars*	*Ruby laser*	He-Ne *laser*
1.	Active medium	$Al_2O_3 : Cr^{3+}$	He : Ne (10 : 1) gas mixture
2.	Entity responsible for laser action	Cr^{3+}-ions	Ne atoms
3.	Pumping agent	Xenon flash lamp	Electric discharge
4.	Properties of the laser beam	Degree of monochromaticity, directionality and coherency are less than that of He-Ne laser.	Degree of monochromacity, directionality and coherence are higher than that of Ruby laser.
5.	Nature of output laser	Pulsed	Continuous.
6.	Maximum output power	A few kW	$\sim$ 100 mW
7.	Wavelength of the emitted laser beam	694.3 nm	632.8 nm
8.	Cooling	Due to heat generated it needs cooling during operation	No such cooling is needed
9.	Spectral width	Wide	Extremely narrow

7.18 Applications of Laser

Dur to its (a) *high temporal and spatial coherence* (b) *monochromaticity* (c) *narrow angular spread* and (d) *high brightness, laser has the following important applications in different fields*:

1. **Communication**: Modulated laser beams are used to transmit messages through optical fibres. More channels of message can be obtained due to its very narrow bandwidth.

The storage capacity in computers is highly improved due to its narrow bandwidth.

2. **Measurement of long distance**: Due to its small angular spread, it is used to measure the distance of the moon from the earth within an error of 0.6 m, which is negligible.

3. **Surgery**: In surgery, laser is used in (i) bloodless surgery, e.g., to weld the detached retina (ii) drilling the teeth and preventing tooth decay by depositing hard materials on the tooth surface (iii) removing tumours and removal of infected cells (iv) the treatment of cataract.

4. **Laser is used in**: (i) holography (ii) cutting of micro-electric circuits (iii) physiotherapy (iv) unblocking the heart valves, breaking kidney stones, etc.

5. **Weather forecasting**: The pictures of the wind movement, clouds, etc. obtained by laser is used in weather forecasting.

6. **Scientific research**: Laser is used to drill holes, fuse and melt metals as it can be focussed in a small area raising the temperature to about 1000 K.

7. **Nuclear fission**: Laser beam can be used to induce nuclear fission, which occurs at $\sim 10^8$ K.

8. **National defense**: For guiding missiles and satellites laser beam is used.

9. **Industry**: (i) In plastic industry laser beam is used in cutting, welding of plastic sheets. (ii) Laser beam is used to detect any flaw in cast iron pipes, etc.

7.19 Answers of Some Important Questions

7.1. *What does the acronym LASER stand for?* [cf. **WBUT 2005; BUH 2000**]

Ans. The acronym LASER stands for Light Amplification by Stimulated Emission of Radiation.

7.2. *What are the main specialities of Laser beam?*

Ans. Laser beam has the following main specialities (a) highly coherent (b) highly directional (c) spectral line width is extremely narrow (d) high energy content, i.e., high brightness.

7.3. *What are the different types of interaction that an atom can have with incident electromagnetic waves?*

Ans. Einstein proposed that an atom having discrete energy levels can interact with the incident electromagnetic radiation in the following three ways (a) Stimulated absorption (b) Spontaneous emission (c) Stimulated emission.

The above three processes form the basic principle of laser.

7.4. *What is stimulated absorption?*

Ans. If an atom having two energy states E_1 and E_2 $(E_2 > E_1)$ be irradiated with an electromagnetic radiation of frequency $\nu = \frac{E_2 - E_1}{h}$, then the atom is pumped to the excited state E_2. This process of absorption of the incident radiation and subsequent transition $E_1 \to E_2$ is called stimulated absorption. The rate of stimulated absorption depends on (i) the intensity of the incident radiation and (ii) the number of atoms in E_1.

7.5. *What is spontaneous emission?*

Ans. The life-time of the excited state (E_2) being very small $(\tau \sim 10^{-8}\,\text{s})$ an atom in the excited state transits to the lower energy state (E_1) with the emission of a photon of frequency $\nu = \frac{E_2 - E_1}{h}$, where h is the Planck's constant.

This type of emission is called the *spontaneous emission* of radiation. The rate of spontaneous transition is (i) proportional to the number of atoms in the excited state (ii) independent on the intensity of the exciting radiation.

The spontaneous emission is random in character and incoherent having broad spectrum.

7.6. *What is stimulated emission?*

Ans. Einstein proposed that an atom in the excited state (E_2) can transit to the lower energy state (E_1) being triggered by an incident photon of appropriate frequency and the emitted radiation is called *stimulated emission.*

The rate of stimulated emission depends on (i) the number of atoms in the excited state and (ii) the intensity of the external radiation.

This leads to the amplification of the incident radiation. This type of radiation is *highly directional* and *coherent.*

7.7. *Distinguish between spontaneous and stimulated emissions of radiation.*
[BUH 2003; cf. CUH 1992; WBUT 2006]

Ans. See Section 7.4 (p. 194).

7.8. *State whether a laser is an amplifier or an oscillator?* **[BUH 1999, 2001, '05]**

Ans. The mirrors in a resonator provide positive feedback to the photons amplified in the active medium. In analogy with electronic oscillator (in which a part of the output voltage is fedback to the input in proper phase, i.e., *positive feedback*) the laser is an *oscillator.*

7.9. *State the characteristic of laser light that differs from those of ordinary light.*
[BUH 2002, cf. 2001; WBUT 2005]

Ans. See Section 7.5 (p. 194).

7.10. *What is population inversion?* **[CUH 1992]**

Ans. If the number of atoms N_2 in the higher energy state E_2 of a medium is greater than the number of atoms N_1 in the lower energy state E_1, then the situation is called *population inversion.* Population inversion is a prerequisite for light amplification in a non-equilibrium distribution of atoms among the various energy levels of the system.

7.11. *What is a metastable state?*

Ans. For population inversion to be established the life-time of the excited state must be $\sim 10^{-6}$ to 10^{-3} s which is 10^3 to 10^6 times the life-time of the ordinary energy levels. Such an energy level is called a *metastable state.* The characteristics of a metastable state are (i) the life-time is very large ($\sim 10^{-3}$ s) (ii) the probability of spontaneous decay is very low (iii) due to large life-time the atoms making transition to the metastable state find sufficient time to be accumulated in this level and population inversion occurs.

7.12. *What are the essential requirements for the construction of a laser?* **[BUH 2001, '03, '05]**

Ans. See Section 7.14 (p. 201).

7.13. *Explain the concept of negative temperature in connection with the operation of a laser.*
[BUH 2002, '05]

Ans. If Ni be the number of atoms in a higher energy state E_i and N_0 be that in the ground state, then according to Boltzmann's law at thermal equilibrium, $N_i = N_0 e^{-E_i/kT}$.

From this relation it is evident that $N_i > N_0$, if T is *negative.* Again, if $N_i > N_0$, then we say that population inversion has occurred in the medium. That is why the state of population inversion is sometimes referred to as *negative temperature state.* It does not mean that temperature below 0 K is attainable. It simply means that population inversion is obtained at ordinary temperature.

7.14. *Show that population inversion is impossible by direct pumping in a two-level system.*

Ans. According to Boltzmann relation to the number of atoms N_i in the excited state E_i at equilibrium temperature is $N_i = N_0 e^{-\frac{E_i}{kT}}$, where k is the Boltzmann constant.

If $kT \gg E_i$, then $e^{-\frac{E_i}{kT}} \to 1$ and then $N_i = N_0$.

In no case N_i can be greater than N_0. In other words, the higher energy level E_i can never be more populated than the ground level E_0 only by means of optical pumping. Hence, by direct pumping population inversion is impossible in a two-level system.

7.15. *Name two lasers in the visible region.* [BUH 2002]

Ans. The two lasers in the visible region are (i) Ruby laser ($\lambda = 694.3\,\text{nm}$), He-Ne laser ($\lambda = 632.8\,\text{nm}$) (ii) Argon ion laser ($\lambda = 450\text{-}520\,\text{nm}$), multiple lines.

7.16. *A laser beam shows high degree of directionality and monochromaticity—why?*

Ans. An incident photon stimulates an atom in the metastable state to emit a photon of same energy, i.e., frequency, phase and direction by transition from higher energy state (E_2) to lower energy state (E_1) so that $h\nu = E_2 - E_1$. Also the *imperfection, thermal distortion* and *scattering*, which are present in a solid-state laser are totally absent in a gas laser. Since the frequency, phase and direction of the stimulated photon are the same as that of the stimulating photon, laser beam shows *high degree of directionality and monochromaticity.*

7.17. *In what respect* He-Ne *laser is superior to Ruby laser?*

Ans. Ruby laser is a *three-level laser system* and no significant emission occurs until at least half of the ground state atoms have been excited to the metastable state. This means population inversion exists only when the metastable state is more densely populated than the ground state, which makes the high threshold for optical pumping and wastage of energy.

But He-Ne laser is a *four-level system* and population inversion begins in the metastable state E_2 and the laser terminal state E_1 with a little pumping. Here the spontaneous emission rate from E_1 to E_0 (ground state) must be faster than the rate from E_2 to E_1. Otherwise, population of E_1 could exceed that of E_2, when the population inversion would be destroyed.

Moreover, He-Ne laser provides a continuous laser beam but the Ruby laser is pulsed.

7.18. *What is resonance transfer of energy?*

Ans. If two different atoms, one excited and the other in normal state of identical nature, collide, then maximum energy transfer occurs from one atom to the other by which the excited atom returns to the ground state and the normal atom is raised to the excited state. This mechanism of energy transfer is called *resonance transfer of energy.*

Example: In He-Ne atoms, the excited Helium atom (He*) collides with normal Ne atom in the ground state and transfers whole of its energy to Ne atom raising it to the excited state according to the scheme $\text{He}^* + \text{Ne} \rightarrow \text{He} + \text{Ne}^*$.

The He atom, then transits to the ground state.

7.19. *What do you mean by an active medium?*

Ans. See Section 7.10 (p. 198).

7.20. *What is threshold inversion density?*

Ans. If the population inversion density is sufficient to compensate loss by the gain, then laser action starts. This population inversion density for which loss is compensated by the gain is called the *threshold inversion density.*

7.21. *Indicate the units of energy density $E(\nu)$, and Einstein's A and B coefficients.*

Ans. Unit of $E(\nu)$ is $\text{J/m}^3\text{/s}$. Unit of A is s^{-1}. Unit of B is $\text{m}^3\text{/J/s}^2$.

7.20 Solved Problems

Ex. 1 *A beam of laser having power* 50 mW *and aperture* 5×10^{-3} m, *wavelength* 700 nm *is focussed with the lens of focal length* 0.2 m. *Calculate the* (i) *areal spread and* (ii) *intensity of the image.*

Soln. (i) The angular spread of the laser beam is

$$d\theta = \frac{\lambda}{d} = \frac{700 \times 10^{-9}}{5 \times 10^{-3}} = 1.4 \times 10^{-4} \text{ radiation.}$$

The areal spread of the given laser beam is

$$A = (d\theta \times f)^2 = (1.4 \times 10^{-4} \times 0.2)^2 = \mathbf{0.078 \times 10^{-8}\ m^2}.$$

(ii) Intensity of the laser beam is

$$I = \frac{\text{Power}}{\text{Area}} = \frac{50 \times 10^{-3}}{0.48 \times 10^{-8}} = \mathbf{104 \times 10^5\ watt/m^2}.$$

Ex. 2 *A* He-Ne *laser beam of wavelength* $\lambda = 632.8\,\text{nm}$ *is focussed on an area* λ^2. *If the laser beam radiates energy at the rate of* 1 mW, *find out the intensity of focussed beam.*

Soln. Area on which the laser beam is focussed is $A = \lambda^2 = (6.328 \times 10^{-7})^2 = 40 \times 10^{-14}\,\text{m}^2$.

Intensity of the beam

$$I = \frac{P}{A} \doteq \frac{10^{-3}\,\text{W}}{40 \times 10^{-14}\,\text{m}^2} = 2.5 \times 10^9\,\text{W/m}^2.$$

Ex. 3 *Find the ratio of spontaneous emission to stimulated emission by an incandescent bulb at* 2500 K. *The frequency in the optical region is* 5.9×10^{14} Hz.

Soln. The required ratio

$$R = \frac{A_{21}}{B_{21}E(\nu)} = (e^{h\nu/kT} - 1) = \exp\left[\frac{(6.6 \times 10^{-34}) \times (5.9 \times 10^{14})}{(1.38 \times 10^{-23}) \times 2500}\right] - 1$$

$$= \exp\left[\frac{6.6 \times 5.9}{1.38 \times 2.5}\right] - 1 = \mathbf{7.9 \times 10^4}.$$

Ex. 4 *In a* He-Ne *laser transition from 3s to 2p-level gives a laser transition of wavelength* 632.8 nm. *If the energy of the 2p-level is* 15.2×10^{-19} J, *how much pumping energy is required, if there is no energy loss in the* He-Ne *laser?* **[WBUT 2008; CU 1997]**

Soln. Here the energy of the emitted laser beam is

$$E = \frac{hc}{\lambda} = \frac{(6.625 \times 10^{-34}) \times (3 \times 10^8)}{632.8 \times 10^{-9}} = 3.1 \times 10^{-19}\,\text{J} = 1.935\,\text{eV}.$$

Again, energy of the $2p$-level is $15.2 \times 10^{-19}\,\text{J} = 9.488\,\text{eV}$.

$\therefore$ the energy of the pumping source is $(9.488 + 1.935)\,\text{eV} = \mathbf{11.42\,eV}$.

Ex. 5 *If the two plane mirrors forming the resonant cavity in a* He-Ne *laser be separated by* 30 cm, *what is the mode of separation?*

Soln. We know that the frequency difference between two consecutive modes is

$$\Delta\nu = \frac{c}{2d},$$

where c = velocity of light in vaccuum = 3×10^8 m/s, $d = 30\,\text{cm} = 0.3\,\text{m}$.

$$\therefore \Delta\nu = \frac{3 \times 10^8}{2 \times 0.3} = \mathbf{5 \times 10^8\ Hz}.$$

Ex. 6 *What is the coherence length of laser beam in vacuum, if the bandwidth of laser beam is* 30 MHz. **[BUH 2001; cf. CUH 1993]**

Soln. Coherence time

$$\tau_c = \frac{1}{\Delta\nu} = \frac{1}{30 \times 10^6}.$$

$\therefore$ coherence length $l_c = c\tau_c = \frac{3\times10^8}{30\times10^6} = \mathbf{10\,m}$.

Ex. 7 *What is the diffraction limited spot size diameter if the beam is focussed by a lens of focal length* $f = 1.5\,\text{m}$? $[\lambda = 1.05 \times 10^{-6}\,\text{m},\ d = 2.5 \times 10^{-2}\,\text{m}]$

Soln. The diffraction limited spot size diameter is

$$D = 2.44 \times \frac{f\lambda}{d} = 2.44 \times \frac{(1.5 \times 2.5 \times 10^{-2}) \times (1.05 \times 10^{-6})}{2.5 \times 10^{-2}} = \mathbf{3.9 \times 10^{-6}\ m.}$$

EXERCISES

7(A)–Objective Type Questions

Select the correct answer:

7.1. Einstein in 1917 theoretically proved the existence of (a) spontaneous emission (b) stimulated emission (c) non-radiative transition (d) thermal emission

7.2. The first laser discovered by Maiman was (a) He-Ne laser (b) Ruby laser (c) CO_2 laser (d) Nd-YAG laser.

7.3. The example of pulsed laser is (a) Ruby laser (b) He-Ne laser (c) Nitrogen laser (d) Argon ion laser.

7.4. Two different laser sources can produce sustained interference because laser is (a) highly coherent (b) incoherent (c) highly bright (d) highly directional.

7.5. Emission of a photon assisted by an extraneous photon of proper frequency is called (a) spontaneous emission (b) stimulated emission (c) stimulated absorption (d) none of these.

7.6. Boltzmann's law gives the density of population of various energy levels in (a) thermodynamical equilibrium (b) thermal equilibrium (c) thermal inequilibrium (d) both in (b) and (c).

7.7. A metastable state has life-time τ of the order of (a) 10^{-8} s (b) 10^{-5} s (c) 10^{-3} s (d) 10^{-10} s.

7.8. Presence of a metastable state is essential for (a) stimulated absorption (b) spontaneous emission (c) stimulated emission (d) non-radiative transition.

7.9. If N_1, N_2 be the number of atoms in energy states E_1, E_2 ($E_2 > E_1$), then the condition for inverse population is (a) $N_1 = N_2$ (b) $N_1 < N_2$ (c) $N_1 > N_2$ (d) $N_1 \geq N_2$.

7.10. For creating population inversion, a system must be at least: (a) two-level (b) three-level (c) four-level (d) any level.

7.11. For laser action to occur the medium must have at least, (a) 4 energy levels (2) 2 energy levels (c) 3 energy levels. **[WBUT 2005]**

7.12. The population inversion in preparing laser beam can be achieved (a) when one of the excited states is more populated than the ground states (b) when one of the excited state is less populated than the ground state (c) when the population of one excited state and the ground state are equal (d) irrespective of any relationship between the population in the excited state and the ground state.

7.13. Stimulated absorption rate depends on (a) intensity of the incident radiation (b) number of atoms in the ground state (c) both (a) and (b) (d) none of these.

7.14. Spontaneous emission rate depends on (a) the number of atoms in the excited state (b) intensity of the exciting radiation (c) the number of atoms in the ground state (d) none of these.

7.15. Stimulated emission rate depends on (a) the number of atoms in the excited state (b) the intensity of the external radiation (c) both (a) and (b) (d) none of these

7.16. An active medium is that (a) which can move (b) in which population inversion is created (c) in which population inversion can never be created (d) which can interact with other media.

7.17. The rate of upward and downward transitions in thermal equilibrium is (a) unequal (b) equal (c) both (a) and (b) (d) none of these.

7.18. For light amplification the responsible mechanism is (a) spontaneous absorption (b) spontaneous emission (c) stimulated emission (d) both (b) and (c).

7.19. The intensity of light passing through an active medium (a) decreases (b) increases (c) remains unaltered (d) first decreases and then increases.

7.20. The probability of spontaneous emission increases as (a) ν (b) ν^2 (c) ν^3 (d) ν^{-1}, where ν is the frequency of the exciting radiation.

7.21. The ratio (ρ) of stimulated emission to the rate of spontaneous emission is given by (a) $\rho = \frac{1}{e^{h\nu/kT}+1}$ (b) $\rho = \frac{1}{e^{h\nu/kT}-1}$ (c) $\rho = e^{h\nu/kT}+1$ (d) $\rho = \left(e^{h\nu/kT}\right)^{-1}$.

7.22. The activator atom in Ruby laser is (a) oxygen (b) aluminum (c) chromium (d) iron.

7.23. In Ruby laser, population inversion is created by (a) optical pumping (b) electric discharge (c) chemical reaction (d) injection current, etc.

7.24. In gas laser population inversion is generally created by (a) optical pumping (b) electric discharge (c) chemical reaction (d) none of these.

7.25. For a fixed set of mirrors, threshold gain of laser beam is (a) $\beta \propto d$ (b) $\beta \propto \frac{1}{d}$ (c) $\beta \propto d^2$ (d) $\beta \propto d^{-2}$, where d is the distance between the mirrors.

7.26. Intensity of laser beam passing through a distance x and amplification β is (a) $I(x) = I_0e^{\beta x}$ (b) $I(x) = I_0e^{-\beta x}$ (c) $I(x) = I_0e^{-\beta/x}$ (d) $I(x) = I_0\frac{1}{e^{\beta x}}$.

7.27. He-Ne laser system gives out (a) always continuous wave (CW) (b) Pulsed wave (c) both (a) and (b) (d) none of these.

7.28. Ruby laser system gives out (a) CW beam (b) Pulsed beam (c) both (a) and (b) (d) none of these.

7.29. The standard wavelength emitted from He-Ne laser is (a) 543 nm (green) (b) 595 nm (yellow-orange) (c) 632.8 nm (red) (d) 612 nm (red-orange).

7.30. According to Boltzmann distribution law (a) $N_i = N_o e^{-E_i/kT}$ (b) $N_i = N_o e^{E_i/kT}$ (c) $N_i = \frac{N_o}{e^{-E_i/kT}}$ (d) $N_i = \frac{1}{N_o}e^{-E_i/kT}$.

7.31. The colour of the laser output from a Ruby laser is (a) green (b) blue (c) red (d) violate.

7.32. The ratio of He and Ne in He-Ne laser is (a) 1 : 100 (b) 10 : 1 (c) 1 : 10 (d) 100 : 1.

7.33. In He-Ne laser, Ne atoms obtain energy (a) on collision with He atoms (b) from chemical reactions (c) from electrical pumping (d) from optical pumping. **[WBUT 2008]**

7.34. In Ruby Laser, the active medium is (a) solid (b) liquid (c) gas (d) a mixture of solid and gas. **[WBUT 2010]**

ANSWERS

7.1. (b) **7.2.** (b) **7.3.** (a), (c) **7.4.** (a) **7.5.** (b) **7.6.** (b) **7.7.** (c) **7.8.** (c) **7.9.** (b) **7.10.** (b) **7.11.** (c) **7.12.** (a) **7.13.** (c) **7.14.** (a) **7.15.** (c) **7.16.** (b) **7.17.** (b) **7.18.** (c) **7.19.** (b) **7.20.** (c) **7.21.** (b) **7.22.** (c) **7.23.** (a) **7.24.** (b) **7.25.** (b) **7.26.** (a) **7.27.** (a) **7.28.** (b) **7.29.** (c) **7.30.** (a) **7.31.** (c) **7.32.** (b) **7.33.** (a) **7.34.** (a).

7(B)–Short Answer Type Questions

7.1. What does the acronym LASER stand for? **[cf. WBUT 2005; BUH 2000]**

7.2. What are the differences between laser light and normal light? **[WBUT 2007; BUH 2002]**

7.3. Explain briefly the characteristics of laser beam. **[WBUT 2005]**

7.4. What do you mean by *spontaneous emission* and *stimulated emission*?

7.5. Distinguish between spontaneous and stimulated emission of radiation.

[**WBUT 2006; BUH 2003; CUH 1992**]

7.6. (a) What is population inversion? Explain with reference to energy level diagram.

(b) Explain the significance of *negative temperature.*

7.7. (a) What are Einstein's A and B coefficients? Explain their significance.

(b) State the units of $E(\nu)$, A and B.

7.8. (a) Why population inversion cannot be created in a two-level system?

(b) Mention the different techniques to create population inversion.

7.9. (a) What is the threshold condition for light amplification?

(b) How does threshold gain (β_{th}) depend on the reflectivity?

7.10. (a) Why the Ruby laser emits pulsed laser beam?

(b) Why the laser beam obtained from He-Ne laser is continuous wave (CW)?

7.11. (a) State some applications of Ruby laser and He-Ne laser.

(b) What are the active centres in the above two lasers.

7.12. What is population inversion in the context of LASER? Does it violate Maxwell-Boltzmann distribution law? [**WBUT 2010**]

7(C)–Numerical Problems

7.1. The life-time for spontaneous emission of the upper state is 10^7 s. Find the inversion density for a He-Ne laser that operates at 633 nm, given that the gain constant = 3% m^{-1}, temperature of discharge is 150 °C. [*Ans.* $3.15 \times 10^4/m^3$]

7.2. From the following data, compute the gain constant of a laser: line width = 0.15 nm, wavelength λ_0 = 650 nm, life-time of spontaneous emission = 2×10^{-4} s and inversion density $N_2 - N_1 = 5 \times 10^{22}/m^3$. [*Ans.* 3.65]

7.3. A normal optical source emits a radiation of wavelength 6000 Å at a temperature 900 K. Show that the stimulated emission in the system is negligible compared to the spontaneous emission.

7.4. A 3 mW laser beam of wavelength $\lambda = 6 \times 10^{-5}$ cm is focussed by a convex lens. The focal length of the lens is 5 cm and radius of the circular aperture is 1 cm. Find the intensity of the focussed spot. If the coherence time of the laser beam is 50 ns, find the monochromaticity ($\Delta\nu/\nu$) of the source. [*Ans.* 4×10^{-8}]

[**CUH 2002**]

7.5. Calculate the gain constant β of a laser beam from the following data:

$N_2 - N_1 = 5 \times 10^{16}\ cm^{-3}$, $\lambda_0 = 650$ nm, $\Delta\lambda = 1.5$ nm and the life-time for spontaneous emission $\tau = 2 \times 10^{-4}$ s. [*Ans.* 3.96/cm]

7.6. A Ruby laser pulse of output beam spot diameter 5 mm and divergence 1 milli radian is directed towards the moon. Calculate the area on the moon illuminated if the distance between the earth and the moon is 376284 km. [*Ans.* 445000 km^2]

7(D)–Broad Answer Type Questions

7.1. (a) Describe with energy level diagrams the phenomena of spontaneous emission, stimulated emission and stimulated absorption in a two-level system. [**CUH 1993**]

(b) Compare spontaneous emission with stimulated emission.

(c) How much higher probability of stimulated emission compared to that of spontaneous emission is achieved in laser? [**WBUT 2005**]

7.2. (a) What is population inversion? Explain. **[WBUT 2005]**

(b) How higher probability of stimulated emission compared to that of spontaneous emission is achieved in LASER? **[WBUT 2005]**

(c) Describe the essential parts of the laser. **[WBUT 2004]**

7.3. (a) Explain briefly the working principle of a LASER. **[WBUT 2006]**

(b) What is an optical resonator? **[WBUT 2006]**

(c) Explain whether a laser is an amplifier or oscillator. **[BUH 1999, 2001, '05]**

7.4. (a) Draw a neat sketch of Ruby laser. **[WBUT 2004]**

(b) Discuss the operation of a Ruby laser with the help of energy level diagram. **[WBUT 2004]**

(c) Mention some important uses of Ruby laser.

7.5. (a) Explain the principle of He-Ne laser with energy level diagram. **[WBUT 2007]**

(b) Discuss some advantages of He-Ne laser in comparison to Ruby laser.

(c) State some applications of laser. **[WBUT 2005]**

7.6. (a) Describe quantitatively the operation of He-Ne laser with rough energy level diagram of relevant atom involved. **[BUH 2002, '03; CUH 2002]**

(b) What do you mean by metastable state? Explain.

(c) Describe different methods of achieving population inversion.

7.7. (a) Find a relation between Einstein's A and B coefficients in connection with laser transition.

(b) What is a cavity resonator? How does it work?

(c) Explain the concept of negative temperature in connection with the operation of a laser. **[BUH 2005, '06]**

7.8. (a) Define Einstein's A and B coefficients and deduce their mutual relation.

(b) Show that the ratio of spontaneous and stimulated emission is proportional to the cube of the frequency.

(c) Explain the construction and action of the optical resonator in a Ruby laser. **[WBUT 2008]**

(d) Explain the importance of metastable state in laser.

(e) How population inversion can be obtained?—Explain.

7.9. (a) Define Einstein A, B coefficients of absorption and emission. Find out the relation among them. **[WBUT 2010]**

(b) Describe the function of an optical resonator. What is the use of such a device in the context of LASER generation? **[WBUT 2010]**

Chapter 8 | Module 3.3

Holography

8.1 Introduction

Holography is the technique of producing three-dimensional images using photography film and coherent light (i.e., without camera or lenses). The term *holography* is derived from the Greek word *holos* meaning the *whole* as it contains the whole information (viz., the phase and amplitude). The principle of holography was discovered by *Dennis Gabor* in 1948.

In ordinary photography by camera lens only the intensity (which is the square of amplitude) of light from the different parts of the object is recorded, the phase distribution on the photo-plate is lost. Thus, the perspective of the image cannot be changed by viewing it from different angles or any unfocus part cannot be focussed. In holography both the amplitude and the phase of light wave are recorded. This is done by the *technique of interference* in which three-dimensional form of the image is produced. The resulting photographic record containing the three-dimensional image is called *hologram,* analogous to the *negative* of an ordinary photograph. The process of holography does not mean the image to the object, rather it records the reflected light wave.

8.2 Principle of Holography

The principle of holography can be explained in the following two ways:

(i) Recording of the hologram and (ii) Reconstruction of the image.

Recording of the hologram: The basic principle of holography depends on the technique of interference of light waves with high degree of coherence.

A coherent laser beam is split up into two beams A and B by a beam splitter S as shown in Fig. 8.1. One part A of the beam is travelling towards the object O and is called the *signal beam.* The other part B of the beam travels in the direction of the incident beam and is called the *reference beam.* The diffracted beam from the object whose hologram is to be recorded and the reference beam interfere on the photo-plate P. This interference is the hologram of the object O. The interference pattern contains information regarding phase and amplitude. The hologram thus prepared contains large number of closely spaced points, which can be viewed by a high power microscope.

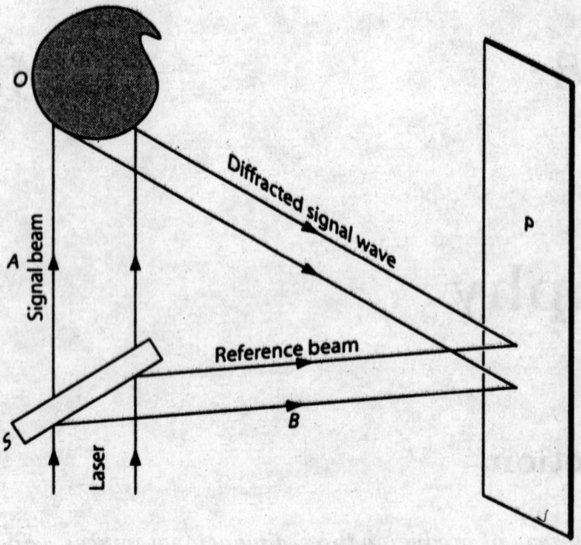

Fig 8.1: Production of hologram.

Reconstruction of the image. If the object O is now removed and the hologram be placed, where it was when formed, then the reference laser beam B, known as the *read out waves*, interacts with the interference pattern on the plate and two images are formed by the diffracted waves. One of the images appears at the original position of the object and is the virtual image. The other image is real, and can be viewed on looking through the hologram (acting like a window), and can be photographed directly [Fig. 8.2].

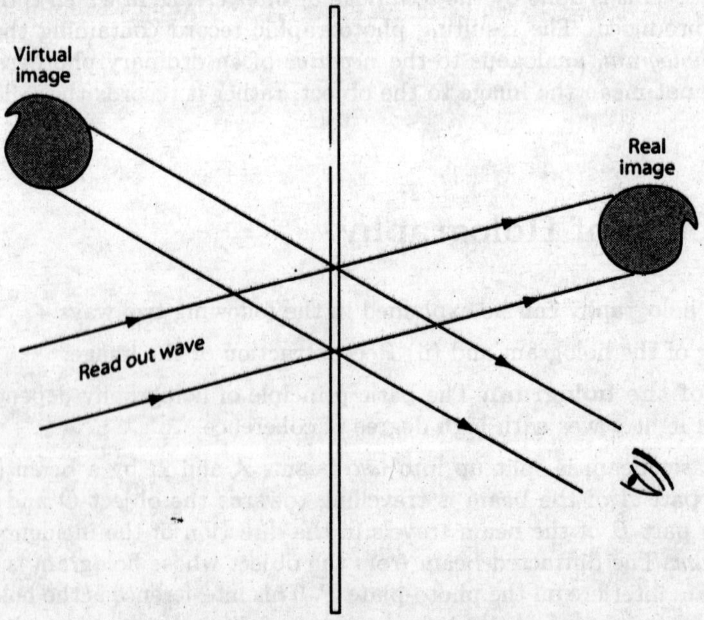

Fig. 8.2: Reconstruction of the image.

If the hologram be illuminated with the reference beam, the object would be seen at the exact location and perspective as if one is looking through a transparent window. This is the virtual image. The image from another angle can be viewed just as a $3D$ object is viewed.

The real image is found between the observer and the plate P, which appears inverted in depth, i.e., the foreground and background are reversed. That is why, the interest of the observer lies with the virtual image.

The reconstruction of the hologram being through the diffraction process, the coherent requirements are not as stringent as the recording set-up. In the process of reconstruction of the hologram, a wave of wave length different from that used in the process of recording can be used resulting magnification and change of location of the image.

8.3 Characteristics of Holograms

Following are some of the important characteristics of hologram:

(i) A hologram is superior over a photograph in that the former has greater information storage capacity than the latter. Many sets of information can be recorded on a hologram by changing the angle at which the reference beam strikes the holographic plate. But in a photographic plate superposition of several images is meaningless. The readout wave, in reconstruction, should be incident at the same angle to exact that particular image stored.

(ii) In a hologram, the information about an object is recorded over the whole area of the hologram and hence destruction of any part of it is an irreparable loss. Each point on the hologram receives light from all parts of the object and as such contains information about the geometrical characteristics of the entire image. Every part of the hologram, no matter how small, is capable of producing the entire image of the object. This makes holography a reliable method for data storage.

(iii) The greatest achievement in holography lies in its capacity to produce $3D$ image of an object, which appears as if the object is being seen. On tilting the head, image from another angle is observed.

Theory of hologram. Let a small object O be illuminated by the source, where most of the light falls undisturbed on the photographic plate P as shown in Fig. 8.3. The scattered or diffracted light from O also falls on P and interfere with the direct beam.

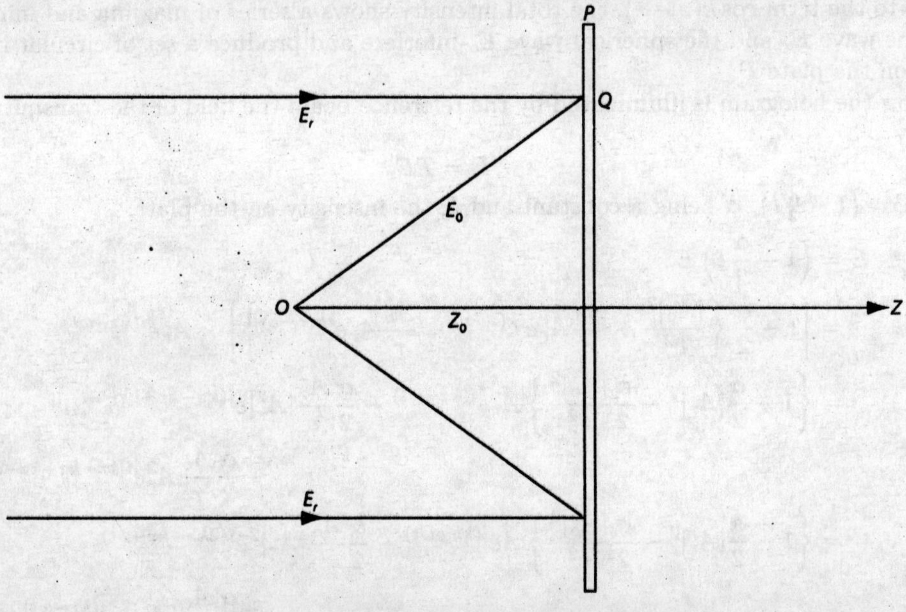

Fig. 8.3: A point source O is illuminated by the source.

To find out an expression for the intensity at Q on the photo-plate, let the electric field of the incident light wave at Q be

$$E = E_r + E_\circ, \tag{8.1}$$

where E_r is the electric field due to reference beam, which is a plane wave given by

$$E_r = A_r e^{i(kz-\omega t)}. \tag{8.2}$$

$E_\circ$ is the electric field of the scattered wave from the object O, whose amplitude and phase vary with position. The reflected wavefronts are spherical and concentric with the point of origin.

The field $E_\circ$ scattered from the object is given by

$$E_\circ = \frac{A_\circ}{r} e^{i(kr-\omega t)}, \tag{8.3}$$

where $r = Z_\circ$ is the distance of the plate from the object O.

The intensity at Q is

$$\begin{aligned}
I &= |E_r + E_\circ|^2 \\
&= \left| A_r e^{i(kz-\omega t)} + \frac{A_\circ}{r} e^{i(kr-\omega t)} \right|^2 \\
&= |A_r|^2 + \left|\frac{A_\circ}{r}\right|^2 + \frac{A_\circ}{r} A_r e^{i(kz-\omega t)} e^{-i(kr-\omega t)} + \frac{A_\circ}{r} A_r e^{-i(kz-\omega t)} e^{i(kr-\omega t)} \\
&= |A_r|^2 + \frac{|A_\circ|^2}{r^2} + \frac{A_\circ}{r} A_r e^{ik(z-r)} + \frac{A_\circ}{r} A_r e^{-ik(z-r)} \\
&= |A_r|^2 + \frac{|A_\circ|^2}{r^2} + \frac{A_\circ}{r} A_r \left[e^{i(kz-r)} + e^{-i(kz-r)} \right] \\
&= |A_r|^2 + \frac{|A_\circ|^2}{r^2} + \frac{A_\circ}{r} A_r \left[\cos k(z-r) + i \sin k(z-r) + \cos(kz-r) - i \sin(kz-r) \right]
\end{aligned}$$

$$\text{or, } I = |A_r|^2 + \frac{|A_\circ|^2}{r^2} + \frac{A_\circ}{r} A_r 2 \cos k(z-r). \tag{8.4}$$

Due to the term $\cos k(z-r)$, the total intensity shows a series of maxima and minima. Thus the plane wave E_r and the spherical wave $E_\circ$ interfere and produce a set of circular interference fringes on the plate P.

If now the hologram is illuminated by the reference beam the field of the transmitted wave is given by

$$E = TE_r, \tag{8.5}$$

where $T = \left(1 - \frac{\alpha}{2} I\right)$, α being a constant and I, the intensity on the plate.

$$\begin{aligned}
\therefore\ E &= \left(1 - \frac{\alpha}{2} I\right) E_r \\
&= \left[1 - \frac{\alpha}{2} \left\{ \frac{|A_\circ|^2}{r^2} + \frac{A_\circ}{r} A_r e^{ik(z-r)} + \frac{A_\circ}{r} A_r e^{ik(r-z)} \right\} \right] \times A_r e^{i(kz-\omega t)} \\
&= \left\{ 1 - \frac{\alpha}{2} |A_r|^2 - \frac{\alpha}{2} \frac{|A_\circ|^2}{r^2} \right\} A_r e^{i(kz-\omega t)} - \frac{\alpha}{2} \frac{A_\circ}{r} |A_r^2| e^{i(kz-kr+kz-\omega t)} \\
&\qquad - \frac{\alpha}{2} \frac{A_\circ}{r} A_r^2 e^{i[kr-kz+kz-\omega t)]} \\
&= \left\{ 1 - \frac{\alpha}{2} |A_r|^2 - \frac{\alpha}{2} \frac{|A_\circ|^2}{r^2} \right\} A e^{i(kz-\omega t)} - \frac{\alpha}{2} \frac{A_\circ}{r} |A_r|^2 e^{i(2kz-kr-\omega t)} \\
&\qquad - \frac{\alpha}{2} \frac{A_\circ}{r} |A_r|^2 e^{i(kr-\omega t)}.
\end{aligned} \tag{8.6}$$

In eqn (8.6), the first term represents the *attenuated reference wave*, the second term represents a *spherical wave surface, which is a replica of the original wave* but has a conjugate or reverse curvature, which converges at a point O' producing a real image at this point [Fig. 8.4].

The third term represents a *diverging spherical wave surface* identical with the wage surface emitted by the object except for a constant factor. This diverging spherical wave when projected back appears to emanate from an apparent object located at the place, where the original object was located. This is the *virtual image* of the object. Hence, the hologram so obtained produces a virtual image O and a *real image* O' of the same object.

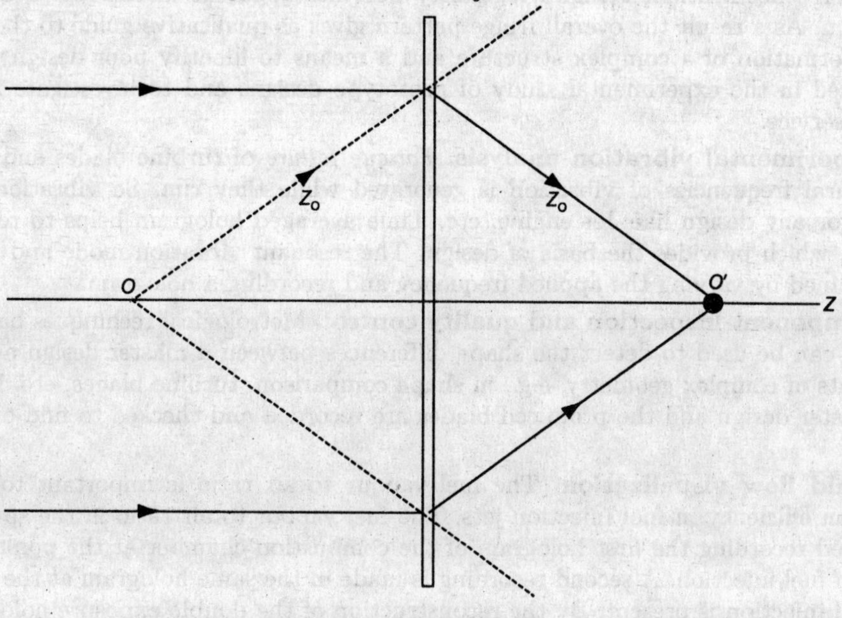

Fig. 8.4: Production of a virtual image *O* and a real image of the object.

If the reference beam be removed, then the blackening of the photo-plate would be proportional to $|E_\circ|^2$, because in that case, only $|E_\circ|$ would be recorded. In other words, in absence of the reference beam, only the amplitude information would be recorded and the phase information would be missed. In presence of reference wave, both the amplitude and phase information are possible to be recorded and the complete reconstruction of the object would be obtained.

8.4 Applications of Holography

Holography is widely used in science and technology in the following fields:

1. **Data storage.** Since each point on a hologram contains information about the geometrical characteristics of the object, a small portion of the hologram is sufficient to reconstruct the whole image of the object. It is thus a reliable method of data storage.

2. By slightly tilting the photograph a number of images can be recorded on a single hologram. Thus, a hologram possesses high information storage capacity.

3. **Measurement of distribution and size of floating fog like particles.** Normally the floating fog like particles in a sample volume are in constant motion and do not remain stationary for the observer to be focussed. The motion of the particles can be *frozen* by a short pulse laser and a hologram is made of such illuminated sample volume. An image of the entire volume is produced in the reconstruction process, and the *particle size, distribution and cross sectional geometry* can be measured by a microscope.

4. **Non-destructive testing (NDT).** The presence of fault in a body can be detected by this method without causing material damage while testing. Interference methods being non-contacting and can detect surface displacements resulting from the application of very small load is extremely suitable for this purpose. Defects appear as a discontinuity in surface displacement. The discontinuity manifests as an irregularity in the fringe pattern. Tin section weld inspection is an example of NDT.

5. **Engineering design investigation.** In the regions of strain concentration, fringes appear with higher concentration. Fringe irregularity becomes apparent in the region of uneven load distribution. As a result the overall fringe pattern gives as qualitative guide to the homogeneity of the deformation of a complex structure and a means to identify poor design characteristic. This is used in the experimental study of prototype designs and to investigate the sources of failure in service.

6. **Experimental vibration analysis.** *Fatigue failure* of turbine blades and rotors due to their natural frequencies of vibration is generated while they run. So vibrational analysis is essential for any design like Jet engine, etc. Time averaged hologram helps to record resonant frequency, which provides the basis of design. The resonant vibration mode and frequency can be determined by varying the applied frequency and recording a hologram.

7. **Component inspection and quality control.** Metrological techniques based on optical technique can be used to detect the shape differences between a master design and production components of complex geometry, e.g., in shape comparison, turbine blades, etc. Fringe pattern of the master design and the produced blades are recorded and checked to find out their shape difference.

8. **Fluid flow visualization.** The fuel vapour to air ratio is important to calculate the combustion efficiency in fuel injection jets. The fuel vapour to air ratio in the spray of the fuel is measured recording the first hologram of the combustion chamber at the point of interest in absence of fuel injection. A second recording is made in the same hologram at the same point at which fuel injection is present. By the reconstruction of the double exposure hologram, a fringe pattern showing the distribution of refractive index and hence the density distribution in the spray is observed. Thus, the desired fuel vapour to air ratio in the fuel spray is determined.

9. **Field of medicine.** Holography finds important applications in the following field of medicine:

(i) Holography of retina is used in the treatment of eye. (ii) It is used to study chest wall motion during systolic and diastolic.

8.5 Distinction Between Holography and Photography

Holography	*Photography*
1. Holography is produced due to the interference of reference beam and the transmitted beam.	1. In photography, the photo-plate is exposed to ordinary light of proper intensity (amplitude) and the photograph of the object is obtained.
2. It produces 3D image of an object.	2. It produces 2D image of an object.
3. It gives depth information of an object and provides a reliable method of data storage.	3. It fails to provide depth information and only a particular plane is pictured.

Holography	*Photography*
4. Hologram contains amplitude and phase variation at different points on it.	4. A photograph contains only the amplitude variation in a particular plane.
5. By slightly tilting the photograph a number of images can be recorded on a single hologram.	5. Superposition of so many images on a photograph is meaningless. Only one image can be recorded on a photography

8.6 Answers of Some Important Questions

8.1. *What is a holography?*
Ans. Holography is the technique of producing three-dimensional image of an object using *photographic film* and *coherent light.* It contains the whole information regarding *phase* and *amplitude.* Holography was derived from the Greek word *holos* meaning *whole* and the technique was discovered by *Danis Gabor* in 1948.
8.2. *What is the technique of holography?*
Ans. The technique of interference of two coherent beams forms the basis of holography. Interference between the reference beam and the diffracted beam (called the signal beam) from the object interfere at a point on the photo-plate. This interference pattern is the *hologram.* The hologram thus produced contains a large number of points, which are closely spaced and can be viewed through a powerful microscope.
8.3. *How the 3D image is reproduced from a hologram?*
Ans. To reproduce the 3D image of an object the hologram is illuminated by a coherent light (laser beam) usually of the same wavelength as the original beam [Fig. 8.4]. Here the hologram behaves like a *diffraction grating* producing two sets of diffracted waves, which are equally inclined with the photo-plate. One set of waves forms a *real image* and the other set forms a *virtual image* on the photographic plate (or screen). The image can be viewed on looking through the hologram acting like a window. Thus, a 3D view is recorded on a two-dimensional photographic film.
8.4. *Is holography possible with other waves except light waves?*
Ans. Yes. In recent years holography with sound waves instead of light waves, has been discovered. *Acoustical hologram* is produced using ultrasonic waves, which is increasingly used in medical science for diagnosis purpose.
8.5. *Mention some important areas of application of holography.*
Ans. Holography is used in the following important areas:
(i) Data storage, (ii) Non-destructive testing (NDT), (iii) Experimental engineering design investigation, (iv) Experimental vibration analysis, (v) Component inspection and quality control, (vi) Fluid flow visualization, (vii) Medicine field.
8.6. *Bring out the fundamental differences between holography and photography.*
Ans. See Section 8.5 (p. 220).

EXERCISES

8(A)–Objective Type Questions

Select the correct answer:

8.1. Holography is based on: (a) the interference of reference and object waves, which are coherent (b) superposition of object and reference waves, which are of slightly different wavelengths (c) recording of superposed images of two different wavelengths (d) recording of phase information of the resultant of the reference and object waves. **[WBUT 2010]**

8.2. In holography we obtain image, which is (a) 2D (b) 3D (c) 1D (d) none of these.

8.3. In holography, which one is recorded? (a) phase information (b) amplitude information (c) both amplitude and phase information (d) nome of these.

8.4. The scattered waves from the object is (a) spherical and concentric around the point of origin (b) plane (c) elliptical (d) hyperbolic.

8.5. The reference wave (a) is a plane wave (b) is a spherical wave (c) may be both (a) and (b) (d) none of these.

8.6. The hologram placed in *readout waves* produces (a) a virtual image (b) a real image (c) a virtual and a real images (d) general illumination.

8.7. Each part of a hologram, no matter how small, can produce (a) a fraction of the entire image (b) the entire image (c) no image (d) none of these.

8.8. A hologram can record (a) only one image (b) many images but slightly tilting the plate (c) no image (d) none of these.

ANSWERS

8.1. (a) **8.2.** (b) **8.3.** (c) **8.4.** (a) **8.5.** (a) **8.6.** (c) **8.7.** (b) **8.8.** (b).

8(B)–Short Answer Type Questions

8.1. Explain the principle of holography.

8.2. Explain how holographic images are reconstructed from the holograms. **[WBUT 2010]**

8.3. What are reference waves, signal waves, diffracted signal waves, readout waves with reference to holography technique (show pictorially).

8.4. What are the basic characteristics of holograms?

8.5. Point out some important applications of holography.

8.6. In what respect holography is superior to photographic technique?

8.7. What is a hologram. How is it produced?

8(C)–Broad Answer Type Questions

8.1. (a) Describe the principle of holography. (b) Discuss with relevant diagrams, the process of recording a hologram and reconstruction of the image. (c) Point out the basic characteristics of a hologram.

8.2. (a) What is holography? (b) Give the theory of holography. (c) Can you produce hologram using ultrasonic waves? Discuss the applications of such holograms.

8.3. (a) How does holography differ from photography? (b) Discuss the various applications of holography. (c) Discuss with diagram how real and virtual images of an object can be obtained by placing a hologram in front of the readout waves.

Chapter 9 | Module 4.1

Special Theory of Relativity

9.1 Introduction

Albert Einstein in 1905, proposed his *special theory of relativity* by introducing new ideas of *space* and *time.* The results of mechanical experiments were correctly predicted over the entire range of speeds from $\frac{u}{c} \to 0$ to $\frac{u}{c} \to 1$. The new theory does not completely discard Newtonian mechanics instead includes the results of old theory (*classical mechanics*) as a special case in the limit $\frac{u}{c} \to 0$. In other words, all the results obtained from special theory of relativity reduces to corresponding classical results in the limit of low speed, i.e., $\frac{u}{c} \to 0$.

Einstein rejected the *absolute* nature of fundamental quantities, viz., *space*, *time* and *mass* as postulated by Newtonian mechanics by introducing new and original ideas.

The special features of Einstein's special theory of relativity are as follows: (i) *the cause that led its formulation* (i.e., *its origin*) (ii) *the postulates* and *principles* (iii) *the conclusions derived from this theory* and (iv) *the experimental evidences in its favour.*

The theory of relativity based on new concepts of *space* and *time* is divided into two parts: (a) *Special theory of relativity* and (b) *General theory of relativity.*

The special theory of relativity deals with inertial systems—the systems that move in uniform rectilinear motion w.r.t. one another. *For inertial systems all systems of coordinates are equally suitable for description of physical phenomenon.*

General theory of relativity deals with accelerated systems—*the systems that move with accelerated velocity w.r.t. each other.* It is applicable to the laws of gravitation in a more defined way than Newtonian approach.

9.2 Frame of Reference

The motion being always relative, we always have to assume the existence of something relative to which the motion is to be described. The motion of an object is meaningless unless it is described relative to a system—called the *frame of reference.* So *a system relative to which the motion of any object is described is called a frame of reference.* The motion observed by observers in different systems is different.

Example. The earth, the walls of the room, the position of stars along a plumb line, a tree, the sun, etc. serve the purpose of frame of reference. According to Newton, *absolute space* is a standard frame of reference.

Frame of reference are of two types, viz., (i) *Inertial frame of reference* and (ii) *Non-inertial frame of reference.*

(i) ***Inertial frame of reference.*** *A frame of reference w.r.t., which a body not acted upon by external force is unaccelerated is called an inertial (or; Newtonian or unaccelerated) frame of reference.* This also includes the state of rest. The laws of mechanics are the same in inertial frames and all inertial frames are equally valid.

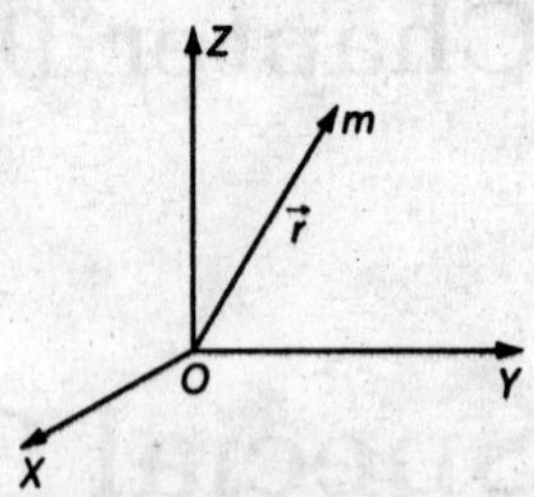

Fig. 9.1: Position of a particle in Cartesian system.

If (x, y, z) be the position of a particle at time t in Cartesian coordinate system, and if no external force is acting on the body, then

$$m\frac{d^2\vec{r}}{dt^2} = 0 \quad \text{or,} \quad m\frac{d^2x}{dt^2} = 0, \quad m\frac{d^2y}{dt^2} = 0, \quad m\frac{d^2z}{dt^2} = 0.$$

$$\therefore \frac{dx}{dt} = v_x = \text{constant}; \quad \frac{dy}{dt} = v_y = \text{constant}; \quad \frac{dz}{dt} = v_z = \text{constant}.$$

Thus, the X-, Y- and Z-components of velocity of the body are constant if there is no external force, i.e., *without the application of external force a body in motion continues its motion with uniform velocity.* This is *Newton's first law of motion.* Thus, an inertial frame is one in which the law of inertial (Newton's first law) is valid. Any reference frame moving with constant velocity relative to an inertial frame is also inertial. *Inter galactic* space is approximated to be an inertial frame of reference.

(ii) ***Non-inertial frame of reference.*** *A frame w.r.t., which a body, not acted upon an external force is accelerated, is called a non-inertial (or accelerated) frame of reference.* In other words, *the accelerated frames are non-inertial frames.*

A rotating frame of reference (viz., the earth's frame) is a non-inertial frame.

In absence of external force, i.e., $\vec{F} = 0$

$$m\frac{d^2x}{dt^2} = ma_x \text{ or, } \frac{d^2x}{dt^2} = a_x; \quad m\frac{d^2y}{dt^2} = ma_y \text{ or, } \frac{d^2y}{dt^2} = a_y; \quad m\frac{d^2z}{dt^2} = ma_z \text{ or, } \frac{d^2z}{dt^2} = a_z.$$

Since the X-, Y- and Z-components of acceleration a_x, a_y and a_z are not zero, it is evident that, a body not acted upon by an external force experiences an acceleration.

9.3 Galilean Transformation

Galilean transformation is a process in which the data of results of observation in one inertial frame is transformed to those in the other inertial frame, when both the inertial frames move with uniform velocity relative to each other, the time interval registered in both the frames being the same.

9.3.1 Galilean Coordinate Transformation

We consider inertial frames S and S', S' moving with uniform velocity v along the common X-axis, At $t = 0$, both origins O and O' of S and S' were in coincidence.

Let an event occurs at P, whose coordinates are (x, y, z, t) and (x', y', z', t') as observed by observers at S and S' respectively. Since the observer in S' moves a distance vt from O in time t, we have

$$\left.\begin{aligned} x' &= x - vt \\ y' &= y \\ z' &= z \\ t' &= t \end{aligned}\right\} \tag{9.1}$$

The eqns (9.1) are called *Galilean transformation of coordinates.* These equations ensure that *the physical laws that one invariant w.r.t. these equations are valid everywhere and at all times.* Also the *space and time coordinates are separate.*

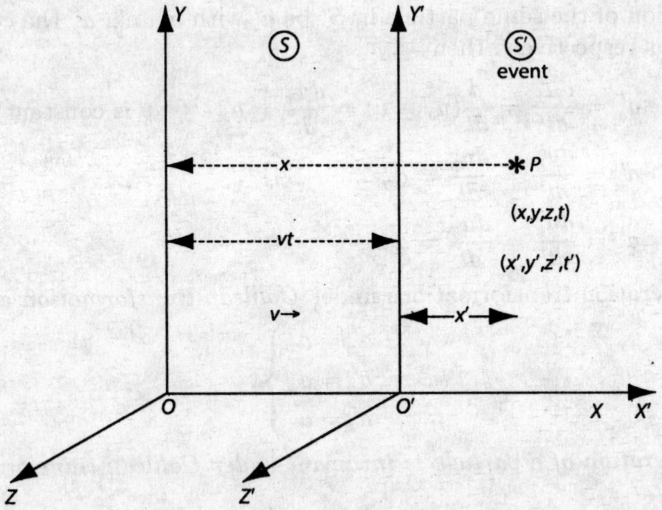

Fig. 9.2: S'-frame moving with uniform velocity v w.r.t. S along the common X-axis.

9.3.2 Galilean Velocity Transformation

If (x, y, z, t) and (x', y', z', t') be the coordinates of the same event observed by observers in S and S' respectively, where S' is moving with constant velocity v relative to S along the common positive X-axis, then

$$x' = x - vt$$
$$y' = y$$
$$z' = z$$
$$t' = t.$$

If (u_x, u_y, u_z) and (u'_x, u'_y, u'_z) are the components of velocity $\vec{u}$ as measured in S and S' respectively, then

$$u'_x = \frac{dx'}{dt'} = \frac{d}{dt}(x - vt)\frac{dt}{dt'} = \left(\frac{dx}{dt} - v\right) \quad (\because t = t')$$

or, $u'_x = u_x - v.$

Also

$$u'_y = \frac{dy'}{dt'} = \frac{dy}{dt}\frac{dt}{dt'} \quad \text{or,} \quad u'_y = u_y.$$

Similarly, $u'_z = u_z$.

Thus, *the velocity transformation equations* are

$$\left.\begin{aligned} u'_x &= u_x - v \\ u'_y &= u_y \\ u'_z &= u_z \end{aligned}\right\}. \tag{9.2}$$

9.3.3 Galilean Acceleration Transformations

If $\vec{a}$ be the acceleration of the particle relative to frame S with a_x, a_y, a_z the components of $\vec{a}$ along X-, Y- and Z-directions respectively, then

$$a_x = \frac{du_x}{dt}, \quad a_y = \frac{du_y}{dt} \quad \text{and} \quad a_z = \frac{du_z}{dt}.$$

If the acceleration of the same particle in S' be $\vec{a}'$ with a'_x, a'_y, a'_z the components along X-, Y- and Z-directions respectively, then

$$a'_x = \frac{du'_x}{dt'} = \frac{d}{dt}(u_x - v) = \frac{du_x}{dt} = a_x \quad (\because v \text{ is constant})$$

$$a'_y = \frac{du'_y}{dt'} = \frac{du_y}{dt} = a_y$$

$$a'_z = \frac{du'_z}{dt'} = \frac{du_z}{dt} = a_z.$$

Thus, the acceleration transformations under *Galilean transformation* are

$$\left.\begin{aligned} a'_x &= a_x \\ a'_y &= a_y \\ a'_z &= a_z \end{aligned}\right\}. \tag{9.3}$$

Thus, *the acceleration of a particle is invariant under Galilean transformation.*

9.3.4 Galilean Transformation of Force

If m and m' be the mass of a particle in S and S'-frame respectively, then the corresponding forces are $\vec{F} = m\vec{a}$ and $\vec{F}' = m'\vec{a}'$.

According to Newton's law, $m = m'$ and $\vec{a} = \vec{a}'$.

$$\therefore \vec{F} = \vec{F}'. \tag{9.4}$$

Thus, *the force is invariant under Galilean transformation.* Thus, Newton's laws of motion are invariant under Galilean transformation. This is also known as the *Galilean invariance.*

Galilean Principle of Invariance

If $\vec{F}$ and $\vec{F}'$ be the forces acting on a body measured by observers in S and S'-frame respectively, then according to Galilean transformation of force, $\vec{F} = \vec{F}'$.

Thus, the Newton's laws of motion are invariant (or, covariant) under Galilean transformation in all inertial frame of reference. This is known as the *principle of Galilean invariance.*

Thus, if a series of mechanical experiments whether conducted in a car at rest w.r.t. the ground or moving smoothly over a straight track at constant velocity, gives the same results. It follows that *the laws of mechanics have the same form in all inertial frames.* This is known as the *Galileo's principle of relativity.* Also Newtonian Mechanics predicts that *all inertial frames in uniform translation relative to one another are equivalent in so far as the mechanical phenomena are concerned.* This equivalence of all inertial frames in respect to the laws of classical mechanics is referred to as *Newtonian relativity.*

It can be shown that the following are invariant under Galilean transformation: (1) *Newton's second law of motion* (2) *Law of conservation of momentum* and (3) *Law of conservation of energy.*

Law of Conservation of Momentum

We consider the collision of two particles of masses m_1 and m_2 in frame S (rest frame). Let u_1, u_2 and v_1, v_2 be the velocities of particles of masses m_1 and m_2 respectively before and after collision. Then according to the law of conservation of momentum in S-frame

$$m_1u_1 + m_2u_2 = m_1v_1 + m_2v_2. \tag{9.5}$$

If m_1 and m_2 remain constant in S'-frame (the moving frame with constant velocity v w.r.t. S along the common positive X-axis), then

$$m_1u_1' + m_2u_2' = m_1v_1' + m_2v_2', \tag{9.6}$$

where u_1', u_2' and v_1', v_2' are the velocities of m_1 and m_2 before and after collision in S'-frame.

Now, according to *Galilean transformation of velocities*, we have

$$\left.\begin{aligned} u_1 = u_1' + v, \quad u_2 = u_2' + v \\ v_1 = v_1' + v, \quad v_2 = v_2' + v \end{aligned}\right\}. \tag{9.7}$$

From eqns (9.5) and (9.7), we get

$$m_1(u_1' + v) + m_2(u_2' + v) = m_1(v_1' + v) + m_2(v_2' + v)$$

$$\text{or,} \quad m_1u_1' + m_2u_2' = m_1v_1' + m_2v_2'.$$

Hence, *the law of conservation of momentum is invariant under Galilean transformation.*

Law of Conservation of Energy

We consider the collision of two particles of masses m_1 and m_2 in rest frame S. Let u_1, u_2 and v_1, v_2 be the velocities of m_1 and m_2 before and after collision. Then according to the law of conservation of energy in S,

$$\tfrac{1}{2}m_1u_1^2 + \tfrac{1}{2}m_2u_2^2 = \tfrac{1}{2}m_1v_1^2 + \tfrac{1}{2}m_2v_2^2 + W, \tag{9.8}$$

where W is that part of initial energy of the particles, which appears in some other form like heat, etc. after collision.

If u_1', u_2' and v_1', v_2' be the velocities of particle of masses m_1 and m_2 (m_1 and m_2 remain constant) before and after collision in moving frame S', then

$$\tfrac{1}{2}m_1u_1'^2 + \tfrac{1}{2}m_2u_2'^2 = \tfrac{1}{2}m_1v_1'^2 + \tfrac{1}{2}m_2v_2'^2 + W, \tag{9.9}$$

where W is assumed to remain constant.

Now, according to Galilean transformation of velocities,

$$\left.\begin{aligned} u_1 = u_1' + v, u_2 = u_2' + v \\ v_1 = v_1' + v, v_2 = v_2' + v \end{aligned}\right\}, \tag{9.10}$$

where v is the velocity of S' w.r.t. S.

Using eqns (9.10) in eqn (9.8), we get

$$\tfrac{1}{2}m_1(u_1' + v)^2 + \tfrac{1}{2}m_2(u_2' + v)^2 = \tfrac{1}{2}m_1(v_1' + v)^2 + \tfrac{1}{2}m_2(v_2' + v)^2 + W$$

$$\text{or,} \ \tfrac{1}{2}m_1u_1'^2 + \tfrac{1}{2}m_2u_2'^2 + v(m_1u_1' + m_2u_2') = \tfrac{1}{2}m_1v_1'^2 + \tfrac{1}{2}m_2v_2'^2 + v(m_1v_1' + m_2v_2') + W \tag{9.11}$$

$$\text{or,} \ \tfrac{1}{2}m_1u_1'^2 + \tfrac{1}{2}m_2u_2'^2 = \tfrac{1}{2}m_1v_1'^2 + \tfrac{1}{2}m_2v_2'^2 + W. \tag{9.12}$$

$$[\because m_1u_1' + m_2u_2' = m_1v_1' + m_2v_2'].$$

Eqn (9.12) shows that *the law of conservation of energy is invariant under Galilean transformation.*

9.4 Michelson-Morley Experiment

A. A. Michelson, first in 1881 and then in collaboration with W. W. Morley in 1887, attempted to determine the velocity of earth relative to hypothetical ether medium supposed to be at rest. Michelson-Morley experiment actually laid the experimental foundation of special theory of relativity proposed by Einstein in 1905.

Michelson interferometer was employed for this purpose, which consists of two optically plane mirrors M_1 and M_2 fixed on two mutually perpendicular arms and a half-silvered glass plate G as shown in Fig. 9.3. The working principle is as follows:

S is a monochromatic source of light placed at the focus of the lens L. The refracted rays from L are rendered parallel and then split-up into two beams by G. The vertical beam travels towards M_2 and then reflected back. The horizontal beam travels towards M_1 and then reflected to G. Since the two beams are derived from the same beam they are coherent and reunite at G and produce *interference pattern*, which can be viewed through the telescope T.

The apparatus is arranged to move with the earth's orbital velocity v along the direction of the earth's orbit round the sun when the drift of ether will change the speed of light. The whole apparatus is mounted on a large stone slab, which floats in a pool of mercury.

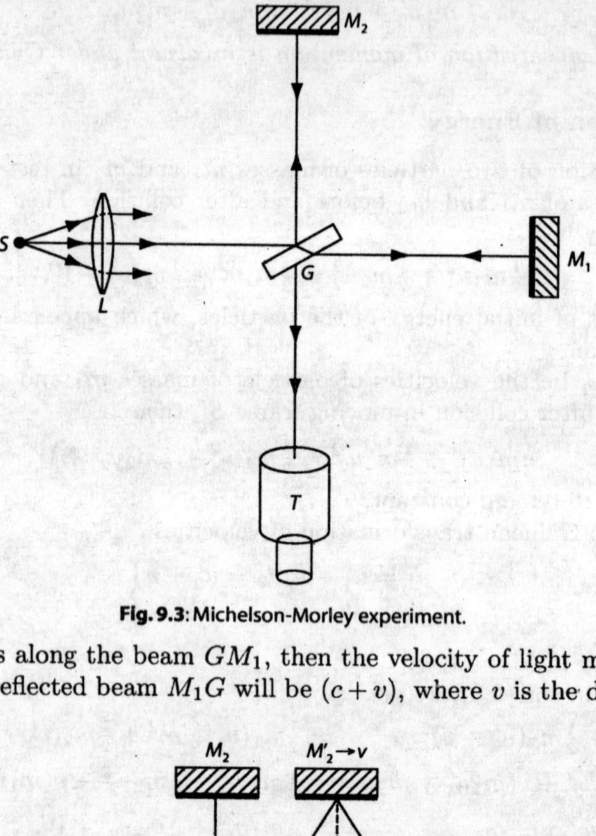

Fig. 9.3: Michelson-Morley experiment.

If the drift of ether is along the beam GM_1, then the velocity of light moving to M_1 will be $(c - v)$. The velocity of reflected beam M_1G will be $(c + v)$, where v is the drift velocity of ether relative to earth.

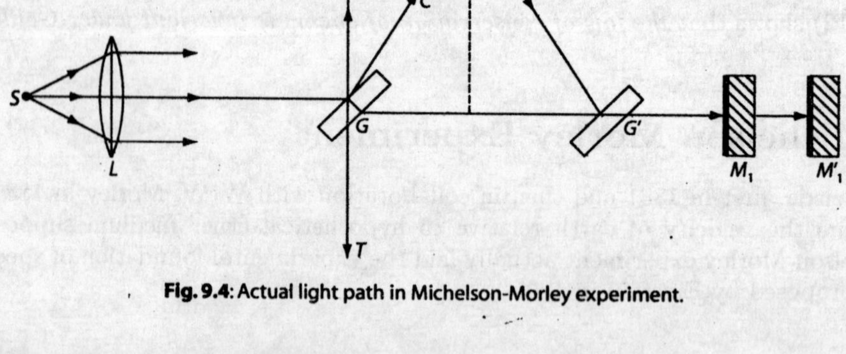

Fig. 9.4: Actual light path in Michelson-Morley experiment.

Time taken by the beam for the round trip GM_1G is

$$T_1 = \frac{d}{c-v} + \frac{d}{c+v}, \quad [\text{where } GM_1 = GM_2 = d]$$

$$= \frac{2dc}{c^2 - v^2} = \frac{\frac{2d}{c}}{1 - \frac{v^2}{c^2}}. \tag{9.13}$$

The vertical beam travels towards M_2 and gets reflected from M_2 only when G has moved to G' as shown in Fig. 9.4. The vertical component of velocity of light is $\sqrt{c^2 - v^2}$.

So the time taken for the round trip GM_2G is

$$T_2 = \frac{2d}{\sqrt{c^2 - v^2}} = \frac{\frac{2d}{c}}{\sqrt{1 - \frac{v^2}{c^2}}}. \tag{9.14}$$

Now, due to the motion of the interferometer relative to ether, the difference of time between the two parts will be

$$\Delta T = T_1 - T_2 = \frac{2d}{c}\left[\frac{1}{1 - \frac{v^2}{c^2}} - \frac{1}{\sqrt{1 - \frac{v^2}{c^2}}}\right]$$

$$= \frac{2d}{c}\left[\left(1 + \frac{v^2}{c^2}\right) - \left(1 + \frac{v^2}{2c^2}\right)\right] \quad \left[\because \frac{1}{1 - \frac{v^2}{c^2}} \simeq 1 + \frac{v^2}{c^2} \text{ and } \frac{1}{\sqrt{1 - \frac{v^2}{c^2}}} \simeq 1 + \frac{v^2}{2c^2}\right]$$

$$\text{or,} \quad \Delta T = \frac{2d}{c}\left[\frac{v^2}{2c^2}\right] = \frac{v^2 d}{c^3}. \tag{9.15}$$

Now, if the instrument is turned through 90°, the role of the horizontal and vertical paths gets interchanged. Then

$$T_1' = \frac{\frac{2d}{c}}{\sqrt{1 - \frac{v^2}{c^2}}} \quad \text{and} \quad T_2' = \frac{\frac{2d}{c}}{1 - \frac{v^2}{c^2}}.$$

$$\therefore \Delta T' = T_1' - T_2' = \frac{2d}{c}\left[\frac{1}{\sqrt{1 - \frac{v^2}{c^2}}} - \frac{1}{1 - \frac{v^2}{c^2}}\right] = -\frac{v^2 d}{c^3}. \tag{9.16}$$

$$\therefore (\Delta T)_{\text{total}} = \Delta T - \Delta T' = \frac{v^2 d}{c^3} - \left(-\frac{v^2 d}{c^3}\right) \quad \text{or,} \quad (\Delta T)_{\text{total}} = \frac{2v^2 d}{c^3}. \tag{9.17}$$

Then the path difference between the two components of the incident beam is

$$\Delta = c(\Delta T)_{\text{total}} = \frac{2v^2 d}{c^2}. \tag{9.18}$$

If now N be the number of *fringe* passing through the vertical cross-wire of the telescope, then

$$N = \frac{\text{Path difference}}{\lambda} = \frac{2v^2 d}{\lambda c^2}. \tag{9.19}$$

Taking wavelength of the monochromatic light $\lambda = 5500$ Å and $v = 30$ km/s and $d = 11$ m, we get

$$N = \frac{2 \times 11 \times (3 \times 10^4)^2}{5500 \times 10^{-10} \times (3 \times 10^8)^2} = 0.4.$$

A *fringe shift* of 0.4 could easily be measured since Michelson instrument was sensitive enough to measure a shift of $\frac{1}{100}$th part of a fringe. But actually no fringe shift was observed. The experiment was repeated at different places at different times and no shift of fringe was detected. Then it was concluded that (i) *the relative velocity between ether and the earth is zero*, (ii) *ether does not exist.*

Explanation of Negative Results

Following interpretations were put forward to account the *negative results* of Michelson-Morley experiment.

(i) *Ether drag hypothesis.* According to Michelson *ether* was dragged along with the moving earth and hence there was no relative velocity between the earth and the ether. This explanation, however, goes against the observed phenomenon of *aberration of star light.*

(ii) *Lorentz-Fitzgerald contraction hypothesis.* Fitzgerald and Lorentz suggested independently that when a material body is in motion, its length undergoes a contraction by a factor $\sqrt{1-\frac{v^2}{c^2}}$ along the direction of motion. This would equalize time T_1 and T_2 in Michelson-Morley experiment leading to no *fringe shift.* This is, however, a purely mathematical concept having no experimental basis.

(iii) *Constancy of speed of light.* Einstein proposed that the velocity of light is an invariant, i.e., it is constant and does not depend upon the motion of the source, observer or medium. Under that condition, $T_1 = T_2 = \frac{2d}{c}$ and hence no fringe shift to be observed.

Thus, the negative results of Michelson-Morley experiment leads to the fact that *velocity of light is invariable and remains constant in all directions.* Effect of ether also remained undetected. Doppler's effect verified that the velocity of light is independent of the motion of source, observer and medium.

The results supports the non-existence of ether medium and hence attempt to take ether as a *fixed frame of reference* goes in vain.

9.5 Postulates of Special Theory of Relativity

Einstein's special theory of relativity is based on the following two postulates:

Postulate I *All the laws of physics are the same (i.e., covariant) in all inertial frames of reference having an uniform translational velocity w.r.t. one another.*

Postulate II *The velocity of light is the same in all inertial frames and is independent of the motion of the source and observer.*

9.6 Insufficiency of Galilean Transformation

We consider two inertial frames S and S', S' moving with velocity v relative to S along the common X-axis. Let their origins O and O' coincide at $t = t' = 0$ and a light wave originates at O and O', when $t = 0$.

The wavefront of light in S-frame would be a sphere, whose equation is

$$x^2 + y^2 + z^2 = c^2t^2, \tag{9.20}$$

where c is the velocity of light.

In view of *Galilean transformation*, eqn (9.20) can be written as

$$(x' + vt)^2 + y'^2 + z'^2 = c^2t'^2 \tag{9.21}$$

in S'-frame. Eqn (9.21) does not represent a sphere indicating that the velocity of light is not invariant, i.e., velocity of light depends on direction, which is not in accord with the postulate of special theory of relativity. The principle of special theory of relativity demands that the wavefront in S' should also be a sphere given by

$$x'^2 + y'^2 + z'^2 = c^2 t'^2. \tag{9.22}$$

Thus, *Galilean transformation is insufficient and a new set of transformation formulae is required.*

9.7 Lorentz Transformation of Space and Time

We consider two inertial frames S and S', S' moving with velocity v relative to S along the common X-axis as shown in Fig. 9.5. Let the origins O and O' of S and S' coincide at $t = 0$.

Let a pulse of light be generated at $t = 0$, which grows in space. At time $t = t$, let the pulse reach the point P, whose coordinates are (x, y, z, t) and (x', y', z', t') in S and S' respectively.

The equation of the spherical wavefront in S and S' at t is given by

$$x^2 + y^2 + z^2 - c^2 t^2 = 0 \tag{9.23}$$

$$\text{and} \quad x'^2 + y'^2 + z'^2 - c^2 t'^2 = 0. \tag{9.24}$$

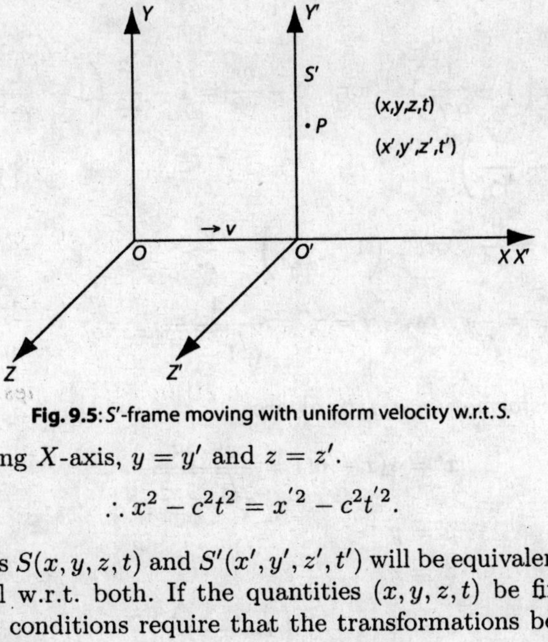

Fig. 9.5: S'-frame moving with uniform velocity w.r.t. S.

Since the motion is along X-axis, $y = y'$ and $z = z'$.

$$\therefore x^2 - c^2 t^2 = x'^2 - c^2 t'^2. \tag{9.25}$$

Now, the inertial frames $S(x, y, z, t)$ and $S'(x', y', z', t')$ will be equivalent if the transformation equations are symmetrical w.r.t. both. If the quantities (x, y, z, t) be finite, then (x', y', z', t') must also be finite. These conditions require that the transformations between $S(x, y, z, t)$ and $S'(x', y', z', t')$ must be *linear.*

Hence, we take the transformations as

$$\left.\begin{aligned} x' &= \gamma(x - vt) \\ x &= \gamma'(x' + vt') \end{aligned}\right\}, \tag{9.26}$$

where γ, γ' are two *undetermined constants.*

Now, from the second equation of eqns (9.26), we have

$$t' = \frac{1}{v}\left(\frac{x}{\gamma'} - x'\right) = \frac{1}{v}\left\{\frac{x}{\gamma'} - \gamma(x - vt)\right\} \quad \text{[from eqn (9.26)]}$$

$$= \gamma\left\{t - \frac{x}{v}\left(1 - \frac{1}{\gamma\gamma'}\right)\right\}. \tag{9.27}$$

Substituting the values of x' and t' from eqns (9.26) and (9.27) in eqn (9.25), we get

$$x^2 - c^2t^2 = \gamma^2(x - vt)^2 - c^2\gamma^2\left\{t - \frac{x}{v}\left(1 - \frac{1}{\gamma\gamma'}\right)\right\}^2. \tag{9.28}$$

Now, equating the coefficients of t^2 and x^2 from both sides of eqn (9.28), we get

$$-c^2 = \gamma^2v^2 - c^2\gamma^2 \tag{9.29}$$

$$\text{and} \quad 1 = \gamma^2 - c^2\gamma^2\frac{1}{v^2}\left(1 - \frac{1}{\gamma\gamma'}\right)^2. \tag{9.30}$$

From eqn (9.29),

$$\gamma^2\left(c^2 - v^2\right) = c^2.$$

$$\therefore \gamma = \frac{c}{\sqrt{c^2 - v^2}} \quad \text{or,} \quad \gamma = \frac{1}{\sqrt{1 - \frac{v^2}{c^2}}}. \tag{9.31}$$

From eqn (9.30),

$$\frac{1}{\gamma^2} = 1 - \frac{c^2}{v^2}\left(1 - \frac{1}{\gamma\gamma'}\right)^2 \quad \text{or,} \quad 1 - \frac{v^2}{c^2} = 1 - \frac{c^2}{v^2}\left(1 - \frac{1}{\gamma\gamma'}\right)^2 \quad \text{[from eqn (9.31)]}$$

$$\text{or,} \quad \frac{v^2}{c^2} = \frac{c^2}{v^2}\left(1 - \frac{1}{\gamma\gamma'}\right)^2$$

$$\therefore \left(1 - \frac{1}{\gamma\gamma'}\right)^2 = \frac{v^4}{c^4} \quad \text{or,} \quad \left(1 - \frac{1}{\gamma\gamma'}\right) = \frac{v^2}{c^2} \tag{9.32}$$

$$\text{or,} \quad \frac{1}{\gamma\gamma'} = 1 - \frac{v^2}{c^2} = \frac{1}{\gamma^2} \quad \text{or,} \quad \gamma = \gamma' = \frac{1}{\sqrt{1 - \frac{v^2}{c^2}}}.$$

$\therefore$ the required *transformation equations* are

$$\left.\begin{aligned} x' &= \gamma(x - vt) = \frac{x - vt}{\sqrt{1 - \frac{v^2}{c^2}}} \\ y' &= y \\ z' &= z \\ t' &= \gamma\left\{t - \frac{x}{v}\left(1 - \frac{1}{\gamma\gamma'}\right)\right\} \\ &= \gamma\left\{t - \frac{x}{v}\cdot\frac{v^2}{c^2}\right\} \quad \text{[from eqn (9.32)]} \\ \text{or,} \quad t' &= \frac{t - \frac{vx}{c}}{\sqrt{1 - \frac{v^2}{c^2}}} \end{aligned}\right\}. \tag{9.33}$$

These are *Lorentz transformation equations from S- to S'-system.*

The *reverse transformation equations from S′ to S* are obtained replacing v by $-v$ as

$$\left.\begin{aligned} x &= \frac{x' + vt}{\sqrt{1 - \frac{v^2}{c^2}}} \\ y' &= y \\ z' &= z \\ t' &= \frac{t + \frac{vx}{c^2}}{\sqrt{1 - \frac{v^2}{c^2}}} \end{aligned}\right\} . \tag{9.34}$$

Special Case. In the low velocity region, i.e., $v \ll c$, $\frac{v}{c} \ll 1$. Lorentz transformation eqns (9.33) reduce to

$$x' = x - vt$$
$$y' = y$$
$$z' = z$$
$$\text{and} \quad t' = t \quad \left(\because \frac{vx}{c^2} \ll 1\right),$$

which are *Galilean transformations.* Hence, *for low velocities, the Lorentz transformations reduce to Galilean transformations.* Thus, Lorentz transformations hold for all velocities but Galilean transformations are valid only for low velocities. *Therefore, Galilean transformations are special cases of Lorentz transformations.*

9.8 Consequences of Lorentz Transformation Equations

9.8.1 Length Contraction (or Lorentz-Fitzgerald Contraction)

We consider two inertial frames S and S', S' moving with velocity v relative to S along the common X-axis [Fig. 9.6]. The origins of S and S' coincide at $t = 0$.

A rod of length l_0 is at rest in S' such that $x'_2 - x'_1 = l_0$, where x'_1 and x'_2 are the x-coordinates of the left and right ends of the rod. l_0 is the length of the rod measured by a stationary observer and is called the *proper length.*

If x_1 and x_2 be the x-coordinates of the same rod in frame S, then $x_2 - x_1 = l$.

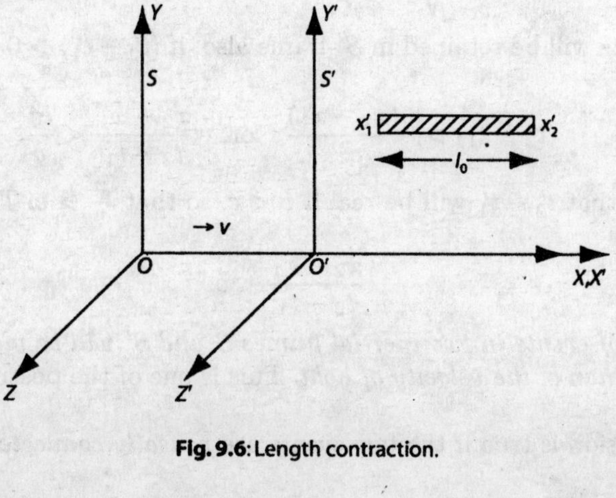

Fig. 9.6: Length contraction.

Now, according to *Lorentz transformation,*

$$x_1' = \frac{x_1 - vt}{\sqrt{1-\frac{v^2}{c^2}}} \quad \text{and} \quad x_2' = \frac{x_2 - vt}{\sqrt{1-\frac{v^2}{c^2}}}.$$

$$\therefore x_2 - x_1 = (x_2' - x_1')\sqrt{1-\frac{v^2}{c^2}} \quad \text{or,} \quad l = l_0\sqrt{1-\frac{v^2}{c^2}}. \tag{9.35}$$

[As the end points of the rod are measured at the same instant, $t = t'$.]

Since $v < c$, $\sqrt{1-\frac{v^2}{c^2}} < 1$. $\therefore l < l_0$.

Thus, *the length* (l_0) *of the rod at rest in the moving frame* S' *appears to be contracted by a factor of* $\sqrt{1-\frac{v^2}{c^2}}$ *to the observer in the* S*-frame. In other words, the length of the rod will appear contracted to the moving observer by a factor* $\sqrt{1-\frac{v^2}{c^2}}$. This is the basis of famous *Lorentz-Fitzgerald contraction hypothesis* to explain the negative result of Michelson-Morley experiment.

Conclusions.

(i) When $v < c$, $\frac{v^2}{c^2} \ll 1$, then $l \simeq l_0$, i.e., the length of the moving rod is the same when it was at rest.

(ii) When $v \simeq c$, $\sqrt{1-\frac{v^2}{c^2}} < 1$. $\therefore l < l_0$, the length of the moving rod appears to be less than the length when it was at rest.

(iii) When $v = c$ or, $v > c$, $\frac{v^2}{c^2} = 1$ or, > 1.

$\therefore l = 0$ or imaginary.

The imaginary value of length is impossible. So *material body can never attain the velocity of light.*

(iv) The contraction of length takes place only along the direction of motion. The contraction along perpendicular direction of the motion is zero.

Order of events. It two events occur at x_1 and x_2 in time t_1 and t_2 respectively in S-frame, and if the second event follows the first, then $t_2 > t_1$. Is this order maintained in S'-system also?

If t_1' and t_2' be the time in which the same two events occur in S'-frame, then

$$t_1' = \frac{t_1 - \frac{vx_1}{c^2}}{\sqrt{1-\frac{v^2}{c^2}}}, \quad t_2' = \frac{t_2 - \frac{vx_2}{c^2}}{\sqrt{1-\frac{v^2}{c^2}}}.$$

$$\therefore t_2' - t_1' = \frac{1}{\sqrt{1-\frac{v^2}{c^2}}}\left[(t_2 - t_1) - \frac{v(x_2 - x_1)}{c^2}\right]. \tag{9.36}$$

The order of events will be retained in S'-frame also, if $(t_2' - t_1') > 0$. The necessary condition for this to happen is

$$(t_2 - t_1) > \frac{v(x_2 - x_1)}{c^2} \quad \text{or,} \quad \frac{x_2 - x_1}{t_2 - t_1} < \frac{c^2}{v}.$$

Eqn (9.36) shows that $t_2' - t_1'$ will be real if $v < c$, so that $\frac{c^2}{v} > c$. Thus, the above condition will hold good, if

$$\frac{x_2 - x_1}{t_2 - t_1} < c.$$

Hence, *the order of events in two inertial frames* S *and* S' *will be maintained if no signal can have a speed greater than c, the velocity of light.* This is one of the postulates of special theory of relativity.

The above conclusion is true if the two events are *causally connected.*

9.8.2 Two Events are said to be Simultaneous if They Occur at the Same Time in a Frame of Reference

We consider two inertial frames of reference S and S', the latter moving with velocity v along the common X-axis. Let two events occur simultaneously in frame S at points $P(x_1, y_1, z_1, t_1)$ and $Q(x_2, y_2, z_2, t_2)$ as observed by the observer O in S-frame. As the two events occur simultaneously in frame S, we have $t_1 = t_2$.

If the same two events occur at t_1' and t_2' as observed by the observer O' in frame S', then using *Lorentz transformation equation,* we get

$$t_1' = \frac{t_1 - \frac{vx_1}{c^2}}{\sqrt{1-\frac{v^2}{c^2}}} \quad \text{and} \quad t_2' = \frac{t_2 - \frac{vx_2}{c^2}}{\sqrt{1-\frac{v^2}{c^2}}}.$$

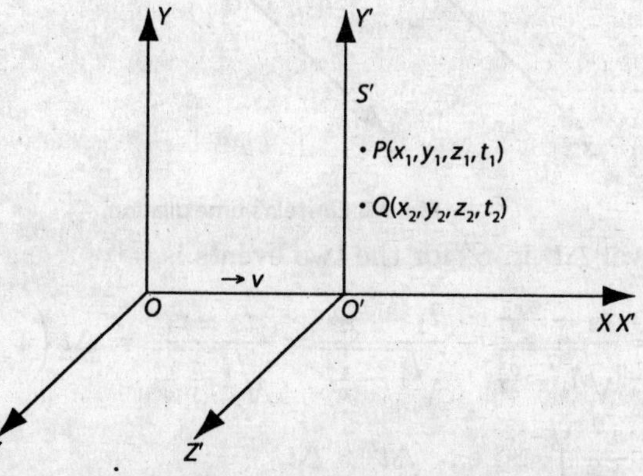

Fig. 9.7: Simultaneity of two events.

Now, the time interval of the events in S'-frame is

$$t_2' - t_1' = \frac{t_2 - t_1}{\sqrt{1-\frac{v^2}{c^2}}} - \frac{\left(\frac{v}{c^2}\right)(x_2 - x_1)}{\sqrt{1-\frac{v^2}{c^2}}}.$$

Since $t_2 = t_1$, we have

$$t_2' - t_1' = -\frac{\frac{v}{c^2}(x_2 - x_1)}{\sqrt{1-\frac{v^2}{c^2}}}. \tag{9.37}$$

$$\because x_2 \neq x_1, \quad t_2' - t_1' \neq 0, \quad \text{i.e.,} \quad t_2' \neq t_1'.$$

Thus, the *two events at different planes P and Q, which are simultaneous for an observer O at rest in frame S are no longer simultaneous to an observer O' in frame S' moving with velocity v relative to S.*

Hence, we conclude that, *simultaneity is not absolute but relative.*

9.8.3 Einstein's Time Dilation (or Apparent Retardation of Clocks)

Let a clock be placed in frame S, which is at rest at x_1. Let this clock give a signal at t_1 in S, which is recorded by the observer O' in S' at t_1'. If the frame S' moves with velocity v relative to S along the common X-axis [Fig. 9.8], then from *Lorentz transformation equation,*

$$t_1' = \frac{t_1 - \frac{vx_1}{c^2}}{\sqrt{1-\frac{v^2}{c^2}}}. \tag{9.38}$$

If the clock gives a second signal at t_2 in S, then let the corresponding time in S' be t'_2, where

$$t'_2 = \frac{t_2 - \frac{vx_2}{c^2}}{\sqrt{1 - \frac{v^2}{c^2}}}. \tag{9.39}$$

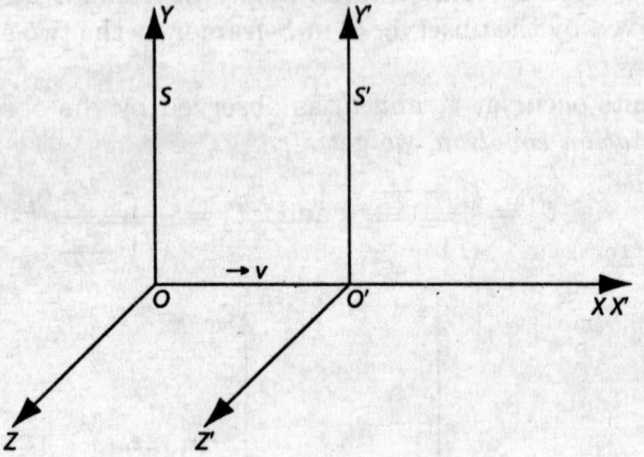

Fig. 9.8: Einstein's time dilation.

Thus, the time interval $\Delta t'$ in S' for the two events is

$$\Delta t' = t'_2 - t'_1 = \frac{t_2 - \frac{vx_2}{c^2}}{\sqrt{1 - \frac{v^2}{c^2}}} - \frac{t_1 - \frac{vx_1}{c^2}}{\sqrt{1 - \frac{v^2}{c^2}}} = \frac{t_2 - t_1}{\sqrt{1 - \frac{v^2}{c^2}}} = \Delta t \left(1 - \frac{v^2}{c^2}\right)^{-\frac{1}{2}} \qquad [\because x_1 = x_2]$$

$$\text{or,} \quad \Delta t' \simeq \Delta t \left(1 + \frac{v^2}{2c^2}\right), \quad \text{i.e.,} \quad \Delta t' > \Delta t. \tag{9.40}$$

Thus, the time interval of the two events appear to be dilated to the observer O' in S' by a factor

$$\frac{1}{\sqrt{1 - \frac{v^2}{c^2}}}.$$

If now the clock be placed at x'_1 in S', then it will be moving with velocity v relative to S. If the clock now gives signals at time t'_1 and t'_2, then by inverse Lorentz transformation,

$$t_1 = \frac{t'_1 + \frac{vx'_1}{c^2}}{\sqrt{1 - \frac{v^2}{c^2}}} \tag{9.41}$$

$$\text{and} \quad t_2 = \frac{t'_2 + \frac{vx'_2}{c^2}}{\sqrt{1 - \frac{v^2}{c^2}}}, \tag{9.42}$$

where t_1 and t_2 are the time recorded by O in S corresponding to t'_1 and t'_2 in S'.

$$\therefore \quad t_2 - t_1 = \frac{t'_2 - t'_1}{\sqrt{1 - \frac{v^2}{c^2}}} \quad \text{or,} \quad \Delta t = \frac{\Delta t'}{\sqrt{1 - \frac{v^2}{c^2}}} \quad [\because x_1 = x_2], \tag{9.43}$$

$$\text{i.e.,} \quad \Delta t > \Delta t'.$$

In this case the clock is at rest in S', and S is moving with velocity $-v$ relative to S'. Here also we get the same result that the time interval Δt recorded by a clock in S appears lengthened to an observer in moving frame by a factor $\frac{1}{\sqrt{1-\frac{v^2}{c^2}}}$.

Hence, a moving clock always appears to go slow. Here we conclude that, *every clock appears to go at its fastest rate when it is at rest relative to the observer but the rate appears to go on slowing by a factor* $\sqrt{1-\frac{v^2}{c^2}}$ *as its velocity v goes on increasing.*

Proper time. *Proper time is the minimum time interval between any two events, which has been measured relative to a frame in which the same two events take place at a certain point.*

9.8.4 Experimental Verification of Time Dilation

The phenomenon of *time dilation* has been verified in case of nuclear particles called *meson.*

In the laboratory, the π^+-mesons are produced with speed $v = 0.99c$, when high energy particles from synchrotron strikes a target. The flux of π^+-meson was measured at two different places separated by 30 metres. A π^+-meson decays into μ^+-meson and neutron ($\pi^+ \to \mu^+ + n$).

The laboratory time interval $\Delta t'$ in travelling a distance of 30 metres can be found out as below:

$$\Delta t' = \frac{30}{0.99 \times 3 \times 10^8} \simeq 10 \times 10^{-8} \text{ s.}$$

The half-life of π^+-meson is $T_{\frac{1}{2}} = 1.8 \times 10^{-8}$ s and $\Delta t' \simeq 5.6 T_{\frac{1}{2}}$.

This suggests that the flux of π^+-meson should decrease to $2^{-5.6}$ or less than 2% of the original flux after describing a distance of 30 metres. But experimentally measured flux of π^+-meson after travelling a distance of 30 metres was found to be $\sim$ 60% of the original flux.

This discrepancy can be understood by calculating the proper time Δt by the relation

$$\Delta t' = \frac{\Delta t}{\sqrt{1-\frac{v^2}{c^2}}}$$

$$\text{or,} \quad \Delta t = \Delta t'\sqrt{1-\frac{v^2}{c^2}} = 10 \times 10^{-8} \times \sqrt{1-(0.99)^2} \quad \left[\because \frac{v}{c} = 0.99\right]$$

$$= 1.4 \times 10^{-8} \text{ s} \simeq 0.78 \times T_{\frac{1}{2}}.$$

Thus, $\Delta t = 0.78 \times T_{\frac{1}{2}}$ s. In this time, the flux of π^+-meson should reduce to $2^{-0.78}$ or, $\sim$ 60%, which is exactly observed.

Here, the laboratory measurement of time $\Delta t'$ taken by π^+-meson in travelling 30 metres is 10×10^{-8} s. But in the moving frame of π^+-meson the time taken Δt to travel a distance of 30 metres is 1.4×10^{-8} s. Thus

$$\frac{\Delta t'}{\Delta t} = \frac{10 \times 10^{-8}}{1.4 \times 10^{-8}} \quad \text{or,} \quad \Delta t' \simeq 7 \times \Delta t,$$

i.e., *7-fold time dilation has occurred.*

9.9 Einstein's Theorem of Velocity Addition

The simple law of velocity addition $\vec{V} = \vec{v}_1 + \vec{v}_2$, which holds good in Newtonian mechanics, breaks down in Einsteiniun relativity.

We consider two inertial frames of reference S and S', S' moving with velocity $\vec{v}'$ relative to S along the common X-axis. Let a particle move with constant velocity $\vec{u} = \hat{i}u_x + \hat{j}u_y + \hat{k}u_z$ in S, then

$$u_x = \frac{dx}{dt}, \quad u_y = \frac{dy}{dt}, \quad u_z = \frac{dz}{dt}.$$

Similarly, in S'-frame,

$$u'_x = \frac{dx'}{dt'}, \quad u'_y = \frac{dy'}{dt'}; \quad u'_z = \frac{dz'}{dt'}.$$

Now, from inverse Lorentz transformation,

$$x = \frac{x' + vt'}{\sqrt{1 - \frac{v^2}{c^2}}}, \quad y = y', \quad z = z' \quad \text{and} \quad t = \frac{t' + \frac{v\,x'}{c^2}}{\sqrt{1 - \frac{v^2}{c^2}}}.$$

$$\therefore dx = \frac{dx' + v\,dt'}{\sqrt{1 - \frac{v^2}{c^2}}}, \quad dy = dy', \quad dz = dz' \quad \text{and} \quad dt = \frac{dt' + \frac{v\,dx'}{c^2}}{\sqrt{1 - \frac{v^2}{c^2}}}$$

$$\therefore u_x = \frac{dx}{dt} = \frac{(dx' + v\,dt')/\sqrt{1 - \frac{v^2}{c^2}}}{\left[dt' + (\frac{v\,dx'}{c^2})\right]/\sqrt{1 - \frac{v^2}{c^2}}} = \frac{(\frac{dx'}{dt'}) + v}{1 + (\frac{v}{c^2})\,(\frac{dx'}{dt'})}$$

$$\text{or,} \quad u_x = \frac{u'_x + v}{1 + (\frac{v}{c^2}\,u'_x)}. \tag{9.44}$$

Similarly,

$$u_y = \frac{dy}{dt} = \frac{dy'}{(dt' + \frac{v\,dx'}{c^2})/\sqrt{1 - \frac{v^2}{c^2}}} = \frac{\frac{dy'}{dt'}}{(1 + \frac{v\,dx'}{c^2 dt'})/\sqrt{1 - \frac{v^2}{c^2}}}$$

$$\text{or,} \quad u_y = \frac{u'_y\sqrt{1 - \frac{v^2}{c^2}}}{1 + \frac{v\,u'_x}{c^2}} \quad \text{and} \quad u_z = \frac{dz}{dt} = \frac{u'_z\sqrt{1 - \frac{v^2}{c^2}}}{1 + \frac{v\,u'_x}{c^2}}. \tag{9.45}$$

Eqns (9.44) and (9.45) give the *law of addition of velocities.*

Special cases

(i) If u'_x and v are smaller compared to c, $\frac{vu'_x}{c^2}$ can be neglected in comparison to unity. The eqn (9.44) then reduces to

$$u_x = u'_x + v, \tag{9.46}$$

which is the *classical law of addition of velocities.*

(ii) When v or, $u'_x = c$, we have from eqn (9.44),

$$u_x = \frac{u'_x + c}{1 + \frac{u'_x c}{c^2}} = \frac{u'_x + c}{c + u'_x} c = c. \tag{9.47}$$

Thus, *if an object moves with velocity c w.r.t. other, then their relative velocity is always c, whatever may be the velocity of the other.*

(iii) When $u'_x = c = v$, then

$$u_x = \frac{c + c}{1 + \frac{c^2}{c^2}} = \frac{2c}{2} = c. \tag{9.48}$$

Thus, *if an object moves with velocity c in S′-frame and S′-frame moves with velocity v = c, then the velocity of the object is $u_x = c$.*

Hence, *addition of any velocity to c, leads to the velocity c only.*

9.10 Relativistic Variation of Mass

Why mass should vary with velocity in relativity?

The law of conservation of mass and conservation of momentum in *classical mechanics* in moving frame S' are expressed as

$$\sum m = k_1, \text{ conservation of mass}$$

$$\text{and} \quad \sum mu' = k_2, \text{ conservation of momentum,}$$

where m is the mass and k_1 and k_2 are constants. In the rest frame S, we have

$$\sum mu = \sum m\frac{u'+v}{1+\frac{u'v}{c^2}} \quad \left[\because u = \frac{u'+v}{1+\frac{u'v}{c^2}}\right]$$

$$= \frac{\sum mu'}{1+\frac{u'v}{c^2}} + \frac{v\sum m}{1+\frac{u'v}{c^2}} \neq \text{a constant.}$$

Thus, the conservation law of momentum, which holds in S'-frame breaks down in S-frame. To restore the law of conservation of momentum in S-frame as well, the mass m must undergo variation as

$$m' = \frac{m}{\sqrt{1-\frac{v^2}{c^2}}}.$$

If m_0 be the rest mass of a body and m the mass when moving with velocity v, then

$$m = \frac{m_0}{\sqrt{1-\frac{v^2}{c^2}}}.$$

This is the well-known *mass variation* relation.

9.10.1 Derivation of Mass Variation Relation

We consider two inertial frames of reference S and S', S' moving with velocity v relative to S along the common X-axis. Let two exactly similar elastic balls of mass m approach each other with velocity u and $-u$ respectively in S'-frame. After collision the two balls coalesce into a single body of mass $2m$.

Now, according to the *law of conservation of momentum*,

$$mu + m(-u) = 2mV',$$

where V' is the velocity of the coalesced ball

$$\text{or,} \quad 2mV' = 0.$$

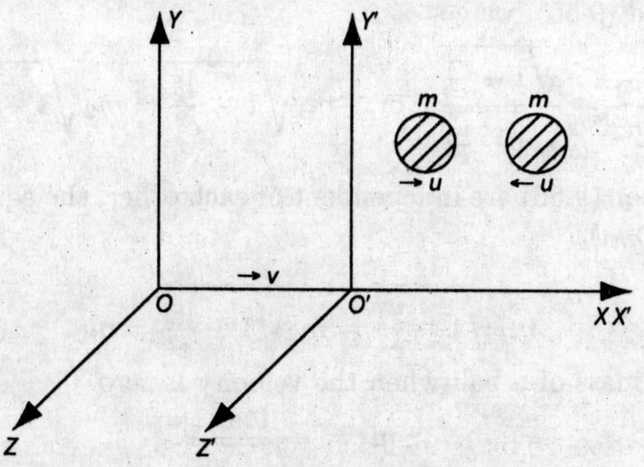

Fig. 9.9: Mass variation with velocity.

But $2m \neq 0$, $\therefore$ $V' = 0$. In other words, the coalesced body of mass $2m$ comes to rest in S'-frame.

We now consider the collision w.r.t. the frame S. Let u_1 and u_2 be the velocities of the balls before collision in S. Then according to the law of addition of velocities,

$$u_1 = \frac{u+v}{1+\frac{uv}{c^2}} \tag{9.49}$$

$$\text{and} \quad u_2 = \frac{-u+v}{1-\frac{uv}{c^2}}. \tag{9.50}$$

After collision let the velocity of the coalesced mass be v in the rest frame S. Considering the conservation of momentum, we get

$$m_1u_1 + m_2u_2 = (m_1+m_2)v, \tag{9.51}$$

where m_1 and m_2 are the masses of the balls moving with velocities u_1 and u_2 respectively.

Substituting the values of u_1 and u_2 from eqns (9.49) and (9.50) in eqn (9.51), we get

$$m_1\left(\frac{u+v}{1+\frac{uv}{c^2}}\right) + m_2\left(\frac{-u+v}{1-\frac{uv}{c^2}}\right) = (m_1+m_2)\,v$$

$$\text{or,} \quad m_1\left(\frac{u+v}{1+\frac{uv}{c^2}} - v\right) = m_2\left(v - \frac{-u+v}{1-\frac{uv}{c^2}}\right)$$

$$\text{or,} \quad m_1\frac{u\left(1-\frac{v^2}{c^2}\right)}{1+\frac{uv}{c^2}} = m_2\frac{u\left(1-\frac{v^2}{c^2}\right)}{1-\frac{uv}{c^2}} \quad \text{or,} \quad \frac{m_1}{m_2} = \frac{1+\frac{uv}{c^2}}{1-\frac{uv}{c^2}}. \tag{9.52}$$

Now,

$$1-\frac{u_1^2}{c^2} = 1 - \frac{\left(\frac{u+v}{c}\right)^2}{\left(1+\frac{uv}{c^2}\right)^2} = \frac{\left(1-\frac{u^2}{c^2}\right)\left(1-\frac{v^2}{c^2}\right)}{\left(1+\frac{uv}{c^2}\right)^2}. \tag{9.53}$$

Similarly

$$1-\frac{u_2^2}{c^2} = \frac{\left(1-\frac{u^2}{c^2}\right)\left(1-\frac{v^2}{c^2}\right)}{\left(1-\frac{uv}{c^2}\right)^2}. \tag{9.54}$$

Now, dividing eqn (9.54) by eqn (9.53), we get

$$\frac{1-\frac{u_2^2}{c^2}}{1-\frac{u_1^2}{c^2}} = \frac{\left(1+\frac{uv}{c^2}\right)^2}{\left(1-\frac{uv}{c^2}\right)^2} \quad \text{or,} \quad \frac{\sqrt{1-\frac{u_2^2}{c^2}}}{\sqrt{1-\frac{u_1^2}{c^2}}} = \frac{1+\frac{uv}{c^2}}{1-\frac{uv}{c^2}}. \tag{9.55}$$

From eqns (9.52) and (9.55), we get

$$\frac{m_1}{m_2} = \frac{\sqrt{1-\frac{u_2^2}{c^2}}}{\sqrt{1-\frac{u_1^2}{c^2}}} \quad \text{or,} \quad m_1\sqrt{1-\frac{u_2^2}{c^2}} = m_2\sqrt{1-\frac{u_2^2}{c^2}}. \tag{9.56}$$

Since both sides of eqn (9.56) are independent of each other, the equation will be valid, if each side is a constant (say, m_0).

Then

$$m_1\sqrt{1-\frac{u_1^2}{c^2}} = m_2\sqrt{1-\frac{u_2^2}{c^2}} = m_0.$$

Here m_0 is the rest mass of a ball when the velocity is zero.

$$\therefore m_1 = \frac{m_0}{\sqrt{1-\frac{u_1^2}{c^2}}}.$$

So if m be the mass of a body moving with velocity v and m_0 be its rest mass, then

$$m = \frac{m_0}{\sqrt{1-\frac{v^2}{c^2}}}. \tag{9.57}$$

Eqn (9.57) represents the *relativistic variation of mass.*

Special cases (i) In the low velocity region, i.e., $v \ll c$, $\frac{v^2}{c^2}$ is negligible compared to unity. $\therefore m = m_0$.

This justifies the constancy of mass of a body in Newtonian mechanics.

(ii) When $v \to c$, $\left(\frac{v}{c}\right)^2 \to 1$ and $m \to \infty$, which is *impossible. Hence, no material particle can assume velocity equal to* (or *greater than*) *c, the velocity of light in vacuum.*

9.11 Einstein's Mass Energy Relation

Let $\vec{ds}$ be the displacement of a body of mass m due to the action of a constant force $\vec{F}$ on the body at rest. The work done is $dW = \vec{F} \cdot \vec{ds}$ and the total work done is

$$W = \int_0^s \vec{F} \cdot \vec{ds} = \int_0^v \frac{d}{dt}(mv)ds = \int_0^v \frac{d}{dt}(mv)\frac{ds}{dt}dt = \int_0^v v\frac{d}{dt}(mv)dt \quad \left[\because \frac{ds}{dt} = v\right]$$

$$= \int_0^v vd(mv), \tag{9.58}$$

where v is the velocity of the body.

According to *work-energy principle*, this work manifests as the KE T of the body.

$$\therefore T = \int_0^v vd(mv) = \int_0^v vd\left(\frac{m_0 v}{\sqrt{1-\frac{v^2}{c^2}}}\right) \quad \left[\because m = \frac{m_0}{\sqrt{1-\frac{v^2}{c^2}}}\right]$$

$$\text{or,}\quad T = m_0 \int_0^v v \frac{\sqrt{1-\frac{v^2}{c^2}} + \frac{v}{2}\left(1-\frac{v^2}{c^2}\right)^{-\frac{1}{2}}\frac{2v}{c^2}}{\left(1-\frac{v^2}{c^2}\right)} dv$$

$$= m_0 \int_0^v v\left(\frac{1}{\sqrt{1-\frac{v^2}{c^2}}} + \frac{\frac{v^2}{c^2}}{\left(1-\frac{v^2}{c^2}\right)^{\frac{3}{2}}}\right) dv = m_0 \int_0^v v\left[\frac{1-\frac{v^2}{c^2}+\frac{v^2}{c^2}}{\left(1-\frac{v^2}{c^2}\right)^{\frac{3}{2}}}\right] dv$$

$$= m_0 \int_0^v v\left[\frac{1}{\left(1-\frac{v^2}{c^2}\right)^{\frac{3}{2}}}\right] dv = m_0 \int_0^v \frac{v\,dv}{\left(1-\frac{v^2}{c^2}\right)^{\frac{3}{2}}} = m_0 c^2 \left[\frac{1}{\left(1-\frac{v^2}{c^2}\right)^{\frac{1}{2}}}\right]_0^v$$

$$= m_0 c^2 \left[\frac{1}{\left(1-\frac{v^2}{c^2}\right)^{\frac{1}{2}}} - 1\right] = \frac{m_0 c^2}{\left(1-\frac{v^2}{c^2}\right)^{\frac{1}{2}}} - m_0 c^2$$

$$= mc^2 - m_0 c^2 \quad \left[\because m = \frac{m_0}{\sqrt{1-\frac{v^2}{c^2}}}\right]$$

$$= (m - m_0)^2$$

$$\text{or,}\quad T = \Delta mc^2 \quad [\Delta m = (m - m_0)].$$

Also

$$T = E - m_0 c^2, \tag{9.59}$$

where $m_0 c^2$ is the *rest mass energy* of the body and $mc^2 = E$ represents its *total energy.*

So

$$E = T + m_0 c^2 \tag{9.60}$$

$$\text{and}\quad E = mc^2. \tag{9.61}$$

Eqn (9.61) represents the *relativistic mass energy relation.*

Eqn (9.61) is widely used in calculating *nuclear fission or fusion energy.* This relation actually opened a new era in human civilisation—called *the atomic* or *nuclear age.*

9.12 Energy and Momentum of Relativistic Particle

If m_0 be the rest mass of a body, then its mass when moving with velocity v is

$$m = \frac{m_0}{\sqrt{1-\frac{v^2}{c^2}}}.$$

Momentum of the body

$$p = mv = \frac{m_0 v}{\sqrt{1-\frac{v^2}{c^2}}} \quad \text{or,} \quad p^2c^2 = \frac{m_0^2 v^2 c^2}{1-\frac{v^2}{c^2}} \tag{9.62}$$

$$\text{or,} \quad p^2c^2 + m_0^2c^4 = \frac{m_0^2 v^2 c^2}{1-\frac{v^2}{c^2}} + m_0^2c^4.$$

[by adding the term $m_0^2c^4$ on both sides of eqn (9.62)]

$$\text{or,} \quad p^2c^2 + m_0^2c^4 = \frac{m_0^2v^2c^2 + m_0^2c^4 - m_0^2v^2c^2}{1-\frac{v^2}{c^2}} = \frac{m_0^2c^4}{1-\frac{v^2}{c^2}} = \left(\frac{m_0c^2}{\sqrt{1-\frac{v^2}{c^2}}}\right)^2 = m^2c^4$$

$$\text{or,} \quad (mc^2)^2 = p^2c^2 + m_0^2c^4 \quad \text{or,} \quad E^2 = p^2c^2 + m_0^2c^4 \tag{9.63}$$

$$\text{or,} \quad E = \sqrt{p^2c^2 + m_0^2c^4}. \tag{9.64}$$

This represents *the relation between energy and momentum of a relativistic particle.*

9.13 Answers of Some Important Questions

9.1. *State the special features of Einstein's special theory of relativity.*

Ans. The special features of Einstein's special theory of relativity are: (i) the cause that led its formulation (i.e., its origin), (ii) the postulates and the principles, (iii) the conclusions derived from this theory, (iv) the experimental evidences in favour of it.

The theory of relativity based on new concepts of space and time is divided into two parts: (a) *Special theory of relativity* and (b) *General theory of relativity.*

9.2. *What do you mean by frame of reference?*

Ans. A frame of reference is a system relative to which the motion of any body is described. The motion observed by observers in different systems is different. For example, the earth, the walls of a room, a tree, the position of a star along a plumb line, the sun, etc. serve the purpose of reference frame.

Frame of reference are of the following two types: (i) Inertial frame (or, Newtonian frame) of reference; (ii) Non-inertial frame (or, accelerated frame) of reference.

9.3. *What are inertial and non-inertial frame of reference?*

Ans. *Inertial frame of reference.* A frame of reference w.r.t., which a body, not acted upon by external force, is unaccelerated is called an *inertial frame* of reference or Newtonian frame of reference. This also includes the state of rest. *The laws of mechanics are the same in all inertial frames.*

If no external force is applied, a body in motion continues its motion with uniform velocity. This is *Newton's first law of motion.*

Intergalactic space is approximated to be an inertial frame of reference.

Non-inertial frame of reference. A frame w.r.t., which a body, not acted upon by an external force, is accelerated is called a *non-inertial frame* of reference. All accelerated frames are non-inertial frames.

In these frames, Newton's laws of motion are not valid in their original form but by introducing some modification, they can be applied in this frame of reference. The earth, a train moving with acceleration, etc. are the common examples of accelerated frame of reference.

9.4. *What is Galilean principle of invariance?*

Ans. The Galilean principle of invariance states that the laws of mechanics have the same form in all inertial frames of reference.

Thus, a series of mechanical experiments whether conducted in a car at rest w.r.t. the ground or moving smoothly over a straight track at constant velocity give the same result.

9.5. *What is Newtonian relativity?*

Ans. Newtonian mechanics predicts that, all inertial frames in uniform translation relative to one another are equivalent so far as the mechanical phenomena are concerned. This equivalence of all inertial frames in respect to the laws of classical mechanics is referred to as *Newtonian relativity.*

9.6. *Discuss the negative result of Michelson-Morley experiment and the conclusions drawn therefrom.*

Ans. The theoretical calculations show that, if there is a relative velocity between the ether and the earth, then a fringe shift of 0.4 would occur. Michelson-Morley experimental set-up is sufficient sensitive to measure this fringe shift. But experimentally no shift of fringe was detected. Then it was concluded that (i) *the relative velocity between ether and the earth is zero* and (ii) *ether does not exist at all.*

9.7. *How the negative result of Michelson-Morley experiment was explained?*

Ans. Following interpretations were put forward to explain the negative result of Michelson-Morley experiment:

(a) *Ether drag hypothesis.* According to Michelson, ether was dragged along with the moving earth and hence there was no relative velocity between the earth and the ether.

(b) *Lorentz-Fitzgerald contraction hypothesis.* Fitzgerald and Lorentz independently suggested that when a material body is in motion its length undergoes a contraction by a factor $\sqrt{1-\frac{v^2}{c^2}}$ along the direction of motion, where v is the velocity of the body and c, the velocity of light. This leads to no time difference $(T_1 - T_2)$ for horizontal and vertical rays and hence no fringe shift.

(c) *Constancy of speed of light.* Einstein proposed that the velocity of light c is independent of the motion of the source, observer or medium. So $T_1 = T_2 = \frac{2d}{c}$, d is the distance of the mirrors from the half silvered glass plate and hence there is no fringe shift.

9.8. *State the postulates of special theory of relativity.*

Ans. Einstein's special theory of relativity is based on the following two postulates:

(a) All the laws of physics are the same (i.e., covariant) in all inertial frames of reference having an uniform translational velocity w.r.t. one another.

(b) The velocity of light is the same in all inertial frames and is independent of the motion of the source and the observer.

9.9. *Discuss the insufficiency of Galilean transformation.*

Ans. *See* Section 9.6 (p. 230).

9.10. *State Lorentz transformation equations.*

Ans. Consider two inertial frames S and S', S' moving with constant velocity v w.r.t. S along the common X-axis. If (x, y, z, t) and (x', y', z', t') be the coordinates of an object as measured by the observer in S and S' respectively, then according to Lorentz transformation,

$$x' = \frac{x - vt}{\sqrt{1 - \frac{v^2}{c^2}}}$$

$$y' = y$$

$$z' = z$$

$$\text{and} \quad t' = \frac{t - \frac{vx}{c}}{\sqrt{1 - \frac{v^2}{c^2}}},$$

where c is the velocity of light.

9.11. *What is proper length? State the Lorentz-Fitzgerald law of length contraction.*

Ans. *The length of a rod measured by a stationary observer is called the proper length.*

The Lorentz-Fitzgerald law of length contraction states that, *the length of a rod will appear contracted to the moving observer by a factor* $\sqrt{1 - \frac{v^2}{c^2}}$.

9.12. *What is proper time? Stat the Einstein's principle of time dilation.*

Ans. *Proper time.* The minimum time interval between any two events, which has been measured relative to a frame in which the same two events take place at a certain point is called *proper time.*

Einstein's principle of time-dilation. The principle states that the time interval between any two events appears to be dilated to the moving observer at O' in S' by a factor $\sqrt{1 - \frac{v^2}{c^2}}$.

If Δt and $\Delta t'$ be the time interval as observed by observers in S and S', S' moving with velocity v relative to S along the common X-axis, then $\Delta t' > \Delta t$.

According to this principle, *every clock appears to go at its fastest rate when it is at rest relative to the observer but the rate appears to go on slowing by a factor* $\sqrt{1 - \frac{v^2}{c^2}}$ *as its velocity v goes on increasing.*

9.13. *Show that addition of any velocity to c leaves the velocity c only, where c is the velocity of light.*

Ans. Let S'-frame moves with velocity v relative to the rest frame S and the velocity of a photon w.r.t. S'-frame be c, then the velocity of the photon in S-frame is

$$u = \frac{c + v}{1 + \frac{cv}{c^2}} = \frac{(c + v)}{(c + v)}c = c.$$

Hence, the result follows.

9.14. *Why the mass of a relativistic particle should vary with velocity?*

Ans. To satisfy the *law of conservation of momentum of a relativistic particle* both in rest frame (S) as well as in moving frame (S') it is essentially needed that the mass of the moving body must vary with velocity.

If m and m_0 be respectively the mass of the body when moving with velocity v and when at rest, then

$$m = \frac{m_0}{\sqrt{1 - \frac{v^2}{c^2}}}, \quad c = \text{the velocity of light.}$$

This is the well-known *mass-variation relation.*

9.15. *State Einstein's mass-energy relation.*

Ans. If m be the mass of a relativistic particle, then its energy is $E = mc^2$. Thus, in view of Einstein, there is *no basic difference between mass and energy.* Actually mass is the *condensed form of energy.*

This relation is widely used to calculate nuclear *fission* or *fusion energy.*

9.14 Solved Problems

Ex. 1 *Show by direct application of Lorentz transformation,* $x'^2 - c^2t'^2 = x^2 - c^2t^2$.

Soln. We consider two inertial frames S and S'. S' moving with velocity v along the common X-axis. Let (x', y', z', t') and (x, y, z, t) be the coordinates of an event in S'- and S-frames respectively.

Then by *Lorentz transformation, equations* are

$$x' = \frac{x - vt}{\sqrt{1 - \frac{v^2}{c^2}}}$$

$$y' = y$$

$$z' = z$$

$$t' = \frac{t - \frac{vx}{c^2}}{\sqrt{1 - \frac{v^2}{c^2}}}.$$

Now,

$$x'^2 = \left(\frac{x - vt}{\sqrt{1 - \frac{v^2}{c^2}}}\right)^2 = \frac{x^2 + v^2t^2 - 2x\,vt}{1 - \frac{v^2}{c^2}}$$

$$\text{and} \quad c^2t'^2 = c^2\left(\frac{t - \frac{vx}{c^2}}{\sqrt{1 - \frac{v^2}{c^2}}}\right)^2 = \frac{c^2t^2 + \frac{v^2x^2}{c^2} - 2xvt}{1 - \frac{v^2}{c^2}}$$

$$\therefore x'^2 - c^2t'^2 = \left(\frac{x^2 + v^2t^2 - 2x\,vt}{1 - \frac{v^2}{c^2}}\right) - \frac{c^2t^2 + \frac{v^2x^2}{c^2} - 2x\,vt}{\left(1 - \frac{v^2}{c^2}\right)}$$

$$= \frac{x^2 + v^2t^2 - 2x\,vt - c^2t^2 - \frac{v^2x^2}{c^2} + 2x\,vt}{1 - \frac{v^2}{c^2}}$$

$$= \frac{x^2\left(1 - \frac{v^2}{c^2}\right) - c^2t^2\left(1 - \frac{v^2}{c^2}\right)}{\left(1 - \frac{v^2}{c^2}\right)} = \mathbf{x^2 - c^2t^2}.$$

Ex. 2 *Two rods of proper length l_0 move lengthwise towards each other parallel to their common axis with the same velocity v relative to laboratory frame. Show that the length of each rod in the reference frame fixed to the other rod is $l = l_0\frac{1-\beta^2}{1+\beta^2}$, where $\beta = \frac{v}{c}$.* **[CUH 2003; WBUT 2007]**

Soln. As the rods are approaching each other, if the velocity of one rod be v, then the velocity of the other is $-v$. Then the relative velocity of one relative to other is

$$u = \frac{v - (-v)}{1 - \frac{v(-v)}{c^2}} = \frac{2v}{1 + \frac{v^2}{c^2}}.$$

Now, according to the *principle of length contraction,* the length of the rod moving with velocity u is

$$l = l_0\sqrt{1-\frac{u^2}{c^2}} = l_0\sqrt{1-\frac{4v^2}{c^2\left(1+\frac{v^2}{c^2}\right)^2}} = l_0\sqrt{1-\frac{4\beta^2}{(1+\beta^2)^2}} \quad \left[\because \beta = \frac{v}{c}\right]$$

$$\therefore l = l_0\frac{1-\beta^2}{1+\beta^2} \text{ (Proved)}.$$

Ex. 3 *A rigid bar of length* 1.5 m *is at rest relative to* S'. *If it makes an angle* $\alpha' = 45°$ *with the* x'*-axis, find the length of the bar and its orientation relative to the frame* S, *when* $v = 0.95$ c.

Soln. If L'_x and L'_y be the x'- and y'-components of the rod of length L' in S', then

$$L'_x = L'\cos\alpha' = 1.5 \times \frac{1}{\sqrt{2}} = 1.06 \text{ m}$$

$$L'_y = L'\sin\alpha' = 1.5 \times \frac{1}{\sqrt{2}} = 1.06 \text{ m}.$$

The vertical component will not suffer length contraction since it is at 90° to the direction of v.

$\therefore L_y = L'_y = 1.06$ m; $L_x = L'_x\sqrt{1-\frac{v^2}{c^2}} = 1.06 \times \sqrt{1-0.95^2} = \mathbf{0.33}$ **m.**

So the length of the bar relative to S is

$$L = \sqrt{L_x^2 + L_y^2} = \sqrt{(0.331)^2 + (1.06)^2} = \mathbf{1.11 \text{ m}}.$$

Orientation of the rod relative to S is

$$\tan\alpha = \frac{L_y}{L_x} = \frac{1}{\sqrt{1-0.95^2}} = 3.20. \quad \therefore \alpha = \mathbf{72°40'}.$$

Ex. 4 *Show that if* L_0^3 *be the rest volume of a cube, then* $L_0^3(1-\beta^2)^{1/2}$ *is the volume as viewed from a reference frame moving with uniform velocity* v *parallel to an edge of the cube.*

Soln. We know that the length of a body undergoes Lorentz contraction by a factor

$$\sqrt{1-\frac{v^2}{c^2}} = \sqrt{1-\beta^2}.$$

$\beta = \frac{v}{c}$ in the direction of its motion. The perpendicular dimensions to the direction of velocity v remain unchanged.

Since the cube has rest length L_0, its rest volume is L_0^3. The length L_0 along the direction of v is changed to $L_0\sqrt{1-\beta^2}$. The other two sides each of length L_0 remain unchanged.

So the volume of the moving cube is

$$L_0\sqrt{1-\beta^2} \times L_0 \times L_0 = L_0^3\sqrt{1-\beta^2} = L_0^3(1-\beta^2)^{1/2} \text{ (Proved)}.$$

Ex. 5 *The length of a rod is found to be half of its length when at rest. What is the speed of the rod relative to the observer?*

[KU 1997]

Soln. Let L_0 and L be the length of the rod when it is at rest and moving with velocity v, then

$$L = L_0\sqrt{1-\frac{v^2}{c^2}}.$$

By the problem, $L = L_0/2$.

$$\therefore \frac{1}{2}L_0 = L_0\sqrt{1-\frac{v^2}{c^2}} \quad \text{or,} \quad \frac{1}{2} = \sqrt{1-\frac{v^2}{c^2}} \quad \text{or,} \quad \frac{1}{4} = 1-\frac{v^2}{c^2}.$$

$$\therefore \frac{v^2}{c^2} = 1-\frac{1}{4} = \frac{3}{4} \quad \text{or,} \quad \frac{v}{c} = \frac{\sqrt{3}}{2}. \quad \therefore v = \frac{\sqrt{3}}{2} \times c = \mathbf{0.866\,c}.$$

Ex. 6 *What are* (i) *proper time* (ii) *non-proper time* (iii) *proper length?*

Soln. (i) *Proper time.* The time interval of two events, which occur at the same position measured by a clock in the inertial frame in which the events occur is called *proper time.*

(ii) *Non-proper time.* The time interval between the same two events measured by an observer from an inertial frame, which is moving w.r.t. the clock is known as *non-proper* or *relativistic time.*

(iii) *Proper length.* The length measured by a scale at rest w.r.t. the object is called *proper length.*

Ex. 7 *Show that the four-dimensional volume element $dx\,dy\,dz\,dt$ is invariant under Lorentz transformations.* **[PU 1995]**

Soln. According to the principle of length contraction and time dilation, we have

$$L = L_0\sqrt{1-\frac{v^2}{c^2}},$$

where L and L_0 are the lengths in S' and S respectively, and

$$T = T_0\sqrt{1-\frac{v^2}{c^2}},$$

where T is the observed time in moving frame S' and T_0 is the proper time in stationary frame S.

If dx' and dx be the observed length in S' and proper length in S, then

$$dx' = dx\sqrt{1-\frac{v^2}{c^2}},$$

where S' moves with velocity $\vec{v}$ relative to S along the common X-axis. Then $dy' = dy$ and $dz' = dz$.

Also if dt' be the observed time in S' and dt the proper time in S, then

$$dt' = dt/\sqrt{1-\frac{v^2}{c^2}}.$$

∴ the four-dimensional volume element is

$$dx'\,dy'\,dz'\,dt' = dx\sqrt{1-\frac{v^2}{c^2}}\,dy\,dz\frac{dt}{\sqrt{1-\frac{v^2}{c^2}}} = \mathbf{dx\,dy\,dz\,dt}.$$

Ex. 8 *A clock in a spaceship emits signal at an interval of* 5s *as observed by an astronaut in the spaceship. If the spaceship travels with a speed of* 1.8×10^8 m/s, *find the interval between the successive signals as received by an observer on the ground.*

Soln. According to relativistic formula for *time dilation,*

$$\Delta t = \frac{\Delta t_0}{\sqrt{1-\frac{v^2}{c^2}}},$$

where Δt_0 is the time interval of the clock in a frame, where it is at rest and Δt is the time interval in the clock of the observer.

$\Delta t_0 = 5$ s and $v = 1.8 \times 10^8$ m/s; $c = 3 \times 10^8$ m/s

$$\therefore \Delta t = \frac{5}{\sqrt{1-\left(\frac{1.8\times10^8}{3\times10^8}\right)^2}} = \frac{5\times3}{\sqrt{3^2-(1.8)^2}} = \frac{15}{\sqrt{4.8\times1.2}} = \frac{15}{2.4} = \mathbf{6.25\,s}.$$

Ex. 9 *Half-life of a particle at rest is* 20 *nanoseconds. What will be the half-life if its speed is* 0.9*c*?

Soln. Here half-life of the particle in the rest frame is $T_0 = 20 \times 10^{-9}$ s.

If T be the half-life of the particle when its speed $v = 0.9c$, then

$$T = \frac{T_0}{\sqrt{1-\frac{v^2}{c^2}}} = \frac{20\times10^{-9}}{\sqrt{1-\left(\frac{0.9c}{c}\right)^2}} = \frac{20\times10^{-9}}{\sqrt{1-0.81}} = \frac{20\times10^{-9}}{\sqrt{0.19}} = \frac{20\times10^{-9}}{0.44} = \mathbf{45.45\times10^{-9}\ s.}$$

Ex. 10 *A clock keeps correct time. With what speed should it travel relative to an observer so that it appears to lose* 1 min *in a day?* **[GNDU 1997]**

Soln. Time measured by the clock $T = 24\times60$ min. When the clock is in motion, time measured by the observer $T_0 = (24\times60-1)$ mins.

Then

$$T = \frac{T_0}{\sqrt{1-\frac{v^2}{c^2}}},$$

where v is the velocity of the clock

$$\text{or,}\quad 1-\frac{v^2}{c^2} = \left(\frac{T_0}{T}\right)^2 = \frac{(24\times60-1)^2}{(24\times60)^2} = \left(\frac{1439}{1440}\right)^2 = 0.9986.$$

$$\therefore \frac{v^2}{c^2} = 1-0.9986 = 0.0014.$$

$$\therefore \frac{v}{c} = 0.0374 \text{ and } v = 0.0374\times c = 0.0374\times3\times10^8\ \text{m/s} = \mathbf{0.1122\times10^8\ m/s.}$$

Ex. 11 *The proper life of* π^+*-meson is* 2.5×10^{-8} s. *If a beam of these mesons of velocity* 0.8*c is produced, calculate the distance, the beam can travel before the flux of the meson beam is reduced to* $1/e^2$ *times the initial flux.*

Soln. According to the principle of time dilation,

$$\Delta t' = \frac{\Delta t}{\sqrt{1-\frac{v^2}{c^2}}}.$$

Here $\Delta t = 2.5\times10^{-8}$ s, $v = 0.8c$.

$$\therefore \Delta t' = \frac{2.5\times10^{-8}}{\sqrt{1-\left(\frac{0.8c}{c}\right)^2}} = \frac{2.5\times10^{-8}}{0.6} = 4.6\times10^{-8}\ \text{s}.$$

If N_0 and N be the flux of π^+-meson at $t=0$ and at $t=\tau$, then

$$N = N_0 e^{-t/\tau}\quad [\tau = \text{the mean time}].$$

But $N = (1/e^2)N_0$,

$$\therefore \frac{1}{e^2}N_0 = N_0 e^{-t/\tau}\quad \text{or,}\quad t = 2\tau = 2\Delta t'.\quad [\because \tau = \Delta t' = 4.16\times10^{-8}\ \text{s}].$$

$\therefore 2\Delta t' = 2\times4.16\times10^{-8}\ \text{s} = 8.32\times10^{-8}$ s.

Now, the distance traversed by the beam before the flux of π^+-meson reduced to $1/e^2$ times the initial flux is $= 2\Delta t'\times0.8c = (8.32\times10^{-8})\times0.8\times(3\times10^8) =$ **19.97 m.**

Ex. 12 *A beam of* μ*-mesons produced at a height of* 20 km *in the earth's atmosphere, travels downwards with a velocity of* 0.99*c. If* 99% *of the original mesons decay before reaching the earth's surface, find the mean life time of the* μ*-mesons.*

Soln. The time T recorded by an observer in the earth's frame for a μ-meson to travel a distance of 20 km with a velocity of $0.99c$ w.r.t. the earth, then we have

$$T = \frac{20 \times 10^3}{0.99 \times 3 \times 10^8} = 67.34 \times 10^{-6}\,\text{s}.$$

If the same time appears to be T_0 to an observer moving with μ-meson (in an inertial frame) in which μ-meson is at rest, then

$$T = \frac{T_0}{\sqrt{1 - \frac{v^2}{c^2}}}$$

$$\text{or,}\quad T_0 = T\sqrt{1 - \frac{v^2}{c^2}} = 67.34 \times 10^{-6} \quad \left[1 - (0.99)^2\right]^{1/2}$$

$$= 67.34 \times 10^{-6} \times 0.141 = 9.495 \times 10^{-6}\,\text{s}.$$

In time T_0, 99% of μ-meson decay in the rest frame.

If N_0 and N be the no. of μ-meson at $t = 0$ and $t = T_0$ and λ, the decay constant, then

$$N = N_0 e^{-\lambda T_0} \quad \text{or,} \quad \frac{N_0}{N} = e^{\lambda T_0} \quad \text{or,} \quad \lambda T_0 = \log_e \frac{N_0}{N} = 2.3026 \times \log_{10} \frac{N_0}{N}.$$

But $N = 1\%$ and $N_0 = 100\%$.

$$\therefore \frac{N_0}{N} = \frac{100}{1} = 100.$$

Hence, $\lambda T_0 = 2.3026 \times \log_{10} 100 = 2 \times 2.3026 = 4.6052$.

$$\therefore \lambda = \frac{4.6052}{T_0} = \frac{4.6052}{9.495 \times 10^{-6}}.$$

So mean life time

$$\tau = \frac{1}{\lambda} = \frac{9.495 \times 10^{-6}}{4.6052} = \mathbf{2.06 \times 10^{-6}\,s}.$$

Ex. 13 *How much younger an astronaut will appear to the earth observer, if he returns after one year having moved with velocity* $0.5c$? [PU 1994]

Soln. Let T_0 = proper time in the spaceship of the astronaut

T = observed time of 1 year on the earth.

Then

$$T = \frac{T_0}{\sqrt{1 - \frac{v^2}{c^2}}} \quad \text{or,} \quad T_0 = T\sqrt{1 - \frac{v^2}{c^2}}$$

or, $T_0 = 1 \times \sqrt{1 - (0.5)^2} = \sqrt{0.75} = 0.866$ year = 10 months and 12 days.

Now, the astronaut will appear younger by

$T - T_0 = 12 -$ (10 months and 12 days) = **1 month and 18 days.**

Ex. 14 *An electron is moving with a speed of* $0.85c$ *in a direction opposite to that of a moving photon. Calculate the relative velocity of the electron and photon.*

Soln. Let the photon and the electron move along the positive and negative X-axis with velocity c and v as shown in Fig. 9.10. The electron is supposed to be in the rest frame S.

Then the system S' (or Laboratory) moves with velocity $-v = 0.85c$ relative to S. Then $v = 0.85c$, $u' = c$.

$$\therefore u = \frac{u' + v}{1 + \frac{u'v}{c^2}} = \frac{c + 0.85c}{1 + \frac{0.85c \times c}{c^2}} = \frac{c + 0.85c}{c + 0.85c}c = c.$$

Fig. 9.10

Thus, the relative velocity of the electron and photon is c.

Ex. 15 *If two photons are approaching each other in free space, then find the relative velocity of one photon w.r.t. the other.* **[WBUT 2007]**

Soln. Let the two photons having velocity c move along the X-direction. As the photons approach each other, we have by *inverse velocity transformation*, the relative velocity of the photons is

$$u = \frac{u' + v}{1 + \frac{u'v}{c^2}} = \frac{c + c}{1 + \frac{c.c.}{c^2}} = \frac{2c}{2} = c. \quad [\because u' = c,\ v = c]$$

Thus, the relative velocity of approach of the photons is c. This shows that *velocity of light is constant in all inertial frames.*

Ex. 16 *Show that addition of any velocity to c leaves the velocity c only.*

Soln. Let the velocity of S'-frame relative to S be v and the velocity of photon is c in S'-frame. Then the velocity of the photon in S-frame is

$$u = \frac{c + v}{1 + \frac{cv}{c^2}} = \left(\frac{c + v}{c + v}\right) c = c.$$

Hence, the result follows.

Ex. 17 *A scientist observes that a certain atom X moving relative to him with velocity* 2×10^{10} cm/s *emits a particle Y, which moves with velocity* 2.8×10^{10} cm/s *w.r.t. the atom. Calculate the velocity of Y relative to the scientist.*

Soln. Let the scientist be in S-frame and the atom in S'-frame.

Then $u' = 2.8 \times 10^{10}$ cm/s is the velocity of Y relative to S'

$v = 2 \times 10^{10}$ cm/s is the velocity of S' relative to S.

Now, the velocity of Y relative to S-frame is

$$u = \frac{u' + v}{1 + \frac{u'v}{c^2}} = \frac{2.8 \times 10^{10} + 2 \times 10^{10}}{1 + \frac{(2.8 \times 10^{10})(2 \times 10^{10})}{(3 \times 10^{10})^2}} = \frac{4.8 \times 10^{10}}{1 + \frac{5.6}{9}} = \frac{4.8 \times 10^{10}}{1.622} = \mathbf{2.95 \times 10^{10}\ cm/s.}$$

Ex. 18 *Starting from the Lorentz transformation show that if* (x_1, y_1, z_1, t_1) *and* (x_2, y_2, z_2, t_2) *represent two events, then the quantity.*

$(x_1 - x_2)^2 + (y_1 - y_2)^2 + (z_1 - z_2)^2 - c^2(t_1 - t_2)^2$ *is invariant in all inertial frames.* **[WBUT 2007]**

Soln. Let S and S' be two inertial frames, S' moving with velocity v relative to S along the positive X-axis. Let the coordinates of the events in S be (x_1, y_1, z_1, t_1) and (x_2, y_2, z_2, t_2). The corresponding coordinates of the same two events in S' be (x_1', y_1', z_1', t_1') and (x_2', y_2', z_2', t_2').

According to *inverse Lorentz transformations*,

$$x_1 = \frac{x_1' + vt_1'}{\sqrt{1 - \frac{v^2}{c^2}}},\ y_1 = y_1',\ z_1 = z_1' \text{ and } t_1 = \frac{t_1' + \frac{v x_1'}{c^2}}{\sqrt{1 - \frac{v^2}{c^2}}}$$

$$\text{and } x_2 = \frac{x_2' + vt_2'}{\sqrt{1 - \frac{v^2}{c^2}}},\ y_2 = y_2',\ z_2 = z_2' \text{ and } t_2 = \frac{t_2' + \frac{v x_2'}{c^2}}{\sqrt{1 - \frac{v^2}{c^2}}}.$$

$$\therefore (x_1 - x_2)^2 + (y_1 - y_2)^2 + (z_1 - z_2)^2 - c^2 (t_1 - t_2)^2$$

$$= \left[\frac{x_1' + vt_1'}{\sqrt{1-\frac{v^2}{c^2}}} - \frac{x_2' + vt_2'}{\sqrt{1-\frac{v^2}{c^2}}}\right]^2 + (y_1' - y_2')^2 + (z_1' - z_2')^2 - c^2 \left[\frac{t_1' + \frac{v}{c^2} x_1'}{\sqrt{1-\frac{v^2}{c^2}}} - \frac{t_2' + \frac{v}{c^2} x_2'}{\sqrt{1-\frac{v^2}{c^2}}}\right]^2$$

$$= \left[\frac{x_1'^2 + 2vx_1't_1' + v^2t_1'^2 - x_2'^2 - 2vx_2't_2' - v^2t_2'^2}{(1-\frac{v^2}{c^2})}\right]$$

$$- 2\frac{(x_1' + vt_1')(x_2' + vt_2')}{(1 - v^2/c^2)} + (y_1' - y_2')^2 + (z_1' - z_2')^2$$

$$- c^2 \left[\frac{t_1'^2 + \frac{v^2}{c^4} x_1'^2 + 2\frac{v}{c^2} x_1' t_1' - t_2'^2 - \frac{v^2}{c^4} x_2'^2 - 2\frac{v}{c^2} x_2' t_2'}{(1-\frac{v^2}{c^2})}\right] + 2\frac{(t_1' + \frac{v}{c^2} x_1')(t_2' + \frac{v}{c^2} x_2')}{1/v^2/c^2}$$

Simplifying and rearranging we get,

$$(x_1' - x_2')^2 + (y_1' - y_2')^2 + (z_1' - z_2')^2 - c^2 (t_1' - t_2')^2$$

Hence, the result follows.

Ex. 19 *When at rest a box is a cube with sides of length* 25 cm. *What is its apparent volume when it moves with a speed* 2×10^{10} cm/s *parallel to one of its sides?* **[BUH 2003]**

Soln. As the cube moves parallel to one of its sides there will be contraction of length along the direction of motion. If l_0 and l be the length of the side of the box when at rest and when moving with velocity $v = 2 \times 10^{10}$ cm/s, then its volume in moving condition is

$$V = l_0^2 l = l_0^3 \left[1 - \frac{v^2}{c^2}\right] = l_0^3 \left[1 - \left(\frac{2 \times 10^{10}}{3 \times 10^{10}}\right)^2\right] = V_0 \left[1 - \left(\frac{2}{3}\right)^2\right] = V_0 \left(1 - \frac{4}{9}\right) = \frac{5}{9} V_0.$$

So the apparent volume of the box is $\mathbf{\frac{5}{9} V_0}$.

Ex. 20 *Calculate the velocity of an electron at which its mass is* $\sqrt{3}$ *times the rest mass.* **[PU 1991]**

Soln. According to relativistic mass variation,

$$m = \frac{m_0}{\sqrt{1 - \frac{v^2}{c^2}}},$$

where m_0 and m are the masses of the electron when at rest and when moving with velocity v respectively

$$\text{or,} \quad \frac{m_0}{m} = \sqrt{1 - \frac{v^2}{c^2}} = \frac{1}{\sqrt{3}} \quad \text{or,} \quad 1 - \frac{v^2}{c^2} = \frac{1}{3} \quad \text{or,} \quad \frac{v^2}{c^2} = 1 - \frac{1}{3} = \frac{2}{3}.$$

$$\therefore v = \sqrt{\frac{2}{3}} c = 0.816 \times 3 \times 10^{10} = \mathbf{2.448 \times 10^{10}\ m/s.}$$

Ex. 21 *In an electric field the speed of an electron changes from* 0.95c *to* 0.98c. *Calculate its change of mass and the work done on the electron to change the velocity.*

Soln. If m_1 and m_2 be the masses of the electron when its velocity is $0.95c$ and $0.98c$ respectively.
Then

$$m_1 = \frac{m_0}{\sqrt{1-(0.95)^2}} = 3.2m_0 \quad \text{and} \quad m_2 = \frac{m_0}{\sqrt{1-(0.98)^2}} = 5.0m_0,$$

where m_0 is the rest mass of the electron $= 9.11 \times 10^{-31}$ kg.

$\therefore$ change in mass of the electron is

$$\Delta m = m_2 - m_1 = (5.0 - 3.2) \times m_0 = 1.8 \times 9.11 \times 10^{-31} = 16.4 \times 10^{-31}\ \text{kg}.$$

Now, the work done on the electron is equal to the change in KE given by

$$\Delta T = T_2 - T_1 = (\Delta m)c^2 = 1.8m_0c^2$$

$$= 1.8 \times 0.51\ \text{MeV} \quad [\because \text{rest mass energy of the electron } m_0c^2 = 0.51\ \text{MeV}]$$

$$= \mathbf{0.92\ MeV}.$$

Ex. 22 *A body of rest mass m_0 breaks up into two parts spontaneously with rest masses m_{01} and m_{02} and speeds v_1 and v_2 respectively. Prove that $m_0 > (m_{01} + m_{02})$.*

Soln. The rest mass energies of the broken parts are

$$\frac{m_{01}c^2}{\sqrt{1-\frac{v_1^2}{c^2}}} \quad \text{and} \quad \frac{m_{02}c^2}{\sqrt{1-\frac{v_2^2}{c^2}}}$$

and the rest mass energy of the body is m_0c^2.

According to mass-energy conservation law,

$$m_0c^2 = \frac{m_{01}c^2}{\sqrt{1-\frac{v_1^2}{c^2}}} + \frac{m_{02}c^2}{\sqrt{1-\frac{v_2^2}{c^2}}} > (m_{01}c^2 + m_{02}c^2), \quad \because \sqrt{1-\frac{v^2}{c^2}} < 1.$$

$$\therefore m_0 > (m_{01} + m_{02}) \quad \textbf{(proved).}$$

Ex. 23 *If one gram of a substance is fully converted into energy, how many calories of heat will be produced?* **[PU 1995]**

Soln. Energy produced,

$$E = m_0c^2 = 10^{-3} \times (3 \times 10^8)^2 = 9 \times 10^{13}\ \text{J}$$

$$[\because m_0 = 1\ \text{gm} = 10^{-3}\ \text{kg and } c = 3 \times 10^8\ \text{m/s}]$$

$$\text{or,} \quad E = \frac{9 \times 10^{13}}{4.2} = \mathbf{2.14 \times 10^{13}\ calories.}$$

$$[\because J = 4.2\ \text{J/cal}].$$

Ex. 24 *Show that for low speed the relativistic KE of a body reduces to classical energy value.*

Soln. The relativistic energy of a particle of mass m moving with velocity v is $E = mc^2$ and the rest mass energy $E_0 = m_0c^2$, m_0 = rest mass of the particle.

The KE of the particle is

$$E_k = E - E_0 = mc^2 - m_0c^2$$

$$\text{or,} \quad E_k = \frac{m_0c^2}{\sqrt{1-\frac{v^2}{c^2}}} - m_0c^2 = m_0c^2\left[\frac{1}{\sqrt{1-\frac{v^2}{c^2}}} - 1\right]$$

$$= m_0c^2\left[\left(1-\frac{v^2}{c^2}\right)^{-1/2} - 1\right] = m_0c^2\left[1 + \frac{v^2}{2c^2} + \frac{3v^4}{8c^4} + \frac{5v^6}{16c^6} + \cdots - 1\right].$$

Now, since $v \ll c$, $\frac{v}{c} \ll 1$, i.e., $\frac{v}{c} \to 0$.

$$\therefore E_k \simeq m_0c^2\left[1 + \frac{v^2}{2c^2} - 1\right] \quad \text{or,} \quad E_k = \frac{1}{2}m_0v^2.$$

This gives the classical value of KE of the particle.

Ex. 25 *Calculate the increase in mass of* 500 gm *of metal of specific heat* 0.15 cal/gm/K *when heated through* 900 °C. **[PU 1993]**

Soln. To heat the metal through 900 °C the heat to be supplied is

$$E = 500 \times 0.15 \times 900 = 6.75 \times 10^4 \text{ cal}$$
$$= 6.75 \times 10^4 \times 4.2 = 28.35 \times 10^4 \text{ J}.$$

Now, the increase in mass is

$$\Delta m = \frac{E}{c^2} = \frac{28.35 \times 10^4}{(3 \times 10^8)^2} = \mathbf{3.15 \times 10^{-12}\ J}.$$

Ex. 26 *Find the rest mass of a photon.*

Soln. Momentum of a particle of energy $h\nu$ is $p = \frac{h\nu}{c}$.
Again, momentum of a particle of mass m moving with velocity v is

$$p = mv = \frac{m_0v}{\sqrt{1 - \frac{v^2}{c^2}}} \quad \text{or,} \quad m_0 = \frac{p\sqrt{1 - \frac{v^2}{c^2}}}{v}.$$

But for a photon, $v = c$, $\therefore$ the rest mass of the photon is

$$m_0 = \frac{p\sqrt{1 - \frac{c^2}{c^2}}}{c} = \mathbf{0}.$$

Ex. 27 *For what value of* $\beta (= \frac{v}{c})$ *will the relativistic mass of a particle exceeds its rest mass by a given fraction* f?

Soln.

$$f = \frac{m - m_0}{m_0} = \frac{m}{m_0} - 1, \quad \therefore \frac{m}{m_0} = \frac{1}{\sqrt{1 - \beta^2}}. \quad \left[\because m = \frac{m_0}{\sqrt{1 - \frac{v^2}{c^2}}}\right]$$

$$\therefore f = \frac{1}{\sqrt{1 - \beta^2}} - 1 \quad \text{or,} \quad f^2 + 1 + 2f = \frac{1}{1 - \beta^2} \quad \text{or,} \quad 1 - \beta^2 = \frac{1}{f^2 + 2f + 1}$$

$$\text{or,} \quad \beta^2 = 1 - \frac{1}{f^2 + 2f + 1} = \frac{f^2 + 2f + 1 - 1}{(1 + f)^2} = \frac{f(f + 2)}{(1 + f)^2}$$

$$\text{or,} \quad \beta = \frac{\sqrt{f(f + 2)}}{1 + f}.$$

Ex. 28 *Calculate the energy equivalent to* 1 amu *in* MeV. [Avogadro Number, $N = 6 \times 10^{23}$/gm-mole

Soln. Here $m = 1\,\text{amu} = \dfrac{1}{6 \times 10^{23}}\,\text{gm} = \dfrac{1}{6 \times 10^{26}}\,\text{kg}.$

$c = 3 \times 10^8\,\text{m/s}$

$$\therefore E = \left(\frac{1}{6 \times 10^{26}}\right) \times (3 \times 10^8)^2 \quad [\because E = mc^2]$$

$$= \frac{9 \times 10^{16}}{6 \times 10^{26}}\,\text{J} = \frac{9 \times 10^{16}}{6 \times 10^{26} \times (1.6 \times 10^{-19})}\,\text{eV}$$

$$= 937 \times 10^6\,\text{eV} = \mathbf{937\,MeV}.$$

EXERCISES

9(A)–Objective Type Questions

Select the correct answer:

9.1. The frame of reference in which a body, not acted upon an external force, experiences an acceleration is (a) inertial frame (b) non-inertial frame (c) moving frame (d) none of these.

9.2. Michelson-Morley experiment was performed to (a) measure speed of light (b) to prove existence of ether (c) to measure speed of the earth relative to ether (d) to test the isotropy of space.
[PU 1994]

9.3. The postulates of special theory of relativity are applicable to (a) stationary frame (b) accelerated frame (c) inertial frame (d) none of these. **[PU 1994]**

9.4. Frame S' moves along the positive X-direction relative to S-frame. A rod placed along Y-axis and observed from S-frame appears (a) contracted (b) elongated (c) unchanged (d) none of these.
[PU 1994]

9.5. Under Galilean transformation the laws of electromagnetism (a) are not valid (b) are found to change their forms (c) are changed in some cases (d) do not change their forms.

9.6. According to Galilean principle of invariance, the laws of mechanics are the same (a) in inertial frames (b) in accelerated frames (c) in stationary frames (d) none of these.

9.7. According to the special theory of relativity, velocity of light in vacuum (a) is independent of the motion of the source observer (b) depends on the motion of the light source (c) depends on the motion of the observer (d) none of these.

9.8. Rest volume L_0^3 of a cubical box corresponds to relativistic volume as
(a) $L_0^3(1-\beta^2)^{1/2}$ (b) $L_0^3(1-\beta^2)^{3/2}$ (c) $L_0^3/\sqrt{1-\beta^2}$ (d) $L_0^3(1-\beta^2)$.

9.9. Lorentz transformation reduces to Galilean transformation when (a) $c = v$ (b) $c > v$ (c) $v \ll c$ (d) none of these.

9.10. Two photons approach each other. Their relative velocity is (a) 0 (b) $< c$ (c) $> c$ (d) $= c$.
[HPU 1993; WBUT 2005]

9.11. If two photons recede from each other, their relative velocity is (a) $c/2$ (b) $2c$ (c) c (d) 0.
[HPU 1993]

9.12. How fast a particle must travel so that its mass becomes twice its rest mass? (a) $0.5c$ (b) $2c$ (c) $\frac{\sqrt{3}}{2}c$ (d) $0.25c$. **[WBUT 2004]**

9.13. The rest mass energy of an electron is (a) 1 MeV (b) 0.51 MeV (c) 100 MeV (d) 5.1 MeV.

9.14. The rest mass of a photon is (a) E/c^2 (b) p/c (c) 0 (d) pc.

9.15. The length of a rod seems contracted when measured by moving observer in the ratio (a) $1:(1-\frac{v^2}{c^2})$ (b) $10:(1-\frac{v^2}{c^2})^{3/2}$ (c) $1:(1-\frac{v^2}{c^2})^{1/2}$ (d) $v:(1-\frac{v^2}{c^2})^{1/2}$.

9.16. A clock goes slower in a (a) rest frame (b) moving frame (c) accelerated frame (d) none of these.

9.17. If a particle moves with velocity $v > c$, then its mass will be (a) real (b) imaginary (c) complex (d) constant.

9.18. If a mass of 1 kg is fully converted into energy, then the energy value in kWh will be (a) 10^{10} (b) 2.5×10^{10} (c) 5×10^{10} (d) 15×10^{10}.

9.19. Relativistic KE of a body reduces to classical KE, if the velocity of the body is such that (a) $v = c$ (b) $v \ll c$ (c) $v > c$ (d) $v = c/2$.

9.20. To increase the velocity of a moving particle, amount of energy to be supplied is (a) 9 (b) ∞ (c) $\frac{mc^2}{2}$ (d) mc^2.

9.21. The relativistic variation of mass of a particle with velocity is a consequence of law of (a) conservation of energy (b) conservation of mass (c) conservation of momentum (d) conservation of mass-energy.

9.22. If the energy of a particle is much higher than its rest energy, then the energy is (a) independent of the momentum (b) proportional to the momentum (c) proportional to the square of the momentum (d) proportional to the square root of the momentum. **[WBUT 2010]**

ANSWERS

9.1. (b) **9.2.** (c) **9.3.** (c) **9.4.** (c) **9.5.** (b) **9.6.** (a) **9.7.** (a) **9.8.** (a) **9.9.** (c) **9.10.** (d) **9.11.** (c) **9.12.** (c) **9.13.** (b) **9.14.** (c) **9.15.** (c) **9.16.** (b) **9.17.** (b) **9.18.** (b) **9.19.** (b) **9.20.** (b) **9.21.** (c) **9.22.** (b).

9(B)–Short Answer Type Questions

9.1. What are inertial and non-inertial frames of reference?

9.2. Write down Galilean transformation equations.

9.3. What is Galilean principle of invariance? Does the law of electromagnetism hold in it?

9.4. What is the negative result of Michelson-Morley experiment? What is the cause of this negative result?

9.5. What are basic postulates of special theory of relativity? **[WBUT 2003, '05]**

9.6. Write the Lorentz transformation relation for time coordinate in a frame moving with velocity v in the Z-direction relative to stationary frame. **[WBUT 2005]**

9.7. When a particle moves with a velocity much less compared to the velocity of light in free space, show that the relativistic expression for KE approaches the classical result. **[WBUT 2007]**

9.8. Two photons approach each other in free space, find the relative velocity of one w.r.t. the other. **[WBUT 2007; BUH 2006]**

9.9. Distinguish between inertial and non-inertial frame of references. **[WBUT 2002]**

9.10. Show that for values of $v \ll c$, Lorentz transformation reduces to the Galilean transformation. **[WBUT 2002]**

9.11. What do you mean by *proper length* and *proper time*?

9.12. Why is it necessary to redefine momentum in relativistic dynamics? **[CUH 1993]**

9.13. A meter ruler is moving parallel to its length. What is its apparent length when its mass is twice its rest mass? **[BUH 2002]**

9.14. Show graphically the variation of relativistic mass with velocity of a particle.

[**Ans.** The relativistic mass of a particle moving with velocity v is given by

$$m = \frac{m_0}{\sqrt{1-\frac{v^2}{c^2}}}, \quad (v < c)$$

where m_0 = rest mass of the particle and c is the velocity of light.

The above relation shows that $m > m_0$ for $0 < v < c$.

If $v = c$, then $m = \frac{m_0}{0} = \infty$, which is impossible. So a material particle cannot have velocity equal to c.

In between $v = 0$ and $v = c$, the mass m of the particle increases enormously as shown in Fig. 9.11.]

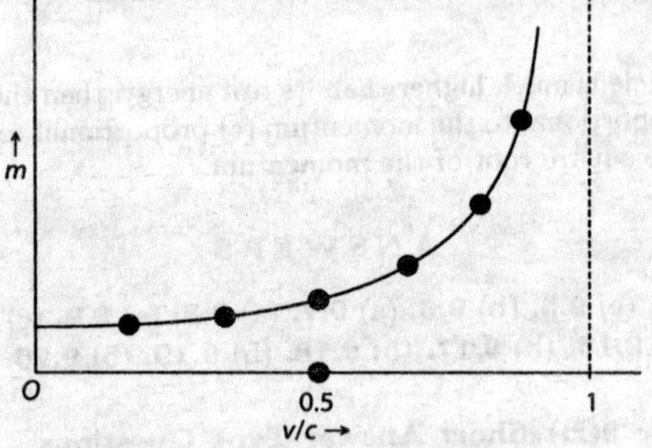

Fig. 9.11: Variation of mass with velocity.

9.15. Is it possible for two objects to have relative velocity larger than c, from the point of view of a third observer? **[BUH 2003]**

9.16. Are two events, which occur simultaneously at x'_1 and x'_2 in S' simultaneous in S? (S and S' are two inertial frames moving relative to each other) **[BUH 2002]**

9.17. A frame S' moves with a uniform velocity relative to S. How will a circle in S' look like in S? **[BUH 2001]**

9.18. Does the slowing down of a moving clock depend upon whether it is moving towards or away from the observer? **[BUH 2001]**

9.19. What is twin paradox in relativity?

[**Ans.** Twin paradox is a consequence of relativistic time dilation

The Twin Paradox. Let twin brothers A and B born at the origin O of an inertial frame on the earth at $t = 0$. Just after his birth the brother B travels along X-axis with a velocity $v = \frac{\sqrt{3}}{2}c$ for 15 years as measured by his own clock and reaches the point P. Then B returns to O with the same speed. So B (in moving frame S') travels for $15+15 = 30$ years from the moment of his birth. So the age of B will be 30 years *by his clock*. $T_0 = 30$ years is the proper time.

Now, the age of B as measured by the clock of A at O will be

$$T = \frac{T_0}{\sqrt{1-\frac{v^2}{c^2}}} = \frac{30}{\sqrt{1-\frac{3}{4}}} = 60 \text{ years.}$$

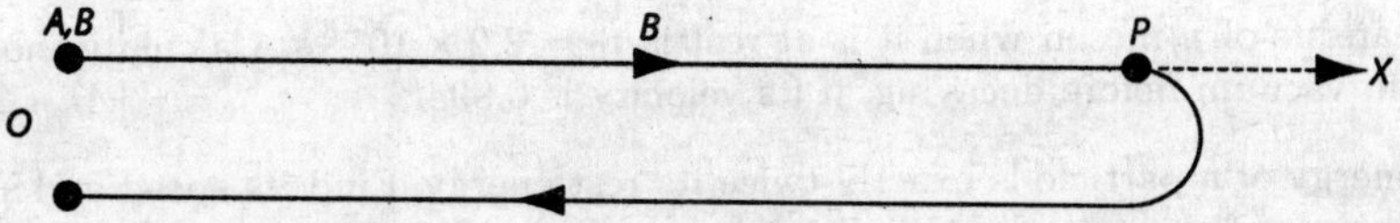

Fig. 9.12

The age of A according to his own clock is also 60 years. But the age of B on return to O is 30 years by his own clock.

Again, A appears to move with $v = \frac{\sqrt{3}}{2}c$ w.r.t. B in opposite direction as B is in moving frame S'. When B returns to O, A will be of 30 years old but B himself will be 60 years. This problem is known as *twin paradox*.

Resolution of the twin paradox. The one-sidedness of the events is the root of the paradox. During the return journey, B had a sudden reversal in direction of velocity with constant magnitude. This leads to arise acceleration and change from one inertial frame to other. In case of B this does not arise. Thus, the complete equivalence of frames of A and B does not prevail. This is a clear contradiction to the basic postulate of relativity. This explains the reason of twin paradox.]

9(C)–Numerical Problems

9.1. Calculate the wavelength of the radiation emitted by the annihilation of an electron with a positron of rest mass 9.1×10^{-31} kg.

[*Hints.* $mc^2 + mc^2 = 2h\nu = \frac{2hc}{\lambda}$.

$$\therefore \lambda = \frac{hc}{mc^2} = \frac{h}{mc} = \frac{6.62 \times 10^{-34}}{9.1 \times 10^{-31} \times 3 \times 10^8} = \mathbf{0.024 \times 10^{-10}\ m}].$$

9.2. If an electron and positron annihilate, calculate energy released, given mass of an electron is 9.1×10^{-31} kg. [*Ans.* 1.02 MeV] **[WBUT 03]**

9.3. Calculate the energy equivalence to one gram of mass. **[WBUT 2005]**

9.4. Two rods of proper length l_0 move lengthwise towards each other parallel to their common axis with the same velocity v relative to the laboratory frame. Show that the length of each rod in the reference frame fixed to the other rod is

$$l = l_0 \frac{1-\beta^2}{1+\beta^2}, \text{ where } \beta = \frac{v}{c}.$$ **[WBUT 2007]**

9.5. Starting from Lorentz transformation show that if (x_1, y_1, z_1, t_1) and (x_2, y_2, z_2, t_2) represent two events, then the quantity

$$(x_1 - x_2)^2 + (y_1 - y_2)^2 + (z_1 - z_2)^2 - c^2(t_1 - t_2)^2$$

is invariant in all inertial frames. **[WBUT 2007]**

9.6. A rigid rod of prime length $L = 1.5$ m is at rest relative to a system S. If the rod makes an angle $\theta = 45°$ w.r.t. X-axis, what will be its length and orientation relative to another frame S moving at $v = 0.98c$? [*Ans.* 1.08 m, $\theta_1 = 78.7°$]

9.7. The total electrical energy generated in a country in a particular year was 7.5×10^{11} kWh. How much mass was converted into energy? [*Ans.* 30 kg] **[DU 85]**

9.8. Two particles of each rest mass 3×10^{-25} kg approaching each other in head on collision. If each particle has an initial velocity of 2×10^{8} m/s, calculate the velocity of one particle as seen by the other. [*Ans.* $u' = 2.769 \times 10^{8}$ m/s, $m = 7.8 \times 10^{-25}$ kg] **[BU 1984]**

9.9. The mean life of μ-meson when it is at rest is $\tau_0 = 2.2 \times 10^{-8}$ s. Calculate the average distance it travels in vacuum before decaying, if its velocity is $0.99c$. [*Ans.* 29.7 m] **[BUH 02]**

9.10. Total energy of a particle is exactly twice its rest energy. Find its speed and show that the result does not depend upon the rest mass of the particle. [*Ans.* $0.866c$]

9.11. Calculate the momentum of a photon, whose energy is 1.00×10^{-12} ergs. [*Ans.* $\frac{1}{3} \times 10^{-27}$ J]

9.12. The density of gold is 19.3×10^{3} kg/m^3 in a rest frame S. Calculate its density that an observer in a frame S' would determine, if the frame S' is moving along the X-axis with a speed of $0.9c$.

[*Hints.* $L = L_0\sqrt{1 - \frac{v^2}{c^2}}$. $\frac{v}{c} = 0.9$. $\therefore L = 0.4359 L_0$. Volume $V_0 = L_0 \times y \times z$, $V = L \times y \times z = 0.4359 V_0$.

Mass $m = \dfrac{m_0}{\sqrt{1 - \frac{v^2}{c^2}}} = \dfrac{m_0}{0.4359}$.

Original density $\rho_0 = \dfrac{m}{V_0}$.

$$\therefore \frac{\rho}{\rho_0} = \frac{m}{V} \times \frac{V_0}{m_0} = \frac{m}{m_0} \times \frac{V_0}{V} = \left(\frac{1}{0.4359}\right)\left(\frac{1}{0.4359}\right).$$

$$\therefore \rho = 101.6 \times 10^{3}\ \text{kg/m}^3].$$

9.13. Show that the rest mass of a particle of momentum p and KE T are related by

$$m = \frac{p^2c^2 - T^2}{2Tc^2}.$$

9.14. Calculate the velocity of electrons accelerated by a potential of 1 MeV. [*Ans.* 2.8×10^{8} m/s]

9.15. How much mass is lost when 1 kg of water at 0 °C turns to ice at 0 °C? [*Ans.* 2.116×10^{-13} kg]

9.16. An electron and a photon both have momenta 2 MeV/c. Find the total energy of each. **[BUH 2001]**

9.17. How much electric energy could theoretically be obtained by annihilation of 1 gm of matter. [*Ans.* $E = 5.5 \times 10^{32}$ eV]

9.18 An electron is observed moving at 50% of the speed of light ($v = 1.5 \times 10^{8}$ m/s): What is the relativistic mass of the electron? What is the KE of the electron? **[WBUT 2010]**

9(D)–Broad Answer Type Questions

9.1. What is a non-inertial frame of reference? What are the basic postulates of special theory of relativity? Using Lorentz transformation derive expression for (a) length contraction (b) time dilation. **[WBUT 2001]**

9.2. (a) What are the basic postulates of special theory of relativity?

(b) If an electron and positron annihilate, calculate the energy released, given that mass of an electron is 9.1×10^{-3} kg.

(c) Deduce Einstein's mass-energy relation $E = mc^2$, considering the mass variation with energy. **[WBUT 2003]**

9.3. (a) Distinguish between *inertial* and *non-inertial frame* of reference.

(b) Show that for values of $v \ll c$, Lorentz transformation reduces to the Galilean transformation.

(c) Deduce the mathematical expression for the law of addition of relativistic velocities.

(d) In laboratory, electrons from two accelerators are projected with the same speed of 2×10^8 m/s in opposite direction. What is their relative velocity? **[WBUT 2002]**

9.4. (a) Write basic postulates of special theory of relativity.

(b) Why is it called special theory of relativity?

(c) Deduce the relation of time dilation from Lorentz transformation.

(d) A plane electromagnetic wave propagates in the medium, which moves with velocity v in X-direction relative to rest frame. Refractive index of light in this medium is n. Find the velocity of wave in the frame moving with constant velocity v relative to rest frame and along X-direction. **[WBUT 2002]**

9.5. (a) What are basic postulates of special theory of relativity?

(b) Establish Einstein's mass energy relation.

(c) Calculate the energy equivalence to 1 gm of mass. **[WBUT 2005]**

9.6. (a) When a particle moves with a velocity much less compared to the velocity of light in free space, show that the relativistic expression of KE approaches the classical limit.

(b) If two photons are approaching each other in free space, find the relative velocity of one photon w.r.t. the other. **[WBUT 2007]**

9.7. Describe the Michelson-Morley experiment and discuss its significance.

[WBUT 2002; CUH 1993, 2003]

9.8. Explain the phenomenon of time dilation. Mention two of its important consequences. **[BUH 2002]**

9.9. Derive Lorentz transformation equations based on the fundamental postulates of the special theory of relativity. **[NBU 2003]**

9.10. Write down Lorentz transformation equations for two inertial frames S and S' moving relative to each other with a velocity v along the common X-axis. **[BU 2003; CUH 1994]**

(a) Show that for small velocities they reduce to Galilean transformation. **[CUH 1991]**

(b) Show that the time interval between two events is minimum in the Lorentz frame in which the events (assuming time like separation) take place at the same place. **[CUH 1994]**

(c) Show that the speed of light is the same in all inertial frames and it cannot be exceeded.

[BUH 2003]

9.11. Show that the KE required by a body is directly proportional to the change in mass due to its motion. **[BHU 2002, '03]**

9.12. Obtain a relation between the total energy and the dynamical mass of a particle. **[BHU 2003]**

9.13. Show that if the variation of mass with velocity is taken into account, the KE of a particle of rest mass m_0 and moving with the velocity v is given by

$$T = m_0c^2\left[\left(1 - \frac{v^2}{c^2}\right)^{-\frac{1}{2}} - 1\right],$$

where c is the velocity of light.

If a particle could move with the velocity of light, how much KE it would possess?

9.14. (a) Derive Einstein relation

$$E^2 = p^2c^2 + m_0^2c^4,$$

where the symbols have their usual significance.

(b) A particle of mass M at rest decays spontaneously into two particles of masses m_1 and m_2. Obtain energies E_1 and E_2 of daughter particles in terms of M, m_1 and m_2.

(c) Show that $E = mc^2$ and $m = \dfrac{m_0}{\sqrt{1-\frac{v^2}{c^2}}}$, the symbols having their usual significances.

9.15. Show that the velocity u of a relativistic particle is given by

$$u = \frac{pc}{\sqrt{p^2 + m_0^2c^4}} \quad \text{and} \quad E = \sqrt{p^2c^2 + m_0^2c^4},$$

where p is the relativistic momentum. Hence, write down formulae for energy and momentum.

[HPU 1994; LU 1995; GNDU 1996]

Chapter 10 | Module 4.2

Quantum Physics–I

10.1 Introduction

Classical physics based on Newton's law of motion successfully explained the motion of objects of *macroscopic dimension* and the related phenomena. Classical theory, however, failed to explain the experimental results of *black body radiation, photoelectric effect, Compton effect, stability of atoms, etc.*

The failure of classical theory to account the above facts led to the discovery of a new theory, which could provide satisfactory explanation to the phenomena related to microscopic bodies like atoms, electrons, etc. The revolutionary new concepts gave birth to *quantum physics*—an entirely new branch of physics. Quantum physics is now regarded as the *foundation of physics.*

10.2 Black Body Radiation

When a radiation passes through matter, a fraction of it is absorbed, some fraction is reflected and the rest is transmitted. If a, b, c represent the respective fraction, then $a + b + c = 1$.

If for a body $b = c = 0$, i.e., the body neither reflects nor transmits, then it absorbs the incident radiation completely and the body appears black whatever be colour of the incident radiation on it. Since a perfect absorber is also a perfect emitter, the body emits radiation of all wavelengths. Such a radiation is called *black body radiation* or *full* (*or total*) *radiation* and the body emitting black body radiation is called a *black body.* In fact, no substance possesses strictly the property of an ideal black body but *lampblack* and *platinum black* behave close to the perfect black body for all practical purposes.

If a black body is placed in an isothermal enclosure, it will attain *thermal equilibrium* with the enclosure and emit *full radiation* characteristic of temperature but independent on the nature of the substance. Hence, the thermal radiation from an isothermal enclosure is identical with that from a black body at the same temperature. That is why black body radiation is sometimes termed as *thermal radiation.*

A hollow copper sphere, coated with lampblack on its inner surface and a fine hole with a pointed projection just in front of the hole serves the purpose of a practical black body [Fig. 10.1]. When radiation enters the hole it suffers multiple reflections and are completely absorbed. Then the body behaves as an *absorption black body.*

When this body is placed in a constant temperature enclosure, the thermal radiation comes out of the whole and the hole acts as an *emission black body* [Fig. 10.2].

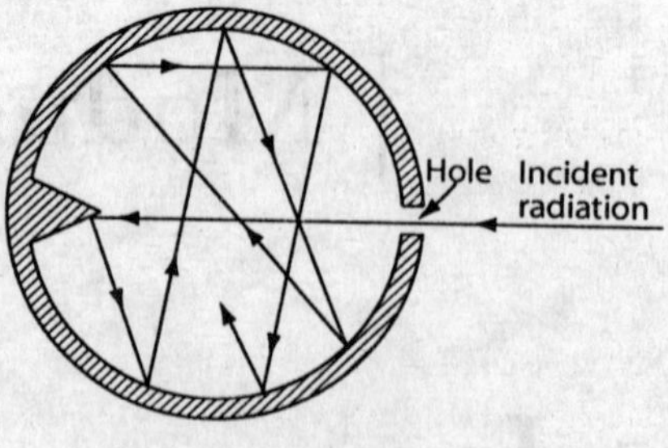

Fig. 10.1: Absorption black body.

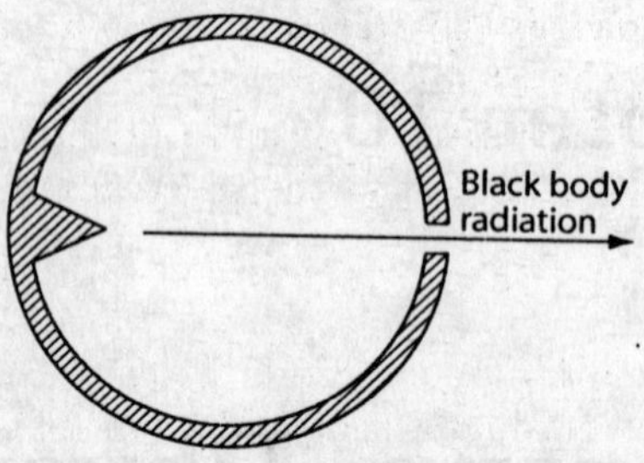

Fig. 10.2: Emission black body.

10.3 Distribution of Energy in the Spectrum of Black Body Radiation

The distribution of thermal energy in the spectrum of black body radiation was experimentally studied by *Paschen, Lummer, Pringsheim,* etc. between the years 1893 to 1899.

The results of their experiments are represented graphically in Fig. 10.3, where the intensity of the emitted radiation (E_λ) is plotted as a function of wavelength (λ) at different temperatures of the black body. The important features of the curves are as below:

(i) *At a given temperature the energy is not uniformly distributed in the radiation spectrum.*

(ii) *At a given temperature the intensity of radiation* (E_λ) *increases with the increase of wavelength* (λ), *attains a maximum value at a particular value of* λ *and then decreases as* λ *increases.*

(iii) *The wavelength corresponding to the maximum* E_λ *decreases with the increase of temperature.*

(iv) *There is an increase in* E_λ *with the increase in temperature at all values of* λ.

(v) *At a particular temperature the area under the curve represents the total energy emitted, which is found to be proportional to the fourth power of temperature* (T), *i.e.,* $E \propto T^4$.

This is *Stefan's law* of total radiation.

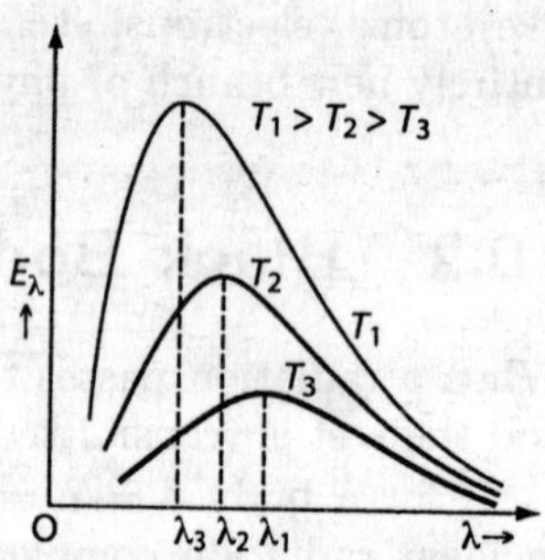

Fig. 10.3: Spectrum of black body radiation.

10.3.1 Wien's Law of Energy Distribution

(1) **Wien's displacement law**: Applying the *laws of thermodynamics,* Wien (in 1893) showed that, the wavelength (λ_m) corresponding to the maximum emission is inversely proportional to the absolute temperature (T) of the black body, i.e.,

$$\lambda_m \propto \frac{1}{T} \quad \text{or,} \quad \lambda_m T = \text{constant } (C). \tag{10.1}$$

The value of the constant $C = 2.898 \times 10^{-3}$ m-K.

The relation (10.1) is known as *Wien's displacement law.* This shows that *with the increase of temperature of the black body the maximum intensity of radiation shifts towards shorter wavelength.*

The total energy density represented by the area under the curve at a particular temperature is

$$E = \int_{\lambda=0}^{\infty} E_\lambda d\lambda = \sigma T^4, \tag{10.2}$$

which is known as the *Stefan-Boltzmann law* of total radiation and $\sigma = 5.57 \times 10^8$ watt/m^2/deg^4 is known as the *Stefan constant.*

(2) **Wien's distribution law**: Applying the *laws of thermodynamics,* Wien (1893) obtained the energy distribution law in the spectrum of black body radiation as

$$E_\lambda d\lambda = \frac{C_1}{\lambda^5} e^{-\frac{C_2}{\lambda T}} d\lambda, \tag{10.3}$$

where C_1 and C_2 are constants. Expression (10.3) is called the *Wien's law* of energy distribution in the spectrum black body radiation. It is to be kept in mind that *Wien's law can explain the energy distribution in the shorter wavelength region alone and breaks down in longer wavelength region* [cf. Fig. 10.4].

10.3.2 Rayleigh-Jeans Law

To overcome the failure of *Wien's distribution law* to account for the experimental results at larger wavelength, *Lord Rayleigh* and *James Jeans* attempted to derive an improved radiation law assuming the following:

(i) *The thermal radiation in the cavity is electromagnetic in nature having perfectly reflecting walls. The incident and reflected waves in the cavity form standing waves.*

(ii) *The number of modes (standing waves) per unit volume in the frequency range* ω *and* $\omega + d\omega$ *is given by*

$$N(\omega)d(\omega) = \frac{\omega^2}{\pi^2 c^3} d\omega. \tag{10.4}$$

$$\because \omega = \frac{2\pi c}{\lambda}, \quad |d\omega| = \frac{2\pi c}{\lambda^2} d\lambda,$$

where λ is the wavelength and c, the velocity of the electromagnetic wave.

$$\therefore N(\lambda)d\lambda = \frac{8\pi}{\lambda^4} d\lambda. \tag{10.5}$$

Again,

$$\because c = \nu\lambda, \lambda = \frac{c}{\nu} \text{ and } |d\lambda| = \frac{c}{\nu^2} d\nu,$$

where $\nu = \frac{2\pi}{\omega}$.

$\therefore$ from eqn (10.5),

$$N(\nu)d\nu = \frac{8\pi}{c^3}\nu^2 d\nu. \tag{10.6}$$

(iii) *According to Maxwell's law of equipartition of energy the average energy per degree of freedom is*

$$\langle\epsilon\rangle = \frac{1}{2}kT,$$

where k is the *Boltzmann constant* and T is the *absolute temperature of the emitter.*

Now, corresponding to the electric field and magnetic field of the electromagnetic wave there are two states of polarization and hence, $\langle\epsilon\rangle = 2 \times \frac{1}{2}kT = kT$.

Thus, the energy density of thermal radiation in the frequency range ω to $\omega + d\omega$ is

$$E(\omega)d\omega = \frac{\omega^2}{\pi^2 c^3} kT d\omega. \tag{10.7}$$

In terms of wavelength,

$$E(\lambda)d\lambda = \frac{8\pi}{\lambda^4} kT d\lambda. \tag{10.8}$$

This is the *Rayleigh-Jeans law* of black body radiation according, which $E(\lambda) \propto \frac{1}{\lambda^4}$, when T is constant.

This law can account the energy distribution only in the long wavelength region and fails in the short wavelength region. Hence, in spite of various attempts the problem remains unsettled. [cf. Fig. 10.4]

10.3.3 Failure of Classical Theory to Explain the Spectral Distribution of Energy

We have seen that Wien's fifth power law $\left(E_\lambda \propto \frac{1}{\lambda^5}\right)$ agrees with experimental results only in the *low wavelength* ($\lambda \to 0$) *region but breaks down in the large wavelength* ($\lambda \to \infty$) *region* [Fig. 10.4].

Rayleigh-Jeans law ($E_\lambda \propto \frac{1}{\lambda^4}$), on the other hand, can account for the distribution of energy *in the long wavelength region and fails completely in the short wavelength.* This result is known as *ultraviolet catastrophe.*

Ultraviolet catastrophe: According to *Rayleigh-Jeans law* the energy density in the black body radiation between wavelengths λ and $\lambda + d\lambda$ at temperature T is

$$E_\lambda d\lambda = \frac{8\pi kT}{\lambda^4} d\lambda \quad \text{or,} \quad E_\lambda = \frac{8\pi kT}{\lambda^4}. \qquad (10.9)$$

Eqn (10.9) shows that, the energy density $E_\lambda \to \infty$, as $\lambda \to 0$, whereas experimental results show that $E_\lambda \to 0$, as $\lambda \to 0$. This implies that *for a given quantity of radiant energy, all the energy is finally confined in the short wavelength (ultraviolet) region.*

This is a *serious discrepancy* between theory and experiment and is known as *ultraviolet catastrophe* or *Jeans' paradox.* This disastrous result is due to the limitation of well established classical laws on the basis of which Rayleigh-Jeans law is derived.

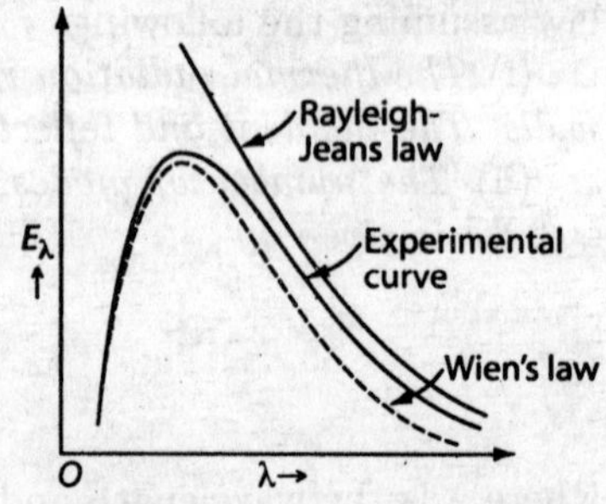

Fig. 10.4: Comparison of Wien's law and Rayleigh-Jeans law with the experimental curve of black body radiation.

10.3.4 Planck's Radiation Law

To provide a satisfactory explanation to the distribution of energy in the spectrum of black body radiation, *Max Planck* (in 1901) put forward his *quantum theory* of radiation.

He assumed that,

(i) *the atoms in the walls of the black body behave like simple harmonic oscillator each having a characteristic frequency of oscillation.*

(ii) *In thermal equilibrium, the emission and absorption of energy by the oscillators occur at equal rate. The emission and absorption of energy take place in the form of discrete packets of energy known as quantum (or, photon). The energy of a photon* (ϵ) *is proportional to its frequency* (ν), *i.e.,* $\epsilon = h\nu$, *where* $h = 6.626 \times 10^{-34}$ J-s *and is called Planck's constant.*

(iii) *The number of oscillators (or degrees of freedom) per unit volume in the frequency range* ν *to* $\nu + d\nu$ *is*

$$N(\nu)d\nu = \frac{8\pi h\nu^2}{c^3} d\nu.$$

Let $N_0, N_1, N_2, \cdots, N_n$ be the number of oscillators with energy $0, \epsilon, 2\epsilon, \cdots, n\epsilon$, then total number of oscillator is

$$N = N_0 + N_1 + N_2 + \cdots + N_n = \sum_{n=0}^{n} N_n \qquad (10.10)$$

and total energy

$$E = 0N_0 + \epsilon N_1 + 2\epsilon N_2 + \ldots + n\epsilon N_n = \sum_{n=0}^{n} n\epsilon N_n. \tag{10.11}$$

According to *Maxwell-Boltzmann distribution function* the number of oscillator in the energy state ϵ_n is $N_n = N_0 e^{-\frac{\epsilon_n}{kT}} = N_0 e^{-\frac{nh\nu}{kT}}$, N_0 being the number of oscillator in the state $\epsilon = 0$, i.e., the *ground state.*

Now, the average energy per mode of standing wave is equal to the average energy $\langle\epsilon\rangle$ of an oscillator given by

$$\langle\epsilon\rangle = \frac{E}{N} = \frac{\sum_{n=0}^{\infty} n\epsilon N_n}{\sum_{n=0}^{\infty} N_n} = \frac{\sum_{0}^{\infty} n\epsilon N_0 e^{-\frac{n\epsilon}{kT}}}{\sum_{0}^{\infty} N_0 e^{-\frac{n\epsilon}{kT}}} = \frac{\sum_{0}^{\infty} nh\nu e^{-\frac{nh\nu}{kT}}}{\sum_{0}^{\infty} e^{-\frac{nh\nu}{kT}}}$$

$$= \frac{h\nu e^{\frac{-n\nu}{kT}} + 2h\nu e^{-\frac{2h\nu}{kT}} + 3h\nu e^{-\frac{3h\nu}{kT}} + \cdots}{e^{-\frac{h\nu}{kT}} + e^{-\frac{2h\nu}{kT}} + e^{-\frac{3h\nu}{kT}} + \cdots}$$

$$= \frac{h\nu e^{-\frac{h\nu}{kT}}\left[1 + 2e^{-\frac{h\nu}{kT}} + 3e^{-\frac{2h\nu}{kT}} + \cdots\right]}{e^{-\frac{h\nu}{kT}}\left[1 + e^{-\frac{h\nu}{kT}} + e^{-\frac{2h\nu}{kT}} + \cdots\right]}$$

$$= \frac{h\nu x[1 + 2x + 3x^2 + 4x^3 + \cdots]}{[1 + x + x^2 + x^3 + \cdots]}, \quad \left[\text{where } x = e^{-\frac{h\nu}{kT}}\right]$$

$$= \frac{h\nu x(1-x)^{-2}}{(1-x)^{-1}} = \frac{h\nu x}{1-x} = \frac{h\nu}{\frac{1}{x} - 1}$$

$$\text{or, } \langle\epsilon\rangle = \frac{h\nu}{e^{\frac{h\nu}{kT}} - 1}. \tag{10.12}$$

Thus, *the average energy per degree of freedom of the oscillator is* $\frac{h\nu}{e^{h\nu/kT}-1}$ *according to quantum theory as against kT according to classical equipartition law.*

Now, the energy density of the black body radiation between frequencies ν and $\nu + d\nu$ is

$$E_\nu d\nu = \frac{8\pi\nu^2 d\nu}{c^3} \times \frac{h\nu}{e^{h\nu/kT} - 1} = \frac{8\pi h\nu^3}{c^3}\frac{1}{e^{h\nu/kT} - 1}d\nu. \tag{10.13}$$

Again, since $c = \nu\lambda$, $\therefore \nu = \frac{c}{\lambda}$ and $|d\nu| = \frac{c}{\lambda^2}d\lambda$. Hence, in terms of wavelength λ, the energy density in the wavelength range λ to $\lambda + d\lambda$ is

$$E_\lambda d\lambda = \frac{8\pi hc}{\lambda^5}\frac{1}{e^{hc/\lambda kT} - 1}d\lambda. \tag{10.14}$$

Eqns (10.13) and (10.14) represent *Planck's radiation law,* which agrees well with experimental energy distribution curve for entire wavelength region of the spectrum.

Deduction of Stefan's law from Planck's Law

According to Planck's law,

$$E_\nu d\nu = \frac{8\pi h\nu^3}{c^3}\frac{1}{e^{h\nu/kT} - 1}d\nu.$$

The total energy density

$$E = \int_0^\infty E_\nu d\nu = \frac{8\pi h}{c^3}\int_0^\infty \frac{\nu^3 d\nu}{e^{h\nu/kT}-1}$$

$$= \frac{8\pi(kT)^4}{c^3h^3}\int_0^\infty \frac{x^3}{e^x-1}dx, \quad \left[\text{putting } \frac{h\nu}{kT} = x\right]$$

$$= \frac{8\pi(kT)^4}{c^3h^3} \times \frac{\pi^4}{15}, \quad \left[\because \int_0^\infty \frac{x^3}{e^x-1} = \frac{\pi^4}{15}\right]$$

$$= \frac{8\pi^5k^4}{15c^3h^3}T^4$$

$$\text{or, } E = AT^4, \tag{10.15}$$

where $A = \frac{8\pi^5k^4}{15c^3h^3}$ = is a constant.

Eqn (10.15) represents *Stefan's law of total radiation.*

Deduction of Wien's displacement law from Planck's law of radiation

Planck's radiation law in terms of λ is

$$E_\lambda d\lambda = \frac{8\pi hc}{\lambda^5}\frac{1}{e^{hc/\lambda kT}-1}d\lambda.$$

According to Planck's law, E_λ is the maximum when $\lambda = \lambda_m$ (say).

Now,

$$E_\lambda d\lambda = \frac{8\pi hc}{\lambda^5(e^{hc/\lambda kT}-1)}d\lambda = \frac{8\pi hc}{y}d\lambda,$$

where $y = \lambda^5(e^{hc/\lambda kT}-1)$.

$$\therefore \frac{dy}{d\lambda} = 5\lambda^4(e^{hc/\lambda kT}-1) - \lambda^5\frac{hc}{\lambda^2kT}\frac{hc}{e^{\lambda kT}}.$$

For $\lambda = \lambda_m$, $\frac{dy}{d\lambda} = 0$.

$$\therefore \left(1 - e^{-\frac{hc}{\lambda_m kT}}\right) = \frac{hc}{5\lambda_m kT}$$

$$\text{or, } 1 - e^{-x} = \frac{x}{5} \quad \left[\text{where } x = \frac{hc}{\lambda_m kT}\right]$$

$$\text{or, } \frac{xe^x}{e^x-1} = 5. \tag{10.16}$$

Eqn (10.16) is a *transcendental equation*, which cannot be solved by analytical method. Solving graphically, we get

$$x = \frac{hc}{\lambda_m kT} = 4.965 \tag{10.17}$$

$$\text{or, } \lambda_m T = 0.29 \text{ cm-K} = \text{a constant.} \tag{10.18}$$

Eqn (10.18) represents *Wien's displacement law.*

Wien's distribution law: We have

$$E_\lambda d\lambda = \frac{8\pi hc}{\lambda^5}\frac{1}{e^{hc/\lambda kT}-1}d\lambda.$$

Now, in the low wavelength region λ is small, so λT is a small quantity.

$$\therefore e^{hc/\lambda kT} - 1 \simeq e^{hc/\lambda kT}.$$

So

$$E_\lambda d\lambda = \frac{8\pi hc}{\lambda^5} e^{-\frac{hc}{\lambda kT}} d\lambda \quad \text{or,} \quad E_\lambda d\lambda = \frac{c_1}{\lambda^5} e^{-\frac{c_2}{\lambda kT}} d\lambda, \tag{10.19}$$

where c_1 and c_2 are two constants.

Deduction of Rayleigh-Jeans Law

In the long wavelength region $\frac{hc}{\lambda kT}$ is small.

$$\therefore e^{hc/\lambda kT} \simeq 1 + \frac{hc}{\lambda kT} \quad \text{[higher powers neglected due to smallness].}$$

$$\therefore E_\lambda d\lambda = \frac{8\pi hc}{\lambda^5} \frac{1}{e^{hc/\lambda kT} - 1} d\lambda = \frac{8\pi hc}{\lambda^5} \frac{1}{\frac{hc}{\lambda kT}} d\lambda = \frac{8\pi hc}{\lambda^5} \times \frac{\lambda kT}{ch} d\lambda$$

$$\text{or,} \quad E_\lambda d\lambda = \frac{8\pi kT}{\lambda^4} d\lambda.$$

This is Rayleigh-Jeans law.

10.4 Photoelectric Effect

Photoelectric effect is a process in which electrons are emitted from a metal surface under the action of a beam of electromagnetic radiation of appropriate frequency. The emitted electrons are called *photoelectrons* and the current produced by the photoelectrons is called *photocurrent.*

The well known photosensitive metals are Na, K, Cs, etc. The photoelectric effect was first discovered by *Heinrich Hertz* in 1887 and subsequent investigations were carried out by *Wallwachs* (1888) and then by *Lenard* (1899).

10.4.1 Experimental Study of Photoelectric Effect

A schematic diagram of the experimental study is shown in Fig. 10.5.

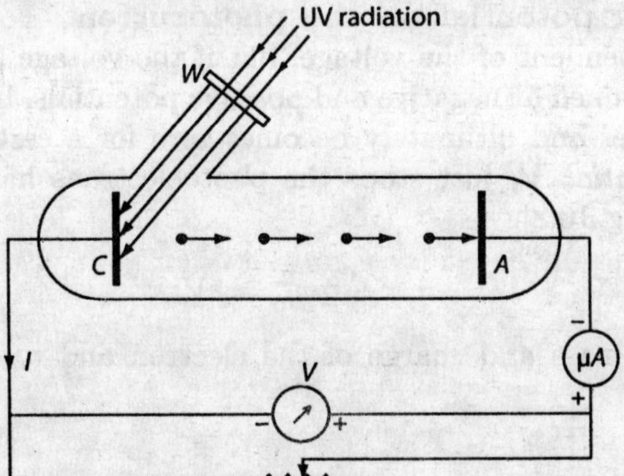

Fig. 10.5: Experimental arrangement for the study of photoelectric effect.

C is a photo sensitive metal plate connected to the negative terminal of battery B and another metal plate A is connected to the positive terminal. A and C thus act as anode and cathode respectively placed in an evacuated quartz tube T fitted with a quartz window W. A and C can be maintained at variable potential difference (P.D.), and ultraviolet radiation falls on C through W.

The emitted photoelectrons are accelerated towards A and constitute a photocurrent (I), which can be measured by the micrometer (μA). The P.D. between A and C can also be measured by the voltmeter (V).

10.4.2 Experimental Observations

(a) **Existence of threshold frequency**: Photoelectrons are not emitted from C if the frequency of the incident radiation falls below a certain minimum frequency ν_0, whatever great the intensity may be. ν_0 is called the *threshold frequency* and is different for different metals [Fig. 10.6].

Definition: *Threshold frequency is the minimum frequency of incident radiation that can cause the emission of photoelectrons with zero KE from the metal surface and below, which no emission occurs.*

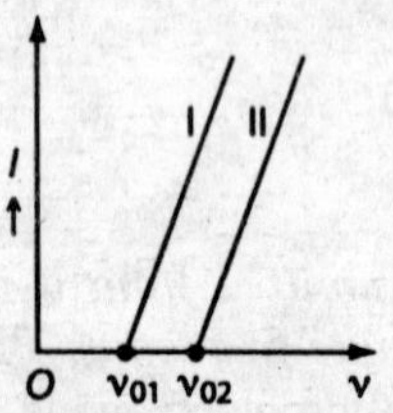

Fig. 10.6: Variation of photocurrent (I) with frequency (ν) of the incident radiation.

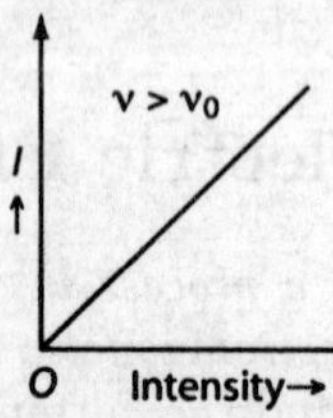

Fig. 10.7: Variation of photocurrent (I) with the intensity of the incident radiation.

(b) **Effect of Intensity on Photocurrent**: If the frequency of the incident radiation is greater than threshold frequency, i.e., $\nu > \nu_0$, then photocurrent I increases linearly with the intensity of the incident radiation [Fig. 10.7].

(c) **Effect of retarding potential on the photocurrent**: For positive anode voltage the photocurrent (I) is independent of the voltage, but if the voltage is reversed, then the anode A and cathode C are maintained at negative and positive potentials. If now the voltage is increased, then photocurrent reduces and ultimately becomes zero for a certain value of voltage, say, V_s, called the *stopping potential*. V_s just stops the photoelectrons having the maximum KE from reaching the anode A [Fig. 10.8].

$$\therefore \tfrac{1}{2}mv_m^2 = eV_s, \tag{10.20}$$

where m and e are the mass and charge of the electron and v_m is the maximum velocity of emission.

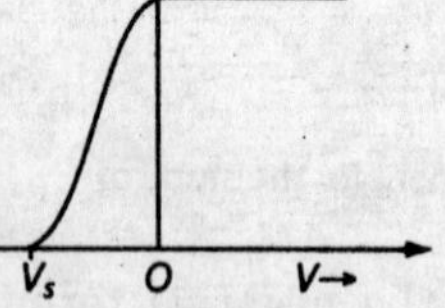

Fig. 10.8: Variation of photocurrent with voltage.

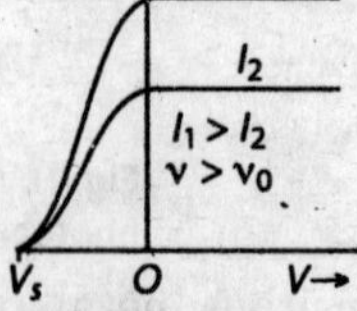

Fig. 10.9: Variation of photocurrent with intensity.

(d) **Effect of intensity on retarding potential**: The retarding potential and hence maximum KE of emission does not depend on intensity of the incident radiation.
If $\nu > \nu_0$, then the photocurrent increases with the increase of intensity [Fig. 10.9].

(e) **Effect of frequency on stopping potential**: The stopping potential of a given metal increases with the increase of frequency $\nu > \nu_0$ of the incident radiation. This shows that the maximum KE of emission increases with frequency ν [Fig. 10.10].

(f) **The photoelectric effect is a spontaneous process,** i.e., there is no time lag between the incidence of light on the metal surface and the emission of photoelectrons.

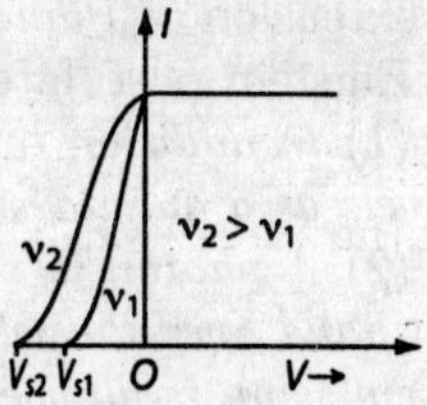

Fig. 10.10: Variation of photocurrent with frequency.

10.4.3 Characteristics (Laws) of Photoelectric Effect

Based on the experimental observations, following characteristics of photoelectric effect are noted:

(1) *There exists a minimum frequency* (ν_0) *of the incident radiation below, which photoelectric effect does not occur.* ν_0 *depends on the nature of emitting metal.*

(2) *If* $\nu > \nu_0$, *then photocurrent is directly proportional to the intensity of the incident radiation.*

(3) *If* $\nu > \nu_0$, *then the velocity and hence the KE of the photoelectrons range from zero to a certain maximum value.*

(4) *The maximum velocity* (ν_m) *is independent of intensity but depends on frequency of the incident radiation.*

(5) *Photoelectric effect is an instantaneous process. There is no time lag between the incidence of radiation and emission of photoelectrons.*

10.4.4 Failure of Classical Theory to Explain Photoelectric Effect

Classical theory assumes that *radiation is electromagnetic waves.* When such a radiation falls on the metal surface, the varying electric intensity acts on the free electrons, which are accelerated and gather energy to be emitted from metal surface.

This *classical theory of radiation fails to explain the main three characteristics of photoelectricity.*

(i) According to wave theory, the kinetic energy (KE) of the emitted electrons should increase with the increase of intensity of the incident radiation. But experimentally KE of the photoelectrons is independent on the intensity.

(ii) If the intensity of the incident radiation be sufficient, then according to wave theory, photoelectrons should be emitted at any frequency. But experiment shows that, if the frequency be smaller than threshold frequency ν_0, photoelectric effect does not occur whatever be the intensity of incident radiation.

(iii) In view of classical wave theory, energy of the incident radiation is spread over the wavefront. Since the energy of the radiation incident on any electron is very small, some time will be required by an electron to absorb required energy for emission. But experimentally, the photoelectric effect is *spontaneous* without any *time lag* between the incidence of radiation and emission of electrons.

Hence, classical theory is *unable to explain the phenomenon of photo-electricity.*

10.4.5 Einstein's Photoelectric Effect

Einstein, in 1905, argued that the observed facts of photo-electricity could be accounted for by an extension of *Planck's quantum theory.*

Einstein postulated that,

(1) *In addition to emission or absorption of radiation as discrete quanta of energy* $\epsilon = h\nu$, *it travels as a quanta with the velocity of light. A photon is a bundle of localized energy in space.*

(2) *A photon is a concentrated form of energy* $h\nu$, *which cannot be split up into fraction. It can hand over the whole of its energy to an electron at a moment and no fractional absorption or emission is allowed.*

(3) *The electrons are bound in the metal and as such require some energy, called the work function* W_0, *to be emitted from it.*

Photoelectric equation: A free electron in the metal gains energy $h\nu$ by absorbing a photon. This energy of the electron is utilized in the following two ways: (i) *A part of the absorbed energy is used to liberate the electron from the metal surface.* (ii) *The remaining part of energy provides KE to the emitted electron.*

Thus, according to conservation of energy,

$$h\nu = W_0 + \tfrac{1}{2}mv_m^2, \tag{10.21}$$

where W_0 = the *work function of the metal* and v_m = *the maximum velocity of the photoelectron.*

If the energy of the incident photon is just sufficient to liberate electron from the metal, then $v_m = 0$ and the electron after liberation will stick to the metal surface.

From eqn (10.21),

$$h\nu_0 = W_0 + 0 \quad \text{or,} \quad W_0 = h\nu_0 \quad [\nu_0 \text{ is the threshold frequency}].$$

$\therefore$ from eqn (10.21), using $W_0 = h\nu_0$, we get

$$h\nu = h\nu_0 + \tfrac{1}{2}mv_m^2 \quad \text{or,} \quad \tfrac{1}{2}mv_m^2 = h(\nu - \nu_0). \tag{10.22}$$

Eqn (10.22) is the *Einstein's photoelectric equation.*

If V_s be the magnitude of the negative potential just to stop the highest energy electron to reach the anode, then

$$eV_s = \tfrac{1}{2}mv_m^2 \quad \text{and so} \quad eV_s = h(\nu - \nu_0), \tag{10.23}$$

V_s is the *stopping potential* and is higher for higher value of ν.

Discussion: From eqns (10.22) and (10.23), we note that

(i) If $\nu < \nu_0$, the KE of the photoelectrons is negative, which is *impossible* and photoelectron emission cannot take place. So for the emission of photoelectrons there must be a threshold frequency ν_0.

(ii) The maximum KE of the photoelectrons depends on the frequency ν and not on intensity of the incident radiation. That is why $\frac{1}{2}mv_m^2$ increases with the increase of ν.

(iii) If keeping $\nu > \nu_0$, the intensity of the incident radiation be increased, then the maximum KE will remain the same, but due to increased number of photons, the number of emitted electrons will increase leading to increased photocurrent.

(iv) As the energy of the incident photon is transferred *instantaneously* to the free electron in the metal, there is *no time lag* between incidence of radiation and emission of photoelectrons.

10.5 The Compton Effect

When a beam of monochromatic X-rays is scattered by loosely bound electrons of light elements (like carbon), the scattered radiation in a particular direction contains two components—one of *lower frequency* called *modified radiation* and the other of the *same frequency* called the *unmodified radiation.* This effect was observed by *A.H. Compton* in 1922 and the explanation was given by him in 1923 and is known as *Compton effect.*

The wavelength difference ($\Delta\lambda$) between the *unmodified* and *modified radiation* is called the *Compton shift.* $\Delta\lambda$ *is independent of the wavelength of the incident radiation and of the nature of the scatter but depends on the angle of scattering.*

Experimental arrangement: The experimental arrangement to study *Compton shift* is diagrammatically shown in Fig. 10.11.

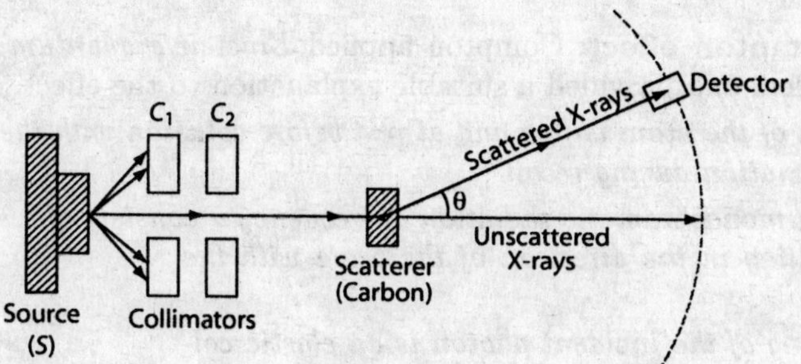

Fig. 10.11: Experimental arrangement for the study of Compton scattering

Monochromatic X-rays from the source S is collimated by the collimating slits C_1 and C_2 and made incident on the carbon scatterer. The intensity of scattered X-rays was measured by the detector for various angle of scattering θ. The intensity variation of scattered X-rays is shown in Fig. 10.12.

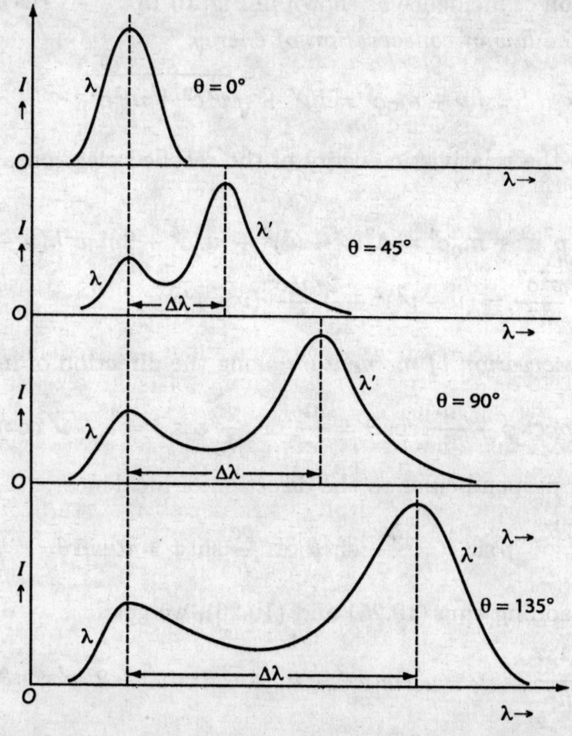

Fig. 10.12: Intensity vs. wavelength plot in Compton scattering.

The results show that

(i) For each value of the angle of scattering, well-defined intensity peaks arise. For $\theta = 0°$, only one peak appears at λ corresponding to incident radiation.

(ii) For $\theta = 45°, 90°$ and $135°$ two peaks appear and the value of $\Delta\lambda$ increases with the scattering angle θ.

(iii) The Compton shift is $\Delta\lambda = \lambda_0(1 - \cos\theta)$, where $\lambda_0 = \frac{h}{m_0 c}$ is called the *Compton wavelength.*

Theory of Compton effect: Compton applied *Einstein's quantum theory of radiation* to the scattering problem and provided a suitable explanation to the effect. He assumed that,

(1) *The electron of the atom is free and at rest before collision with the photon and undergoes relativistic mass variation during recoil.*

(2) *The incident monochromatic radiation of frequency ν consists of a stream of photons, each of energy $h\nu$ travelling in the direction of the beam with the speed of light.*

(3) *The scattering of the incident photon is an elastic collision of the photon with the electron of the target atom and as such the law of conservation of momentum and energy holds good.*

Let a photon of energy $h\nu$ collides with an electron at rest. The photon will hand over some of its energy to the electron during collision and will be scattered with less energy $h\nu'$ along a direction making an angle θ with the direction of incidence. The electron will recoil in a direction making an angle ϕ with the direction of incidence as shown in Fig. 10.13.

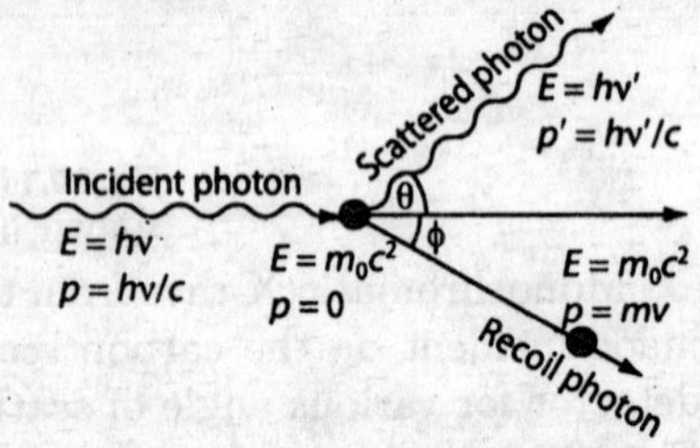

Fig. 10.13: Compton scattering.

Then according to the *law of conservation of energy,*

$$h\nu + m_0c^2 = h\nu' + \sqrt{p^2c^2 + m_0^2c^4},$$

where $\sqrt{p^2c^2 + m_0^2c^4}$ is the relativistic energy of the recoiled electron and m_0, the rest mass of the electron.

$$\text{or, } p^2c^2 + m_0^2c^4 = h^2(\nu - \nu')^2 + m_0^2c^4 + 2m_0c^2h(\nu - \nu')$$

$$\text{or, } \frac{p^2c^2}{h^2} = (\nu - \nu')^2 + \frac{2m_0c^2}{h}(\nu - \nu'). \tag{10.24}$$

From the *law of conservation of momentum* along the direction of incidence gives

$$p\cos\phi + \frac{h\nu'}{c}\cos\theta = \frac{h\nu}{c} \text{ or, } \frac{pc}{h}\cos\phi = \nu - \nu'\cos\theta \tag{10.25}$$

and along the direction perpendicular to the direction of incidence,

$$p\sin\phi = \frac{h\nu'}{c}\sin\theta \text{ or, } \frac{pc}{h}\sin\phi = \nu'\sin\theta. \tag{10.26}$$

Now, squaring and adding eqns (10.25) and (10.26), we get

$$\frac{p^2c^2}{h^2} = (\nu - \nu'\cos\theta)^2 + \nu'^2\sin^2\theta = \nu^2 - 2\nu\nu'\cos\theta + \nu'^2$$

$$\text{or, } \frac{p^2c^2}{h^2} = (\nu - \nu')^2 + 2\nu\nu'(1 - \cos\theta). \tag{10.27}$$

Now, equating the right-hand side eqns (10.24) and (10.27), we get

$$\frac{2m_0c^2}{h}(\nu - \nu') = 2\nu\nu'(1 - \cos\theta)$$

$$\text{or,}\ \frac{\nu - \nu'}{\nu\nu'} = \frac{h}{m_0c^2}(1 - \cos\theta)$$

$$\text{or,}\ \frac{1}{\nu'} - \frac{1}{\nu} = \frac{h}{m_0c^2}(1 - \cos\theta) \qquad (10.28)$$

$$\text{or,}\ \frac{\lambda'}{c} - \frac{\lambda}{c} = \frac{h}{m_0c^2}(1 - \cos\theta) \quad [\because c = \nu\lambda]$$

$$\text{or,}\ \lambda' - \lambda = \Delta\lambda = \frac{h}{m_0c}(1 - \cos\theta) \qquad (10.29)$$

$$\text{or,}\ \Delta\lambda = \frac{2h}{m_0c}\sin^2\frac{\theta}{2}. \qquad (10.30)$$

Eqn (10.29) or (10.30) gives the *Compton shift* $\Delta\lambda$.
Now,

$$\frac{h}{m_0c} = \frac{6.626 \times 10^{-34}}{(9.11 \times 10^{-31}) \times (2.99 \times 10^8)} = 0.02433\ \text{Å}.$$

$$\therefore\ \Delta\lambda = 0.02433(1 - \cos\theta). \qquad (10.31)$$

The experimental value of $\Delta\lambda$ is in excellent agreement with the theoretical value.
The *Compton wavelength* is thus

$$\lambda_0 = \frac{h}{m_0c} = 0.02433\ \text{Å}.$$

Now,

$$E = h\nu = \frac{hc}{\lambda_0} = \frac{hc}{h/m_0c} = m_0c^2.$$

Thus, λ_0 is the wavelength of a photon whose energy is equal to the rest mass energy (m_0c^2) of the scattering electron.

Conclusion: (i) *$\Delta\lambda$ is the maximum for $\theta = 180°$ and $\Delta\lambda = 0$ for $\theta = 0°$*. (ii) *$\Delta\lambda$ is independent on λ and the nature of the scattering material.* (iii) *If the target electron is bound to the carbon atom, then m_0 should be replaced by M, the mass of the atom in eqn* (10.30), *i.e.*,

$$\Delta\lambda = \frac{2h}{Mc}\sin^2\frac{\theta}{2}.$$

Since $M = 12 \times 1840 \times m_0$ (for carbon),

$$\therefore\ \Delta\lambda = \frac{2}{12 \times 1840} \times \left(\frac{h}{m_0c}\right)\sin^2\left(\frac{180°}{2}\right) \quad [\text{for maximum value of } \Delta\lambda,\ \theta = 180°]$$

$$= 9.058 \times 10^{-5} \times 0.02433\ \text{Å}$$

$$= 2.2 \times 10^{-6}\ \text{Å, which is negligible.}$$

Hence, the *Compton shift is negligible if the target is a bound electron.*

10.5.1 The Direction of the Recoil Electron

In Compton effect the *scattered photon* of energy $h\nu'$ and the *recoil electron* of energy $E = mc^2$ move along direction making angle θ and ϕ respectively w.r.t. to the direction of incidence of photon of energy $h\nu$ after impact.

According to the *law of conservation of momentum along and perpendicular to the direction of incidence*, we have

$$\frac{h\nu}{c} + 0 = \frac{h\nu'}{c}\cos\theta + mv\cos\phi \quad \text{or,} \quad mvc\cos\phi = h\nu - h\nu'\cos\theta \tag{10.32}$$

and

$$0 = \frac{h\nu}{c}\sin\theta - mv\sin\phi \quad \text{or,} \quad mvc\sin\phi = h\nu\sin\theta. \tag{10.33}$$

Dividing eqn (10.33) by (10.32), we get

$$\frac{\sin\phi}{\cos\phi} = \tan\phi = \frac{h\nu'\sin\theta}{h\nu - h\nu'\cos\theta} = \frac{\frac{hc}{\lambda'}\sin\theta}{\frac{hc}{\lambda} - \frac{hc}{\lambda'}\cos\theta} = \frac{\sin\theta}{\frac{\lambda'}{\lambda} - \cos\theta}$$

$$\text{or,} \quad \tan\phi = \frac{\sin\theta}{\left[\frac{\frac{h}{m_0c}(1-\cos\theta)+\lambda}{\lambda}\right] - \cos\theta} \qquad \left[\because \lambda' - \lambda = \frac{h}{m_0c}(1-\cos\theta)\right]$$

$$= \frac{\sin\theta}{\frac{h\nu}{m_0c^2}(1-\cos\theta) + (1-\cos\theta)} \qquad \left[\because \lambda = \frac{c}{\nu}\right]$$

$$= \frac{2\sin\frac{\theta}{2}\cos\frac{\theta}{2}}{\left(\frac{h\nu}{m_0c^2}+1\right)(1-\cos\theta)} = \frac{2\sin\frac{\theta}{2}\cos\frac{\theta}{2}}{\left(\frac{h\nu}{m_0c^2}+1\right)2\sin^2\frac{\theta}{2}}.$$

$$\therefore \tan\phi = \frac{\cot\frac{\theta}{2}}{\left(1+\frac{h\nu}{m_0c^2}\right)}. \tag{10.34}$$

Relation (10.34) gives the *direction of the recoil electron after scattering.*

10.5.2 Kinetic Energy of the Recoil Electron

We have the KE of the incident photon $= h\nu$
and the KE of the scattered photon $= h\nu'$.
$\therefore$ KE of the recoil electron is

$$E_r = h\nu - h\nu' = \frac{hc}{\lambda} - \frac{hc}{\lambda+\Delta\lambda} \qquad [\because \lambda' = \lambda + \Delta\lambda]$$

$$= hc\left[\frac{\lambda+\Delta\lambda-\lambda}{\lambda(\lambda+\Delta\lambda)}\right] = hc\left[\frac{\Delta\lambda}{\lambda(\lambda+\Delta\lambda)}\right]$$

$$= hc\left[\frac{\frac{h}{m_0c}(1-\cos\theta)}{\lambda\left\{\lambda+\frac{h}{m_0c}(1-\cos\theta)\right\}}\right]$$

$$= \frac{hc}{\lambda}\left[\frac{\frac{h}{m_0c}(1-\cos\theta)}{\lambda\left\{1+\frac{h}{m_0c\lambda}(1-\cos\theta)\right\}}\right]$$

$$\text{or,} \quad E_r = h\nu\left[\frac{\left(\frac{h\nu}{m_0c^2}\right)(1-\cos\theta)}{\left\{1+\frac{h\nu}{m_0c^2}(1-\cos\theta)\right\}}\right] \qquad \left[\because \lambda = \frac{c}{\nu}, \therefore c\lambda = \frac{c^2}{\nu}\right]$$

$$= h\nu\frac{\alpha(1-\cos\theta)}{1+\alpha(1-\cos\theta)}, \tag{10.35}$$

where $\alpha = \dfrac{h\nu}{m_0c^2}$.

Eqn (10.35) gives the *expression for the KE of the recoil electron.*
E_r is the maximum E_{rm}, when $\theta = 180°$.
So

$$E_{rm} = \frac{2\alpha h\nu}{1+2\alpha}. \tag{10.36}$$

Again, expressing θ in terms of ϕ with the help of eqns (10.34) and (10.35), we get

$$E_r = h\nu \frac{2\alpha \cos^2\phi}{(1+\alpha)^2 - \alpha^2\cos^2\phi}. \tag{10.37}$$

For $\phi = 0$,

$$E_r = E_{rm} = 2\alpha h\nu/(1+2\alpha).$$

10.6 Failure of Classical Theory to Explain Compton Effect

In view of classical theory, the incident X-rays are nothing but a wave train of electromagnetic waves. The electric intensity $\left(\vec{E} = \vec{E}_0 \sin\omega t\right)$ of the incident wave acts on the scattering target electron of the material and *forced vibration* is set up. The electron oscillates with the frequency ω of the incident radiation. The electron being an accelerated charged particle, radiates *electromagnetic radiation* uniformly in all directions and the scattering is *spherically symmetric.* Thus, the scattered radiation is independent of θ and of the same wavelength (λ) as the incident radiation.

Thus, the *existence of modified radiation of wavelength (λ') and hence the Compton shift remained unexplained according to classical theory.*

10.7 Explanation of Compton Effect on the Basis of Quantum Theory

In view of quantum theory the incident X-radiation is a shower of photon of energy $\epsilon = h\nu = \frac{hc}{\lambda}$, where λ is its wavelength. The photon on collision with the target electron at rest hands over some of its energy to it and scatters with reduced energy $\epsilon' = h\nu' = \frac{hc}{\lambda'}$, λ' being the wavelength of the scattered radiation.

The Compton shift is given by

$$\Delta\lambda = \lambda' - \lambda = \frac{h}{m_0 c}(1-\cos\theta).$$

(a) **Scattering by free electron**: When the incident photon collides with a free electron, we get modified radiation of wavelength

$$\lambda' = \lambda + \frac{h}{m_0 c}\ (1-\cos\theta)$$

along with unmodified radiation of wavelength λ.

(b) **Scattering by tightly bound electron**: If the target electron is tightly bound to the atom, then the whole atom is involved in the collision process and the mass m_0 should be replaced by the mass of the atom $M(M \gg m_0)$ in the expression of Compton shift. Calculation shows that the Compton shift $\Delta\lambda$, then becomes too small to be detected and $\lambda' = \lambda$, i.e., there is no unmodified radiation.

(c) **Scattering by low atomic number elements**: In low atomic number elements the number of electrons is less and are loosely bound to the nucleus. Such electrons may be supposed to be free electrons. The scattering by these elements (like carbon) gives rise to both modified and unmodified radiation.

(d) **Scattering by heavy elements**: In heavy elements the number of tightly bound electrons is more than the loosely bound electrons. Scattering with tightly bound electrons give rise to unmodified radiation and loosely bound electrons give rise to modified radiation. The intensity of the unmodified radiation in heavy elements is, therefore, more than that of the light elements.

10.8 Dual Nature Light (Wave Particle Dualism)

The optical phenomena like *interference, diffraction* of light, etc. are the result of superposition of light waves, i.e., *interaction of light with light.* The interpretation of the experimental results of interference, diffraction is provided by assuming the wave nature of light. But the phenomena like *black body radiation, photoelectric effect, Compton effect,* which are the result of interaction of *radiation with matter* cannot be understood on the basis of wave theory, or the electromagnetic theory of radiation. In order to provide satisfactory interpretation to the above phenomena, the quantum theory is the only help. According to quantum theory, radiation consists of *shower of quanta or photons* each of energy $\epsilon = h\nu$.

We thus conclude that light plays *dual role,* sometimes wave, sometimes particle depending on the situation. This is what is called *dual nature of light or wave-particle duality.*

Luis de Broglie of France showed in 1924 that this dualism is not confined to radiation only. It holds good for matter as well. According to him a *material particle of mass m moving with velocity v having momentum $p = mv$ is associated with a wave of wavelength λ given by $\lambda = \frac{h}{p}$, where h is the Planck's constant.* The associated wave with the material particle is called *matter wave* and the relation is known as *de Broglie relation.*

10.9 Bohr's Principle of Complementarity

The particle and wave characters of photons are *complimentary* and are *mutually exclusive.* In some experiments like *black body radiation, photoelectric effect, Compton effect,* the particle character is revealed, the wave character is totally absent, whereas in other experiments like, *interference, diffraction, polarization, etc.* the wave character is revealed and the particle character is totally absent. This mutual exclusiveness of particle nature and wave nature is known as the *principle of complementarity.* This principle was first enunciated by *N. Bohr* and is known as the *Bohr's principle* of complementarity.

10.10 Answers of Some Important Questions

10.1. *What led to the discovery of quantum theory?*

Ans. Classical theory based on *Newton's laws of motion* failed to explain the experimental results of *black body radiation, photoelectric effect, Compton effect, stability of atom,* etc. The failure of classical theory to account for the above experimental results led to the discovery of a new theory—the *quantum theory. Max Planck* in 1901 proposed *quantum theory,* which could successfully explain the distribution of energy in the *spectrum of black body radiation* and thus the door of modern physics was opened.

10.2. *What do you mean by black body and black body radiation?* **[CUH 1992, '95, 2002]**

Ans. A black body is a substance, which can absorb radiations of all wavelengths incident on it and emits radiations of all wavelengths when heated.

A perfect black body is an *ideal concept.*

The radiation emitted by a black body consists of *all wavelengths* and is called a *black body radiation* (or *full radiation* or *total radiation*).

10.3. *What are the characteristics of black body radiation?*

Ans. Following are the important characteristics of black body radiation:

(i) Black body radiation consists of nearly all wavelengths.

(ii) Black body radiation resembles in many ways a perfect gas. Perfect gas consists of molecules of KE from 0 to ∞ and black body radiation consists of photons of energy $E = 0$ to ∞.

(iii) Black body radiation is isotropic and homogeneous.

(iv) The energy density of black body radiation depends only on the temperature of the emitter.

(v) The character of the emitted and absorbed black body radiation is the same.

10.4. *Define emissive power and absorptive power of a black body. What is Kirchhoff's law of black body radiation?* **[CUH 1992]**

Ans. Emissive power: It is defined as the energy emitted normally from unit area of the surface of the emitter per unit time through unit solid angle.

If $u_\lambda d\lambda$ be the energy radiated normally from an elementary area dS in elementary solid angle $d\omega$ in time dt, then

$$e_\lambda d\lambda = \frac{u_\lambda d\lambda}{dS d\omega dt}.$$

Here, e_λ represents the *emissive power* in the wavelength range λ to $\lambda + d\lambda$.

Absorptive power: It is defined as the ratio of the absorbed radiation (Q'_λ) to the incident (Q_λ) on the body, i.e., absorptive power

$$a_\lambda = \frac{Q'_\lambda}{Q_\lambda}.$$

Kirchhoff's law: It is defined as the ratio of the emissive power to the absorptive power of a body at any temperature and for the same wavelength and is equal to the emissive power of a perfect black body, i.e.,

$$\frac{e_\lambda}{a_\lambda} = E_b,$$

where E_b = emissive power of black body.

10.5. *What are the basic differences between Wien's law and Rayleigh-Jeans law of black body radiation?*

Ans. Following are the basic differences between the above two laws:

(i) Wien's law was deduced by considering black body radiation to be longitudinal waves. But in the derivation Rayleigh-Jeans law, radiation was supposed to be transverse electromagnetic waves.

(ii) According to Wien's theory the energy density of wavelength λ, i.e., $E_\lambda \propto \frac{1}{\lambda^5} e^{-\frac{c_2}{\lambda T}}$, where C_2 is a constant and T is the absolute temperature of the emitter.

But according to Rayleigh-Jeans law

$$E_\lambda \propto \frac{1}{\lambda^4}.$$

(iii) Wien's law can expiain the distribution of energy in the short wavelength region of the black body energy distribution curve and fails in the long wavelength region.

Rayleigh-Jeans law is valid only in the long wavelength region and leads to ultraviolet catastrophe in the short wavelength region of the black body spectrum.

10.6. *Explain what is meant by ultraviolet catastrophe.*

Ans. See Section 10.3.3 (p. 264).

10.7. *State Stefan-Boltzmann law.*

Ans. The law states that, the total thermal energy (Q) emitted per unit area per second from a perfect black body is proportional to the fourth power of its absolute temperature (T), i.e., $Q = \sigma T^4$, where $\sigma = 5.672 \times 10^5$ erg/cm^2/K^4 is the Stefan's constant.

If a body at temperature T be surrounded by another black body at absolute temperature T_1, then the amount of heat lost per second per unit area of the former is $Q = \sigma(T^4 - T_1^4)$.

10.8. *Clarify the basic difference in the formulation point of view of Wien's law, Rayleigh-Jeans law and Planck's law of radiation.*

Ans. The basic difference in the formulation point of view of the above three laws are as follows:

(i) *Wien's law* and *Rayleigh-Jeans law* were derived on the classical approach that energy of black body radiation is *continuous*, whereas *Planck's law* assumed that energy is *discrete*.

(ii) Wien assumed thermal radiation to be *longitudinal wave*, whereas Rayleigh and Jeans assumed it to be *transverse electromagnetic wave*.

But Planck assumed that thermal radiation is *discrete and consists of photons*. Photons are particles of concentrated energy $\epsilon = h\nu$, where ν is the frequency of radiation and h is the Planck's constant. These are unbreakable in any situation.

Planck's law of radiation corrected the *radical wrong* in Wien's law and Rayleigh-Jeans' law and removed *ultraviolet catastrophe*.

10.9. *What is photoelectric effect? How does classical theory fail to explain this effect?*

[cf. BUH 2005; CUH 1990; NBUH 2003]

Ans. See Sections 10.4 (p. 267) and 10.4.4 (p. 269).

10.10. *Discuss how the photon concept explains the observed facts of photo-electricity.*

[CUH 1990]

Ans. See Section 10.4.5 (p. 270).

10.11. *Explain why electron cannot be scattered at an angle greater than* 90°.

[NBUH 2003; CUH 1996]

Ans. The relation between the angle of recoil of the electron (ϕ) and angle of scattering of photon (θ) is

$$\tan\phi = \frac{\cot\frac{\theta}{2}}{\left(1 + \frac{h\nu}{m_0c^2}\right)}.$$

From the above relation, if $\theta = 0$, then $\tan\phi = \infty$ and $\phi = \pi/2$.

And if $\theta = \pi$, $\tan\phi = 0$ and $\phi = 0$.

Since $0 \leq \theta \leq \pi$, the value of ϕ must be within 0 and $\pi/2$. Hence, the electron cannot be scattered at an angle greater than 90°.

10.12. *Why Compton effect cannot be observed with visible light?* **[WBUT 2008]**

Ans. Compton effect is the phenomenon of scattering of photon with *free electron* (or, loosely bound electron). Now, the energy of a photon of visible light ($\lambda = 4000 \times 10^{-10}$ m to 7500×10^{-10} m) is extremely smaller than the binding energy of an electron even for the lightest element hydrogen. The energy of a photon of $\lambda = 6000 \times 10^{-10}$ m is 2 eV, whereas the binding energy of the electron in hydrogen is 13.6 eV. That is why this electron cannot be regarded as a free electron. On the other hand, the energy of an X-ray photon is a few thousand times greater than that of a visible light photon and as such the outer electron of light elements like carbon can be regarded as free electrons. For this reason Compton effect is not possible with visible light photon.

Further more, the maximum value of Compton shift $\Delta\lambda \sim 1.1 \times 10^{-8}$ m and to observe Compton shift with optical instrument it must have resolving power

$$\frac{\lambda}{\Delta\lambda} = \frac{6000 \times 10^{-10}}{1.1 \times 10^{-8}} \simeq 54,$$

which is beyond the limit of resolution of an optical instrument.

10.13. *Why Compton effect cannot be observed with macroscopic bodies?*
Ans. Compton shift, $\Delta\lambda = \frac{h}{m_0 c}(1 - \cos\theta)$. For a particular angle of scattering of the photon θ is constant and $\Delta\lambda \propto \frac{1}{m_0}$, where m_0 is the rest mass of the free electron.

For macroscopic bodies the mass $M \to \infty$ compared to the mass $m_0 = 9.1 \times 10^{-31}$ kg. So if macroscopic bodies be the target, then $\Delta\lambda \to 0$. That is why *Compton effect cannot be expected with macroscopic bodies.*

10.14. *Discuss the limit of applicability of classical theory.*
Ans. The dimensions of a physical quantity is called *action.* Planck's constant ($h = 6.626 \times 10^{-34}$ J-s) has the dimensions [energy] × [times] = [length] × [momentum]. The Planck's constant is the *fundamental quantum of action.*

If the value of dimension of action of any physical system is of the order of Planck's constant (h), then the system is to be dealt with *quantum mechanics.* On the other hand, the system is to be dealt with *classical mechanics* if the value of dimension of action is greater than the value of h.

10.15. *Can cent per cent energy of the incident photon be transferred to the target electron in Compton effect? Discuss.*
Ans. See solution of Example 22 (Sec. 10.11, p. 285).

10.16. *Compare photoelectric effect and Compton effect.*
Ans. (i) Both photoelectric effect and Compton effect result due to the action of a photon with an electron.

(ii) The laws of conservation of momentum and energy are valid in both the effects.

(iii) In photoelectric effect the energy of the incident photon is completely transferred to the free electron in metal but in Compton effect only fraction of the energy of the incident photon is transferred to the loosely bound electron.

(iv) If the energy of the incident photon (ϵ) be greater than the work function (W_0) of the metal, then the electron is elected from it leaving a positive charge. But in Compton effect the electron is recoiled after the scattering of the photon.

(v) Threshold energy of photoelectric effect is lower than that in Compton effect.

10.11 Solved Problems

Ex. 1 *Calculate the Compton wavelength* (in Å) *for an electron.* **[WBUT 2005]**

Soln. The rest mass of an electron $m_0 = 9.11 \times 10^{-28}$ gm, Planck's constant $h = 6.625 \times 10^{-27}$ erg-sec and velocity of light in vacuum $c = 3 \times 10^{10}$ cm/sec. Now, the Compton wavelength of an electron is

$$\lambda_0 = \frac{h}{m_0 c} = \frac{6.625 \times 10^{-27}}{(9.11 \times 10^{-28}) \times (3 \times 10^{10})} = \mathbf{0.0242\ \text{Å}}.$$

Ex. 2 *For heating metals, a very small hole in an electric furnace acting as a black body is used. If the area of the hole is* 100 mm^2 *and it is desired to maintain the metal at* 1100 °C, *how much energy travels per* sec *through this hole.* [$\sigma = 5.67 \times 10^{-8}$ W/m^2/K^4]

Soln. Heat energy passing out of the hole per sec is

$$\begin{aligned} E &= \sigma A T^4 \\ &= (5.67 \times 10^{-8})(1 \times 10^{-4}) \times (1373)^4 \quad [\because T = 273 + 1100 = 1373\ \text{K}] \\ &= \mathbf{20.13\ J/sec}. \end{aligned}$$

Ex. 3 *An electric heater emits* 1500 W *of thermal radiation. Assuming that the coils of the heater radiates like black body, determine its temperature.*
[*Given: Surface area of the coil* = $0.030\,\text{m}^2$ *and* $\sigma = 6.00 \times 10^{-8}\,\text{W/m}^2/\text{K}^4$]

Soln. Energy radiated by the surface of the heater coil

$$E = A\sigma T^4 \quad \text{or,} \quad 1500 = (0.03) \times (6.0 \times 10^{-8}) \times T^4$$

$$\text{or,} \quad T^4 = \frac{1500}{0.03 \times 6 \times 10^{-8}} = \frac{1500 \times 10^8}{0.18} \quad \therefore \quad T = \mathbf{955.4\ K.}$$

Ex. 4 *Calculate the black body temperature from of following data: Stefan's constant* $\sigma = 1.37 \times 10^{-12}$ cal/cm²/s. *Solar constant* $S = 2.3$ cal/cm²/min, *distance between the sun and the earth* = 1.5×10^{13} cm. **[PU 1966; DU 1971]**

Soln. Energy radiated per sec per unit area

$$E = \frac{4\pi R^2}{4\pi r^2} S = \left(\frac{R}{r}\right)^2 S = \left(\frac{1.5 \times 10^{13}}{7 \times 10^{10}}\right)^2 \times \frac{2.3}{60}\ \text{cal/s.}$$

Again, $E = \sigma T^4$ or, $\sigma T^4 = E$.

$$\therefore 1.37 \times 10^{-12} \times T^4 = \left(\frac{1.5 \times 10^{13}}{7 \times 10^{10}}\right) \times \frac{2.3}{60} \quad \text{or,} \quad T^4 = \frac{\left(\frac{1.5\times10^{13}}{7\times10^{10}}\right)^2 \times \frac{2.3}{60}}{1.37 \times 10^{-12}}.$$

$$\therefore T = \mathbf{5987\ K.}$$

Ex. 5 *Taking sun's temperature to be* 5902 K, *calculate the wavelength corresponding to the maximum emission of sun.* [*Given: the constant* $c = 0.2892$]

Soln. From Wien's displacement law

$$\lambda_m T = \text{constant } (c)$$

$$\text{or,} \quad \lambda_m = \frac{c}{T} = \frac{0.2892}{5902} = 4900 \times 10^{-8}\ \text{cm}$$

$$\text{or,} \quad \lambda_m = \mathbf{4900\ \text{Å}.}$$

Ex. 6 *What is the energy in* electron volt *of a photon of wavelength* 1 Å? *What is its momentum?* [$h = 6.62 \times 10^{-34}$ J-s]

Soln. The energy of a photon of frequency ν is

$$E = h\nu = \frac{hc}{\lambda}, \quad [\text{where } c = \nu\lambda].$$

Now, $h = 6.62 \times 10^{-34}$ J-s, $c = 3 \times 10^8$ m/s and $\lambda = 1\,\text{Å} = 10^{-10}$ m.

$$\therefore E = \frac{6.62 \times 10^{-34} \times 3 \times 10^8}{10^{-10}} = 19.86 \times 10^{-16}\ \text{joules}$$

$$= \frac{19.86 \times 10^{16}}{1.6 \times 10^{-19}}\ \text{eV} \quad [\because\ 1\,\text{eV} = 1.6 \times 10^{-19}\,\text{J}]$$

$$= \mathbf{12412.5\,eV.}$$

Ex. 7 *Calculate the number of photons emitted per* second *by a* 100 watt *sodium lamp.* [λ = 5893 Å *for sodium*].

Soln. The energy of a photon of frequency ν is

$$E = h\nu = \frac{hc}{\lambda} \quad [\because c = \nu\lambda].$$

Now, $h = 6.63 \times 10^{-34}$ J-s, $c = 3 \times 10^8$ m/s, $\lambda = 5893 \times 10^{-10}$ m.

$$\therefore E = \frac{(6.63 \times 10^{-34}) \times (3 \times 10^8)}{5893 \times 10^{-10}} = 3.375 \times 10^{-19} \text{ J}.$$

According to the problem, the energy emitted per sec by 100 watt 1 A is 100 J/sec. So the number of photons emitted per sec is

$$N = \frac{100}{3.37 \times 10^{-19}} = \mathbf{2.96 \times 10^{20}}.$$

Ex. 8 *The work function of a photosensitive metal is* 2.1 eV. (i) *Calculate the threshold wavelength,* (ii) *What is the wavelength of the incident radiation, if the stopping potential is* 0.43 V? [*Given*: $h = 6.63 \times 10^{-34}$ J-s *and* $e = 1.6 \times 10^{-19}$ C]

Soln. (i) The work function of the metal $W_0 = 2.1\text{ eV} = 2.1 \times 1.6 \times 10^{-19}$ J.

If λ_0 be the threshold wavelength, then

$$W_0 = h\nu_0 = \frac{hc}{\lambda_0} \quad \text{or,} \quad \lambda_0 = \frac{hc}{W_0} = \frac{(6.63 \times 10^{-34}) \times (3 \times 10^8)}{2.1 \times 1.6 \times 10^{-19}} = 5920 \times 10^{-10}\text{ m} = \mathbf{5920\ \text{Å}}.$$

(ii) If λ be the wavelength corresponding to stopping potential $V_s = 0.43$ V, then we have

$$h\nu = \text{eV}_s + W_0$$

$$\text{or,} \quad \frac{hc}{\lambda} = \text{eV}_s + W_0$$

$$\text{or,} \quad \lambda = \frac{hc}{(\text{eV}_s + W_0)}$$

$$= \frac{(6.63 \times 10^{-34}) \times (3 \times 10^8)}{(0.43 \times 1.6 \times 10^{-19}) + (2.1 \times 1.6 \times 10^{-19})}$$

$$= \frac{6.63 \times 3 \times 10^{-26}}{(0.43 + 2.1) \times 1.6 \times 10^{-19}}$$

$$= 4913 \times 10^{-10}\text{ m} = \mathbf{4913\ \text{Å}}.$$

Ex. 9 *If the threshold wavelength of sodium metal is* 6800 Å, *what is the photoelectric work function of sodium in electron volt.* [$h = 6.62 \times 10^{-34}$ J-s]

Soln. Threshold frequency

$$\nu_0 = \frac{c}{\lambda_0} = \frac{3 \times 10^8}{6800 \times 10^{-10}}\text{ Hz} \quad [\because \lambda_0 = 6800 \times 10^{-10}\text{ m}].$$

The work function

$$W_0 = h\nu_0 = \frac{(6.62 \times 10^{-34}) \times 3 \times 10^8}{6800 \times 10^{-10}}\text{ J} = \frac{(6.62 \times 10^{-34}) \times 3 \times 10^8}{(6800 \times 10^{-10}) \times (1.6 \times 10^{-19})}\text{ eV} = \mathbf{1.82\text{ eV}}.$$

Ex. 10 *Light of wavelength* 2000 Å *falls on a photosensitive material having work function* 4.2 eV. *What is the KE of the fastest and slowest photoelectron? What is the stopping potential?* [*Given*: $h = 6.62 \times 10^{-34}$ J-s; $e = 1.6 \times 10^{-19}$ C] **[HPU 1994]**

Soln. The energy of the incident photon

$$E = h\nu = \frac{hc}{\lambda} = \frac{(6.62 \times 10^{-34}) \times (3 \times 10^8)}{2000 \times 10^{-10}} = 9.93 \times 10^{-19}\,\text{J}.$$

Work function $W_0 = 4.2\,\text{eV} = 4.2 \times 1.6 \times 10^{-19} = 6.72 \times 10^{-19}$ J.

$\therefore$ KE of the fastest electron is

$$= \tfrac{1}{2}mv_m^2 = h\nu - W_0 = E - W_0 = (9.93 \times 10^{-19} - 6.72 \times 10^{-19})\,\text{J}$$

$$= 3.21 \times 10^{-19}\,\text{J} = \frac{3.21 \times 10^{-19}}{1.6 \times 10^{-19}} = \mathbf{2\ eV}.$$

KE of the electron is lowest when its velocity is zero.

So lowest KE $= \frac{1}{2}m \cdot 0^2 = \mathbf{0}$.

The stopping potential is given by

$$h\nu = eV_s + W_0 \text{ or, } 9.93 \times 10^{-19} = V_s \times 1.6 \times 10^{-19} + 6.72 \times 10^{-19} \text{ or, } V_s = \frac{3.21 \times 10^{-19}}{1.6 \times 10^{-19}} = \mathbf{2\ volts}.$$

Ex. 11 *If V_{s_1} and V_{s_2} be the stopping potentials for a metal corresponding to wavelengths λ_1 and λ_2, then find out an expression for Planck's constant.*

Soln. From Einstein's photoelectric equation

$$h\nu_1 = eV_{s_1} + W$$

$$\text{and}\quad h\nu_2 = eV_{s_2} + W$$

$$\text{or,}\quad \frac{hc}{\lambda_1} = eV_{s_1} + W_0$$

$$\text{and}\quad \frac{hc}{\lambda_2} = eV_{s_2} + W_0.$$

Subtracting, we have

$$hc\left(\frac{1}{\lambda_1} - \frac{1}{\lambda_2}\right) = e(V_{s_1} - V_{s_2}) \text{ or, } hc\left(\frac{\lambda_2 - \lambda_1}{\lambda_1\lambda_2}\right) = e(V_{s_1} - V_{s_2}) \text{ or, } h = \frac{e(V_{s_1} - V_{s_2})\lambda_1\lambda_2}{c(\lambda_2 - \lambda_1)},$$

where e is the electronic charge and c is the velocity of light.

Ex. 12 *Light of a wavelength* 2000 Å *falls on an aluminium surface having work function* 4.2 eV. *Find out threshold wavelength and stopping potential.*

Soln. Threshold wavelength

$$\lambda_0 = \frac{hc}{W_0} = \frac{12400}{W_0(\text{eV})}\,\text{Å} = \frac{12400}{4.2}\,\text{Å} = \mathbf{2952\ Å}.$$

Stopping potential V_s is given by

$$eV_s = \frac{hc}{\lambda} - W_0 = \frac{12400}{\lambda(\text{Å})} - W_0(\text{eV}) = \frac{12400}{2000} - 4.2 = 2\,\text{eV}.$$

$\therefore V_s = \mathbf{2\ volts}$.

Ex. 13 *Show that a photon cannot hand over its full energy to a free electron during collision.*

Soln. First we assume that it is possible for a photon to hand over its full energy to the electron during collision: Then from the law of conservation of energy and momentum, we get

$$h\nu + m_0c^2 = 0 + \sqrt{p^2c^2 + m_0^2c^4}, \tag{10.38}$$

where $h\nu' = 0$, if whole of the energy be given to the free electron and $\sqrt{p^2c^2 + m_0^2c^4}$ is the energy of the recoil electron and

$$\frac{h\nu}{c} + 0 = 0 + p \quad \text{or,} \quad h\nu = pc. \tag{10.39}$$

$$\therefore pc + m_0c^2 = [p^2c^2 + m_0^2c^4]^{1/2} \quad \text{[from eqns (10.38) and (10.39)]}$$

or, $p^2c^2 + 2pm_0c^3 + m_0^2c^4 = p^2c^2 + m_0^2c^4$

or, $2pm_0c^3 = 0$

or, $2\dfrac{h\nu}{c}m_0c^3 = 0$

or, $\mathbf{2m_0h\nu c^2 = 0.}$ (10.40)

For eqn (10.40) to be satisfied either $m_0c^2 = 0$ or, $h\nu = 0$, which is *impossible*. Hence, a *photon on collision with a free electron cannot hand over its energy to the electron.*

Ex. 14 *An X-ray photon of wavelength* 0.3 Å *is scattered through an angle* 45° *by a loosely bound electron. Find the wavelength of the scattered photon.*

Soln. The wavelength of the scattered photon

$$\lambda' = \lambda + \frac{h}{m_0c}(1 - \cos\theta).$$

Here $\lambda = 0.3$ Å, $h = 6.63 \times 10^{-34}$ J-s, $m_0 = 9.11 \times 10^{-31}$ kg and $c = 3 \times 10^8$ m/s.

Now, $\frac{h}{m_0c} = \frac{6.63\times10^{-34}}{9.11\times10^{-31}\times3\times10^8} = \frac{6.63\times10^{-11}}{9.11\times3}$ m.

$$\therefore \lambda' = 0.3 + 0.02426(1 - \cos 45°)$$
$$= 0.3 + 0.02426(1 - 0.7071)$$
$$= 0.3 + 0.02426 \times 0.2929$$
$$= \mathbf{0.3071\ Å}.$$

Ex. 15 *An X-ray photon is found to have doubled its wavelength on being scattered by* 90°*. Find the energy and wavelength of the incident photon.* **[LU 1996]**

Soln. The Compton shift $\Delta\lambda = \frac{h}{m_0c}(1 - \cos\theta)$.

Here $\theta = 90°$.

$$\therefore \Delta\lambda = \frac{h}{m_0c} = \frac{6.63 \times 10^{-34}}{(9.1 \times 10^{-31}) \times 3 \times 10^8} = \mathbf{0.02425 \times 10^{-10}\ m.}$$

If λ be the wavelength of the incident photon, then the wavelength of the scattered photon is $\lambda' = 2\lambda$.

$\therefore \lambda + \Delta\lambda = 2\lambda$ or, $\lambda = \Delta\lambda = 0.02425 \times 10^{-10}$ m $=$ **0.02425 Å**.

Again, energy of the incident photon is

$$E = h\nu = \frac{hc}{\lambda} = \frac{(6.63 \times 10^{-34}) \times (3 \times 10^8)}{0.02425 \times 10^{-10}} = \frac{19.89 \times 10^{-16}}{0.02425}\ \text{J} = \frac{19.89 \times 10^{-16}}{0.02425 \times 1.6 \times 10^{-19}}\ \text{eV}$$
$$= 0.51 \times 10^6\ \text{eV} = \mathbf{0.51\ MeV}.$$

Ex. 16 *Show that the Compton shift of the scattered radiation varies from* 0 *to* 0.0484 Å, *if the angle of scattering varies from* 0 *to* 180°. **[LU 1992]**

Soln. The Compton shift $\Delta\lambda = \frac{h}{m_0 c}(1 - \cos\theta)$.

(i) When $\theta = 0, \cos\theta = 1. \therefore \Delta\lambda = \frac{h}{m_0 c}(1 - 1) = 0$.

(ii) When $\theta = 90°, \cos\theta = 0. \therefore \Delta\lambda = \frac{h}{m_0 c}(1 - 0) = \frac{h}{m_0 c} = \mathbf{0.0242\ \text{Å}}$

(iii) When $\theta = 180°, \cos 180° = -1. \therefore \Delta\lambda = \frac{h}{m_0 c}(1 + 1) = 0.0242 \times 2 = \mathbf{0.0484\ \text{Å}}$

Hence, when θ varies from 0 to 180°, **$\Delta\lambda$ varies from 0 to 0.0484 Å.**

Ex. 17 *An X-ray photon of frequency* 3×10^{19} Hz *collides with a free electron at rest and is scattered at* 60°. *Find the direction in which the electron will move.* [*Given:* $(\frac{h}{m_0 c}) = 0.02426$ Å]

Soln. The frequency of the incident photon $\nu = 3 \times 10^{19}$ Hz, angle of scattering $\theta = 60°$ and $h = 6.63 \times 10^{-34}$ J-s.

The angle of recoil is given by $\cot\frac{\theta}{2} = (1 + \frac{h\nu}{m_0 c^2})\tan\phi$.

Now, $\frac{h\nu}{m_0 c^2} = \left(\frac{h}{m_0 c}\right)\frac{\nu}{c} = 0.02426 \times 10^{-10} \times \frac{3\times10^{19}}{3\times10^{8}} = 0.2426$.

So $\cot\left(\frac{60°}{2}\right) = (1 + 0.2426)\tan\phi$ or, $\cot 30° = (1.2426)\tan\phi$.

$\therefore \tan\phi = \frac{1}{1.2426} \times \sqrt{3} = \frac{1.7320}{1.2426} = 1.3938$.

$\therefore \phi = \mathbf{54.34°}$.

Ex. 18 *If a light of wavelength* 6000 Å *falls on a metal surface and emits photoelectrons with velocity* 4×10^5 m/s, *what is the threshold wavelength?* [*Given:* $c = 3 \times 10^8$ m/s] **[WBUT 2007]**

Soln. Einstein's photoelectric equation,

$$h\nu = \frac{1}{2}mv^2 + W_0 \quad \text{or,} \quad \frac{hc}{\lambda} = \frac{1}{2}mv^2 + W_0.$$

$\therefore$ the work function,

$$W_0 = \frac{hc}{\lambda} - \frac{1}{2}mv^2. \qquad (10.41)$$

Here, the wavelength of incident light $= 6000$ Å $= 6000 \times 10^{-10} = 6 \times 10^{-7}$ m.

Velocity of the ejected photoelectron $v = 4 \times 10^5$ m/s.

Mass of the electron $m = 9.11 \times 10^{-31}$ kg.

Planck's constant $h = 6.62 \times 10^{34}$ J-s.

Velocity of light in vacuum $c = 3 \times 10^8$ m/s.

Now, from eqn (10.41)

$$W_0 = \frac{(6.62 \times 10^{-34}) \times (3 \times 10^8)}{6 \times 10^{-7}} - \frac{1}{2} \times 9.11 \times 10^{-31} \times (4 \times 10^5)^2$$

$$= 3.31 \times 10^{-19} - 72.88 \times 10^{-21} = 258.12 \times 10^{-21}\ \text{J}.$$

If λ_0 be the threshold wavelength, then

$$W_0 = \frac{hc}{\lambda_0}$$

$$\text{or,} \quad \lambda_0 = \frac{hc}{W_0} = \frac{(6.62 \times 10^{-34}) \times (3 \times 10^8)}{258.12 \times 10^{-21}} = \frac{19.86}{258.12} \times 10^5$$

$$= 7694.09 \times 10^{-10}\ \text{m} = \mathbf{7694.09\ \text{Å}}.$$

Ex. 19 *If* m_0 *be the rest mass of an electron and* λ_c *the Compton wavelength, then prove that, the maximum recoil energy of a free electron when struck by a photon of wavelength* λ *is* $E_{max} = \frac{2m_0 c^2 \lambda_c^2}{\lambda^2 + 2\lambda_c\lambda}$.

Soln. If ν and ν' be the frequencies of the incident photon and the scattered photon respectively, then the KE of the recoil electron is

$$E = h\nu - h\nu' = hc\left(\frac{1}{\lambda} - \frac{1}{\lambda'}\right).$$

If λ' is maximum, then E becomes maximum,

$$\therefore E_{\text{max}} = hc\left(\frac{1}{\lambda} - \frac{1}{\lambda'_{\text{max}}}\right).$$

But Compton shift, $\lambda' - \lambda = \frac{h}{m_0 c}(1 - \cos\theta) = \lambda_c(1 - \cos\theta)$

$$\therefore \lambda' = \lambda_c(1 - \cos\theta) + \lambda \quad \text{or,} \quad \lambda'_{\text{max}} = 2\lambda_c + \lambda \quad [\text{when } \theta = \pi].$$

$$\therefore E_{\text{max}} = hc\left[\frac{1}{\lambda} - \frac{1}{(\lambda + 2\lambda_c)}\right] = \frac{2hc\lambda_c}{\lambda(\lambda + 2\lambda_c)} = \frac{2m_0c^2\lambda_c^2}{\lambda^2 + 2\lambda_c\lambda} \textbf{ (Proved).}$$

$$[\because h = m_0 c\lambda_c]$$

Ex. 20 *A photon strikes a tightly bound electron to the nucleus. Prove that the Compton shift would be smaller.*

Soln. When a photon strikes a free electron the Compton wavelength is $\lambda_c = \frac{h}{m_0 c}$.

But when the electron belongs to a very heavy element with a large atomic number, the binding energy E_b comes into play. Under such condition, the Compton wavelength can be shown to be

$$\lambda_c = \frac{h/c}{(m_0 + E_b/c^2)} = \frac{h}{m_0 c + \frac{E_b}{c}}.$$

This shows that if E_b be greater, then λ_c will be smaller.

Ex. 21 *When a metal surface is irradiated by light of wavelength λ, the maximum energy of the photoelectron is β. If the same metal be irradiated with photon of wavelength λ_2, what would be the maximum energy of the photoelectrons?*

Soln. The maximum KE of photoelectrons

$$(E_k)_{\text{max}} = h\nu - W_0 = \frac{hc}{\lambda} - W_0. \tag{10.42}$$

When $\lambda = \lambda_1, (E_k)_{\text{max}} = \beta$,

$$\therefore \beta = \frac{hc}{\lambda_1} - W_0 \quad \text{or,} \quad W_0 = \frac{hc}{\lambda_1} - \beta. \tag{10.43}$$

From eqns (10.42) and (10.43), when $\lambda = \lambda_2$,

$$(E_k)_{\text{max}} = \frac{hc}{\lambda_2} - \frac{hc}{\lambda_1} + \beta = \boldsymbol{hc\left(\frac{1}{\lambda_2} - \frac{1}{\lambda_1}\right) + \beta}. \tag{10.44}$$

Ex. 22 *What fraction of the KE of the incident photon can be transferred to the recoil electron? Comment on the result obtained.*

Soln.

$$E_r = h\nu\frac{\alpha(1 - \cos\theta)}{1 + \alpha(1 - \cos\theta)}, \quad \left[\text{where } \alpha = \frac{h\nu}{m_0c^2}\right].$$

E_r is the maximum, when $\theta = 180°$.

$$\therefore E_{rm} = \frac{2\alpha}{1 + 2\alpha}.$$

Comment on the result: We have $\frac{2\alpha}{1+2\alpha} = \frac{1}{1+\frac{1}{2\alpha}} < 1$.

So *the incident photon can never transfer the whole of its energy to the electron.*

Ex. 23 *When electromagnetic radiation of wavelength λ is incident on a photosensitive surface, the KE of the photoelectrons emitted from the surface is* 2 eV. *When the wavelength of the incident radiation is* 2λ, *the KE of the photoelectrons emitted from the surface is* 0.5 eV. *Then find the threshold wavelength (maximum wavelength) for photoelectric emission from the surface in terms of* λ. **[WBUT 2008]**

Soln. If λ_0 be the threshold wavelength, then from Einstein's phtoelectric equation $\frac{hc}{\lambda} = (\text{KE}) + \frac{hc}{\lambda_0}$.

When wavelength of the incident radiation is λ,

$$\frac{hc}{\lambda} = 2 + \frac{hc}{\lambda_0} \tag{1}$$

and when wavelength of incident radiation is 2λ,

$$\frac{hc}{2\lambda} = 0.5 + \frac{hc}{\lambda_0} \tag{2}$$

$$\text{or,} \quad \frac{2hc}{\lambda} = 2 + 4\frac{hc}{\lambda_0}. \tag{3}$$

Subtracting (10.45) from (10.47), we get $\frac{hc}{\lambda} = \frac{3hc}{\lambda_0}$ or, $\lambda_0 = \mathbf{3\lambda}$.

EXERCISES

10(A)–Objective Type Questions

Select the correct answer:

10.1. The wavelength of which the spectral energy density of emitted radiation at temperature T from a black body attains maximum value is proportional to (a) $\frac{1}{T}$ (b) T (c) T^4 (d) $T^{3/2}$. **[WBUT 2005]**

10.2. If visible light is used to study the Compton scattering, then the Compton shift will be (a) negative (wavelength of the scattered light will be lesser) (b) more positive than what is observed with X-ray (c) zero (d) positive but not detectable in the visible window. **[WBUT 2007]**

10.3. The failure of classical theory to explain the energy distribution in the spectrum of black body radiation is due to the assumption that (a) energy of black body radiation is constant (b) energy is discrete in nature (c) energy is continuous in nature (d) energy is infinite.

10.4. Black body radiation is termed as thermal radiation because (a) the quality of the radiation depends on temperature (b) the quality and quantity of radiation depend on temperature (c) the quality and quantity of radiation is independent of temperature (d) none of these.

10.5. Wien's distribution formula accounts for the energy distribution in the spectrum of black body radiation in the (a) high wavelength region (b) low wavelength region (c) entire wavelength region (d) none of these.

10.6. Rayleigh-Jeans law is correct only in the (a) low wavelength region of black body radiation spectrum (b) high wavelength region (c) entire wavelength region (d) none of these.

10.7. Planck's radiation law can account for the energy distribution in the spectrum of black body radiation in the (a) high wavelength region (b) low wavelength region (c) entire wavelength region (d) none of these.

10.8. Wien's distribution law of black body radiation is given by (a) $E_\lambda d\lambda = \frac{8\pi hc}{\lambda^5}\frac{1}{e^{\frac{hc}{\lambda kT}}-1}d\lambda$ (b) $E_\kappa d\lambda = \frac{8\pi kT}{\lambda^4}d\lambda$ (c) $E_\lambda d\lambda = \frac{c_1}{\lambda^5}e^{-\frac{c_2}{\lambda T}}$ (d) $E_\kappa d\lambda = \frac{8\pi kT}{\lambda^5}e^{-\frac{\lambda T}{hc}}d\lambda$.

10.9. The phenomenon of photoelectric effect involves (a) bound electrons in the atom (b) free electrons in metals (c) both bound and free electrons (d) secondary electrons.

10.10. At stopping potential the photocurrent becomes (a) infinite (b) constant (c) zero (d) none of these.

10.11. For photoelectric effect to occur the frequency of the incident light must be (a) below threshold frequency (b) above threshold frequency (c) equal to threshold frequency (d) any frequency.

10.12. Photoelectric current is directly proportional to the intensity of the incident light, if (a) the wavelength (λ) of incident light is greater than threshold wavelength (λ_0) (b) $\lambda < \lambda_0$ (c) $\lambda = \lambda_0$ (d) $0 < \lambda < \infty$.

10.13. If the frequency of the incident radiation (ν) be equal to the threshold frequency (ν_0), then Einstein's photoelectric equation reduces to (a) $h\nu = \frac{1}{2}mv^2 - h\nu_0$ (b) $h\nu = h\nu_0$ (c) $h\nu = \frac{1}{2}mv^2 + h\nu_0$ (d) $h\nu = \frac{1}{2}mv^2$.

10.14. Classical theory fails to account for the instantaneity of photo-electricity because it assumes that (a) energy is discrete packet called *photon* (b) energy is distributed over the entire wavefront of the incident radiation (c) energy is constant (d) none of these.

10.15. Stopping potential in photoelectric effect depends on (a) intensity of incident radiation (b) frequency of the incident radiation (c) volume of the tube containing the anode and cathode (d) pressure in the above tube.

10.16. The energy of the scattered photon in Compton effect is (a) less than the energy of the incident photon (b) more than the energy of the incident photon (c) equal to the energy of the incident photon (d) infinite.

10.17. The Compton shift $\Delta\lambda$ is equal to Compton wavelength (λ_0), when (a) the angle of scattering $\theta = 0$ (b) $\theta = 90°$ (c) $\theta = 180°$ (d) $\theta = 45°$.

10.18. Compton shift $\Delta\lambda$ is more for (a) heavy elements (b) light elements (c) for both light and heavy elements (d) none of these.

10.19. For scattering angle $\theta = 90°$, Compton shift reduces to (a) $\Delta\lambda = \frac{h}{m_0 c}(1 + \cos\theta)$ (b) $\Delta\lambda = \frac{h}{m_0 c}(1 - \cos\theta)$ (c) $\Delta\lambda = 0$ (d) $\Delta\lambda = \frac{h}{m_0 c}$.

10.20. In Compton effect a photon can transfer (a) negligible energy to the electron (b) whole of its energy to the electron (c) less than its energy to the electron (d) none of these.

10.21. Compton wavelength (λ_0) is (a) large for heavy elements (b) small for light elements (c) independent of elements (d) depends on the frequency of the incident photon.

10.22. With increase in wavelength of incident radiation the KE of the emitted photoelectron (a) increases (b) decreases (c) remains constant. **[WBUT 2005]**

10.23. The photoelectric current depends on (a) frequency of the incident radiation (b) intensity of the incident radiation (c) none of these. **[WBUT 2005]**

10.24. Given that the temperature of younger star is higher than that an older one, then (a) a blue star is younger than a red star (b) a blue star is older than a red star (c) both of them are of same age (d) colour of the star cannot be correlated to the age of the star.

ANSWERS

10.1. (a) **10.2.** (a) **10.3.** (c) **10.4.** (b) **10.5.** (b) **10.6.** (b) **10.7.** (c) **10.8.** (c) **10.9.** (b) **10.10.** (c) **10.11.** (b) **10.12.** (b) **10.13.** (b) **10.14.** (b) **10.15.** (b) **10.16.** (a) **10.17.** (b) **10.18.** (b) **10.19.** (a) **10.20.** (c) **10.21.** (c) **10.22.** (b) **10.23.** (b) **10.24.** (a).

10(B)–Short Answer Type Questions

10.1. What are the features of photoelectric effect, which cannot be explained from wave theory of light? **[WBUT 2007]**

10.2. Show graphically how the energy density vs. frequency plot of black body radiation is changed if the temperature is increased? **[WBUT 2007]**

10.3. State Stefan-Boltzmann's law. **[WBUT 2003]**

10.4. Write down Einstein's photoelectric equation. **[BUH 2002; WBUT 2005]**

10.5. Write short notes on Planck's radiation law. **[WBUT 2005]**

10.6. State Wien's displacement law. **[WBUT 2002]**

10.7. Explain why electron cannot be scattered at an angle greater than 90°.**[CUH 2091; NBU 2002)]**

[*Hint.* The angle of scattering of the electron ϕ is given by $\tan\phi = \frac{\cot\frac{\theta}{2}}{1+\frac{h\nu}{m_0c^2}}$.

For a constant frequency ν of the incident photon, $1 + \frac{h\nu}{m_0c^2} = C$ (a constant).

$\therefore C\tan\phi = \cot\frac{\theta}{2}$.

The maximum value of $\cot\frac{\theta}{2} = \infty$.

Therefore, the maximum value of $\tan\theta = \infty$. $\therefore \theta = 90°$.

Therefore, the electron cannot be scattered at an angle greater than 90°.]

10.8. What is the magnitude of momentum carried by a photon of frequency ν? **[BUH 2002]**

10.9. Why does photoelectric effect take place only when photons strike free electrons within the metal? **[BUH 2002]**

10.10. What is the energy of a photon carrying a momentum p? **[BUH 2004]**

10.11. Show that Planck's constant has the dimension of angular momentum. **[BUH 2001]**

10.12. What is Compton effect? Do you observe Compton effect with visible light? Give reasons for your answer. **[BUH 1993]**

[*Hint.* No, because visible light photon energy is comparable to the binding energy of the outer shell electrons of the atoms.]

10.13. What do you understand by *Wave-particle duality*? **[NBUH 2003]**

10.14. Discuss the importance of Compton scattering in connection with the photon concept of radiation. **[CUH 2003]**

10.15. What is photoelectric effect? How does classical theory fail to explain this effect? **[NBUH 2003; BUH 2005; NBUP 2002; cf. CUH 1990]**

10.16. What is ultraviolet catastrophe? Explain.

10.17. What do you mean by black body and black body radiation? **[CUH 1992, '95, 2002]**

10.18. Why is it important to have a clean surface while doing photoelectric experiment? **[HPU 1995]**

[*Hint.* The impurity layer on the metal surface (if it is not clean) have higher work function. The incident photon will have to cross the layer of material having higher work function and ultimately the photoelectron emission may not occur.]

10.19. Why the photoelectrons do not have much energy as the quantum of light, which causes its ejection? [BU 1995]

10.20. Does every photon incident on a photo cathode cause emission of an electron? [HPU 1997]

10.21. Does Compton scattering expression explain unmodified scattering? Explain. [NU 2000, '02]

10.22. State the basic postulates of Planck's quantum theory. How did Einstein modify this theory in the light of experimental results? [NU 2000]

10.23. Explain how the wave theory failed to explain photoelectric effect and Compton effect? [NU 2002]

10.24. Show that the temperature dependence in Stefan's law can be derived from Planck's radiation law. [WBUT 2008]

10(C)–Numerical Problems

10.1 If the incident radiation is of $\lambda = 1.372$ Å, find the wavelength of (a) scattered radiation at an angle of 30° and (b) velocity of the recoiled electron. [*Ans.* 1.375 Å, 2.63×10^6 m/s]

10.2 The photoelectric work function for lithium is 2.3 eV. (a) Find the threshold wavelength λ_0 for the photoelectric effect. (b) If ultraviolet light of wavelength $\lambda = 2000$ Å is incident on lithium surface, obtain the maximum KE of the photoelectrons and the value of the stopping voltage. [CUH 1997]

[*Hint.* (a) $W_0 = h\nu_0$, $\therefore \nu_0 = \frac{c}{\lambda_0} = \frac{W_0}{h} = \frac{2.3\times1.6\times10^{-19}}{6.62\times10^{-34}} = \frac{3.68}{6.62} \times 10^{15} = 0.55589 \times 10^{15}$ Hz.

$\therefore$ threshold wavelength $\lambda_0 = \frac{c}{\nu_0} = \frac{3\times10^8}{0.55589\times10^{15}} = 5.39675 \times 10^{-7}$ m $=$ **5396.75 Å**.

Maximum KE

$$E_{max} = h(\nu - \nu_0) = hc\left(\frac{1}{\lambda} - \frac{1}{\lambda_0}\right)$$

$$= hc\left(\frac{\lambda_0 - \lambda}{\lambda \times \lambda_0}\right) = \frac{(6.62 \times 10^{-34}) \times (3 \times 10^8) \times (5396.75 - 2000) \times 10^{-10}}{5396.75 \times 2000 \times 10^{-20}}$$

$$= \frac{6.62 \times 3 \times 3396.75}{5396.75 \times 2} \times \frac{10^{-36}}{10^{-17}}$$

$$= \frac{67459.455}{10793.5} \times 10^{-19} = 6.25 \times 10^{-19}\text{ J}$$

$$= \frac{6.25 \times 10^{-19}}{1.6 \times 10^{-19}} = 3.9\,\text{eV}.$$

$\therefore$ stopping potential $V_0 = \frac{E_{max}}{e} = \mathbf{3.9\ V}$]

10.3 Electrons are emitted with zero velocity from a certain metal surface when it is exposed to radiation of $\lambda = 6800$ Å. Calculate the threshold frequency and work function of the metal. [*Ans.* 0.44×10^{-15} Hz; 1.82 eV]

10.4 The photoelectric threshold for a certain metal is 3600 Å. Determine the maximum energy of the electrons ejected by radiation of 2000 Å. [*Ans.* $h = 6.6 \times 10^{-34}$ J-s; 4.4×10^{-19} J]

10.5 10 eV photon is incident on molybdenum whose work function is 4.15 eV. What is the stopping potential? [*Ans.* 5.85 V]

10.6 The photoelectric threshold wavelength of tungsten is 2300 Å. The tungsten surface is illuminated with ultraviolet light of wavelength 1800 Å. Calculate (a) the maximum KE of the ejected electrons (b) the maximum velocity of the ejected electrons (c) the stopping potential in volts for the electrons. [*Ans.* (a) 2.387×10^{-19} J = 1.49 eV (b) 7.239×10^5 m/s (c) 1.49 eV]

10.7 The long wavelength limit at which radiation is just able to eject photoelectrons from tantalum is 2974 Å. Find the KE that must be imparted to an electron in order that it may just escape from the metal.
[*Ans.* 4.2 eV]

10.8 The stopping potential for photoelectrons emitted from a sodium photocathode illuminated by light of wavelength $\lambda = 300$ nm is 1.85 V while that for light of $\lambda = 400$ nm is 0.82 V. Determine the value of Planck's constant, work function for sodium and threshold wavelength for sodium.

[BUH 1993]

10.9 Find the wavelength of X-rays, which produce recoil electrons of energy 5000 eV in Compton effect. Assume that the photon is scattered through an angle of 180° and the direction of recoil electron is in the direction of incident photon. [*Ans.* 0.325 Å]

10.10 X-rays with $\lambda = 1$ Å are scattered from a carbon block. The scattered radiation is viewed at 90° to the incident beam. Calculate (a) the Compton shift $\Delta\lambda$ (b) KE imparted to the recoil electron.
[*Ans.* $\Delta\lambda = 0.0243$ Å, KE = 0.295 eV]

10.11 X-rays of $\lambda = 0.709$ Å suffer Compton scattering from a carbon scatter. What is the wavelength of the photons scattered at $\theta = 30°$?

10.12 In Compton scattering the incident photons have $\lambda = 3 \times 10^{-8}$ cm. Calculate λ of the scattered radiation, if they are viewed at an angle of 60° to the direction of incidence. **[AU 1975]**

10.13 Write down the conservation laws in Compton scattering. Show that in Compton scattering while the photon can be scattered at any angle between 0° to 180°, the recoil electron can be emitted at angles between 0° and 90°.
[WBUT 2010]

10(D)–Broad Answer Type Questions

10.1 (a) State and explain Einstein's equation for explaining photoelectric emission. (b) Explain the phenomenon of Compton shift $\Delta\lambda$.
[WBUT 2002]

10.2 (a) Derive an expression for the Compton shift in wavelength for a photon scattered from a free electron at an angle θ. (b) At which angle the shift will be maximum. (c) Calculate the Compton wavelength (in Å) for an electron.
[WBUT 2005]

10.3 What are the conservation laws satisfied in Compton scattering experiment? Derive an expresson for the Compton wavelength shift.
[NBUP 2002; CUH 1996]

10.4 A beam of monochromatic X-rays of wavelength λ is scattered by a light element. Obtain an expression for the wavelength of the scattered radiation. Which apparatus will you set up to verify your result?
[cf. CUH 1994]

10.5 Describe an experimental arrangement to verify Einstein's photoelectric equation. What are the main features that can be observed in such an experiment, which do not admit any explanation by classical theories.
[BUH 2003]

10.6 What is photoelectric effect? How does one explain the fact hat the maximum KE of the ejected electrons is independent of the intensity of the incident light?
[BUH 2001]

10.7 (a) What is Compton effect? Derive an expression for Compton shift and wavelength of scattered photon. (b) What is Compton wavelength. Determine its value. Distinguish between Compton shift and Compton wavelength.
[PU 1997; HPU 1996; PU 1991, '93]

10.8 A photon strikes an electron tightly bound to the nucleus. What is the effect on Compton shift? **[PU 1995; GNDU 1995]**

10.9 Derive Einstein's photoelectric equation. How does it explain the laws of photoelectric emission? Why all the photoelectrons do not have the same energy? **[HPU 1997; PU 1996; LU 1995, CU 1992]**

10.10 How do you account for the presence of unmodified component of X-rays in the study of Compton effect? **[LU 1996; HPU 1996]**

10.11 What is the maximum KE of the recoil electron in Compton effect? Find the maximum energy transferred by the photon to the electron. Is it possible for a photon to transfer its entire energy to the electron? **[PU 1990; GNDU 1996]**

10.12 Explain how the wave theory of light failed to explain photoelectric effect and Compton effect? How the quantum theory explained these phenomena? How energy and momentum conservation occurs in photoelectric effect and Compton effect? **[NU 2002]**

10.13 (a) What is Compton effect? Explain how it confirms the photon nature of radiation? (b) Prove that Compton shift is given by $\Delta\lambda = \frac{h}{m_0 c}(1 - \cos\theta)$, where the symbols have their usual meaning. **[NU 1996]**

10.14 Give suitable reasons of the following: (a) Both photoelectric effect and Compton effect arise due to the action of photons on electrons. Linear momentum is not conserved in photoelectric effect while it is conserved in Compton effect. (b) In Compton effect when a photon collides with a bound electron its wavelength does not change. **[NU 1996]**

10.15 Prove that if during Compton scattering the recoil electron has very low energy, then the change in wavelength of the photon scattered at θ is $\Delta\lambda = \frac{\lambda_0}{2}\left(\frac{\lambda'}{\lambda} + \frac{\lambda}{\lambda'} - 2\cos\theta\right)$.

Chapter 11 | Module 4.3

Quantum Physics–II

11.1 Introduction

We have seen in Chapter 10 that radiation possesses dual character, viz., *wave* and *particle.* The phenomena like *interference* and *diffraction* established the wave nature of radiation, whereas *black body radiation, photoelectric effect* and *Compton effect* established the particle nature.

Now, matter has particle nature. Therefore, due to symmetry of nature matter should also have the same *dual nature,*—wave and particle. The close parallelism between *Maupertian principle of least action* and *Fermat's principle of least time* supports this concept. de Broglie of France first in 1924 predicted wave property of matter, which was verified experimentally by *C.J. Davisson and L.H. Germer in* 1927 and also by *G.P. Thomson in* 1927. The wave property and particle property may be reconciled by a *wave packet* formed due to the superposition of monochromatic waves of different wavelengths. A wave packet is confined to a small region of space and as such suitable to represent a particle, which is also confined to a small region during motion.

11.2 Phase Velocity and Group Velocity

(a) **Phase velocity**: A plane simple harmonic wave travelling along the positive X-direction is represented by

$$y = a \sin \omega \left(t - \tfrac{x}{v}\right), \tag{11.1}$$

where y is the displacement of a particle at time t at a distance x from the origin, v is the velocity of the wave and $\omega = 2\pi\nu$ is the angular frequency.

The phase of the wave represented by eqn (11.1) is $\phi(x,t) = \omega(t - \frac{x}{v})$.

$$\therefore \frac{\partial \phi}{\partial t} = \omega\left(1 - \frac{1}{v}\frac{dx}{dt}\right),$$

where ϕ is constant w.r.t. time, then

$$\frac{\partial \phi}{\partial t} = 0, \quad \therefore 1 - \frac{1}{v}\left(\frac{dx}{dt}\right)_{\phi} = 0 \quad (\because \omega \neq 0).$$

$$\text{or,} \quad \left(\frac{dx}{dt}\right)_{\phi} = v. \tag{11.2}$$

Eqn (11.2) represents the velocity with which the displacement of a given phase moves and is termed as the *phase velocity.*

∴ the phase velocity of the wave is

$$v_p = \left(\frac{dx}{dt}\right)_\phi$$

and is defined as *the velocity with which the displacement of a given phase moves forward.*

Now, replacing v by v_p in eqn (11.1),

$$y = a \sin \omega \left(t - \frac{x}{v_p}\right) = a \sin \left(\omega t - \frac{\omega x}{v_p}\right), = a \sin (\omega t - kx), \tag{11.3}$$

where $k = \frac{\omega}{v_p}$ is the *wave propagation* constant.

$$\therefore v_p = \frac{\omega}{k}. \tag{11.4}$$

Eqn (11.4) gives the expression of v_p in terms of ω and k.

The phase velocity v_p in a *non-dispersive* medium is independent of wavelength (i.e., $\frac{dv_p}{d\lambda} = 0$) but in a *dispersive medium* v_p varies with wavelength (i.e., $\frac{dv_p}{d\lambda} \neq 0$).

(b) **Group velocity**: We consider a group of waves consisting of two components of equal amplitude but slightly different frequencies ω_1 and ω_2. Let two such component waves travelling along the X-direction in a dispersive medium be represented by

$$y_1 = a \sin(\omega t - kx) \tag{11.5}$$

$$\text{and} \quad y_2 = a \sin[(\omega + \Delta\omega)t - (k + \Delta k)x]. \tag{11.6}$$

The resultant displacement at time t and position x is given by

$$y = a \sin(\omega t - kx) + a \sin[(\omega + \Delta\omega)t - (k + \Delta k)x]$$

$$= 2a \cos \tfrac{1}{2}(t\Delta\omega - x\Delta k) \sin \left[\left(\frac{2\omega + \Delta\omega}{2}\right) t - \left(\frac{2k + \Delta k}{2}\right) x\right]. \tag{11.7}$$

$\Delta\omega$ and Δk being compared to ω and k are neglected [Fig. 11.1]. Then

$$y = 2a \cos \tfrac{1}{2}(t\Delta\omega - x\Delta k) \sin(\omega t - kx). \tag{11.8}$$

Eqn (11.8) represents a group of waves

$$\text{or,} \quad y = A \sin(\omega t - kx), \tag{11.9}$$

where the resultant amplitude

$$A = 2a \cos \tfrac{1}{2}(t\Delta\omega - x\Delta k). \tag{11.10}$$

Eqn (11.9) represents the carrier wave travelling with velocity $v_p = \omega/k$.

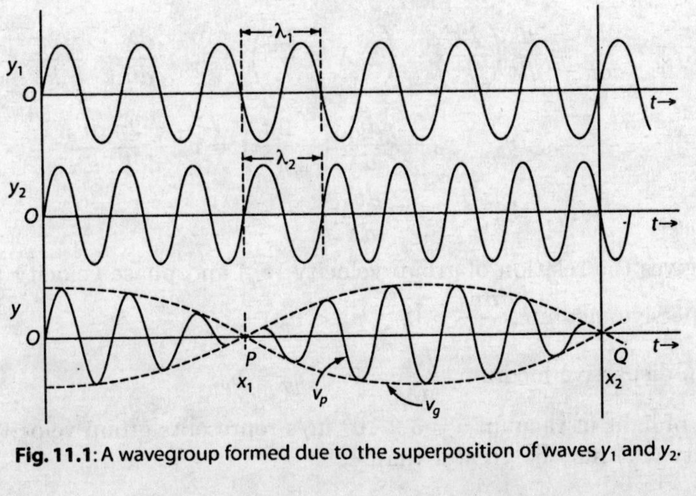

Fig. 11.1: A wavegroup formed due to the superposition of waves y_1 and y_2.

Length of the wave group: Let $P(x_1, 0)$ and $Q(x_2, 0)$ be two successive points at which the resultant amplitude A is zero at time t.

Then from eqn (11.10) at $P(x_1, 0)$,

$$\tfrac{1}{2}(t\Delta\omega - x_1\Delta k) = (2n+1)\tfrac{\pi}{2} \tag{11.11}$$

and at $Q(x_2, 0)$,

$$\tfrac{1}{2}(t\Delta\omega - x_2\Delta k) = (2n-1)\tfrac{\pi}{2}, \tag{11.12}$$

where $n = 0, 1, 2, \cdots$

From eqns (11.11) and (11.12), we have by subtraction

$$\frac{1}{2}(x_2 - x_1)\Delta k = [(2n+1) - (2n-1)]\frac{\pi}{2} \quad \text{or,} \quad x_2 - x_1 = \frac{2\pi}{\Delta k}. \tag{11.13}$$

This represents the length of the wave packet.

$$\therefore\ x_2 - x_1 = \frac{2\pi}{\Delta(\frac{2\pi}{\lambda})} = -\frac{\lambda^2}{\Delta\lambda}. \tag{11.14}$$

Thus, *the spread of the wave packet will be small, if $\Delta\lambda$ is large, i.e., a large number of waves of varying wavelength superpose to produce the wave packet.*

Group velocity: At $t = 0$ and $x = 0$, eqn (11.10) shows that the amplitude $A = 2a$, the maximum.

If at a later time t' and position x', A is again maximum, then

$$\frac{1}{2}(t'\Delta\omega - x'\Delta k) = 0 \quad \text{or,} \quad \frac{x'}{t'} = \frac{\Delta\omega}{\Delta k}.$$

In the limit, when $\Delta k \to 0$,

$$\frac{x'}{t'} = \frac{d\omega}{dk} \quad \text{or,} \quad v_g = \frac{d\omega}{dk}. \tag{11.15}$$

Eqn (11.15) represents the group velocity of the wave group. Thus, *the group velocity v_g may be defined as the velocity with which the maximum amplitude of the wave group moves.*

Relation between v_g and v_p: We have the group velocity

$$v_g = \frac{d\omega}{dk} = \frac{d}{dk}(v_p k) \quad \left(\because v_p = \frac{\omega}{k}\right)$$
$$= v_p + k\frac{dv_p}{dk} = v_p + k\frac{dv_p}{d\lambda}\frac{d\lambda}{dk}.$$

But

$$\frac{d\lambda}{dk} = \frac{d}{dk}\left(\frac{2\pi}{k}\right) \quad \left(\because \lambda = \frac{2\pi}{k}\right) \quad \text{or,} \quad \frac{d\lambda}{dk} = -\frac{2\pi}{k^2}.$$

$$\therefore\ v_g = v_p + k\frac{dv_p}{d\lambda}\left(-\frac{2\pi}{k^2}\right) = v_p - \frac{2\pi}{k}\frac{dv_p}{d\lambda}$$

$$\text{or,} \quad v_g = v_p - \lambda\frac{dv_p}{d\lambda}. \tag{11.16}$$

Eqn (11.16) gives the relation of group velocity (v_g) and phase velocity (v_p).

(i) For a dispersive medium, $\dfrac{dv_p}{d\lambda} > 0$. $\therefore v_g < v_p$.

(ii) For a non-dispersive medium, $\dfrac{dv_p}{d\lambda} = 0$. $\therefore v_g = v_p$.

The velocity of light in vacuum $c = 3 \times 10^8$ m/s represents group velocity (v_g), in some cases the phase velocity (v_p) may be greater than c.

11.3 de Broglie Hypothesis

In view of de Broglie a moving particle in space is associated with a wave called *de Broglie wave* or *matter wave*. The wavelength of the matter wave is given by

$$\lambda = \frac{h}{p} = \frac{h}{mv}, \tag{11.17}$$

where h is the Planck's constant and $p = mv$ is the momentum of the particle of mass m moving with velocity v. It should be kept in mind that *de Broglie waves* and *de Broglie wave packets* are only *theoretical devices* used to explain the experimental observations.

To derive an expression for de Broglie wavelength we recall that energy of a *photon* is

$$E = h\nu = \frac{hc}{\lambda} \quad [\because c = \nu\lambda]. \tag{11.18}$$

Again, in view of Einstein's mass-energy relation,

$$E = mc^2. \tag{11.19}$$

Now, from eqns (11.20) and (11.21), we get

$$\frac{hc}{\lambda} = mc^2 \quad \text{or,} \quad \lambda = \frac{h}{mc},$$

where mc is the momentum of the *photon.*

The above concept is also valid in case of subatomic particles like electron, proton, neutron, etc. If m be the mass and v the velocity of such a material particle, then de Broglie wavelength

$$\lambda = \frac{h}{mv} = \frac{h}{p}.$$

The above relation can also be expressed in terms of KE of the particle.

The KE of the particle

$$E_k = \frac{1}{2}mv^2 = \frac{p^2}{2m}, \quad \therefore p = \sqrt{2mE_k}.$$

So

$$\lambda = \frac{h}{\sqrt{2mE_k}}. \tag{11.20}$$

If E be the total relativistic energy of the particle of rest-mass m_0, then

$$E = E_k + m_0c^2$$

$$\text{or,} \quad \sqrt{p^2c^2 + m_0^2c^4} = E_k + m_0c^2$$

$$\text{or,} \quad p^2c^2 + m_0^2c^4 = E_k^2 + 2E_km_0c^2 + m_0^2c^4$$

$$\text{or,} \quad p = \frac{\sqrt{E_k(E_k + 2m_0c^2)}}{c}.$$

$$\therefore \lambda = \frac{h}{p} = \frac{hc}{\sqrt{E_k(E_k + 2m_0c^2)}} \quad \text{or,} \quad \lambda = \frac{h}{\sqrt{2m_0E_k\left(1 + \frac{E_k}{2m_0c^2}\right)}}. \tag{11.21}$$

11.3.1 de Broglie Wavelength of a Charged Particle Accelerated by an Electric Field V volt/m

If v be the velocity of a charged particle of charge q and mass m, when accelerated by a field V volt/m, then

$$\frac{1}{2}mv^2 = qV \quad \text{or,} \quad v = \sqrt{\frac{2qV}{m}}.$$

∴ de Broglie wavelength associated with the charged

$$\lambda = \frac{h}{mv} = \frac{h}{m\sqrt{\frac{2qV}{m}}} \quad \text{or,} \quad \lambda = \frac{h}{\sqrt{2mqV}}. \tag{11.22}$$

Eqn (11.22) represents the de Broglie wavelength of the *charged particle*.
Now,

$$h = 6.625 \times 10^{-34} \text{ J-s}$$
$$q = 1.602 \times 10^{-19} \text{ C (for electron)}$$
$$m = 9.1 \times 10^{-31} \text{ kg.}$$

Using these standard values in eqn (11.22), we get the de Broglie wavelength associated with an electron.

$$\lambda = \frac{6.625 \times 10^{-34}}{\sqrt{2 \times (9.1 \times 10^{-31}) \times (1.602 \times 10^{-19})V}} \quad \text{or,} \quad \lambda = \frac{12.27}{\sqrt{V}} \text{ Å.} \tag{11.23}$$

11.3.2 Distinction between Matter Waves and Electromagnetic Waves

1. Matter waves are generated due to the motion of particles *whether charged* or *uncharged.* But the electromagnetic waves are produced only by the motion of the *charged particles.*

2. The velocity of matter wave depends on the velocity of the particle, i.e., velocity is not constant, it may be greater than the velocity of light. But velocity of electromagnetic waves is equal to the velocity of light.

3. de Broglie waves are only *theoretical concepts* employed to explain the experimental results but electromagnetic waves have their *real existence.*

Characteristics of matter wave

Following are the few important characteristics of matter waves:

(i) Matter waves are invisible, they are only theoretical concept.

(ii) The wavelength of matter wave is inversely proportional to the mass of the moving article (i.e., $\lambda \propto \frac{1}{m}$). So λ is larger if the particle is smaller.

(iii) de Broglie wave guide the associated moving particle and hence are called the *pilot waves.*

(iv) The probability of finding a particle in space at a time is proportional to the square of the modulus of its amplitude. Hence, the probability of finding the particle is the maximum, where the amplitude is the maximum.

(v) Matter wave is independent of charge of the associated particle.

11.3.3 Bohr's Quantization Condition in the Light of de Broglie's Hypothesis

According to Bohr's quantization condition of angular momentum, the angular momentum of an atomic electron revolving in a stationary orbit is $L = mvr = n\hbar$ [where $\hbar = \frac{h}{2\pi}$], where h is the Planck's constant and $n = 1, 2, 3, \cdots$

In view of de Broglie, the electrons in the stationary orbits are waves of wavelength λ. *A stationary Bohr orbit must certain integral multiple of electron waves.*

$$\therefore 2\pi r = n\lambda = n\frac{h}{mv} \quad \text{or,} \quad mvr = n\frac{h}{2\pi} \quad \text{or,} \quad L = n\hbar, \tag{11.24}$$

which is the *Bohr's quantization condition of angular momentum. Hence, de Broglies hypothesis is in accordance with Bohr's theory.*

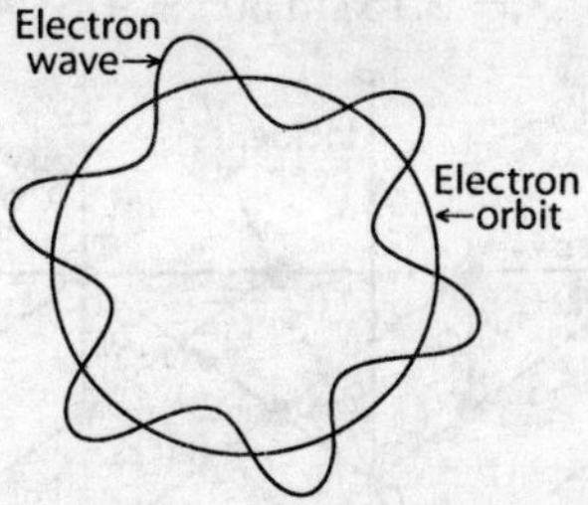

Fig. 11.2: A stationary Bohr orbit containing integral multiple of electron waves.

11.3.4 Davisson and Germer Experiment

C.J. Davisson and L.H. Germer of America performed an experiment (in 1927), which conclusively demonstrated the *wave nature of electrons.*

The experiment: The experimental arrangement is shown in Fig. 11.3.

The emitted electrons from a heated filament F are accelerated by the anode A. The accelerated electrons, then pass through a small hole in A and are incident on a nickel crystal S. The scattered electrons in different directions are then collected by the *Faraday cup C*, which is so designed that it can collect only those electrons, which are of the same energy as the incident beam. The scattered intensity is measured by a galvanometer. The intensity of the scattered beam in different direction was determined by rotating C about an axis through S on a graduated circular scale. The observation is repeated for different accelerating potential (V).

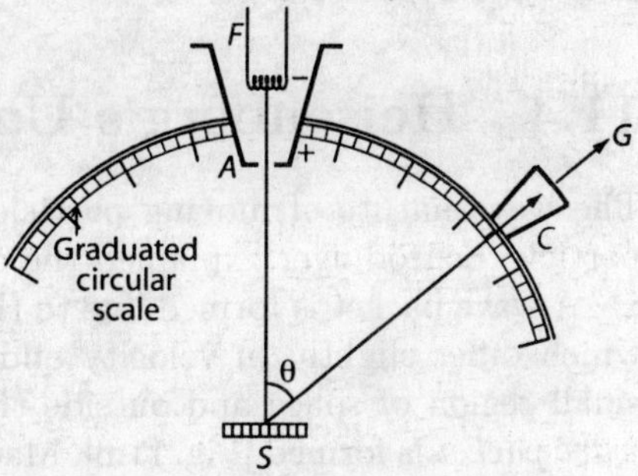

Fig. 11.3: Experimental arrangement of Davisson and Germer.

From the intensity of the scattered beam when plotted against the angle of scattering, we get a curve as shown in Fig. 11.4. The maximum of intensity of the scattered beam for accelerating voltage $V = 54\,\text{V}$ is found to occur at $\theta = 50°$.

$\therefore$ the calculated value of matter wave of electron is

$$\lambda = \frac{12.28}{\sqrt{V}} = \frac{12.28}{\sqrt{54}} = \mathbf{1.67\,\text{Å}}. \tag{11.25}$$

The experimental value of λ can be calculated as follows:

Let the electron beam be incident normally upon the surface of the crystal making an angle ϕ with the set of reflecting planes [Fig. 11.5].

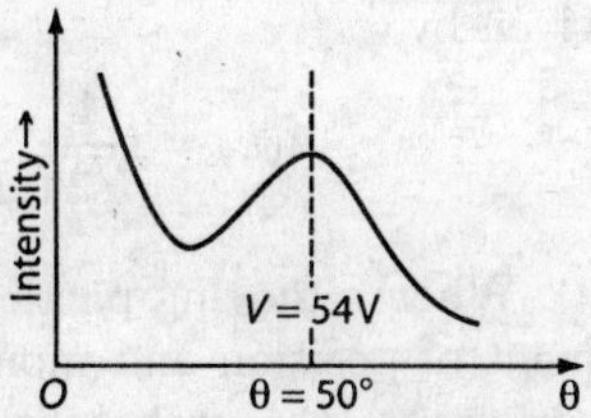

Fig. 11.4: Variation of intensity with θ.

According to *Bragg's law,* the reinforcement of electron waves will take place, if

$$n\lambda = 2D\sin\psi = 2D\cos\phi, \tag{11.26}$$

where $D =$ the spacing of the atomic layers and $\psi = (\frac{\pi}{2} - \phi)$.

Again, if d be the interatomic distance, then $D = d\sin\phi$.

$$\therefore n\lambda = 2d\sin\phi\cos\phi = d\sin 2\phi \quad \text{or,} \quad n\lambda = d\sin\theta, \tag{11.27}$$

where $2\phi = \theta$ and $n = 1, 2, 3, \cdots$ and d is the distance between two reflecting planes.

From X-ray study, the value of d is known as $d = 2.15$ Å.
$\therefore$ for the first order ($n = 1$) diffraction $\theta = 50°$, we get

$$\lambda = 2.15 \sin 50° = 1.65 \text{ Å}. \tag{11.28}$$

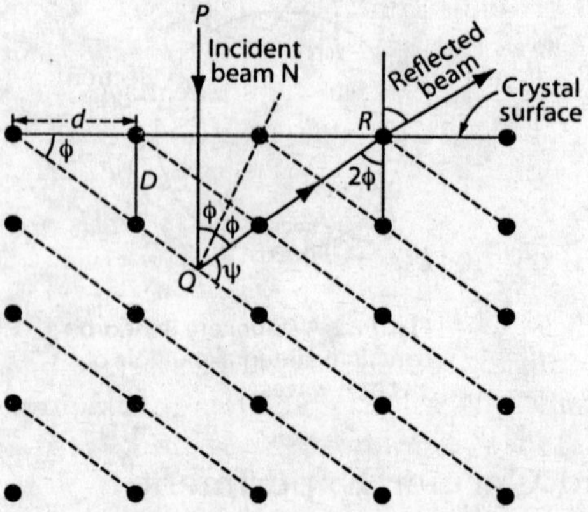

Fig. 11.5: Bragg reflection of electron waves for normal incidence.

Thus, the theoretically calculated value and experimental value of λ are in good agreement as seen from eqns (11.25) and (11.28). Hence, electron shows diffraction, which is a wave property. The experiment of Davisson and Germer, therefore, *provides a direct experimental evidence of de Broglie hypothesis.*

11.4 Heisenberg's Uncertainty Relation

The wave nature of moving particle introduces *uncertainty* in position and momentum of the particle. Schrödinger explained the dual nature (waves and particle) in terms of *wave packet.*

A wave packet is formed due to the superposition of group of infinitely large number of waves, which differ slightly in velocity and wavelength. These waves interfere constructively within a small region of space and outside they interfere destructively giving zero amplitude. Thus, the *wave packet* is formed [Fig. 11.6]. Mathematically, a wave packet is represented by

$$\psi(x,t) = \frac{1}{\sqrt{2\pi}} \int_{-\infty}^{+\infty} A(k) e^{i(kx-\omega t)} dk,$$

where $A(k)$ is a function of k and is known as *Fourier transform* given by

$$A(k) = \frac{1}{\sqrt{2\pi}} \int_{-\infty}^{+\infty} \psi(x,t) e^{-i(kx-\omega t)} dx.$$

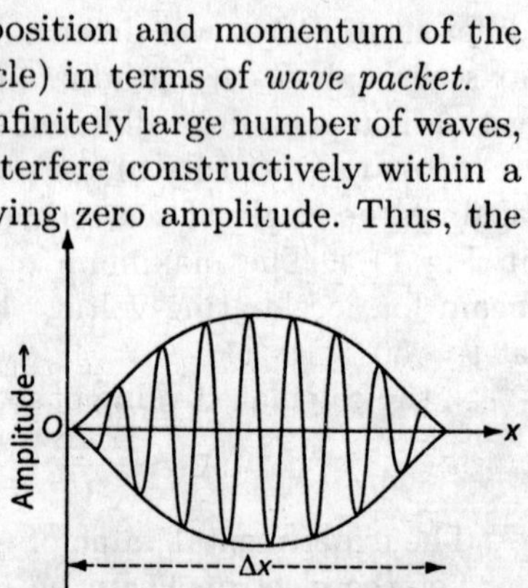

Fig. 11.6: A wave packet.

When a moving particle is conceptualized as a de Broglie wave packet, the precise measurement of position and momentum is disturbed, because when the particle is located within the region Δx, the probability of finding the particle is maximum at the centre of the wave packet and reduces to zero towards the ends. This introduces an uncertainty (Δx) in the position of the particle. Again, the spread of wavelength ($\Delta\lambda$) of the waves forming the wave packet is related to Δx. Also according to de Broglie relation, $p = \frac{h}{\lambda}$.

$$\therefore |\Delta p| = \frac{h}{\lambda^2} \Delta\lambda,$$

i.e., the spread of momentum is related to the spread of wavelength, which in turn depends on the spread (Δx) of the wave packet. Hence, uncertainty in position (Δx) is related to uncertainty in momentum Δp. This is *Heisenberg's uncertainty principle.*

Heisenberg stated his uncertainty principle as below:

It is impossible to specify accurately and simultaneously the value of both members of particular pairs of physical variables (viz., position and momentum; energy and time; angular momentum and angle) that describe the dynamical behaviour of an atomic system, the product of their uncertainties being always of the order of Planck's constant (h).

11.4.1 Statement of Uncertainty Principle

The principle states that, *in any simultaneous determination of the position and momentum of a particle, the product of the uncertainties is equal to or greater than Planck's constant (h).*

If Δx and Δp_x be the fundamental uncertainties in simultaneous measurement of *position and momentum* of a particle, then according to *uncertainty principle,*

$$\Delta x \cdot \Delta p_x \geq \frac{\hbar}{2} \cdot \left[\hbar = \frac{h}{2\pi}\right]. \tag{11.29}$$

In terms of energy (E) of the particle in time (t) in which it is measured, the uncertainty relation is

$$\Delta E \cdot \Delta t \geq \frac{\hbar}{2}, \tag{11.30}$$

where ΔE and Δt are the uncertainties in energy and time respectively.

In terms of angular momentum (L) and angular position (θ), the uncertainty relation is

$$\Delta L \cdot \Delta\theta \geq \frac{\hbar}{2}. \tag{11.31}$$

$\Delta\theta$ and ΔL are the uncertainties in *angular position* and *angular momentum* respectively.

11.4.2 Simple Proof of Heisenberg's Uncertainty Relation

If λ and p_x be the de Broglie wavelength and X-component of momentum of a particle travelling along X-direction, then the *de Broglie relation* gives

$$\lambda = \frac{h}{p_x}, \quad \therefore\ p_x = \frac{h}{\lambda} = \frac{h}{2\pi} \cdot \frac{2\pi}{\lambda} = \hbar k. \tag{11.32}$$

If the wave packet representing the particle is formed due to the superposition of two simple harmonic waves with propagation constant k and $k + \Delta k$, then the length of the wave packet is

$$\Delta x = \frac{2\pi}{\Delta k}, \tag{11.33}$$

Δx represents the uncertainty in position corresponding to the uncertainty Δk in propagation constant k.

From eqn (11.32),

$$\Delta p_x = \hbar \Delta k = \hbar \frac{2\pi}{\Delta x} \quad \text{or,} \quad \Delta x \Delta p_x = \hbar \cdot 2\pi = h. \tag{11.34}$$

The equality (=) is replaced by $\geq$, if the wave packet differs in shape from that shown in Fig. 11.6.

$$\therefore\ \Delta x \Delta p_x \geq h. \tag{11.35}$$

According to *new quantum theory,*

$$\Delta x \Delta p_x \geq \frac{\hbar}{2}. \tag{11.36}$$

Physical Significance of Uncertainty Relation

(1) From relation (11.34) we note that if $\Delta x = 0$, then

$$\Delta p_x = \frac{h}{\Delta x} = \frac{h}{0} = \infty,$$

which means, if the position of a moving particle is determined accurately in simultaneous measurement, then the uncertainty in momentum will be infinite.

Similarly, if the momentum (p_x) of the particle is accurately determined at some instant, then the uncertainty in the position coordinate (x) will be infinite, i.e., $\Delta x = \infty$, if $\Delta p_x = 0$.

Thus, in any simultaneous measurement of position and momentum the accuracy is limited by the uncertainty relation.

(2) From relation (11.36)

$$\Delta x(m\Delta v_x) \geq \frac{\hbar}{2} \quad \text{or,} \quad \Delta x \Delta v_x \geq \frac{\hbar}{2m} \quad [\because\ p_x = mv_x], \tag{11.37}$$

where v_x is the velocity of the particle along X-direction.

$\because \frac{\hbar}{m} \to 0$, $\Delta x \Delta v_x \to 0$, for such particle the position and velocity can be determined almost accurately.

If mass $m = \infty$, i.e., for heavy bodies, $\frac{\hbar}{m} = 0$, then position and momentum of such bodies can be determined with perfect accuracy. This is the limiting case of classical mechanics. Hence, *classical mechanics gives correct result for macroscopic bodies but quantum mechanics is true for microscopic bodies like electrons, protons, etc.*

11.4.3 Simple Proof of Uncertainty Relation between Energy and Time

The group velocity of a wave packet (v_g) representing a moving particle is equal to the particle velocity (v_x) along X-direction. If Δx be the width of the wave packet along X-direction, then the time required to move through Δx is

$$\Delta t = \frac{\Delta x}{v_g} = \frac{\Delta x}{v_x}. \tag{11.38}$$

Now, the KE of the particle is $E = \frac{p_x^2}{2m}$.

$$\therefore \Delta E = \frac{2p_x \Delta p_x}{2m} = \frac{p_x}{m}\Delta p_x = v_x \Delta p_x, \tag{11.39}$$

where ΔE is the uncertainty in energy E.

From eqns (11.38) and (11.39), we get $\Delta E \cdot \Delta t = \Delta x \cdot \Delta p_x$.

$$\therefore \Delta E \Delta t \geq \frac{\hbar}{2}. \tag{11.40}$$

Eqn (11.40) represents uncertainty relation in energy and time.

Physical significance of energy-time uncertainty relation: From relation (11.40) we note that $\Delta E \to \infty$, if $\Delta t \to 0$, i.e., *if the systems remains in energy state E for a minimum time, then the uncertainty in the determination of energy E will be maximum and the converse is also true.*

11.5 Illustration of Heisenberg's Uncertainty Principle by Thought Experiments

11.5.1 Diffraction of Electrons through Single-Slit

Let a mono-energetic electron beam pass through a slit of width Δy, along positive X-direction. Since the moving electron is associated with de Broglie wavelength, $\lambda = \frac{h}{p}$, where p is the momentum of the electron, the passing electron beam may be thought as a monochromatic bunch of waves [Fig. 11.7].

After diffraction through the slit, single-slit diffraction pattern will be observed on the screen (S). If the first minimum (A) after the central maximum be formed at an angle of diffraction (θ), then the condition is

$$\Delta y \sin\theta = \lambda \quad \text{or,} \quad \Delta y = \frac{\lambda}{\sin\theta}. \tag{11.41}$$

[$\because n = 1$ for the first minimum.]

As the electrons pass through the slit of width Δy, they can pass through any point in it. So the uncertainty in position of the electrons is Δy.

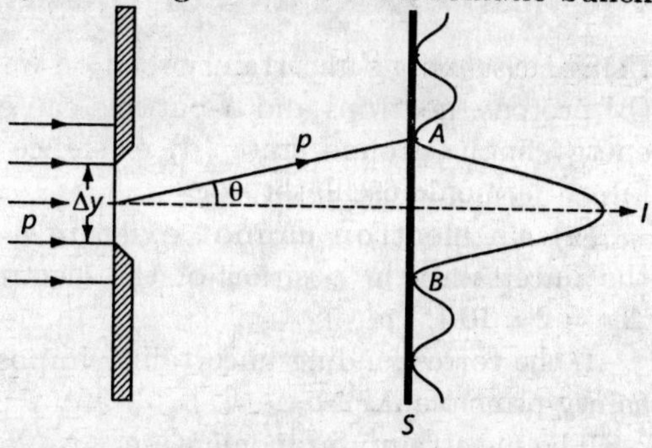

Fig. 11.7: Diffraction of electron through single slit.

Before diffraction the momentum of an electron is p but after diffraction it has momentum components ($p\sin\theta$) to ($-p\sin\theta$) along Y-axis. This implies that the uncertainty in momentum of the electron is

$$\Delta p_y = p\sin\theta - (-p\sin\theta) = 2p\sin\theta = 2\frac{h}{\lambda}\sin\theta \quad \left[\because \lambda = \frac{h}{p}\right]. \tag{11.42}$$

Now, multiplying eqns (11.41) and (11.42), we get

$$\Delta y \Delta p_y = \frac{\lambda}{\sin\theta} \times \frac{2h}{\lambda}\sin\theta = 2h \quad \text{or,} \quad \Delta y \Delta p_y \simeq h. \tag{11.43}$$

Thus, the truth of uncertainty principle is established.

11.5.2 Gamma Ray Microscope (Thought) Experiment

Let the *position* and *momentum* of an electron have to be determined simultaneously using a γ-ray microscope [Fig. 11.8]. If θ be the semi-vertical angle of the objective of the microscope and λ the wavelength of γ-ray, then the resolving power of the microscope is

$$\Delta x = \frac{\lambda}{2\sin\theta}. \tag{11.44}$$

To reduce the uncertainty in position Δx, we must reduce λ and this can be obtained by using γ-rays of short λ.

To determine the position of the electron at least one γ-ray photon scattered by the electron must enter the objective. Now, the scattered photon can enter the objective between $+\theta$ to $-\theta$. Then the recoil electron will have uncertainty in momentum. The X-component of momentum of the electron will lie between $\left(+\frac{h\nu}{c}\sin\theta\right)$ and $\left(-\frac{h\nu}{c}\sin\theta\right)$ [Fig. 11.8].

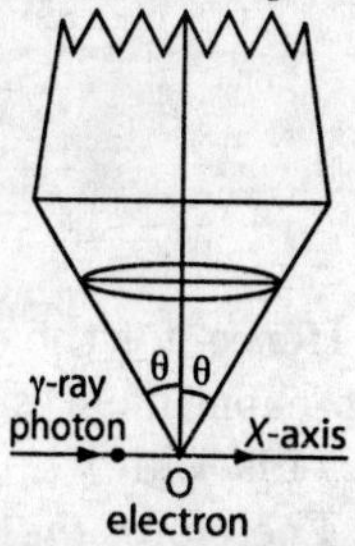

Fig. 11.8: γ-ray microscope experiment.

Thus, the uncertainty in the X-component of momentum of the electron

$$\Delta p_x = \frac{h\nu}{c}\sin\theta - \left(-\frac{h\nu}{c}\sin\theta\right) = \frac{2h\nu}{c}\sin\theta = \frac{2h}{\lambda}\sin\theta. \tag{11.45}$$

In view of conservation of momentum, this is also the uncertainity in X-component of momentum of the photon.

Multiplying eqns (11.44) and (11.45), we get

$$\Delta x \Delta p_x = \frac{\lambda}{2\sin\theta} \times \frac{2h}{\lambda}\sin\theta \quad \text{or,} \quad \Delta x \Delta p_x = h. \tag{11.46}$$

A rigorous deduction using new quantum theory shows that $\Delta x \Delta p_x \geq \frac{\hbar}{2}$.

N.B. For details see *Perspective of Quantum Mechanics* (p. 174) by the author.

11.6 Applications of Uncertainty Relation

Using Heisenberg's uncertainty relation we can show that (a) an electron cannot exist in a nucleus (b) protons, neutrons and α-particle can exist within the nucleus (c) size of hydrogen atom and energy in the ground state (d) existence of finite zero point energy (e) minimum energy of a linear harmonic oscillator.

(a) **An electron cannot exist in a nucleus**: Let the electron exist in the nucleus. Then the uncertainty in position of the electron will be the same as the diameter of the nucleus $\Delta x = 2 \times 10^{-14}$ m.

If the corresponding uncertainty in position of the electron be Δp_x, then according to uncertainty principle $\Delta x \Delta p_x \geq \hbar$.

The uncertainty in momentum

$$\Delta p_x \geq \frac{\hbar}{\Delta x}.$$

$\therefore$ the minimum uncertainty in Δp_x is

$$\Delta p_x = \frac{\hbar}{\Delta x} = \frac{6.63 \times 10^{-34}}{2 \times 3.14 \times (2 \times 10^{-14})} = 5.278 \times 10^{-21} \text{ kg-m/s}.$$

So if the electron exists in the nucleus, then its minimum momentum $p_{\text{min}} = 5.278 \times 10^{-21}$ kg-m/s.

Now, for the electron

$$\begin{aligned} E_{\text{min}}^2 &= p_{\text{min}}^2 c^2 + m_0^2 c^4 \\ &= [5.278 \times 10^{-21} \times 3 \times 10^8] + (9.11 \times 10^{-31})^2 \times (3 \times 10^8)^4 \\ &= \left[2.5 \times 10^{-24} + 6.72 \times 10^{-27}\right] \text{ J}^2. \end{aligned}$$

Neglecting the second term due to smallness,

$$\begin{aligned} E_{\text{min}} &= \sqrt{2.5 \times 10^{-24}} = 1.58 \times 10^{-12} \text{ J} \\ &= \frac{1.58 \times 10^{-12}}{1.6 \times 10^{-19}} \text{ eV} \quad [\because 1 \text{ eV} = 1.6 \times 10^{-19} \text{ J}] \\ &= 9.875 \times 10^6 \text{ eV} = \mathbf{9.875\ MeV}. \end{aligned}$$

Hence, for the electron to be present in the nucleus its minimum energy should be $\sim$ 10 MeV. But experiment shows that the maximum energy of the β-particles emitted from radioactive nuclei is 4 MeV.

Therefore, *the electron cannot exist within the nucleus.*

(b) **Protons, neutrons and α-particles can exist within the nucleus**: If proton, neutron or, α-particle exist in the nucleus, then its position uncertainty must not exceed the nuclear diameter $\sim 2 \times 10^{-14}$ m, i.e., $\Delta x = 2 \times 10^{-14}$ m.

According to uncertainty relation, $\Delta x \Delta p_x \approx \hbar$.

$\therefore$ the uncertainty in the X-component of momentum is

$$\Delta p_x = \frac{\hbar}{\Delta x} = \frac{6.63 \times 10^{-34}}{2\pi \times (2 \times 10^{-14})} = 5.3 \times 10^{-21} \text{ kg-m/s}.$$

Thus, the momentum of the proton, neutron or, α-particle must have the minimum value $p_{\text{min}} = 5.3 \times 10^{21}$ kg-m/s.

Now, the minimum energy corresponding to p_{min} is

$$(E_k)_{\text{min}} = \frac{p_m^2}{2m_0} = \frac{(5.3 \times 10^{-21})^2}{2 \times 1.67 \times 10^{-27}} = 8.4 \times 10^{-15} \text{ J},$$

where the rest-mass of the proton or neutron is

$$m_0 = 1.67 \times 10^{-27} \text{ kg} \quad \text{or,} \quad (E_k)_{\min} = \frac{8.4 \times 10^{-15}}{1.6 \times 10^{-19}} = \mathbf{52\ keV}.$$

The mass of an α-particle being 4 times m_0, the minimum energy of an α-particle for the existence in the nucleus is 13 keV.

Experiments show that the energy carried by a proton or a neutron emitted from a nucleus is greater than 52 keV, their existence within the nucleus is *quite justified.* Also the energy with which an α-particle is emitted from the nucleus is greater than 13 keV and hence *it can exist in the nucleus.*

(c) **Size of hydrogen atom and its energy in the ground state**: If the electron in the H-atom moves round the nucleus in a circular path of radius r, then the position uncertainty of the electron is $\Delta x = r$.

From the uncertainty principle,

$$\Delta x \Delta p_x = \hbar \quad \text{or,} \quad \Delta p_x = \frac{\hbar}{\Delta x} = \frac{\hbar}{r},$$

where Δp_x is the corresponding uncertainty in the X-component of momentum.

Note that according to new quantum mechanics h in Eqn (11.34) can be be replaced by $\hbar$ (or, $\hbar/2$) if a large number of wave superpose to produce wave packet.

The minimum momentum of the electron is

$$p = \Delta p_x = \frac{\hbar}{r}.$$

$\therefore$ the KE of the electron

$$E_k = \frac{p^2}{2m} = \frac{\hbar^2}{2mr^2}.$$

The PE of the electron is $V = -\frac{e^2}{4\pi\epsilon_0 r}$.

$\therefore$ the total energy of the electron is

$$E = E_k + V = \frac{\hbar^2}{2mr^2} - \frac{e^2}{4\pi\epsilon_0 r}.$$

$$\therefore \quad \frac{dE}{dr} = -\frac{\hbar^2}{mr^3} + \frac{e^2}{4\pi\epsilon_0 r^2}.$$

For E to be minimum when the atom is in the ground state, $\frac{dE}{dr} = 0$ and $\frac{d^2E}{dr^2} = -ve.$

$$\therefore -\frac{\hbar^2}{mr^3} + \frac{e^2}{4\pi\epsilon_0 r^2} = 0 \quad \text{or,} \quad r = \frac{4\pi\epsilon_0\hbar^2}{me^2} = \frac{4\pi\epsilon_0}{me^2}\left(\frac{h}{2\pi}\right)^2 \quad \text{or,} \quad r = \frac{\epsilon_0 h^2}{\pi me^2}. \tag{11.47}$$

Now,

$$\frac{d^2E}{dr^2} = \frac{3\hbar^2}{mr^4} - \frac{e^2}{2\pi\epsilon_0 r^3} = \frac{1}{r^3}\left[\frac{3\hbar^2}{mr} - \frac{e^2}{2\pi\epsilon_0}\right]$$

$$= \frac{1}{r^3}\left[\frac{3\hbar^2}{m} \cdot \frac{me^2}{4\pi\epsilon_0\hbar^2} - \frac{e^2}{2\pi\epsilon_0}\right] \quad \text{[using value of } r \text{ from eqn (11.47)]}$$

$$= \frac{1}{r^3}\left[\frac{3e^2}{4\pi\epsilon_0} - \frac{e^2}{2\pi\epsilon_0}\right] = +\text{ve quantity.}$$

So when $r = \frac{\epsilon_0 h^2}{\pi me^2}$, the energy is minimum and the H-atom is in the ground state.

Putting $h = 6.63 \times 10^{-34}$ J-s

$m = 9.11 \times 10^{-31}$ kg

$e = 1.6 \times 10^{-19}$ C

and $\epsilon_0 = 8.85 \times 10^{-12}$ C^2/N-m^2

eqn (11.47), we get

$$r = \frac{(8.85 \times 10^{-12}) \times (6.63 \times 10^{-34})^2}{3.14 \times (9.11 \times 10^{-31}) \times (1.6 \times 10^{-19})^2} = 5.31 \times 10^{-11} \text{ m.}$$

Hence, the size of H-atom is like a sphere of radius 5.31×10^{11} m.

Energy in the ground state: The ground state energy is

$$E = \frac{\hbar^2}{2mr^2} - \frac{e^2}{4\pi\epsilon_0 r} = \frac{1}{r}\left[\frac{\hbar^2}{2m} \cdot \frac{1}{r} - \frac{e^2}{4\pi\epsilon_0}\right]$$

$$= \frac{me^2}{4\pi\epsilon_0\hbar^2}\left[\frac{\hbar^2}{2m} \cdot \frac{me^2}{4\pi\epsilon_0\hbar^2} - \frac{e^2}{4\pi\epsilon_0}\right] \quad \text{[using the value of } r \text{ from eqn (11.47)]}$$

$$= -\frac{me^4}{32\pi^2\epsilon_0^2\hbar^2} = -\frac{me^4}{32\pi^2\epsilon_0^2}\left(\frac{2\pi}{h}\right)^2 = -\frac{me^4}{8\epsilon_0^2 h^2}. \qquad (11.48)$$

Substituting the values of m, e, ϵ_0 and h in eqn (11.48), we find $E = -13.6$ eV, which is the ground state energy of the H-atom.

(d) **Existence of finite zero point energy**: If a particle of rest-mass m_0 be confined to move in a potential of linear dimension l, then the position uncertainty is $\Delta x = l/2$.

According to *uncertainty principle*, $\Delta p_x \Delta x = \hbar/2$ or the uncertainty in the corresponding momentum $\Delta p_x = \frac{\hbar}{2\Delta x} = \hbar/l$.

If the minimum value of momentum p_x is Δp_x, i.e., $p_x = \Delta p_x = \hbar/l$, then the lowest value of KE of the particle is

$$E_{\min} = \frac{p^2}{2m_0} = \frac{p_m^2}{2m_0} = \frac{\hbar^2}{2m_0 l^2}.$$

So the energy of a particle confined in a space cannot have zero energy, instead it will have a minimum energy ($E_{\min}$).

(e) **Minimum energy of a linear harmonic oscillator**: If x be the position of the oscillator of mass m and ω its angular velocity, then the total energy at any instant t is

$$E = \frac{p^2}{2m} + \frac{1}{2}m\omega^2 x^2.$$

The maximum uncertainty in its position is $\Delta x = x$, then the minimum uncertainty Δp in momentum is

$$\Delta p = \frac{\hbar}{\Delta x} = \frac{\hbar}{x} \quad \text{[from uncertainty principle].}$$

The minimum value of momentum of p is $p = \hbar/x$.

$$\therefore E = \frac{1}{2m}\left(\frac{\hbar}{x}\right)^2 + \frac{1}{2}m\omega^2 x^2 = \frac{\hbar^2}{2mx^2} + \frac{1}{2}m\omega^2 x^2. \qquad (11.49)$$

$$\therefore \frac{dE}{dx} = -\frac{\hbar^2}{mx^3} + m\omega^2 x.$$

For E to be minimum when the oscillator is in the ground state $\frac{dE}{dx} = 0$.

$$\therefore -\frac{\hbar^2}{mx^3} + m\omega^2 x = 0 \quad \text{or,} \quad m\omega^2 x = \frac{\hbar^2}{mx^3} \quad \text{or,} \quad x^4 = \frac{\hbar^2}{m^2\omega^2} \quad \text{or,} \quad x^2 = \frac{\hbar}{m\omega}.$$

If this value of x^2 is put in the expression of $\frac{d^2E}{dx^2}$, it comes out to be a negative quantity. So E is the minimum energy of the oscillator, from eqn (11.49) is

$$E_{\min} = \frac{\hbar^2}{2m} \cdot \left(\frac{m\omega}{\hbar}\right) + \frac{1}{2}m\omega^2\left(\frac{\hbar}{m\omega}\right) = \hbar\omega.$$

Taking $\Delta x \Delta p_x = \frac{\hbar}{2}$, we get $E_{\min} = \frac{1}{2}\hbar\omega$.

11.7 Heisenberg's Uncertainty Principle from de Broglie Wave Concept

According to de Broglie wave concept (i) a particle in motion can be taken as a group of waves (ii) the group velocity of the wave group (v_g) is equal to the particle velocity (v).

Let a group of waves consist of only two component waves with equal amplitude and frequencies ω_1 and $\omega_2 (\omega_1 \simeq \omega_2)$. The equations of the waves are

$$\psi_1 = a\cos(\omega_1 t - k_1 x)$$
$$\text{and} \quad \psi_2 = a\cos(\omega_2 t - k_2 x),$$

where ω_1/k_1 and ω_2/k_2 are the respective phase velocities.

The resultant motion is

$$\begin{aligned}\psi &= \psi_1 + \psi_2\\ &= a[\cos(\omega_1 t - k_1 x) + \cos(\omega_2 t - k_2 x)]\\ &= 2a\cos(\omega t - kx)\cos\left(\frac{\Delta\omega}{2}t - \frac{\Delta k}{2}t\right),\end{aligned}$$

where $\omega = \frac{\omega_1+\omega_2}{2}$, $k = \frac{k_1+k_2}{2}$, $\Delta\omega = \omega_1 - \omega_2$, $\Delta k = k_1 - k_2$.

The resultant motion shows loops [cf. Fig. 11.1], which moves with group velocity $v_g = \frac{\Delta\omega}{\Delta k}$. The loop is equivalent to the position of the particle. So there is uncertainty in position of the particle as it can be found at any point within the loop. The error in the measurement in position of the particle is thus equal to the distance between two consecutive nodes.

Now, the condition for the formation of nodes at x is given by

$$\cos\left(\frac{\Delta\omega}{2}t - \frac{\Delta k}{2}x\right) = 0 \quad \text{or,} \quad \frac{\Delta\omega}{2}t - \frac{\Delta k}{2}x = (2n+1)\frac{\pi}{2} \quad [n = 0, 1, 2, \cdots].$$

If x_1 and x_2 be the positions of two consecutive nodes from the origin, then

$$\frac{\Delta\omega}{2}t - \frac{\Delta k}{2}x_1 = (2n+1)\frac{\pi}{2}$$
$$\text{and} \quad \frac{\Delta\omega}{2}t - \frac{\Delta k}{2}x_2 = (2n+3)\frac{\pi}{2}$$
$$\text{or,} \quad \frac{\Delta k}{2}(x_1 - x_2) = \pi$$
$$\text{or,} \quad \Delta x = \frac{2\pi}{\Delta k}. \tag{11.50}$$

Also

$$k = \frac{2\pi}{\lambda} = \frac{2\pi p}{h} \qquad \left[\because \lambda = \frac{h}{p}\right]$$
$$\text{or,} \quad \Delta k = \frac{2\pi}{h}\Delta p$$
$$\text{or,} \quad \Delta p = \frac{h\Delta k}{2\pi}. \tag{11.51}$$

$\therefore$ from eqns (11.50) and (11.51),

$$\Delta x \cdot \Delta p = \frac{2\pi}{\Delta k} \times \frac{(\Delta k)h}{2\pi} = h. \tag{11.52}$$

Eqn (11.52) is the uncertainty principle, where Δp is the uncertainty in momentum measurement of the particle

For a large number of component waves forming the wave packet and having continuous varying wavelengths

$$\Delta x \Delta p \geq \tfrac{1}{2}\hbar. \qquad (11.53)$$

11.8 Answers of Some Important Questions

11.1. *Justify that an electron can be treated more as a wave than a particle.*
Ans. The de Broglie wavelength associated with an electron of energy say 100 eV is 1.33 Å, which is much greater than the size of an electron. It may, therefore, be treated as a wave than a particle.
11.2. *Show that the de Broglie wave process does not affect the interaction between macroscopic bodies.*
Ans. For a macroscopic body, say a cricket ball of mass 500 gm, moving with a velocity of 50 km/hr, has de Broglie wavelength

$$\lambda = \frac{h}{mv} = \frac{6.62 \times 10^{-34} \text{ J-s}}{0.5 \text{ kg} \times 13.9 \text{ m/s}} = 10^{-34} \text{ m} = 10^{-24} \text{ Å}, \qquad (11.54)$$

which is insignificant in comparison to the size of the ball. Hence, we can say that *de Broglie wave process does not affect the interacton between macroscopic bodies.*
11.3. *Comment on the nature of de Broglie wave.*
Ans. de Broglie waves are generated due to the motion of particles, *whether charged or uncharged.* The velocity of the matter wave depends on the velocity of the particle and is greater than the velocity of light. The waves associated with the particles are *not real three-dimensional waves* like sound waves, but are *probability waves* related to the probabilities of finding the particles in various places and with various properties. Thus, de Broglie waves are only theoretical concepts employed to explain the experimental results.
11.4. *What is de Broglie model of atom?*
Ans. According to de Broglie, the orbit of an electron contains an illegal number of waves associated with the electron so that the angular momentum is quantized, i.e., $L = n\hbar$, ($\hbar = h/2\pi$). In view of this we can picture the electron in two ways, (i) a particle moving in an orbit with a certain quantized value of angular momentum (L) or (ii) as a standing de Broglie wave occupying a certain region around the nucleus.
11.5. *Show that only a wave packet and not a monochromatic wave can represent a particle moving in space.*
Ans. If possible let us suppose that a particle (an electron) is described by a wave $\psi(x,t) = A\sin(kx - \omega t)$, where A is the amplitude, ω the angular frequency and k the wave vector. The above equation represents a pure sine wave extending from $-\infty$ to $+\infty$. Though it has a precise wavelength λ and momentum $\hbar k$ it is of *infinite extent* and *completely non-localized.* But an electron is confined in a small volume. So a monochromatic wave $\psi(x,t)$ cannot represent a particle. Hence, the de Broglie waves are *not harmonic waves.*

Several monochromatic waves of slightly different frequency can produce wave packet on superposition. These wave packets are confined to a small extension and can represent particles. The separation between the successive maxima in a wave packet is λ_{av} and it has a particle-like location in space. Hence, *only a wave packet and not a monochromatic wave can represent a moving particle.*
11.6. *Distinguish between matter waves and electromagnetic waves.*
Ans. See Section 11.3.2 (p. 296).
11.7. *State Heisenberg's uncertainty principle. What is the origin of this principle?*

Ans. The uncertainty principle states that, in any simultaneous measurement of position and momentum of a particle, the product of uncertainties is equal to or greater than Planck's constant.

If Δx and Δp_x be the uncertainties of position and momentum, then according to this principle $\Delta x \Delta p_x \geq \hbar/2$, where $\hbar = h/2\pi$, h being the Planck's constant.

The origin of the uncertainty principle is due to the wave nature of particle. Because of this wave nature the position and momentum of a moving particle cannot be determined accurately.

11.8. *In γ-ray microscope experiment, justify the use of γ-rays.*

Ans. The limit of resolution of a microscope is given by $\Delta x = \frac{\lambda}{2\sin\theta}$, where λ is the wavelength of light and θ is the semi-vertical angle of the light entering the objective. Here, Δx is the uncertainty in position of the electron and $\Delta x \propto \lambda$, when θ is constant for a particular microscope. Since λ for γ-ray is smaller than ordinary light, γ-ray is used to reduce Δx. Of course γ-ray does not have visibility producing capacity and hence the experiment is called the *thought experiment*.

11.9. *Prove that the mass of a photon is zero. What is its momentum?*

Ans. If λ be the de Broglie wavelength and m_0, the rest-mass of the photon, then its phase velocity is given by

$$v_p = c\left[1 + \frac{m_0^2 c^2 \lambda^2}{h^2}\right]^{1/2}.$$

But for a photon, $v_p = c$, the velocity of light.

$$\therefore c = c\left[1 + \frac{m_0^2 c^2 \lambda^2}{h^2}\right]^{1/2} \quad \text{or,} \quad \frac{m_0^2 c^2 \lambda^2}{h^2} = 0. \tag{11.55}$$

As c, λ and h can never be zero, $m_0 = 0$.

Again, the relativistic energy E of the photon is $E = \sqrt{p^2c^2 + m_0^2 c^4}$.

$$\because \; E = h\nu,$$

$$\therefore h\nu = \sqrt{p^2c^2 + m_0^2c^4} = pc \quad [\because m_0 = 0].$$

$\therefore$ the momentum of the photon is $p = \frac{h\nu}{c} = \frac{\hbar\omega}{c}$.

11.10. *Justify that in macroscopic world, classical mechanics gives correct result.*

Ans. If Δx and Δp_x be the uncertainties in position and momentum of a moving particle, then according to the uncertainty relation

$$\Delta x \Delta p_x \geq \frac{\hbar}{2} \quad \text{or,} \quad \Delta x \Delta v_x \geq \frac{\hbar}{2m}. \tag{11.56}$$

For heavy bodies the mass $m = \infty$ and hence $\hbar/m = 0$, then $\Delta x \Delta v_x = 0$.

Then the position and momentum of the body can be correctly determined. Thus, classical mechanics is an appropriate mechanics in classical world, which can give correct result.

11.11. *Derive an expression for the thermal de Broglie wavelength of a particle.*

Ans. If m and E_k be the mass and KE of an electron, then its momentum $p = \sqrt{2mE_k}$ [for non-relativistic motion].

From the theorem of equi-partition of energy $E_k = \frac{3}{2}kT$, where k is the Boltzmann constant and T is the absolute temperature.

$$\therefore p = \sqrt{2m \times \frac{3}{2}kT} = \sqrt{3mkT}.$$

Now, the de Broglie wavelength of the particle is

$$\lambda = \frac{h}{p} = \frac{h}{\sqrt{3mkT}}.$$

11.9 Solved Problems

Ex. 1 *Calculate de Broglie wavelength of a neutron of energy* 1 MeV. **[PU 1993]**

Soln. If m and E be the mass and energy of the neutron, then its de Broglie wavelength is given by

$$\lambda = \frac{h}{\sqrt{2mE}}.$$

Here $m = 1.67 \times 10^{-27}$ kg, $E = 1$ MeV $= 10^6$ eV $= 10^6 \times 1.6 \times 10^{-19}$ J $= 1.6 \times 10^{-13}$ J.

$$\therefore \lambda = \frac{6.62 \times 10^{-34}}{\sqrt{2 \times (1.6 \times 10^{-27}) \times (1.6 \times 10^{-13})}} = \mathbf{2.93 \times 10^{-14}\ m.}$$

Ex. 2 *What is the de Broglie wavelength of an electron of energy* 1 MeV.

Soln. Since the electron energy is 1 MeV, i.e., greater than 0.5 MeV, relativistic formula is to be applied.

Now, KE of the electron $E_k = 1$ MeV $= 10^6 \times 1.6 \times 10^{-19}$ J $= 1.6 \times 10^{-13}$ J.

$$m_0c^2 = (9.11 \times 10^{-31}) \times (3 \times 10^8)^2 = 9.11 \times 9 \times 10^{-15}\ \text{J}$$

$$= \frac{9.11 \times 9 \times 10^{-15}}{1.6 \times 10^{-19}}\ \text{eV} = 5.12 \times 10^5\ \text{eV}.$$

So the de Broglie wavelength of the electron is

$$\lambda = \frac{hc}{\sqrt{E_k(E_k + 2m_0c^2)}} = \frac{6.63 \times 10^{-34} \times 3 \times 10^8}{\sqrt{1.6 \times 10^{-13} \times (1.6 \times 10^{-13} + 2 \times 1.6 \times 0.512 \times 10^{-13})}}$$

$$= \frac{6.63 \times 3 \times 10^{-13}}{1.6 \times 1.4227} = 0.008739 \times 10^{-10}\ \text{m} = \mathbf{0.008739\ Å}.$$

Ex. 3 *Show that if a particle moves with velocity* 0.707c, *where c is the velocity of light in vacuum; then its de Broglie wavelength and Compton wavelength become equal.*

Soln. A particle of rest-mass m_0 moving with velocity v has momentum

$$p = mv = \frac{m_0 v}{\sqrt{1 - \frac{v^2}{c^2}}},$$

where the mass of the particle moving with velocity v is

$$m = \frac{m_0}{\sqrt{1 - \frac{v^2}{c^2}}}.$$

Now, the de Broglie wavelength of the particle is

$$\lambda = \frac{h}{p} = \frac{h}{\frac{m_0 v}{\sqrt{1 - \frac{v^2}{c^2}}}} = \frac{h}{m_0 c}\,\frac{\sqrt{1 - \frac{v^2}{c^2}}}{v/c}.$$

For $v = 0.707c = \frac{c}{\sqrt{2}}$, we have

$$\lambda = \frac{h}{m_0 c}\,\frac{\sqrt{1 - \frac{1}{2}}}{\frac{1}{\sqrt{2}}} = \frac{h}{m_0 c}$$

$$\text{or,}\quad \lambda = \lambda_c \quad \left[\text{where } \lambda_c = \frac{h}{m_0 c} \text{ is the Compton wavelength}\right].$$

Ex. 4 *What voltage must be applied to an electron microscope to produce electrons of wavelength* 0.40?

Soln. The de Broglie wavelength of an electron moving under a voltage V volt is

$$\lambda = \frac{h}{\sqrt{2m\,eV}}.$$

Here, $\lambda = 0.40 \times 10^{-10}$ Å, $h = 6.62 \times 10^{-34}$ J-s, $m = 9.1 \times 10^{-31}$ kg.

$$\therefore \lambda = \frac{6.62 \times 10^{-34}}{\sqrt{2 \times (9.1 \times 10^{-31}) \times (1.60 \times 10^{-19})}} \cdot \frac{1}{\sqrt{V}} \quad [\because\ 1 \text{ eV} = 1.60 \times 10^{-19} \text{ J}]$$

$$\text{or,} \quad \sqrt{V} = \frac{6.62 \times 10^{-34}}{(0.4 \times 10^{-10}) \times \sqrt{2 \times (9.1 \times 10^{-31}) \times (1.6 \times 10^{-19})}} = 30.6$$

$$\therefore \quad V = (30.6)^2 = \mathbf{936.3\ V.}$$

Ex. 5 *What is the de Broglie wavelength of an electron, which has been accelerated from rest through a potential difference of* 1000 V.

Soln. de Broglie wavelength of an electron moving under a potential of V volt is $\lambda = \frac{12.25}{\sqrt{V}}$ Å.

Here $V = 100$ volt, $\therefore \lambda = \frac{12.25}{\sqrt{100}} = \mathbf{1.225\ Å.}$

Ex. 6 *Show that an electron moving through a potential difference of* 150 V *has a de Broglie wavelength of* 1 Å.

Soln. Now, the de Broglie wavelength of the electron moving through a P.D. of V volt is

$$\lambda = \frac{h}{\sqrt{2m\,eV}} = \frac{12.25}{\sqrt{V}} = \frac{12.25}{\sqrt{150}} = 1 \text{ Å } \textbf{(Proved).}$$

Ex. 7 *An electron moves in a circular path of radius* 2 cm, *under the influence of a magnetic field of* 0.1 *tesla applied perpendicular to the plane of motion. Calculate the de Broglie wavelength of the electron.* **[BUH 2001]**

Soln. The *Lorentz force* on the electron is $\vec{F} = e(\vec{v} \times \vec{B})$ or, $F = evB$ $[\because \vec{B} \perp \vec{v}]$.

Again, for equilibrium $F = \frac{mv^2}{r}$, where m is the mass of the electron moving with velocity v in a circular path of radius r.

$$\therefore evB = \frac{mv^2}{r} \quad \text{or,} \quad mv = p = erB \quad [p = \text{the momentum of the electron}]]$$

$$\text{or, } \lambda = \frac{h}{mv} = \frac{h}{erB}.$$

Here $h = 6.62 \times 10^{-34}$ J, $r = 2$ cm $= 0.02$ m, $e = 1.6 \times 10^{-19}$ C and $B = 0.1$ tesla.

$$\therefore \lambda = \frac{6.62 \times 10^{-34}}{(1.6 \times 10^{-19}) \times (0.02) \times (0.1)} = \mathbf{0.02\ Å.}$$

Ex. 8 *Calculate the de Broglie wavelength of electrons of energy* 10^4 eV *and compare it with the wavelength of em radiation for which the photon has the same energy.* **[CUH 1997]**

Soln. The de Broglie wavelength of the electron of energy 10^4 eV is given by

$$\lambda = \frac{h}{mv} = \frac{h}{\sqrt{2m\,eV}} = \frac{6.6 \times 10^{-34}}{\sqrt{2 \times (9.1 \times 10^{-31}) \times 10^4 \times (1.6 \times 10^{-19})}} = 0.122 \times 10^{-10} \text{ m} = 0.122 \text{ Å},$$

where $h = 6.6 \times 10^{-34}$ J-s, $m = 9.1 \times 10^{-31}$ kg, $eV = 10^4 \times (1.6 \times 10^{-19})$.

The wavelength of a photon of energy $\epsilon = 10^4$ eV is

$$\lambda_{ph} = \frac{hc}{\epsilon} = \frac{(6.6 \times 10^{-34}) \times (3 \times 10^8)}{10^4 \times (1.6 \times 10^{-19})}$$
$$= \frac{19.8 \times 10^{-26}}{1.6 \times 10^{-15}} = 12.375 \times 10^{-11}\,\text{m}$$
$$= 1.2375 \times 10^{-10}\,\text{m} = 1.2375\,\text{Å}.$$
$$\therefore \frac{\lambda}{\lambda_{ph}} = \frac{0.122}{1.2375} = 0.0985 \simeq \mathbf{0.01.}$$

Ex. 9 *Calculate de Broglie wavelength of electrons of KE* 0.1 MeV *using both relativistic and non-relativistic expression for* λ.

Soln. Non-relativistic value of λ for 0.1 MeV electron is

$$\lambda_{nr} = \frac{h}{\sqrt{2m\,eV}} = \frac{12.26}{\sqrt{V}} = \frac{12.26}{\sqrt{0.1 \times 10^6}} = 0.0388\,\text{Å}.$$

The relativistic value of λ is

$$\lambda_r = \frac{ch}{e\sqrt{V(V + 1.02 \times 10^6)}} = \frac{12.41 \times 10^3}{\sqrt{V(V + 1.02 \times 10^6)}} = \frac{12.41 \times 10^3}{\sqrt{0.1 \times 10^6 (0.1 + 1.02) \times 10^6}} = 0.0371\,\text{Å}.$$

Thus, $\lambda_{nr} > \lambda_r$.

Ex. 10 *What is the de Broglie wavelength of a neutron of KE* 10 MeV? **[BUH 2002]**

Soln. The rest-mass energy of the neutron is

$$m_\circ c^2 = (1.675 \times 10^{-27}) \times (3 \times 10^8)^2\ \text{J} = 942\,\text{MeV}.$$

The KE of the given neutron is 10 MeV, which is much less than its rest-mass energy. So relativistic effect needs to be considered.

Now, the de Broglie wavelength of the neutron is

$$\lambda = \frac{h}{p} = \frac{h}{\sqrt{2m\,eV}} = \frac{6.62 \times 10^{-34}}{\sqrt{2 \times (1.675 \times 10^{-27}) \times (10.0 \times 10^6)}}$$
$$= \frac{6.62 \times 10^{-34}}{\sqrt{3.350 \times 10^{-20}}} = \frac{6.62}{\sqrt{3.35}} \times 10^{-24}\,\text{m}$$
$$= \frac{6.62}{1.83} \times 10^{-24} = 3.614 \times 10^{-24}\,\text{m} = \mathbf{3.614 \times 10^{-14}\,Å}.$$

Ex. 11 *Find the de Broglie wavelength of a* 15 keV *electron.* **[BUH 1998]**

Soln. Energy of the electron $= 15$ keV $= 15 \times 10^{-3}$ MeV $= 0.015$ MeV.

This energy being very small compared to the rest-mass energy of the electron 0.51 MeV, we must use non-relativistic formula.

So the de Broglie wavelength of the electron is

$$\lambda = \frac{h}{p} = \frac{h}{\sqrt{2m\,eV}} = \frac{6.62 \times 10^{-34}}{\sqrt{2 \times (9.11 \times 10^{-31}) \times (1.6 \times 10^{-19}) \times (15 \times 10^3)}}$$
$$= \frac{6.62 \times 10^{-34}}{\sqrt{437.28 \times 10^{-47}}} = \frac{6.62 \times 10^{-34}}{(\sqrt{4372.8}) \times 10^{-24}}$$
$$= \frac{6.62 \times 10^{-10}}{66.127} = 0.1 \times 10^{-10} = \mathbf{0.1\,Å}.$$

Ex. 12 *Find the de Broglie wavelength of* 15 keV *protons (rest-mass of the proton is* 1.67×10^{-27} kg).

Soln. For the proton $m_0c^2 = 9.39 \times 10^8$ eV.

The KE of the proton 15×10^3 eV is small compared to m_0c^2. So non-relativistic formula must be applied.

Hence,

$$\lambda = \frac{h}{\sqrt{2m_0\,eV}} = \frac{6.62 \times 10^{-34}}{\sqrt{2 \times (1.67 \times 10^{-27}) \times (1.6 \times 10^{-19}) \times 15 \times 10^3}} = \mathbf{0.0007396\ Å.}$$

Ex. 13 *Calculate the glancing angle at which electrons of* 100 eV *should fall on the lattice planes of a crystal to obtain a Bragg reflection in the first order. [The d-value (the lattice spacing) =* 2.15 Å.]

Soln. The de Broglie wavelength of 100 eV electron is

$$\lambda = \frac{12.26}{\sqrt{V}}\ Å = \frac{12.26}{\sqrt{100}} = 1.226\ Å.$$

From Bragg's law,

$$2d\sin\theta = n\lambda \quad \text{or,} \quad \sin\theta = \frac{\lambda}{2d}\ (\because n = 1) = \frac{1.226}{2 \times 2.15} = 0.285.$$

$$\therefore \theta = \mathbf{16.5^\circ}.$$

Ex. 14 *An electron of mass* 9.1×10^{-31} kg *has a speed of* 1 km/s *with an accuracy of* 0.05%. *Calculate the uncertainty with which the position of the electron can be located.* **[PU 1994]**

Soln. Error in velocity of the electron $= \frac{0.05}{100}$.

$\therefore$ error in the measurement of momentum

$$\Delta p = \Delta(mv) = m\Delta v = 9.1 \times 10^{-31} \times \left(1000 \times \frac{0.05}{100}\right) = 4.55 \times 10^{-31}\ \text{kg-m/s}.$$

So error in the measurement of position

$$\Delta x = \frac{h}{\Delta p} = \frac{6.62 \times 10^{-34}}{4.55 \times 10^{-31}} = \mathbf{1.45 \times 10^{-3}\ m.}$$

Ex. 15 *The life time of an excited state of an atom is* 10^{-8} sec. *Calculate the minimum uncertainty in the determination of energy of the excited state.*

Soln. If ΔE and Δt be the uncertainty in the determination of energy E and time t, then according to uncertainty relation,

$$\Delta E \cdot \Delta t \geq \frac{h}{2\pi} \text{ or, } \Delta E \geq \frac{h}{2\pi \times \Delta t} \text{ or, } \Delta E \geq \frac{6.62 \times 10^{-34}}{2\pi \times 10^{-8}} \text{ or, } \Delta E \geq 10^{-26}\ \text{J} = \mathbf{6.25 \times 10^{-8}\ eV}.$$

Ex. 16 *The uncertainty in position of a particle is equal to its de Broglie wavelength, what is the uncertainty in its velocity?*

Soln. We have

$$\Delta p \Delta x = h \text{ or, } \Delta(mv)\Delta x = h.$$

$$\therefore \Delta v = \frac{h}{m(\Delta x)} = \frac{h}{m\lambda}\ (\because \Delta x = \lambda) = \frac{p}{m} = v.$$

Thus, the uncertainty in velocity is the velocity itself.

Ex. 17 *An electron is confined to a box of length* 10^{-9} m. *Calculate the minimum uncertainty in its velocity* [*mass of the electron* $= 9 \times 10^{-31}$ kg, $h = 6.6 \times 10^{-34}$ J-s].

Soln. If Δx, Δp be the uncertainty in position and momentum of the electron, then according to uncertainty principle, $\Delta x \Delta p = h$.

Here $\Delta x = 10^{-9}$ m.

$$\therefore \Delta p = \frac{h}{\Delta x} = \frac{6.6 \times 10^{-34}}{10^{-9}} = 6.6 \times 10^{-25} \text{ kg-m/s.}$$

Again, $\Delta p = m(\Delta v)$.

$$\therefore \Delta v = \frac{\Delta p}{m} = \frac{6.6 \times 10^{-25}}{9 \times 10^{-31}} = \mathbf{7.3 \times 10^5 \text{ m/s.}}$$

Ex. 18 *The speed of a bullet of mass* 50 g *is measured to be* 300 m/s *with an uncertainty of* 0.01%. *With what accuracy can we locate the position of the bullet if it is measured simultaneously with its speed.* **[HPU 1995]**

Soln. The uncertainty in the momentum measurement of the bullet is

$$\Delta p = \Delta(mv) = m\Delta v = \left(\frac{50}{1000}\right) \times 300 \times \frac{0.01}{100} = 15 \times 10^{-4} \text{ kg-m/sec.}$$

$\therefore$ uncertainty in the measurement of position

$$\Delta x \geq \frac{h}{\Delta p} \text{ or, } \Delta x \geq \frac{6.62 \times 10^{-34}}{15 \times 10^{-4}} \text{ or, } \mathbf{\Delta x \geq 4.41 \times 10^{-31} \text{ m.}}$$

[If we use $\Delta x \Delta p_x \geq \hbar$, then $\Delta x \geq 7 \times 10^{-32}$ m.]

Ex. 19 *Using the expression of phase velocity*

$$v_p = c\sqrt{1 + \frac{m_0^2 c^2 \lambda^2}{h^2}},$$

prove that the rest-mass of the photon $m_0 = 0$. *Determine its momentum.*

Soln. The phase velocity

$$v_p = c\sqrt{1 + \frac{m_0^2 c^2 \lambda^2}{h^2}}.$$

But the phase velocity of a photon v_p is equal to the velocity of light c in vacuum.

$$\therefore c = c\sqrt{1 + \frac{m_0^2 c^2 \lambda^2}{h^2}} \text{ or, } 1 = 1 + \frac{m_0^2 c^2 \lambda^2}{h^2}.$$

$$\therefore \frac{m_0^2 c^2 \lambda^2}{h^2} = 0 \text{ or, } m_0 = 0. \quad \left[\because \frac{c^2 \lambda^2}{h^2} \neq 0\right]$$

So the rest-mass of a photon $\mathbf{m_0 = 0}$.

Again, the relativistic expression for energy is $E = \sqrt{p^2c^2 + m_0^2 c^4}$.

Also $E = h\nu = \hbar\omega$ for photon.

$\therefore pc = \hbar\omega$ [$\because m_0 = 0$] or the momentum of the photon $\boldsymbol{p = \frac{\hbar\omega}{c}}$.

Ex. 20 *The angular momentum of the electron in the hydrogen atom is* $2\hbar$ *with an error of* 5%. *Show that its angular position in a circular orbit around the nucleus cannot be specified at all.*

Soln. From the uncertainty relation $\Delta\theta\Delta L \geq \hbar$.

But the uncertainty in angular momentum is

$$\Delta L = 2\hbar \times \frac{5}{100} = \frac{\hbar}{10}.$$

$\therefore$ the uncertainty in angular position is

$$\Delta\theta \leq \frac{\hbar}{\Delta L} = \frac{\hbar}{\hbar/10} = 10 \text{ radians.}$$

Since the angular position in a plane can never be greater than $2\pi = 6.283$ radians, with such a larger uncertainty (10 radians) the angular position of the electron cannot be specified.

Ex. 21 *To what accuracy the mass of a meson with life time of* 10^{-23} sec *can be determined?*

Soln. We have the time-energy uncertainty relation $\Delta E\Delta t \geq \hbar$ or, $\Delta(mc^2)\Delta t \geq \hbar$ or, $c^2\Delta m\Delta t \geq \hbar$.

Now, the minimum uncertainty in the mass determination of the meson is

$$(\Delta m)_{\min} = \frac{\hbar}{c^2\Delta t} = \frac{1.054 \times 10^{-34}}{(3 \times 10^8)^2 \times 10^{-23}} = \mathbf{1.171 \times 10^{-28}\ kg.}$$

Ex. 22 *An electron and proton has same de Broglie wavelength. Prove that the energy of the electron is greater.* **[WBUT 2008]**

Soln. If λ_e, λ_p be the de Broglie wavelength of the electron and proton and E_e, E_p the energy of the electron and proton respectively, then

$$\lambda_e = \frac{h}{\sqrt{2m_eE_e}} \text{ and } \lambda_p = \frac{h}{\sqrt{2m_pE_p}} \text{ or, } E_e = \frac{h^2}{2m_e\lambda_e} \text{ and } E_p = \frac{h^2}{2m_p\lambda_p},$$

where m_e and m_p are the mass of the electron and proton respectively.

$$\therefore \frac{E_e}{E_p} = \frac{h^2}{2m_e\lambda_e}\bigg/\frac{h^2}{2m_p\lambda_p} = \frac{m_p}{m_e} > 1.$$

$$\therefore E_e > E_p.$$

EXERCISES

11(A)–Objective Type Questions

Select the correct answer:

11.1. de Broglie's wavelength is associated with particles in motion only, if it is (a) charged (b) uncharged (c) both charged or uncharged (d) none of these.

11.2. de Broglie wavelength for a particle of mass m moving with velocity v is (a) $\lambda = \frac{mv}{h}$ (b) $\lambda = \frac{h}{mv}$ (c) $\lambda = \frac{hv}{m}$ (d) $\frac{m}{hv}$.

11.3. de Broglie wavelength of a body of mass m and KE E is (a) $\frac{2mh}{\sqrt{E}}$ (b) $\frac{h}{\sqrt{2mE}}$ (c) $\frac{h}{2mE}$ (d) $\frac{\sqrt{2mh}}{E}$. **[WBUT 2005]**

11.4. If v_g be the group velocity of the wave group representing a particle moving with velocity v, then (a) $v_g > v$ (b) $v > v_g$ (c) $v_g = v$ (d) $v_g = \frac{1}{v}$.

11.5. If v_p be the phase velocity of de Broglie wave and c the velocity of light, then (a) $v_p < c$ (b) $v_p > c$ (c) $v_p = c$ (d) $v_p = \frac{1}{c}$.

11.6. An electron being accelerated through a potential difference V volt, is associated with de Broglie wavelength (a) $\lambda = 12.27\sqrt{V}$ Å (b) $\lambda = \frac{12.27}{V}$ Å (c) $\lambda = \frac{12.27}{\sqrt{V}}$ (d) $\lambda = 12.27V$ Å.

11.7. de Broglie wavelength of a neutron of energy 1 MeV is (a) 2.86×10^{-14} m (b) 2.86×10^{-14} Å (c) 2.86×10^{14} Å (d) 2.86×10^{-10} m.

11.8. The de Broglie wave representing a moving particle is a (a) monochromatic wave (b) a wave group of continuously varying wavelength (c) a progressive wave (d) a stationary wave.

11.9. Davisson and Germer experiment concerns (a) polarization (b) diffraction (c) interference (d) electron diffraction.

11.10. The uncertainty principle was proposed for the first time by (a) N. Bohr (b) Schrödinger (c) Heisenberg (d) Max Born.

11.11. The product of uncertainties between position (x) and momentum (p) is (a) $\Delta x \Delta p_x = \lambda$ (b) $\Delta x \Delta p_x = h$ (c) $\Delta x \Delta p_x = nh$ (d) $\Delta x \Delta p_x = \frac{1}{\lambda h}$.

11.12. The uncertainty principle is a consequence of (a) wave nature of matter (b) particle nature of wave (c) dual nature of radiation (d) none of these.

11.13. Which is incorrect statement of uncertainty principle? (a) $\Delta x \Delta \dot{p}_x \geq \frac{\hbar}{2}$ (b) $\Delta\theta \Delta L \geq \frac{\hbar}{2}$ (c) $\Delta E \Delta t \geq \frac{\hbar}{2}$ (d) $\Delta\theta \Delta x > \frac{\hbar}{2}$.

11.14. The phase velocity (v_p) of a monochromatic de Broglie wave is (a) $v_p = c^2 v$ (b) $v_p = \frac{c^2}{v}$ (c) $v_p = \frac{v}{c^2}$ (d) $v_p = \frac{1}{c^2 v}$.

11.15. de Broglie wavelength associated with a 15 kV electron is (a) 10 Å (b) 0.10 Å (c) 1 Å (d) 100 Å.

11.16. Davisson and Germer studied diffraction with Nickel crystal and found a first order peak at 65° with electron beam of 54 eV. If instead a 216 eV beam were used, then the peak would have been at (a) 27° (b) 54° (c) 130° (d) 260°. **[WBUT 2007]**

11.17. The energy of an electron having de Broglie wavelength 1 Å is (a) 8.1×10^{-2} eV (b) 8.8×10^{-2} eV (a) 8.1×10^{-12} eV (d) 8.1×10^{-5} eV.

11.18. A stone is dropped from the top of the building. What happens to the de Broglie wavelength of the stone as it falls? (a) It increases (b) It decreases (c) It remains constant (d) de Broglie wavelength cannot be defined. **[WBUT 2008]**

11.19. If the quantum mechanical state of a system is described by $\psi(x) = \psi_0 e^{ik(x-a)}$, for $|x| \leq L$ and $\psi(x) = 0$, otherwise, then (a) the particle has a definite position but uncertainty in momentum (b) the particle has a definite momentum but uncertainty in position (c) the particle has uncertainties both in position and momentum (d) the particle has no uncertainties in either position or momentum.

11.20. If a wave packet is described by $\phi(x) = Ae\left(-\frac{x^2}{2\sigma^2}\right)$, then the momentum uncertainty in proportional to (a) $h\sigma$ (b) $\frac{h}{\sigma}$ (c) $h\sigma^2$ (d) $\frac{h}{\sigma^2}$. **[WBUT 2010]**

ANSWERS

11.1. (c) **11.2.** (b) **11.3.** (b) **11.4.** (c) **11.5.** (b) **11.6.** (c) **11.7.** (a) **11.8.** (b) **11.9.** (d) **11.10.** (c) **11.11.** (b) **11.12.** (a) **11.13.** (d) **11.14.** (b) **11.15.** (b) **11.16.** (b) **11.17.** (a) **11.18.** (c) **11.19.** (c) **11.20.** (d).

11(B)–Short Answer Type Questions

11.1. Differentiate between phase velocity and group velocity of a wave packet.

11.2. Write down the relation of phase velocity and group velocity of a wave packet. When are they equal?

11.3. Why do not we observe de Broglie wavelength with a fast moving cricket ball?

[*Sol.* According to de Broglie's hypothesis, the wavelength of a particle of m moving with velocity v is $\lambda = \frac{h}{p} = \frac{h}{mv}$.

So if the momentum p of the particle is large, then the wavelength λ will be small. If $\lambda \sim 10^{-10}$ m (i.e., the size of an atom), then λ is measurable. If λ is smaller than 10^{-10} m, then de Broglie wavelength cannot be measured.

The *kgr*, which is

Now, for macroscopic bodies (like a cricket ball or a marble) $p \sim 10^{-5}$ kg-m/s because, if we assume $m = 1$ gm $= 10^{-3}$ kg and $v = 1$ cm/s $= 10^{-2}$ m/s, then $p = mv = 10^{-3} \times 10^{-2} = 10^{-5}$ kg-m/s. Then the value of $\lambda = \frac{6.6\times10^{-34}}{10^{-5}} = 6.6 \times 10^{-29}$ m, which is far below $\lambda = 10^{-10}$ m and *hence the de Broglie wavelength associated with a fast moving cricket ball cannot be observed.*]

11.4. Justify that de Broglie wavelength associated only with subatomic particle is observable.

[**Ans.** For an electron $m = 9 \times 10^{-31}$ kg and if it moves with velocity $v = 10^7$ m/s, then its momentum $p = 9 \times 10^{-31} \times 10^7 = 9 \times 10^{-24}$ kg-m/s. Then the de Broglie wavelength, $\lambda = \frac{h}{p} = \frac{6.6\times10^{-34}}{9\times10^{-24}} = 0.73\times10^{-10}$ m, which is measurable. So only subatomic particles like electron, proton, neutron, α-particle can have observable de Broglie wavelength.]

11.5. Show that the particle velocity is equal to the group velocity of the wave packet.

11.6. What do you mean by *Wave particle duality*? **[NBUH 2003]**

11.7. State de Broglie's postulate on the wave like property of particles. **[BUH 2000]**

11.8. State and explain Heisenberg's uncertainty principle. **[WBUT 2003]**

11.9. Justify that an atomic nucleus cannot harbour a free electron.

11.10. Prove that for rotational motion of a particle, the uncertainty principle can be stated in the form $\Delta L \cdot \Delta\theta \geq \hbar$, where ΔL is the uncertainty in the angular momentum of the particle and $\Delta\theta$ is the uncertainty in its angular position.

[**Ans.** The KE of a rotational motion about the axis of rotation is $E = \frac{1}{2}I\omega^2$, where I is the moment of inertia about the axis of rotation and ω, the angular velocity.

Change in KE is

$$\Delta E = \tfrac{1}{2} \times 2\omega\Delta(I\omega) = \omega\Delta L \text{ or, } \omega\Delta L = \Delta E \;[\because L = I\omega]. \tag{1}$$

The angular position at time t is $\theta = \omega t$.

$$\therefore \frac{\Delta\theta}{\omega} = \Delta t. \tag{2}$$

From eqns (i) and (ii), we get $\Delta L\Delta\theta = \Delta E\Delta t$.

But $\Delta E\Delta t \geq \hbar$,

$$\therefore \Delta L\Delta\theta \geq \hbar.]$$

11.11. A beam of short wavelength gives accurately the position of a particle. Justify.

[GNDU 1997, '96]

[**Ans.** To determine the position of a particle accurately Δx must be small. Again, from uncertainty relation, $\Delta p = \frac{\hbar}{2\Delta x}$ should be large for accurate determination of position of the particle.

Now, according to de Broglie relation,

$$\lambda = \frac{h}{p} \text{ or, } p = \frac{h}{\lambda} \text{ or, } |\Delta p| = \frac{h}{\lambda^2}\Delta\lambda.$$

Hence, if $\lambda \to 0$, then $|\Delta p| \to \infty$ and in that case, $\Delta x \to 0$, i.e., the position can be accurately determined if the wavelength is small]

11.12. Show that de Broglie phase velocity exceeds velocity of light as well as particle velocity.

[PU 1990, '91; LU 1996]

11.13. An electron and a proton has the same de Broglie wavelength. Prove that the energy of the electron is greater. **[PU 1997]**

[**Ans.** If M, V be the mass and velocity of proton and m, v those for the electron, then the de Broglie wavelength of the electron is $\lambda_e = \frac{h}{mv}$ and that of the proton is $\lambda_p = \frac{h}{MV}$.

$$\because \lambda_e = \lambda_p, \quad \therefore \frac{h}{mv} = \frac{h}{MV} \text{ or, } mv = MV.$$

$\therefore$ energy of the electron

$$= \frac{1}{2}mv^2 = \frac{1}{2}(mv)v = \frac{1}{2}(MV)v = \frac{1}{2}(MV)\left(\frac{MV}{m}\right) \quad [\because mv = MV]$$

$$= \frac{1}{2}MV^2\frac{M}{m} = \frac{M}{m} \times \text{energy of the proton.}$$

$\because M > m$, the energy of the electron is greater than that of the proton for the same de Broglie wavelength.]

11.14. Show that de Broglie wavelength of a particle of rest-mass m_0 and KE k is given by $\lambda = \frac{hc}{\sqrt{k(k+2m_0c^2)}}$. **[WBUT 2010]**

11.15. Why do not we apply uncertainty principle in case of macroscopic bodies?

11.16. Show that the phase velocity of de Broglie matter wave can exceed the speed of light; explain why it does not imply a violation of special relativity. **[BUH 2006]**

11.17. State de Broglie on the wave like property of a particle. What is the approximate energy of neutrons suitable for crystallographic studies? **[BUH 2000]**

11.18. Describe an experiment, which verified de Broglie's hypothesis. **[WBUT 2010]**

11.19. State and explain de Broglie's hypothesis. Show that the relativistic de Broglie's wavelength is given by $\lambda_{\text{relativistic}} = \frac{h}{\sqrt{E_k(E_k+2m_0c^2)}}$. [The notations have their usual significance] **[WBUT 2010]**

11(C)–Numerical Problems

11.1. A proton and a deuteron have the same KE. Which has longer wavelength? **[PU 1993, '96]**

[$\lambda_d/\lambda_p = \frac{1}{\sqrt{2}}$, $\therefore \lambda_p > \lambda_d$]

11.2. Find the ground state of hydrogen atom using uncertainty principle. **[BUH 2006]**

11.3. Estimate the minimum uncertainty in the momentum of an electron confined in a box of length 1 μm. **[BUH 2004]**

11.4. Assume that an electron is inside a nucleus of radius 10^{-15} m. Calculate from the uncertainty principle the minimum KE of the electron.

[Given: $h = 6.63 \times 10^{-34}$ J-s, electron mass $= 9.1 \times 10^{-31}$ kg] **[WBUT 2007]**

[**Ans.** Here the position uncertainty of the electron is $\Delta x = 10^{-15}$ m. From uncertainty relation $\Delta x \Delta p_x \simeq \frac{\hbar}{2}$, the uncertainty in the corresponding momentum is

$$\Delta p_x \simeq \frac{h}{2\pi\Delta x} = \frac{6.63 \times 10^{-34}}{2 \times 2 \times 3.14 \times 10^{-15}} = \frac{6.63 \times 10^{-19}}{12.56} = 0.5278 \times 10^{-19} \text{ kg-m/s}.$$

Taking the minimum momentum of the electron $p = \Delta p_x$, we have its minimum KE.

$$(\text{KE})_{\min} = \frac{p^2}{2m} = \frac{(0.5278 \times 10^{-19})^2}{2 \times 9.1 \times 10^{-31}} = \frac{0.2786 \times 10^{-38}}{18.2 \times 10^{-31}}$$

$$= 0.0153 \times 10^{-7} \text{ J} = \frac{0.0153 \times 10^{-7}}{1.6 \times 10^{-19}} \text{ eV}$$

$$= 0.009566 \times 10^{+12} = 9.566 \times 10^{9} \text{ eV}.]$$

11.5. Calculate the de Broglie wavelength corresponding to the most probable velocity of thermal neutrons at 300 K. Can you detect it? [Given: mass of a neutron $= 1.676 \times 10^{-27}$ kg.] **[CUH 1992]**

[*Hint.* $v = \sqrt{\frac{2kT}{m}} = 2.22 \times 10^3$ m/s. $\therefore \lambda = \frac{h}{mv} = 1.779$ Å. It can be detected by neutron diffraction through a crystal.]

11.6. Calculate de Broglie wavelength of electrons accelerated through a P.D. of 150 V. [Given: $h = 6.62 \times 10^{-27}$ erg-sec, $m = 9.11 \times 10^{-28}$ gm, $e = 4.8 \times 10^{-10}$ esu] **[CU 1986]**

[*Ans.* 1.001 Å]

11.7. The average period that elapses between the excitation of an atom and the time it emits radiation is 10^{-8} sec. Find the uncertainty in the energy emitted and the uncertainty in the frequency of light emitted. [Given: $\hbar = 1.054 \times 10^{-34}$ J-s.] **[NU 1981]**

[*Ans.* 1.054×10^{-27} J; 1.59×10^{7} Hz]

11.8. What is the de Broglie wavelength of an automible of mass 1000 kg, moving with a velocity of 80 km/hr? Can you detect it? Explain your answer. **[CU 1984]**

[*Ans.* 4.76×10^{-38} m; cannot be detected]

11.9. Consider a beam of 100 eV electron incident normally on the surface of a nickel target as in Davisson and Germer experiment. Assuming lattice constant of nickel to be 2.15 Å, find the angle at which the first order Bragg maximum should occur. **[BUH 2003]**

[*Hint.* $\lambda = \frac{h}{\sqrt{2m_0 \text{eV}}} = \frac{6.62 \times 10^{-34}}{\sqrt{2 \times (9.11 \times 10^{-31}) \times (1.6 \times 10^{-19}) \times 100}} = 1.23 \times 10^{-10}$ m.

From Bragg equation, $2d \sin\theta = \lambda$ (for $n = 1$).

$$\sin\theta = \frac{1.23 \times 10^{-10}}{2 \times 2.15 \times 10^{-10}} \left[\because \sin\theta = \frac{\lambda}{2d}\right] = 0.286. \quad \therefore \theta = 16.6°.]$$

11.10. Calculate the wavelength of an electron of energy (a) 10 eV (b) 100 kV. [*Ans.* 3.87 Å; 0.039 Å]

11.11. Prove that the de Broglie wavelength of a particle of mass m and moving with the most probable speed of a Maxwellian distribution at temperature T is $\lambda = \frac{h}{\sqrt{2mkT}}$.

11.12. Compare the uncertainties in the velocities of an electron and a proton confined in a 10 Å box. [*Ans.* 7.3×10^5 m/s, 4.0×10^2 m/s]

11(D)–Broad Answer Type Questions

11.1. Starting from de Broglie's hypothesis show that the group velocity associated with a particle is the same as the particle velocity. **[WBUT 2007]**

11.2. (a) State and explain Heisenberg's uncertainty principle. (b) Give an account of γ-ray microscope experiment and discuss the result. **[WBUT 2003]**

11.3. (a) What is the experiment evidence in favour of de Broglie's hypothesis of matter waves? (b) Why in case of moving electrons quantum mechanics is used while for moving car we use Newtonian mechanics? Explain. (c) Show that there cannot be any existence of free electrons inside the nucleus of an atom. **[WBUT 2005]**

11.4. Describe the thought experiment with γ-ray microscope. Could the experiment be done with optical microscope? **[CU 1987]**

11.5. Describe Davisson and Germer experiment to demonstrate the wavelike behaviour of moving electrons. Discuss how the dual character of electrons can be reconciled on the basis of de Broglie's theory.

11.6. Describe Davisson-Germer experiment giving emphasis on the interpretation of the result of the experiment. **[BUH 1998]**

11.7. State de Broglie's hypothesis of matter wave. Bring out the similarity between Einstein's relations for photon and de Broglie's relations for material particles.

11.8. Describe an electron diffraction experiment and give an interpretation of your observations. What is the significance of the experiment in the light of wave-particle dualism?

11.9. State Heisenberg's uncertainty relation. Explain it with two simple experiments. How and why the concept of Bohr orbits violates the principle of uncertainty? Why the principle does not reveal itself while working with macroscopic objects.

11.10. (a) What do you understand by wave-particle duality? Explain the significance of the Heisenberg's uncertainty principle. (b) State and explain Heisenberg's uncertainty relation. Discuss how the wave-particle dualism can be reconciled on the basis of this principle.

11.11. Deduce the expression for the de Broglie wavelength of a particle of mass m and energy ϵ. How is it modified if the energy is very high? **[CUH 1997, NBUH 03]**

11.12. Discuss Bohr's principle of complementarity to show that no experiment can be performed, which would reveal the wave and particle characteristics of the electrons simultaneously. Explain how the uncertainty relation can be used to decide about the applicability of classical and quantum concepts.

11.13. (a) What is de Broglie wavelength? Find out an expression for the de Broglie wavelength of an electron having velocity $v \approx c$, where c is the velocity of light in vacuum.

(b) Describe an experiment, which verifies de Broglie's hypothesis about the wave nature of particles.

[WBUT 2008]

(c) Show that the product of group velocity and phase velocity of the de Broglie wave is constant.

[WBUT 2008]

Chapter 12 | Module 5.1

Crystallography

12.1 Introduction

Matter exists in nature in the *solid, liquid* and *gaseous states.* The atoms or molecules in solids are bound to one another with strong forces of attraction and as such solids have definite *shape* and *size.* Solids are classified broadly in two classes, viz., *crystalline* or *noncrystalline* (or *amorphous*).

In crystalline solids, the atoms are arranged in a periodic manner in all three directions. The crystalline state being low energy state most of the solids prefer to be in crystalline state. This is because energy released during the formation of ordered structure is more than that in disorder structure. The crystalline solids are again sub-divided into *single crystalline* and *micro-crystalline* (or *polycrystalline*). The periodicity of atoms extends throughout the volume of the matter in single crystal like diamond, quartz, etc. In polycrystalline material there is an aggregate of so many small crystallites oriented randomly and separated by *grain boundaries.* The crystallites are called *grains.*

In *amorphous solids* the atoms or molecules are randomly arranged and the solid exhibits *short range order.* Amorphous solids are formed when the atoms do not get sufficient time to be arranged in periodic manner. Glass is an example of amorphous solid. The classification of solids are as below:

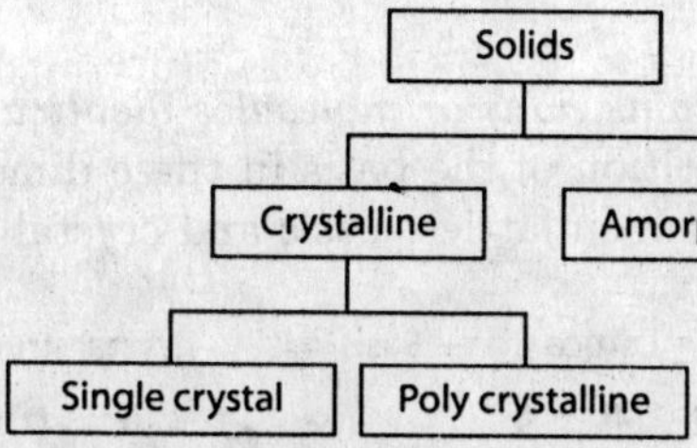

The physical properties of solids depend on the surroundings of atoms, ions or molecules forming the solid. As the amorphous solids have no directional properties they are known as *isotropic* substances. Whereas a crystal has different periodic arrangements in all the three direction, its physical properties vary with direction and, therefore, a crystal is called *anisotropic* substance. The branch of science in which the properties of solids are discussed is called *solid state physics* or *physics of condensed matter.* But the branch of science dealing with the study of geometrical forms and physical properties of crystalline solid is called *crystallography.*

12.1.1 Properties of Crystalline Solids

Crystalline solids like *calcium carbonate ammonium chloride, quartz,* etc. have the following specific properties: (a) *a crystalline solid has a definite melting point* (b) *a crystalline material has a definite geometrical shape and size* (c) *crystalline solids are anisotropic* (d) *the strength of all the bonds between the ions, atoms or molecules being the same, their chemical composition is uniform* (e) *the crystalline solids have definite colour (some are transparent) in visible light.*

12.1.2 Properties of Amorphous Solids

Amorphous solids like *glass,* etc., have the following properties: (a) *Amorphous solids have no definite shape* (b) *they are isotropic in nature* (c) *their chemical composition is not uniform as the bonds between the atoms are not of equal strength* (d) *their melting point is not fixed due to the variation of bond strength.*

12.2 Space Lattice (or Point Lattice)

An infinite regular and periodic array of points in three dimensions in which every point has surroundings identical to that of every other point in the array is called a space lattice.

The term *identical surroundings* means that the lattice appear to have the same view when looked from a point $P(\vec{r})$ as from another point $Q(\vec{r}')$ in the lattice with reference to an arbitrary origin O, Fig. 12.1.

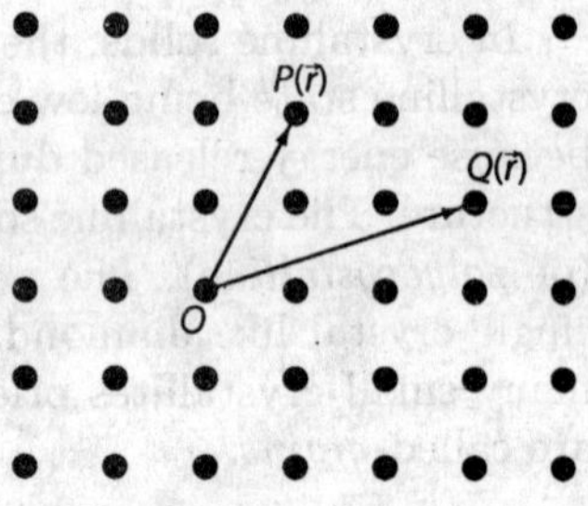

Fig. 12.1: Two-dimensional lattice.

12.3 Crystal Lattice

It is defined as a space lattice in which the lattice sites are occupied by atoms or group of atoms known as basis.

12.4 Basis

A basis is defined as a cluster of atoms, ions or molecules identical in composition, arrangement and orientation. The periodic repetition of the basis in three dimensions forms the structure of the crystal. The simple relation between lattice, basis and crystal structure can be expressed as in Fig. 12.2.

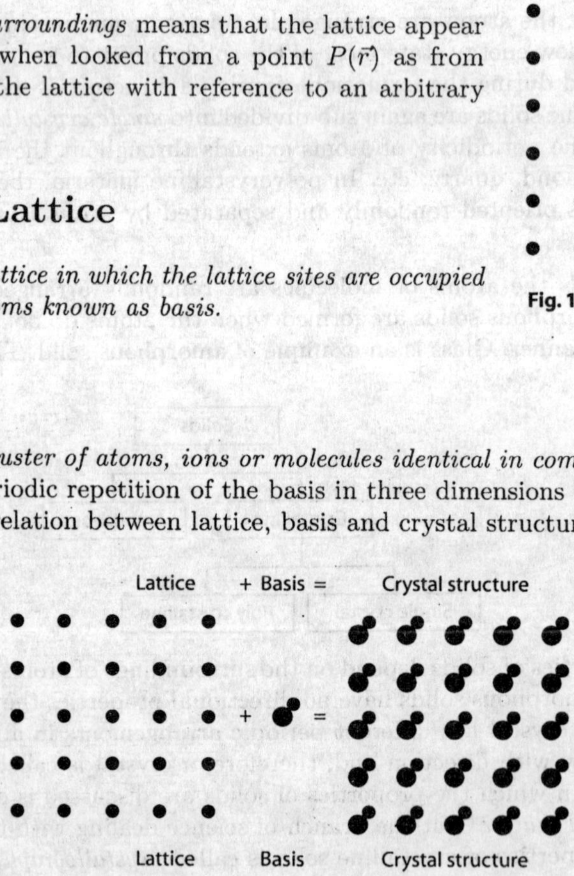

Fig. 12.2: Lattice, Basis and crystal structure.

12.5 Some Crystallographic Terms

We consider a two-dimensional crystal lattice as shown in Fig. 12.3, where the parallelogram $ABCD$ defined by the vectors $\vec{a}$ and $\vec{b}$ may be taken as a unit cell of the crystal. It is evident that the whole crystal may be reproduced by translating parallelogram $ABCD$ along $\vec{a}$ and $\vec{b}$ directions merely by suitably multiplying the vectors $\vec{a}$ and $\vec{b}$. Then any lattice point in the crystal may be represented by a vector $\vec{R}$ such that, $\vec{R} = h\vec{a} + k\vec{b}$, where h and k are integers. For a three-dimensional lattice, $\vec{R} = h\vec{a} + k\vec{b} + l\vec{c}$. The three-dimensional crystal lattice may be reproduced by repeated translation of the non-coplanar vectors $\vec{a}, \vec{b}$ and $\vec{c}$ known as the *basis vectors.*

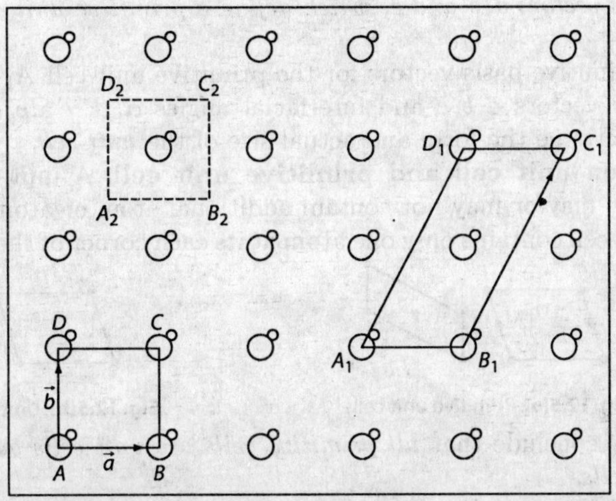

Fig. 12.3: Unit cells in two-dimensional crystal lattice.

12.5.1 Unit Cell

The unit cell of a crystal lattice is defined as *region of the crystal represented by the basis vectors $\vec{a}, \vec{b}$ and $\vec{c}$ such that a translation of this region by any integral multiple of these vectors will lead to a similar region of the crystal.*

It is also defined as *the fundamental elementary pattern of minimum of atoms, molecules or group of molecules, which represent fully all the characteristics of the crystal.*

The choice of a unit cell is not unique.

12.5.2 Wigner-Seitz Primitive Cell

Wigner-Seitz primitive unit cell may be constructed as: (a) *Connect a given lattice point to all the nearby lattice points.* (b) *Draw normals at the midpoints of the lines connecting the lattice points as shown in* Fig. 12.4.

The smallest volume enclosed by the normal planes is the *Wigner-Seitz primitive unit cell.*

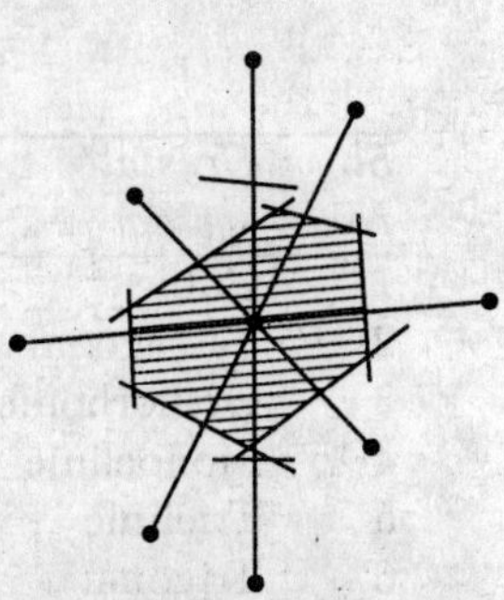

Fig. 12.4: Wigner-Seitz cell.

12.5.3 Basis Vectors

The *three non-coplanar vectors $\vec{a}, \vec{b}$ and $\vec{c}$ used to define a unit cell are called the basis vectors.*

12.5.4 Primitive Unit Cell

The unit cell occupying the smallest volume in the crystal lattice is termed as a primitive unit cell. The unit cells $ABCD$, $A_1B_1C_1D_1$, $A_2B_2C_2D_2$ are all primitive unit cells [Fig. 12.3].

12.5.5 Primitive Basis Vectors

The three non-coplanar vectors $\vec{a}, \vec{b}$ and $\vec{c}$, which define a primitive unit cell are called primitive basis vectors.

$\vec{a}', \vec{b}', \vec{c}'$ are also primitive basis vectors for the primitive unit cell $A_1B_1C_1D_1$.

The primitive basis vectors $\vec{a}, \vec{b}, \vec{c}$ and interfacial angles α, β, γ are called the *basic lattice parameters* as they determine the form and actual size of the *unit cell.*

Difference between unit cell and primitive unit cell: A unit cell, in addition to an atom at its each corner, may or may not contain additional atom or atoms.

But a primitive unit cell contains only one atom at its each corner of the unit cell [Fig. 12.5(a)].

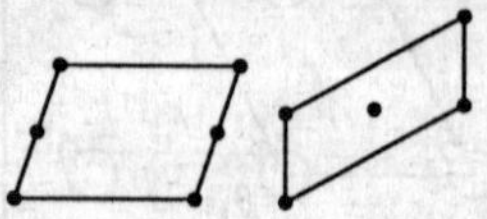

Fig. 12.5(a): Primitive unit cells.

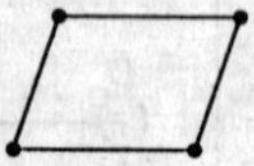

Fig. 12.5(b): Unit cells.

From above we may conclude that *all primitive cells are unit cells but all unit cells may or may not be primitive cells.*

Lattice constants: The length a, b and c specifying the size of a unit cell are called *lattice constants.* For a cubic crystal, $a = b = c$ and a is the lattice constant.

12.6 Bravais Lattice

Bravais in 1884 proposed the concept of the *space lattice. A space lattice is a mathematical concept, which represents infinite number of points in space such that the arrangement of points about a given point is similar to that about any other point.*

There are 14 distinguishable ways in which points can be arranged in three dimensional space. Therefore, Bravais lattices are 14 in number as shown in Fig. 12.6 [Figs 12.6(i) to 12.6(xiv)] and belong to 7-crystal systems of Table 12.1. The 14 Bavais lattices and their names are listed in Table 12.2.

Table 12.1: Crystal systems

Sl. No.	*Crystal system*	*Lattice constants*	α, β, γ	*Example*
1	Cubic	$a = b = c$	$\alpha = \beta = \gamma = 90°$	$NaCl, CaF_2, NaClO_2$
2	Tetragonal	$a = b \neq c$	$\alpha = \beta = \gamma = 90°$	$NiSO_4, SnO_2$
3	Orthorhombic	$a \neq b \neq c$	$\alpha = \beta = \gamma = 90°$	$KNO_3, BaSO_4, MgSO_4$
4	Monoclinic	$a \neq b \neq c$	$\alpha = \beta = 90°, \gamma \neq 90°$	$Na_2SO_4, FeSO_4$
5	Triclinic	$a \neq b \neq c$	$\alpha \neq \beta \neq \gamma \neq 90°$	$CuSO_4, K_2Cr_2O_7$
6	Trigonal	$a = b = c$	$\alpha = \beta = \gamma \neq 90°$	$CaSO_4$
7	Hexagonal	$a = b \neq c$	$\alpha = \beta = 90°, \gamma = 120°$	SiO_2, AgI

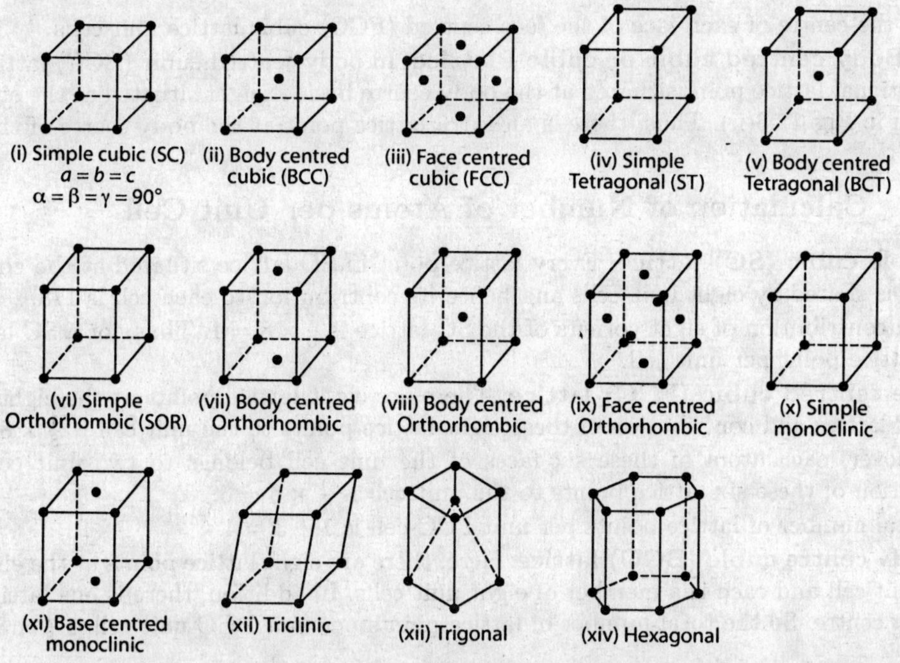

Fig. 12.6: The Bravais lattices in 3-dimensions.

Table 12.2: Bravais lattice and crystal systems

Crysal systems	*Bravais lattices*	*Symbols*
1. Cubic	Simple	P
	Body centred	I
	Face centred	F
2. Tetragonal	Simple	P
	Body centred	I
3. Orthorhombic	Simple	P
	Base centred	C
	Body centred	I
	Face centred	F
4. Monoclinic	Single	P
	Body centred	C
5. Triclinic	Simple	P
6. Trigonal	Simple	P
7. Hexagonal	Simple	P

12.7 Space Lattices of Cubic Systems

There are *three Bravais lattices of cubic system* as follows:

(i) **Simple cubic** (or *primitive cubic*) or *cubic P-type lattice*: In simple cubic (SC) lattice, there are eight lattice points at the corners of the unit cell as shown in Fig. 12.6(i). There is no other lattice point.

(ii) **Face-centred cubic** or *cubic F lattice*: In addition to the lattice points at the eight corners it contains lattice points at the centre of each face [Fig. 12.6(iii)]. There is an extra lattice point at the centre of each face of the face centred (FCC) cubic lattice unit cells.

(iii) **Body centred cubic or cubic I lattice**: In body centred cubic (BCC) lattice, there is one additional lattice point situated at the body centre besides eight situated at the eight corners as shown in Fig. 12.6(ii). Thus, there is an extra lattice point at the body centre of the unit cell.

12.7.1 Calculation of Number of Atoms per Unit Cell

(i) **Simple cubic (SC) lattice**: Every lattice point in SC lattice situated at the corner of the unit cell is shared by eight unit cells and hence its contribution to each cell is 1/8.

Total contribution of eight corners of the SC lattice is $\frac{1}{8} \times 8 = 1$. Thus, for a SC lattice there is one lattice point per unit cell.

(ii) **Face-centred cubic (FCC) lattice**: There are eight lattice points at the eight corners of the FCC lattice and contribution of these eight lattice points to the unit cell is $\frac{1}{8} \times 8 = 1$.

Moreover, each atom of these six faces of the unit cell belongs to two unit cells. So the contribution of these six lattice points to the unit cell is $\frac{1}{2} \times 6 = 3$.

$\therefore$ total number of lattice points per unit FCC cell is $1 + 3 = 4$.

(iii) **Body centre cubic (BCC) lattice**: Here, there are eight lattice points at the eight corners of the unit cell and each is a member of eight unit cells. In addition, there is one lattice point at the body centre. So the total number of lattice points per unit BCC unit cell is $\frac{1}{8} \times 8 + 1 = 2$.

12.7.2 Coordination Number

For a particular lattice, *coordination number* (CN) *is defined as the number of nearest neighbours to a particular atom in the crystal lattice.*

12.7.3 Nearest Neighbour Distance

It is defined as the distance between the centres of two nearest neighbouring atoms in the crystal.

12.7.4 Atomic Radius (r)

Atomic radius is defined as half of the distance between nearest neighbour atoms in the crystal.

Here the atoms are supposed to be *spherical in size* and any two nearest neighbouring atoms touch each other [Fig. 12.7(a)].

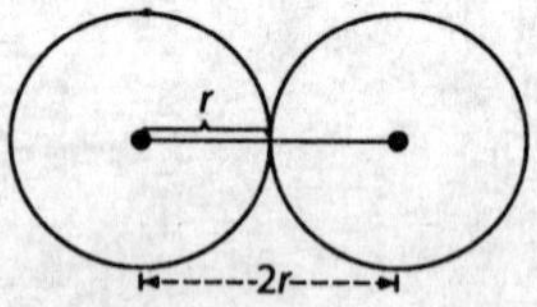

Fig. 12.7(a): Nearest neighbour distance in SC lattice.

(i) **SC lattice**: In Fig. 12.7(b), AB represents the radius r of an atom and $r = a/2$.

The nearest neighbour distance $d = 2r = 2\frac{a}{2} = a$.

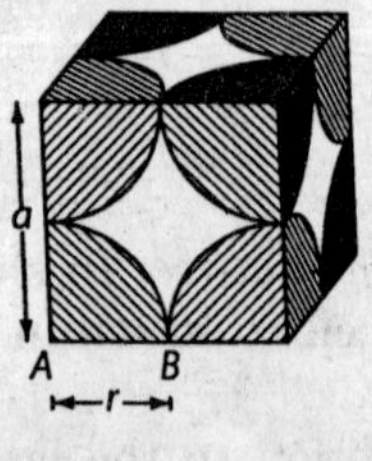

Fig. 12.7(b): Geometry of a SC unit cell.

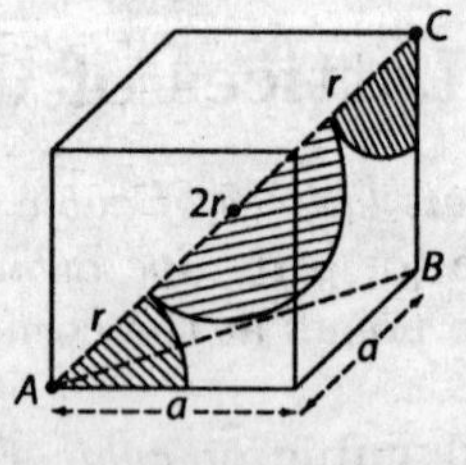

Fig. 12.8: Geometry of a BCC unit cell.

(ii) **BCC lattice**: In Fig. 12.8, a BCC unit cell is shown.

$$AB^2 = a^2 + a^2, \quad \therefore AB = \sqrt{2}a$$

$$\text{and} \quad AC^2 = (\sqrt{2}a)^2 + a^2 = 3a^2.$$

$$\therefore \quad AC = \sqrt{3}a.$$

$\therefore$ atomic radius $r = \frac{\sqrt{3}a}{4}$.

So the atomic radius of a BCC crystal is $r = \frac{\sqrt{3}a}{4}$. ($\because 4r = \sqrt{3}a$).

The nearest neighbour distance $d = 2r = \frac{\sqrt{3}}{2}a$.

(iii) **FCC lattice**: Fig. 12.9 shows an FCC unit cell whose face diagonal is AC.

Face diagonal AC $= \sqrt{2}a = 4r$.

$\therefore$ atomic radius $r = \frac{\sqrt{2}a}{4}$ or, $r = \frac{a}{2\sqrt{2}}$.

The nearest neighbour distance $d = 2r = \frac{a}{\sqrt{2}}$.

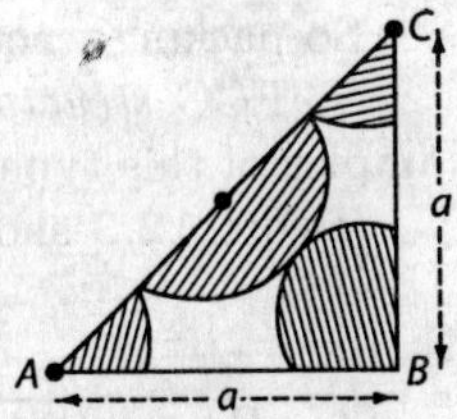

Fig. 12.9: Face diagonal of an FCC unit cell.

12.7.5 Atomic Packing Factor

The atomic packing factor of a unit cell is defined as the ratio of volume occupied by the atoms in the unit cell to the total volume of the unit cell.

If there are n atoms each of volume v and V, the volume of the unit cell, then

$$\text{Packing factor (PF)} = \frac{nv}{V}.$$

Sometimes packing factor (PF) is expressed in %, then % PF = PF × 100.

Atomic packing factor is sometimes called *packing fraction.*

Calculation of packing factor

(i) **Simple cubic (SC) structure**: In simple cubic structure, there is one atom in the unit cell.

Volume of the atom of radius r is $v = \frac{4}{3}\pi r^3$.

But $r = \frac{a}{2}$, where a is the lattice constant of SC structure.

$$\therefore v = \frac{4}{3}\pi\left(\frac{a}{2}\right)^3 = \frac{\pi a^3}{6}.$$

Volume of the unit cell $V = a^3$.

$\therefore$ packing factor $= \frac{1 \times v}{V} = \frac{\pi a^3}{6a^3} = \frac{\pi}{6} = 0.523$.

$\therefore$ packing fraction $= \frac{\pi}{6} \times 100 = 0.523 \times 100 = 52.3\%$.

The structure is thus *loosely packed.* Example of this structure is Polonium.

(ii) **Body centred cubic (BCC) structure**: Here the number of atoms per unit cell is 2.

$\therefore$ volume of these two atoms $= 2 \times \frac{4}{3}\pi r^3$.

But atomic radius $r = \frac{\sqrt{3}}{4}a$.

$\therefore$ total volume occupied by the two atoms $= 2 \times \frac{4}{3}\pi\left(\frac{\sqrt{3}}{4}a\right)^3 = \frac{\sqrt{3}\pi}{8}a^3$.

Volume of the unit cell $V = a^3$.

$\therefore$ packing factor (PF) $= \frac{\sqrt{3}\pi a^3}{8 \times a^3} = \frac{\sqrt{3}\pi}{8} = 0.68$.

$\therefore$ packing fraction $= \frac{\sqrt{3}\pi}{8} \times 100 = 0.68 \times 100 = 68\%$.

Examples of this type of structure are *tungsten, sodium, iron* and *chromium.*

(iii) **Face centred cubic (FCC) structure**: Here the number of atoms per unit cell = 4.

$\therefore$ volume of 4 atoms $= 4 \times v = 4 \times \frac{4}{3}\pi r^3$.

But atomic radius $r = \frac{\sqrt{2}}{4}a$.

$\therefore$ total volume occupied by the 4 atoms $= 4 \times \frac{4}{3}\pi\left(\frac{\sqrt{2}}{4}a\right)^3 = \frac{\sqrt{2}\pi}{6}a^3$ and volume of the unit cell $V = a^3$.

$\therefore$ packing factor $= \frac{\sqrt{2}\pi a^3}{6\times a^3} = \frac{\sqrt{2}\pi}{6} = 0.74$.

So packing fraction (PF) $= \frac{\sqrt{2}\pi}{6} \times 100 = 0.74 \times 100 = 74\%$.

So FCC *structure has the highest packing fraction compared to SC and BCC structure.* Examples of this type of structure are most of the metals like *copper, aluminum, lead* and *silver.*

Table 12.3 shows some crystal structure with example.

Table 12.3: Some crystal structures with examples

Sl. No.	*Parameters*	*Simple cubic* (SC)	*Body centred cubic* (BCC)	*Face centred cubic* (FCC)	*Hexagonal close packed structure* (HCP)	*Diamond cube* (DC)
1.	Atomic radius	$\frac{a}{2}$	$\frac{\sqrt{3}a}{4}$	$\frac{a}{2\sqrt{2}}$	$\frac{a}{2}$	$\frac{\sqrt{3}a}{8}$
2.	Nearest neighbour distance	a	$\frac{\sqrt{3}a}{2}$	$\frac{a}{\sqrt{2}}$	a	$\frac{\sqrt{3}a}{4}$
3.	Coordination number	6	8	12	12	4
4.	Volume of the unit cell	a^3	a^3	a^3	$\frac{3\sqrt{3}}{2}a^2c$	a^3
5.	No. of atoms per unit cell	1	2	4	6	8
6.	Packing factor	0.52	0.68	0.74	0.74	0.34
7.	Example	Polonium	Sodium, Lithium and Chromium	Aluminum, Copper, Silver and Lead	Magnesium, Zinc and Cadmium	Germanium, Silicon and Diamond

12.8 Calculation of Lattice Constant

We consider a cubic crystal of lattice constant a.

Then volume of the unit cell $= a^3$.

If ρ be the density of the crystal, then mass in each unit cell of the crystal $= \rho a^3$.

Again, if there are n molecules per unit cell, M, the molecular weight and N, the Avogadro's number, then

$$\text{Mass of a single molecule} = \frac{M}{N}. \tag{12.1}$$

Mass in each unit cell = (no. of a molecules in a unit cell) $\times$ (mass of a single molecule)

$$= n \times \frac{M}{N}. \tag{12.2}$$

Now, from eqns (12.1) and (12.2), we have

$$\rho a^3 = \frac{nM}{N} \quad \text{or,} \quad a^3 = \frac{nM}{\rho N} \quad \text{or,} \quad \text{Lattice constant } a = \left(\frac{nM}{\rho N}\right)^{1/3} \tag{12.3}$$

Thus, if n, M, ρ and N are known, then the value of the lattice constant a of a crystal can be determined from eqn (12.3).

12.9 Lattice Planes and Miller Indices

A crystal lattice may be considered as an aggregate of a set of parallel equidistant planes passing through the lattice points. These set of parallel planes are called *lattice planes*, which can be chosen in a number of ways like (1), (2), (3) and (4) in Fig. 12.10. Miller discovered a method to designate a plane in a crystal using three numbers h, k and l known as *Miller indices.*

Miller indices are the three smallest possible integers, which have the same ratios as the reciprocals of the intercepts of the plane concerned on the three axes.

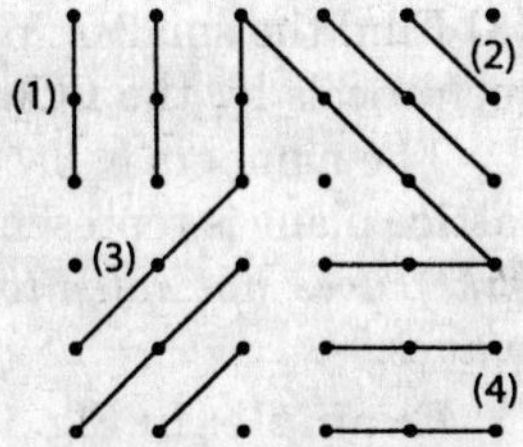

Fig. 12.10: Set of lattice planes.

12.10 Crystallographic Directions

In a crystal, the concentration of atoms in different direction is different, which leads to the *anisotropy* of it. For this reason it is necessary to locate different directions and planes for crystallographic analysis.

Definition: *The crystallographic direction in a crystal is defined as a line joining any two points in the lattice.*

In two dimensions, the crystallographic direction is expressed as a line passing through a point $A(x, y)$ and any other lattice point taken as the origin $O(0, 0)$ as shown in Fig. 12.11(a), the coordinates x and y being whole numbers.

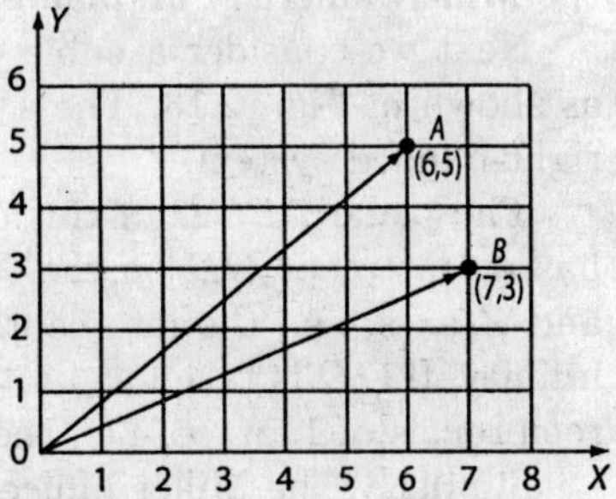

Fig. 12.11(a): Representation of crystallographic directions in two dimensions.

In Fig. 12.11(a), the direction of the points A and B with O as the origin are $(6, 5)$ and $(7, 3)$ respectively.

In three dimensions, the crystallographic direction of any point A having coordinates (x, y, z), where x, y, z are whole numbers, is conventionally denoted by putting x, y, z in a square bracket, i.e., $[x, y, z]$.

The crystallographic directions of the corner points of a unit cube are shown in the table below:

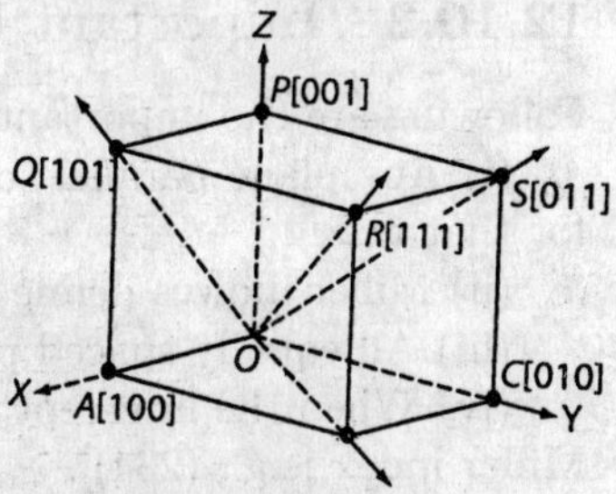

Fig. 12.11(b): Representation of crystallographic directions in three dimensions.

Table 12.4

Points	$A(100)$	$B(110)$	$C(010)$	$P(001)$	$Q(101)$	$R(111)$	$S(011)$
Direction	[100]	[110]	[010]	[001]	[101]	[111]	[011]

12.10.1 Rules to Find the Miller Indices of a Lattice Plane

(1) First, find the intercepts of the plane on the three crystal axes (OX, OY, OZ). Let these be pa, qb and rc, where a, b, c are the corresponding primitives and p, q, r are the intercepts along $\vec{a}, \vec{b}$ and $\vec{c}$ respectively [Fig. 12.12].
(2) Take the reciprocals of the numbers p, q, r in the form $\frac{1}{p}, \frac{1}{q}, \frac{1}{r}$.
(3) Find the smallest possible integers h, k, l by multiplying the reciprocals by the LCM of p, q, r such that $h : k : l = \frac{1}{p} : \frac{1}{q} : \frac{1}{r}$.

The numbers h, k, l are the Miller indices, the corresponding lattice plane is represented by (hkl). It is to be kept in mind that (hkl) *does not refer to a single plane instead a set of parallel planes.*

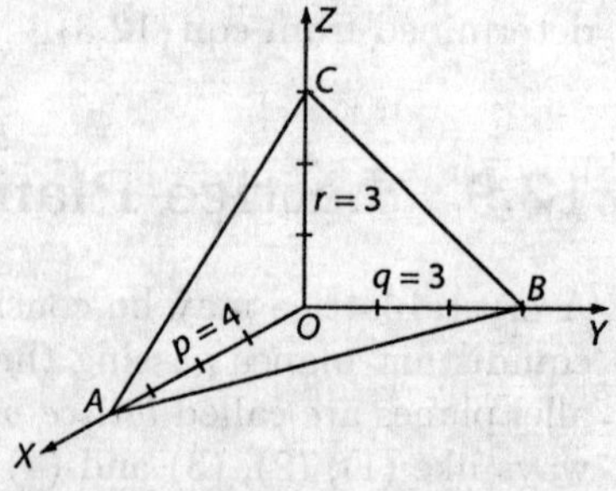

Fig. 12.12: A lattice plane specified by Miller indices (344).

Example: In Fig. 12.12, $p = 4, q = 3$ and $r = 3$.
Reciprocals of p, q, r are $\frac{1}{p}, \frac{1}{q}, \frac{1}{r} - \frac{1}{4}, \frac{1}{3}, \frac{1}{3}$.
Now, LCM of 4, 3, 3 is 12.
$\therefore h : k : l = \frac{12}{4} : \frac{12}{3} : \frac{12}{3} = 3 : 4 : 4.$
So Miller indices of the plane ABC is (344). (344) is also the Miller indices of all planes parallel to ABC.

Next we consider a cubic crystal whose unit cell is a cube as shown in Fig. 12.13. The axes OX, OY and OZ now form a right-handed system.

The plane $ABCD$ of this cube (or any plane parallel to it) has an intercept OD on the X-axis but it is parallel to the Y and Z-axes, i.e., the intercepts on the Y- and Z-axes are each infinity. If OD is taken as unity, then the Miller indices are the reciprocals of $1, \infty, \infty$, i.e., they are (100).

Similarly, the Miller indices of the planes $GADE$ are (110) and those for the plane FED are (111) [Fig. 12.12].

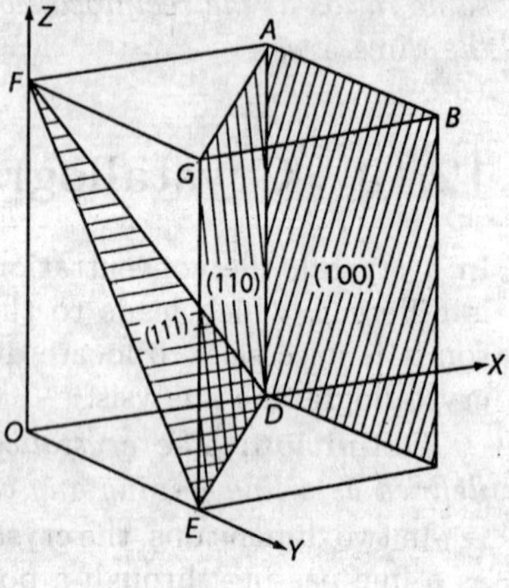

Fig. 12.13: Miller indices of lattice planes in a cubic lattice.

12.10.2 Important Features of Miller Indices

Following are the important features of the Miller indices:

(i) Any plane parallel to one of the coordinate axes has intercept of infinity and index zero for that axis.

(ii) Miller indices define a set of parallel planes and not a particular plane.

(iii) All equally spaced parallel planes have the same index number (hkl).

(iv) When the intercept of a plane on any axis is negative, a bar is put on the corresponding Miller index [e.g., $(2\bar{3}4)$].

(v) If a plane passes through the origin, it is defined in terms of parallel plane having non-zero intercept.

(vi) The complete set of crystallographical equivalent directions of (hkl) is expressed by enclosing the direction indices in *angle brackets* like $\langle hkl \rangle$.

12.11 Inter-planar Spacing

The spacing d_{hkl} between a family of parallel adjacent planes (hkl) can be calculated by taking any lattice point as an origin and erecting coordinate axes in the $\vec{a}, \vec{b}$ and $\vec{c}$ directions and then determining the perpendicular distance between this origin and the plane nearest to this origin.

The intercepts of this (hkl) plane on $\vec{a}, \vec{b}$ and $\vec{c}$ axes are respectively $\frac{a}{h}, \frac{b}{k}$ and $\frac{c}{l}$ as shown in Fig. 12.14.

From Fig. 12.13,

$$d_{hkl} = OP = \tfrac{a}{h}\cos\alpha = \tfrac{b}{k}\cos\beta = \tfrac{c}{l}\cos\gamma, \tag{12.4}$$

where α, β and γ are the angles between the normal to the plane and the $\vec{a}, \vec{b}$ and $\vec{c}$ axes respectively.

If $\hat{n}$ be the unit vector normal to the plane, then

$$\hat{n}\cdot\vec{a} = a\cos\alpha$$
$$\hat{n}\cdot\vec{b} = b\cos\beta$$
$$\hat{n}\cdot\vec{c} = c\cos\gamma.$$

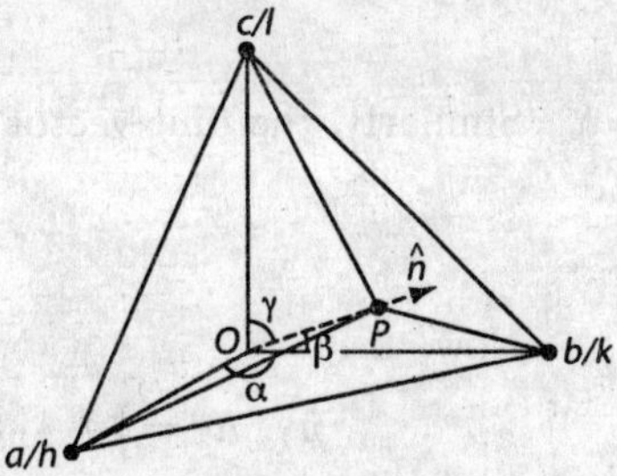

Fig. 12.14: Direction angles of the normal to the (hkl) plane.

$\therefore$ from eqn (12.4),

$$d_{hkl} = \frac{\hat{n}\cdot\vec{a}}{h} = \frac{\hat{n}\cdot\vec{b}}{k} = \frac{\hat{n}\cdot\vec{c}}{l}. \tag{12.5}$$

Now, in an orthogonal lattice, taking X-, Y- and Z-axes along $\vec{a}, \vec{b}$ and $\vec{c}$ directions respectively, the equation to the (hkl) plane having intercepts $\frac{a}{h}, \frac{b}{k}$ and $\frac{c}{l}$ is

$$\frac{x}{a/h} + \frac{y}{b/k} + \frac{z}{c/l} = 1 \quad \text{or,} \quad \frac{hx}{a} + \frac{ky}{b} + \frac{lz}{c} = 1 \quad \text{or,} \quad \phi(x,y,z) = 1 \text{ (constant)}. \tag{12.6}$$

$$\therefore \nabla\phi = \hat{i}\frac{\partial\phi}{\partial x} + \hat{j}\frac{\partial\phi}{\partial y} + \hat{k}\frac{\partial\phi}{\partial z} = \hat{i}\frac{h}{a} + \hat{j}\frac{k}{b} + \hat{k}\frac{l}{c} = |\nabla\phi|\,\hat{n}, \tag{12.7}$$

where $\hat{n}$ is the unit vector along the normal to the (hkl) plane and

$$|\nabla\phi| = \sqrt{(h/a)^2 + (k/b)^2 + (l/c)^2}.$$

$$\therefore \hat{n} = \frac{\nabla\phi}{|\nabla\phi|} = \frac{\frac{h}{a}\hat{i} + \frac{k}{b}\hat{j} + \frac{l}{c}\hat{k}}{\sqrt{(h/a)^2 + (k/b)^2 + (l/c)^2}}.$$

Now, from eqn (12.5),

$$d_{hkl} = \frac{\hat{n}\cdot\vec{a}}{h} = \frac{1}{h}\left[\frac{\left(\frac{h}{a}\right)\hat{i} + \left(\frac{k}{b}\right)\hat{j} + \left(\frac{l}{c}\right)\hat{k}}{\sqrt{\left(\frac{h}{a}\right)^2 + \left(\frac{k}{b}\right)^2 + \left(\frac{l}{c}\right)^2}}\cdot\hat{i}a\right] = \frac{1}{h}\left[\frac{\frac{h}{a}a}{\sqrt{\left(\frac{h}{a}\right)^2 + \left(\frac{k}{b}\right)^2 + \left(\frac{l}{c}\right)^2}}\right].$$

$$\therefore d_{hkl} = \frac{1}{\sqrt{\left(\frac{h}{a}\right)^2 + \left(\frac{k}{b}\right)^2 + \left(\frac{l}{c}\right)^2}}, \tag{12.8}$$

where $\hat{i}, \hat{j}$ and $\hat{k}$ are the usual *triad of unit vectors along* X, Y *and* Z *directions respectively.*

In case of a cubic crystal, $a = b = c$,

$$\therefore d_{hkl} = \frac{a}{\sqrt{h^2 + k^2 + l^2}}. \tag{12.9}$$

12.11.1 Angle between Two Planes having Miller Indices $(h_1k_1l_1)$ and $(h_2k_2l_2)$

We have the unit vector normal to the plane $(h_1k_1l_1)$ is

$$\hat{n}_1 = \frac{\hat{i}\frac{h_1}{a} + \hat{j}\frac{k_1}{b} + \hat{k}\frac{l_1}{c}}{\sqrt{\left(\frac{h_1}{a}\right)^2 + \left(\frac{k_1}{b}\right)^2 + \left(\frac{l_1}{c}\right)^2}}.$$

For a cubic system, $a = b = c$

$$\therefore \hat{n}_1 = \frac{\hat{i}h_1 + \hat{j}k_1 + \hat{k}l_1}{\sqrt{h_1^2 + k_1^2 + l_1^2}}.$$

Similarly, the unit vector normal to the plane $(h_2k_2l_2)$ of a cubic system is

$$\hat{n}_2 = \frac{\hat{i}h_2 + \hat{j}k_2 + \hat{k}l_2}{h_2^2 + k_2^2 + l_2^2}.$$

Now,

$$\hat{n}_1 \cdot \hat{n}_2 = |\hat{n}_1||\hat{n}_2|\cos\theta \quad \text{[where } \theta \text{ is the angle between the planes]}$$

$$= \frac{\hat{i}h_1 + \hat{j}k_1 + \hat{k}l_1}{\sqrt{h_1^2 + k_1^2 + l_1^2}} \cdot \frac{\hat{i}h_2 + \hat{j}k_2 + \hat{k}l_2}{\sqrt{h_2^2 + k_2^2 + l_2^2}}$$

$$= \frac{h_1h_2 + k_1k_2 + l_1l_2}{\sqrt{(h_1^2 + k_1^2 + l_1^2)(h_2^2 + k_2^2 + l_2^2)}}.$$

$$\therefore \theta = \cos^{-1}\left[\frac{h_1h_2 + k_1k_2 + l_1l_2}{\sqrt{(h_1^2 + k_1^2 + l_1^2)(h_2^2 + k_2^2 + l_2^2)}}\right]. \tag{12.10}$$

12.11.2 Inter-planar Spacing between Lattice Planes of Different Cubic Lattice

(i) **SC lattice**: Here the unit cell is a cube of side a and there is only one lattice point at each corner of the cube.

The inter-planar spacing of (hkl) planes for such lattice is

$$d_{hkl} = \frac{a}{\sqrt{h^2 + k^2 + l^2}}.$$

The Miller indices of the three family of planes are (100), (110) and (111).

For (100) planes,

$$d_{100} = \frac{a}{\sqrt{1^2 + 0^2 + 0^2}} = a.$$

For (110) planes,

$$d_{110} = \frac{a}{\sqrt{1^2 + 1^2 + 0^2}} = \frac{a}{\sqrt{2}}.$$

For (111) planes,

$$d_{111} = \frac{a}{\sqrt{1^2 + 1^2 + 1^2}} = \frac{a}{\sqrt{3}}.$$

(ii) **BCC lattice**: In this cubic lattice apart from one atom at each corner, there is an additional atom at the body centre.

The family of the lattice planes are represented by (100), (110) and (111).

For (100) planes

$$d_{100} = \frac{a}{2}, \tag{12.11}$$

since there is an additional plane (100) just half way between (100) planes.

For (110) planes

$$d_{110} = \frac{a}{\sqrt{2}}, \tag{12.12}$$

as the (110) plane of a BCC lattice is similar to (110) plane of a SC lattice.

For (111) planes,
in a BCC lattice, the plane (111) is just midway between (111) planes of a SC lattice.

$$\therefore \quad d_{111} = \frac{1}{2}\frac{a}{\sqrt{3}}. \tag{12.13}$$

(iii) **FCC lattice**: For (100) lattice,

$$d_{100} = \frac{a}{2},$$

since there is an additional plane (100) half way between (100) planes of a SC lattice.

For (110) planes,

$$d_{110} = \frac{1}{2}\frac{a}{\sqrt{2}}, \tag{12.14}$$

since there is an additional plane (110) midway and parallel to the first set between (110) planes of a SC lattice.

For (111) planes

$$d_{111} = \frac{a}{\sqrt{3}},$$

as the (111) planes of a SC lattice pass through the centre of the faces of FCC lattice.

12.11.3 Comparison of SC, BCC, FCC Structure

Table 12.5

Sl. No.	*Properties*	SC	BCC	FCC
1.	Position of atoms	Corners of the cube	Corners and body centre of the cube	Corners and face centres of the cube
2.	Volume of the unit cell	a^3	a^3	a^3
3.	Number of atoms per unit cell	1	2	4
4.	Coordination number (CN)	6	8	12
5.	Nearest neighbour distance ($2r$)	a	$\frac{a\sqrt{3}}{2}$	$\frac{a}{\sqrt{2}}$
6.	Atomic radius	$\frac{a}{2}$	$\frac{a\sqrt{3}}{4}$	$\frac{a}{2\sqrt{2}}$
7.	Atomic packing fraction (APF)	0.52	0.68	0.74
8.	Example	KI	NaCl, KCl	$AlCl_2$, $CuSo_4$

12.12 Answers of Some Important Questions

12.1. *What do you mean by single crystal and micro-crystalline (or, polycrystalline)?*

Ans. If in a solid the periodicity extends throughout the volume of the matter, then the solid is a single crystal, e.g., *diamond, quartz,* etc.

If there is an aggregate of so many small crystallites oriented at random and separated by *grain boundaries*, then the corresponding solid is called a micro-crystallite. The crystallites are called *grains.*

12.2. *Amorphous solids are isotropic but crystals are anisotropic. Why?*

Ans. Amorphous solids have no directional properties. Hence, they are isotropic. But crystals have different periodic arrangements in all three directions, which lead to the variation of its physical properties with direction. That is why a crystal is *anisotropic*.

12.3. *All primitive cells are unit cells but all unit cells may or may not be primitive cell—justify.*

Ans. In a primitive cell there is one atom at each corner of the unit cell, i.e., the number of atoms per unit cell is one. But in a unit cell there may or may not be additional atom or atoms in addition to an atom at each corner.

Hence, all primitive cells are unit cells but all unit cells may or may not be primitive cell.

12.4. *What are the basic lattice parameters?*

Ans. In a unit cell, the intercepts a, b and c made by three adjacent faces with the crystallographic axes are called *primitives* or *lattice constants.* The angles α, β and γ between the crystallographic axes are known as *interfacial angles.* The quantities a, b, c and α, β, γ are the *basic lattice parameters.* These determine the form and actual size of the unit cell.

12.5. *What do you mean by space lattice?*

Ans. A space lattice is a mathematical concept, which represents infinite number of points in space such that the arrangement of points in space about a given point is similar to that about any other point.

12.6. *What are Bravais lattice?*

Ans. The points may be arranged in 14 distinguishable ways in three dimensional space maintaining perfect periodicity. These 14 space lattices are called *Bravais lattice* and belong to 7-crystal system.

12.7. *Define space lattice and crystal lattice.*

Ans. A *space lattice* is an infinite array of points in three dimensions in which every point has surroundings identical to that of every other point in that array.

A *crystal lattice* is a space lattice in which the lattice sites are occupied by atoms or cluster of atoms.

12.8. *What do you mean by lattice constant?*

Ans. In a lattice a unit cell has the length of the edges and the angles between the edges are of definite values. Each edge of the unit cell is the distance between the atoms of the same kind and is known as the *lattice constant.*

12.9. *Define coordination number.*

Ans. Coordination number (CN) is defined as the number of nearest neighbours to a particular atom in the crystal lattice.

12.10. *What is atomic packing factor?*

Ans. The atomic packing factor of a unit cell is defined as the ratio of volume occupied by the atoms in the unit cell to the total volume of the unit cell.

If there are n atoms each of volume v, and V, the volume of the unit cell, then

$$\text{Packing Factor (or Fraction) (PF)} = \frac{nv}{V}.$$

12.11. *Find an expression for the density of crystalline material in terms of lattice constant for a cubic lattice.*

Ans. Let ρ be the density of a cubic crystalline material of lattice constant a. If M_A be the atomic weight, then the volume of one mole of the substance is $v_m = \frac{M_A}{\rho}.v_M$ is also the volume of Avogadro's number (N_A) of atoms.

$\therefore$ the volume of a single atom $v = \frac{M_A}{N_A\rho}$ and volume of n number of atoms $v_n = nv = \frac{nM_A}{N_A\rho}$.

But the effective volume of n atoms is equal to the volume of the unit cell a^3.

So

$$a^3 = \frac{nM_A}{N_A\rho} \quad \text{or,} \quad \rho = \frac{nM_A}{a^3 N_A}.$$

12.12. *What are Miller indices?*

Ans. Miller indices are the three smallest possible integer (hkl), which have the same ratios as the reciprocals of the intercepts of the plane concerned on the three axes. A particular set of values of (hkl) represents a family of parallel crystal planes.

12.13. *What are the specialties of the Miller indices?*

Ans. The Miller indices have the following specialties:

(a) Miller indices (hkl) represent a set of parallel planes and not a particular plane.

(b) If the intercepts are at infinity, the corresponding Miller indices are zero. If a plane passes through the origin, it is defined in terms of parallel plane having non-zero intercept.

(c) The miller indices (hkl) have the same value if the planes are equispaced and parallel.

(d) The Miller index $\bar{h}$ represents that the plane cuts the axis on the negative side of the origin.

12.14. *What are crystallographic directions?*

Ans. The crystallographic direction is a line joing any two points of the space lattice relative to a specific point taken as the origin.

12.15. *If* $(h_1k_1l_1)$ *and* $(h_2k_2l_2)$ *be the Miller indices of two planes, then what is the angle between them?*

Ans. The angle between the two lattice planes is given by

$$\theta = \cos^{-1}\left[\frac{h_1h_2 + k_1k_2 + l_1l_2}{(h_1^2 + k_1^2 + l_1^2)(h_2^2 + k_2^2 + l_2^2)}\right].$$

12.16. *Find the angle between the plane* (111) *and direction* [111].

Ans. The (111) plane in a cubic unit cell is represented by PAC. PA is on the (111) plane and the direction of PA is represented by $[10\bar{1}]$. The angle between the direction $[10\bar{1}]$ and [111] is the angle between the plane (111) and the direction [111].

If θ be the angle between (111) and [111], then

$$\theta = \frac{h_1h_2 + k_1k_2 + l_1l_2}{(h_1^2 + k_1^2 + l_1^2)(h_2^2 + k_2^2 + l_2^2)}.$$

Here

$$h_1 = 1 \qquad h_2 = 1$$
$$k_1 = 0 \quad \text{and} \quad k_2 = 1$$
$$l_1 = -1 \qquad l_2 = 1.$$

$$\therefore \quad \theta = \frac{1\times 1 + 0 \times 1 + (-1\times 1)}{\sqrt{1^2 + 0^2 + (-1)^2}\sqrt{1^2+1^2+1^2}} = 0.$$

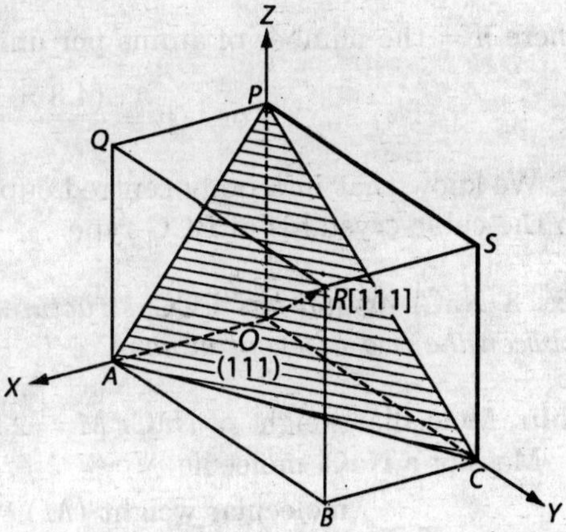

Fig. 12.15

N.B. If a plane and the direction are represented by same value of h, k, l, then they are always perpendicular to each other.

12.13 Solved Problems

Ex. 1 *Copper has* FCC *structure and the atomic radius is* 0.1278 nm. *Calculate the density and the inter-planar spacing for* (321) *planes. Take the atomic weight of copper as* 63.5.

[WBUT 2005]

Soln. For a cubic crystal having lattice constant a and number of atoms n (= 4 for FCC lattice)

$$a = \left(\frac{nM_4}{\rho N}\right)^{1/3},$$

where M_A, ρ and N are respectively the atomic mass, density and Avogadro's number respectively.

Again, for an FCC crystal, $a = \frac{4}{\sqrt{2}}r$, where $r = 0.1278\,\text{nm} = 0.1278 \times 10^{-9}$ m is the atomic radius of the crystal.

Now,

$$\rho = \frac{nM_A}{Na^3} = \frac{4 \times 63.5}{6.023 \times 10^{23} \times \left(\frac{4 \times 0.1278 \times 10^{-9}}{\sqrt{2}}\right)^3}\ \text{gm-cm}^{-3} = \mathbf{8.926\ gm\text{-}cm^{-3}} \quad (12.15)$$

and d_{hkl} = inter-planar spacing for (321) plane

$$= \frac{a}{\sqrt{(h^2 + k^2 + l^2)}} = \frac{\frac{4 \times 0.1278 \times 10^{-9}}{\sqrt{2}}}{\sqrt{3^2 + 2^2 + 1^2}}\ \text{m} = \mathbf{0.0966 \times 10^{-9}\ m.}$$

Ex. 2 *A cubic crystal has lattice constant* 43 Å *and density* 963 kg/m³. *If its atomic weight is* 63, *then what type of cubic unit cell does it form?*

Soln. Here, edge of the unit cell, $a = 4.3\,\text{Å} = 4.3 \times 10^{-10}$ m.

Density of the material of the crystal, $\rho = 963$ kg/m³.

Atomic weight, $M_A = 23$ kg/k-mol.

Then

$$n = \frac{a^3 \rho N}{M_A},$$

where n = the number of atoms per unit cell and N = the Avogadro's number

$$\text{or,} \quad n = \frac{(4.3 \times 10^{-10})^3 \times 963 \times 6.02 \times 10^{26}}{23} = \mathbf{2.}$$

We know that in a body centred cubic (BCC) crystal the number of atoms per unit cell is 2. So the cubic crystal is of BCC type.

Ex. 3 NaCl *crystal has* FCC *structure and is of density* 2.18 gm/cm³. *Calculate the distance between the two adjacent atoms.*

Soln. Molecular weight of NaCl $M = 23 + 35.5 = 58.5$.

Mass of a NaCl molecule

$$m = \frac{\text{molecular weight } (M)}{\text{Avogadro's number } (N)} = \frac{58.5\ \text{gm/mol}}{6.023 \times 10^{23}\ (\text{mol}^{-1})} = 9.713 \times 10^{-23}\ \text{gm.}$$

Number of molecules per unit volume

$$= \frac{\text{mass of unit volume of NaCl } (\rho)}{\text{mass of a NaCl molecule } (m)} = \frac{2.18}{9.713 \times 10^{-23}} = 2.244 \times 10^{22}\ \text{molecule/cm}^3.$$

Now, number of atoms per unit volume of NaCl = $2 \times 2.244 \times 10^{22}$ ($\because$ NaCl is diatomic).

If a and n represent respectively the distance between adjacent atoms and number of atoms along the edge of 1 cm cube, then the length of the edge of a unit cube = na.

Volume of the unit cube = $n^3 a^3 = 1$ cm³

$$\text{or,} \quad (4.49 \times 10^{22})\,(a)^3 = 1 \quad [\because n^3 = 4.49 \times 10^{22}].$$

$$\therefore a = \left(\frac{1}{4.49 \times 10^{22}}\right)^{1/3} = 2.81 \times 10^{-8}\ \text{cm} = \mathbf{2.81\ Å.}$$

Ex. 4 *Calculate the lattice constant of rock-salt from the following data: Molecular weight of* NaCl = 58.45 kg-mol. *Density of* NaCl = 2.17 × 10^3. kg/m^3. *Avogadro's number* N = 6023 × 10^{23}/gm-mole = 6.023 × 10^{26}/kg-mole.

Soln. The number of lattice points (n) per unit cell of NaCl = 4 (as NaCl is FCC crystal).

Now, lattice constant

$$a = \left(\frac{4M}{\rho N}\right)^{1/3} = \left[\frac{4 \times 58.45}{2.17 \times 10^3 \times 6.023 \times 10^{26}}\right]^{1/3} \text{ m} = \mathbf{5.64\ \text{Å}}.$$

Ex. 5 KCl *is of* FCC *crystal structure having density* 1.98 × 10^3 kg/m^3. *If its molecular weight is* 75.6, *calculate,* (i) *the distance from one atom to the next atom of same kind and* (ii) *the distance between adjacent atoms.* **[cf. BUH 1995]**

Soln. If a be the lattice constant of the crystal, then

$$a = \left(\frac{nM_A}{\rho N}\right)^{1/3},$$

where n = number of ions in a unit cell = 4, M_A = molecular weight = 74.6 kg/k mole, ρ = density = 963 kg/m^3 and N = Avogadro's number = 6.02 × 10^{26} atoms/k-mol.

$$\therefore a = \left(\frac{4 \times 74.6}{1.98 \times 10^3 \times 6.023 \times 10^{26}}\right)^{1/3} \simeq 6.3\ \text{Å}.$$

Thus, distance between two atoms of the same kind is a = **6.3 Å**.

Hence, the distance between two adjacent atoms is $a/2$ = **3.15 Å**.

Ex. 6 *Calculate the lattice constant of potassium bromide. Given: density of potassium bromide* = 2.7 × 10^3 kg/m^3 *and belongs to FCC lattice. Molecular weight of potassium bromide* = 119 *and Avogadro's No.* = 6.02 × 10^{23}/gm-mole = 6.02 × 10^{26}/kg-mole. **[CUH 1992 cf. 1991]**

Soln. Mass of a unit cell = $\frac{4\times 119}{6.02\times 10^{26}}$ kg.

[∵ KBr is a FCC crystal having 4 atoms/cell.

Again, mass of a unit cell of KBr can be written as $a^3 \times 2.7 \times 10^3$ kg [a = lattice constant].

So

$$a^3 \times 2.7 \times 10^3 = \frac{4 \times 119}{6.02 \times 10^{26}} \quad \text{or,} \quad a = \left[\frac{4 \times 119}{2.7 \times 10^3 \times 6.02 \times 10^{26}}\right]^{1/3} = \mathbf{6.64\ \text{Å}}.$$

Ex. 7 *In a crystal, a lattice plane cuts intercepts of* $2a$, $3b$ *and* $6c$ *along the three axes, where* a, b *and* c *are primitive vectors of the unit cell. Determine the Miller indices of the given plane.*

Soln. Let pa, qb and rc be respectively the intercepts of the plane on the three crystal axes, then

$$pa : qb : rc = 2a : 3b : 6c$$

$$\text{or,} \quad p : q : r = 2 : 3 : 6$$

$$\text{or,} \quad \frac{1}{p} : \frac{1}{q} : \frac{1}{r} = \frac{1}{2} : \frac{1}{3} : \frac{1}{6} = 3 : 2 : 1.$$

Hence, the Miller-indices of the given plane are **(321)**.

Ex. 8 *In a orthorhombic crystal a lattice plane cuts intercepts of lengths* $3a$, $-2b$ *and* $\frac{3c}{2}$ *along three axes. Deduce the Miller indices of the plane;* a, b, c *are vectors of the unit cell.*

Soln. If the intercepts of the plane in the three crystal axes are pa, qb and rc respectively, then

$$pa : qb : rc = 3a : (-2b) : \frac{3c}{2}$$

$$\text{or,} \quad p : q : r = 3 : (-2) : \frac{3}{2}$$

$$\text{or,} \quad \frac{1}{p} : \frac{1}{q} : \frac{1}{r} = \frac{1}{3} : \left(-\frac{1}{2}\right) : \frac{2}{3} = 2 : (-3) : 4.$$

$\therefore$ the Miller indices of the plane are $(\mathbf{2\bar{3}4})$.

Ex. 9 *Find the Miller indices of a plane having intercepts of* $8a$, $4b$ *and* $2c$ *on the* a, b, c *axes respectively.* **[BUH 2001]**

Soln. If the intercepts of the plane on a, b, c axes respectively be pa, qb are rc, then

$$pa : qb : rc = 8a : 4b : 2c$$

$$\text{or,} \quad p : q : r = 8 : 4 : 2$$

$$\text{or,} \quad \frac{1}{p} : \frac{1}{q} : \frac{1}{r} = \frac{1}{8} : \frac{1}{4} : \frac{1}{2} = 1 : 2 : 4.$$

$\therefore$ the Miller indices of the plane are (**124**).

Ex. 10 *In an orthorhombic crystal a plane makes intercepts* 2.93 mm, 4.47 mm *and* 2.35 mm *along the three crystallographic axes, the corresponding primitives being* 3.05 Å, 6.99 Å *and* 4.90 Å. *Deduce the Miller indices of the cleavage plane.*

Soln. $p \times 3.05 : q \times 6.99 : r \times 4.90 = 2.93 : 4.47 : 2.35$

$$\text{or,} \quad p : q : r = \frac{2.93}{3.05} : \frac{4.47}{6.99} : \frac{2.35}{4.90} = 0.96 : 0.64 : 0.48 = 6 : 4 : 3.$$

$$\therefore \frac{1}{p} : \frac{1}{q} : \frac{1}{r} = \frac{1}{6} : \frac{1}{4} : \frac{1}{3} = 2 : 3 : 4.$$

Hence, Miller indices of the plane are (**234**).

Ex. 11 *Obtain the Miller indices of a plane having intercepts of* a, $b/2$ *and* ∞ *on the* a, b, c *axes respectively.* **[BUH 2004]**

Soln. Let pa, qb and rc be the intercepts on the a, b, c axes respectively, then

$$pa : qb : rc = a : \tfrac{b}{2} : \infty \quad \text{or,} \quad p : q : r = 1 : \tfrac{1}{2} : \infty.$$

$$\therefore \tfrac{1}{p} : \tfrac{1}{q} : \tfrac{1}{r} = 1 : 2 : 0.$$

So the Miller indices of the plane is (**120**).

Ex. 12 *Draw the following planes and directions in the case of an* FCC *structure* (i) (100), (ii) (110), (iii) (112).

Soln. (i) For (100) plane, $h = 1$, $k = 0$, $l = 0$. The reciprocals are $\frac{1}{1}, \frac{1}{0}, \frac{1}{0}$, i.e., $1, \infty, \infty$. The plane is shown in Fig. 12.16(a).

The direction [100] is given by the normal to (100) plane and passing through the origin, i.e., SR.

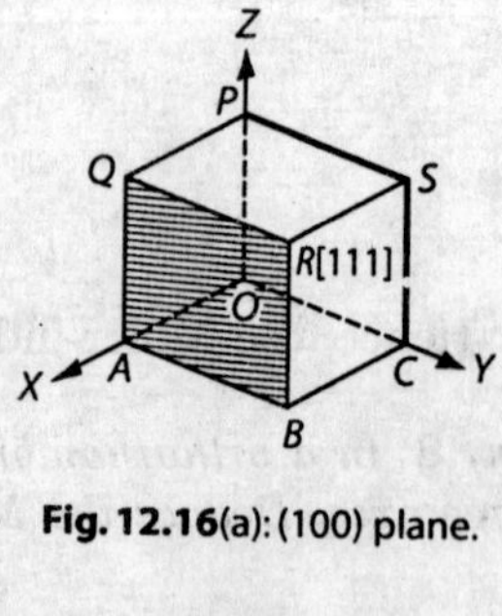

Fig. 12.16(a): (100) plane.

(ii) For (110) plane $h = 1, k = 1, l = 0$.

$$\therefore \tfrac{1}{h} = 1, \tfrac{1}{k} = 1, \tfrac{1}{l} = \infty.$$

The plane is shown in Fig. 12.16(b).

The direction [110] is the perpendicular to (110) and passing through O, i.e., OB.

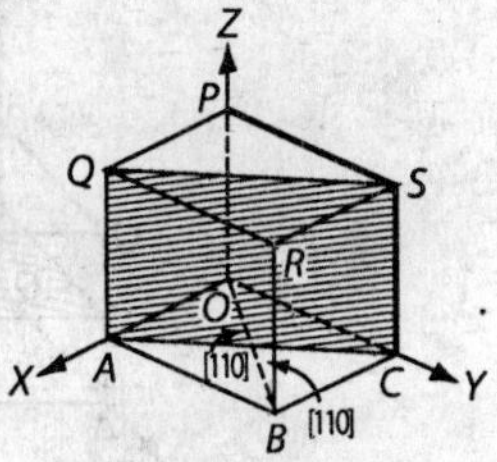

Fig. 12.16(b): (110) plane.

(iii) For (112) plane, $h = 1, k = 1, l = 2$.

$$\therefore \tfrac{1}{h} = 1, \tfrac{1}{k} = 1, \tfrac{1}{l} = 0.5.$$

The plane is shown in Fig. 12.16(c).

The direction [112] is the perpendicular to (112) plane and passing through O, i.e., OR.

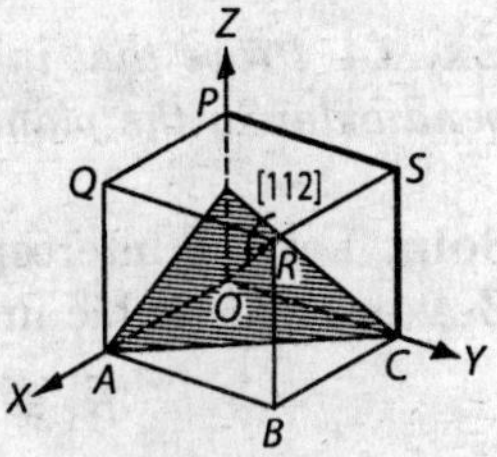

Fig. 12.16(c): (112) plane.

Ex. 13 *Draw the crystal planes* (002), (012) *and* (022) *in a* FCC *lattice.*

Soln. (002) plane: Here

$$\begin{aligned} h &= 0 \quad \therefore \tfrac{1}{h} = \tfrac{1}{0} = \infty \\ k &= 0 \quad \therefore \tfrac{1}{k} = \tfrac{1}{0} = \infty \\ l &= 2 \quad \therefore \tfrac{1}{l} = \tfrac{1}{2} = 0.5. \end{aligned}$$

So the (002) plane has intercepts $\infty, \infty, 0.5$ along X-, Y-, Z-directions respectively as shown in Fig. 12.17(b).

(012) plane: Here

$$\begin{aligned} h &= 0 \quad \therefore \tfrac{1}{h} = \tfrac{1}{0} = \infty \\ k &= 1 \quad \therefore \tfrac{1}{k} = \tfrac{1}{1} = 1 \\ l &= 2 \quad \therefore \tfrac{1}{l} = \tfrac{1}{2} = 0.5. \end{aligned}$$

The (012) plane thus has intercepts ∞, 1 and 0.5 along the X-, Y- and z-directions respectively [Fig. 12.17(b)].

(022) plane: Here

$$\begin{aligned} h &= 0 \quad \therefore \tfrac{1}{h} = \tfrac{1}{0} = \infty \\ k &= 2 \quad \therefore \tfrac{1}{k} = \tfrac{1}{2} = 0.5 \\ l &= 2 \quad \therefore \tfrac{1}{l} = \tfrac{1}{2} = 0.5. \end{aligned}$$

The (022) plane has intercepts ∞, 0.5 and 0.5 along the X-, Y- and Z-directions respectively [Fig. 12.17(c)].

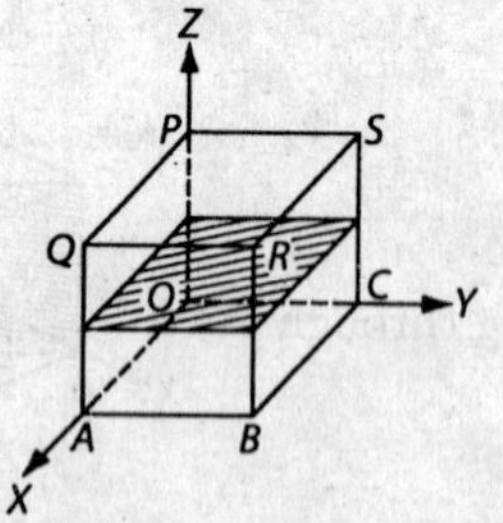

Fig. 12.17(a): (002) plane.

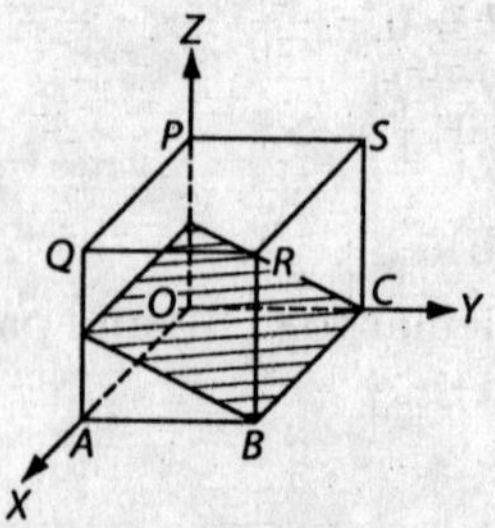

Fig. 12.17(b): (012) plane.

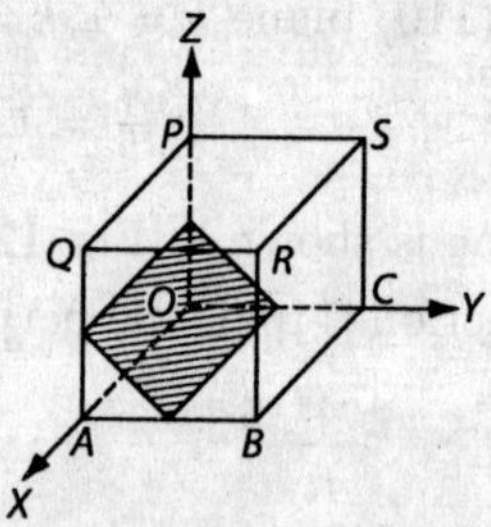

Fig. 12.17(c): (022) plane.

Ex. 14 *Prove that in a cubic crystal the direction [hkl] is perpendicular to the plane (hkl).*

Soln. Let the intercepts of the (hkl) plane along X-, Y- and Z-axes of the cubic unit cell be

$$\vec{R}_1 = \frac{\vec{a}}{h}, \quad \vec{R}_2 = \frac{\vec{b}}{k}, \quad \vec{R}_3 = \frac{\vec{c}}{l},$$

where $\vec{a}, \vec{b}$ and $\vec{c}$ are basis vectors along X-, Y- and Z-directions respectively, Fig. 12.18.

The vector $\vec{H} = h\vec{a} + k\vec{b} + l\vec{c}$ represents the direction $[hkl]$ and $\vec{R}_{12}, \vec{R}_{13}$, lie on the plane (hkl).

Now,

$$\hat{H} \cdot \hat{R}_{12} = (h\vec{a} + k\vec{b} + l\vec{c}) \cdot (\vec{R}_2 - \vec{R}_1) = (h\vec{a} + k\vec{b} + l\vec{c}) \cdot \left(\frac{\vec{b}}{k} - \frac{\vec{a}}{h}\right)$$
$$= \vec{b} \cdot \vec{b} - \vec{a} \cdot \vec{a} = \vec{a} \cdot \vec{a} - \vec{a} \cdot \vec{a}$$
$$= 0 \quad [\because a = b = c \text{ for a cubic crystal}].$$

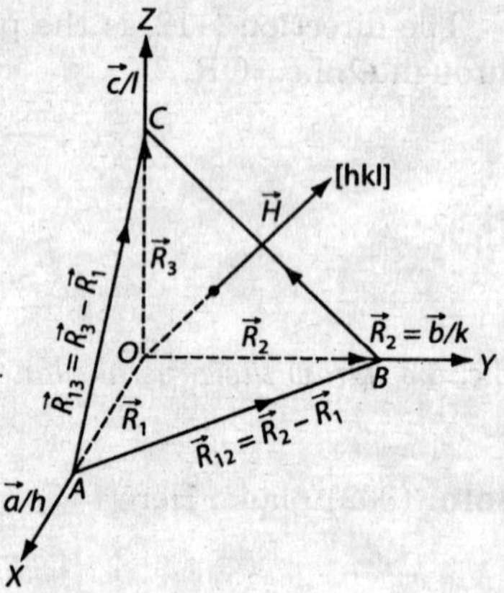

Fig. 12.18: (hkl) plane in a cubic crystal.

Also

$$\vec{H} \cdot \vec{R}_{13} = (h\vec{a} + k\vec{b} + l\vec{c}) \cdot \left(\frac{\vec{c}}{l} - \frac{\vec{a}}{h}\right) = \vec{c} \cdot \vec{c} - \vec{a} \cdot \vec{a} = \vec{a} \cdot \vec{a} - \vec{a} \cdot \vec{a} = 0.$$

Thus, $\vec{H}$ is perpendicular to $\vec{R}_{12}$ as well as $\vec{R}_{13}$, which lie on (hkl) plane. Hence, $\vec{H}$ is perpendicular to (hkl) plane.

Ex. 15 *Calculate the inter-planar spacing d of planes* (111) *in a simple cubic lattice of side a.*

[WBUT 2003, '04, '07]

Soln. We know that the inter-planar spacing d_{hkl} between the consecutive (hkl) planes is

$$d_{hkl} = \frac{1}{\sqrt{\frac{h^2}{a^2} + \frac{k^2}{b^2} + \frac{l^2}{c^2}}}.$$

In this case, $a = b = c$ and $h = k = l = 1$.

$$\therefore d_{hkl} = \frac{1}{\sqrt{\frac{1}{a^2} + \frac{1}{a^2} + \frac{1}{a^2}}} = \frac{\mathbf{a}}{\sqrt{\mathbf{3}}}.$$

Ex. 16 *The Miller indices of a plane in a simple cubic crystal are* (321). *Find the intercepts of the plane and sketch it.*

[BUH 2000]

Soln. Let pa, qb and rc be the intercepts along X-, Y- and Z-axes respectively, where a, b, c are the primitive vectors of the unit cell.

Then $pa : qb : rc = 3a : 2b : c$ or, $p : q : r = 3 : 2 : 1$.

As the crystal is SC, $a = b = c$ and the intercepts on the X-, Y- and Z-axes are **3, 2** and **1** respectively as shown in Fig. 12.19.

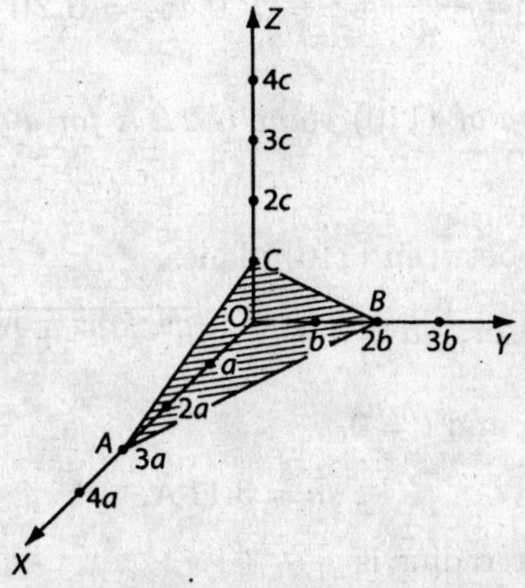

Fig. 12.19: The intercepts of (321) plane.

Ex. 17 *Find the inter-planar spacing for a* (321) *plane in a simple cubic lattice whose lattice constant is* 4.2×10^{-8} cm.

Soln. Here $a = b = c = 4.2 \times 10^{-8}$ cm $= 4.8 \times 10^{-10}$ m and

$$d_{hkl} = \frac{a}{\sqrt{h^2 + k^2 + l^2}} = \frac{4.2 \times 10^{-10}}{\sqrt{3^2 + 2^2 + 1^2}} \quad [\because h = 3, k = 2 \text{ and } l = 1]$$

$$= \mathbf{1.1 \times 10^{-10}\ m.}$$

Ex. 18 *In a tetrahedral lattice* $a = b = 2.5$ Å *and* $c = 1.8$ Å. *Find the lattice spacing between* (111) *planes.*

Soln. For a given set of parallel planes (hkl) the lattice spacing is given by

$$d_{hkl} = \left[\frac{h^2}{a^2} + \frac{k^2}{b^2} + \frac{l^2}{c^2}\right]^{-1/2}.$$

In this case, $h = 1$, $k = 1$, $l = 1$ and $a = b = 2.5$ Å, $c = 1.8$ Å.

$$\therefore d_{hkl} = \left[\frac{1^2}{(2.5)^2} + \frac{1^2}{(2.5)^2} + \frac{1^2}{(1.8)^2}\right]^{-1/2} = \left[\frac{2}{6.25} + \frac{1}{3.24}\right]^{-1/2} = \mathbf{1.26\ Å.}$$

Ex. 19 *The distance* d_{100} *plane in* BCC *structure is* 0.232 nm. *What is the size of unit cell? What is the radius of the atom?* **[WBUT 2006]**

Soln. For a BCC crystal the radius of the atom is given by

$$r = \frac{\sqrt{3}a}{4},$$

where a is the lattice constant.

Now, for a BCC structure,

$$d_{100} = \frac{a}{2} \quad \text{or,} \quad 0.232 = \frac{a}{2} \quad [\because d_{100} = 0.232 \text{ nm}]$$

$$\text{or,} \quad a = 0.464 \text{ nm.}$$

$$\therefore r = \frac{\sqrt{3}a}{4} = \frac{\sqrt{3}}{4} \times 0.464 = \mathbf{0.20 \text{ nm.}}$$

Ex. 20 *The inter-planar spacing of* (110) *plane is* 2.2 Å *for a cubic crystal. Calculate the atomic radius.*

Soln. The inter-planar spacing between (110) planes,

$$d_{100} = \frac{a}{\sqrt{h^2+k^2+l^2}} \quad \text{or,} \quad a = d_{100} \times \sqrt{h^2+k^2+l^2} = 2.2 \times \sqrt{1^2+1^2+0^2}.$$

$\because d_{100} = 2.2$ Å and $h = 1$, $k = 1$ and $l = 0$,

$$\therefore a = 3.11 \text{ Å}.$$

Now, for a SC crystal atomic radius is

$$r = \frac{a}{2} = \frac{3.11}{2} = \mathbf{1.555 \text{ Å}.}$$

For a BCC crystal, $r = \frac{\sqrt{3}}{4}a = 3.11 \times \frac{\sqrt{3}}{4} = 1.347$ Å and for a FCC crystal, $r = \frac{\sqrt{2}}{4}a = 3.11 \times \frac{\sqrt{2}}{4} = \mathbf{1.099}$ **Å.**

Ex. 21 *In a simple cubic crystal, find the ratio of* (a) *spacing of the* (110) *and* (111) *planes* (b) *nearest neighbour distance to the next nearest neighbour distance.*

Soln. (a) For a SC lattice,

$$d_{hkl} = \frac{a}{\sqrt{h^2+k^2+l^2}}.$$

For (110) planes,

$$d_{110} = \frac{a}{\sqrt{1^2+1^2+0^2}} = \frac{a}{\sqrt{2}}.$$

For (111) planes,

$$d_{111} = \frac{a}{\sqrt{1^2+1^2+1^2}} = \frac{a}{\sqrt{3}}.$$

$$\therefore d_{110} : d_{111} = \frac{a}{\sqrt{2}} : \frac{a}{\sqrt{3}} = \mathbf{\frac{1}{\sqrt{2}} : \frac{1}{\sqrt{3}}}.$$

(b) In a SC lattice the nearest neighbour distance $d_1 = a$, where a is the side of the unit cube.

The next nearest neighbour is situated at the end of the face diagonal. Hence, the next nearest neighbour distance

$$d_2 = \sqrt{a^2 + a^2} = a\sqrt{2}.$$

$$\therefore d_1 : d_2 = a : a\sqrt{2} = \mathbf{1 : \sqrt{2}.}$$

Ex. 22 *If the radius of* N_i *atom is* 1.245 Å, *calculate the inter-planar distance of* (100) *and* (111) *planes of* N_i *crystal.*

Soln. If a and r be representatively the lattice constant and radius of the N_i atom, then,

$$a = 2\sqrt{2}r = \frac{4}{\sqrt{2}}r = \frac{(4 \times 1.245)}{\sqrt{2}} = 3.52\,\text{Å}.$$

Now, the inter-planar distance between (100) planes is

$$d_{100} = \frac{a}{\sqrt{1^2 + 0^2 + 0^2}} = a = \mathbf{3.52\,Å}$$

and the inter-planar distance between (111) planes is

$$d_{111} = \frac{a}{\sqrt{1^2 + 1^2 + 1^2}} = \frac{3.52}{\sqrt{3}} = \mathbf{2.03\,Å}. \tag{12.16}$$

Ex. 23 *Copper has an* FCC *structure and the atomic radius is* 0.1278 nm. *Calculate the inter-planar spacing for* (111) *planes.* **[BUH 1998]**

Soln. If a the lattice constant and r the atomic radius, then for FCC cubic crystal

$$a = \frac{4}{\sqrt{2}}r = \frac{4}{\sqrt{2}} \times 0.1278 \times 10^{-9}\text{ m}.$$

$$[\because r = 0.1278\text{ nm} = 0.1278 \times 10^{-9}\text{ m}]$$

Now, inter-planar spacing for (hkl) planes is

$$d_{hkl} = \frac{a}{\sqrt{h^2 + k^2 + l^2}}.$$

Here, $h = k = l = 1$ and $a = \frac{4}{\sqrt{2}} \times 0.1278 \times 10^{-9}$ m.

$$\therefore d_{111} = \frac{\frac{4}{\sqrt{2}} \times 0.1278 \times 10^{-9}}{\sqrt{1^2 + 1^2 + 1^2}} = \frac{4 \times 0.1278 \times 10^{-9}}{\sqrt{6}} = \frac{0.5112 \times 10^{-9}}{2.4495}$$
$$= 0.209 \times 10^{-9}\text{ mm} = \mathbf{0.209\text{ nm}}.$$

Ex. 24 *Find the Miller indices for a plane with the intercepts* $6\vec{a}, 2\vec{b}, 3\vec{c}$. **[BUH 1993]**

Soln. If pa, qb and rc be the intercepts of the plane along X-, Y- and Z-axes respectively, then

$$pa : qb : rc = 6a : 2b : 3c$$
$$\text{or,}\quad p : q : r = 6 : 2 : 3$$
$$\text{or,}\quad \frac{1}{p} : \frac{1}{q} : \frac{1}{r} = \frac{1}{6} : \frac{1}{2} : \frac{1}{3} = 1 : 3 : 2$$
$$\text{or,}\quad h : k : l = 1 : 3 : 2.$$

$\therefore$ the required Miller indices of the plane are **(132)**.

Ex. 25 *The density of* BCC *iron is* 7900 kg-m^{-3} *and its atomic weight is* 56. *Calculate the side of the unit cell and the inter-atomic spacing.* **[BUH 1993]**

Soln. The side of a unit cell of a BCC lattice is

$$a = \left(\frac{nM}{\rho M}\right)^{1/3}$$

where n = number of atoms per unit cell, M = atomic weight, ρ = the density of the material and N = Avogadro's number.

Here, $n = 2, M = 56, \rho = 7900$ kg-m^{-3} and $N = 6.023 \times 10^{-26}$/kg-mole.

$$\therefore a = \left[\frac{2 \times 56}{7900 \times 6.023 \times 10^{26}}\right]^{1/3} = \left[\frac{112}{47581.7 \times 10^{26}}\right]^{1/3} = 2.86 \times 10^{-10} \text{ m} = 2.86 \text{ Å}.$$

Now, inter-atomic spacing of a BCC lattice is

$$a_1 = \frac{\sqrt{3}}{2}a = \frac{\sqrt{3}}{2} \times 2.86 = \frac{1.732 \times 2.86}{2} = \mathbf{2.48 \text{ Å}}.$$

Ex. 26 *Write down Miller indices of a plane having intercepts* $2a$, ∞ *and* $c/2$ *on the* a, b, c *axes respectively.* **[BUH 2002]**

Soln. Let pa, qb and rc be the intercepts of the plane along a, b and c axes respectively. Then

$$pa : qb : rc = 2a : \infty : \tfrac{c}{2}$$
$$\text{or, } \quad p : q : r = 2 : \infty : \tfrac{1}{2}$$
$$\text{or, } \quad \tfrac{1}{p} : \tfrac{1}{q} : \tfrac{1}{r} = \tfrac{1}{2} : \tfrac{1}{\infty} : 2$$
$$\text{or, } \quad h : k : l = 1 : 0 : 4.$$

Hence, the Miller indices of the plane is **(104)**.

Ex. 27 *Find the Miller indices of a plane having intercepts of* $8a$, $4b$ *and* $2c$ *on the* a, b, c *axes respectively.* **[BUH 2001]**

Soln. If pa, qb and rc be the intercepts of the plane along a, b and c axes respectively, then

$$pa : qb : rc = 8a : 4b : 2c$$
$$\text{or, } \quad p : q : r = 8 : 4 : 2$$
$$\text{or, } \quad \tfrac{1}{p} : \tfrac{1}{q} : \tfrac{1}{r} = \tfrac{1}{8} : \tfrac{1}{4} : \tfrac{1}{2} = 1 : 2 : 4$$
$$\text{or, } \quad h : k : l = 1 : 2 : 4.$$

So the Miller indices of the plane is **(124)**.

Ex. 28 *Show that the maximum radius of the sphere that can just fit into the void at the body centre of the* FCC *structure coordinated by the facial atoms is* $0.414r$, *where* r *is the radius of the atom.*

Soln. Let R be the radius of the sphere that can just fit into the void as shown in Fig. 12.20.

$$\therefore 2r + 2R = a \quad \text{or,} \quad r + R = \tfrac{a}{2} \quad \text{or,} \quad R = \tfrac{a}{2} - r.$$

For FCC structure, $a = \frac{4r}{\sqrt{2}}$.

$$\therefore R = \frac{4r}{2\sqrt{2}} - r = \frac{2r}{\sqrt{2}} - r = r(\sqrt{2} - 1) = 0.414r \text{ (Proved).}$$

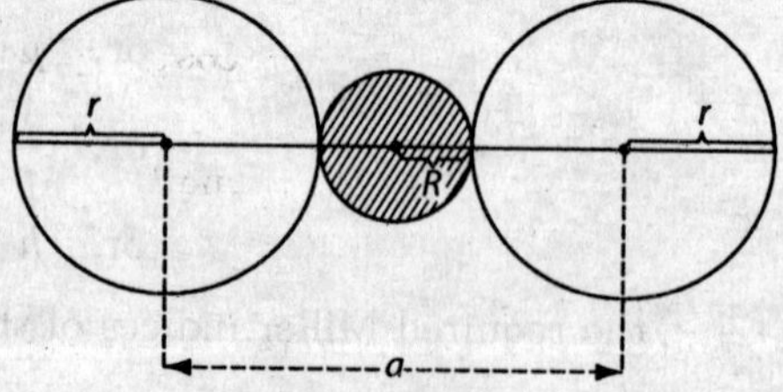

Fig. 12.20

Ex. 29 *Copper has* FCC *structure and the atomic radius is* 0.1278 nm. *Calculate the inter-planar spacing for* (110) *and* (212) *planes.*

Soln. We know that inter-planar spacing d_{hkl} of a lattice having cubic structure of lattice constant a is

$$d = \frac{a}{(h^2 + k^2 + l^2)^{1/2}}.$$

For FCC structure,

$$a = \frac{4r}{\sqrt{2}},$$

where the atomic radius $r = 0.1278 \times 10^{-9}$ m.

$$\therefore a = \frac{4 \times 0.1278 \times 10^{-9}}{\sqrt{2}} = 0.3615 \text{ nm}.$$

Now, inter-planar spacing for (110) plane is

$$d_{110} = \frac{0.3615}{(1^2+1^2+0^2)^{1/2}} = \frac{0.3615}{\sqrt{2}} \text{ nm} = \mathbf{0.2556\ nm}.$$

Inter-planar spacing for (212) plane is

$$d_{212} = \frac{0.3615}{(2^2+1^2+2^2)^{1/2}} = \frac{0.3615}{9^{1/2}} = \mathbf{0.1205\ nm}.$$

Ex. 30 *Show that in simple cubic lattice* $d_{100} : d_{110} : d_{111} = 1 : 0.71 : 0.58$.

Soln. Inter-planar spacing for (hkl) planes in a SC structure of lattice constant a is

$$d = \frac{a}{(h^2+k^2+l^2)^{1/2}}. \quad \therefore d_{100} = \frac{a}{1} = a.$$

$$d_{110} = \frac{a}{(1^2+1^2+0^2)^{1/2}} = \frac{a}{\sqrt{2}} \quad \text{and} \quad d_{111} = \frac{a}{(1^2+1^2+1^2)^{1/2}} = \frac{a}{\sqrt{3}}.$$

$$\therefore \quad d_{100} : d_{110} : d_{111} = 1 : \frac{1}{\sqrt{2}} : \frac{1}{\sqrt{3}} = 1 : 0.71 : 0.58 \text{ (Proved)}.$$

Ex. 31 *Calculate the number of atoms per square millimetre of* (111) *plane in a* SC *structure.*

Soln. Area of the (111) plane is $= \frac{1}{2} \times$ base $\times$ height $= \frac{1}{2} \times a\sqrt{2} \times EO$.
Now, from Fig. 12.21(b),

$$\cos \angle DEO = \cos 30° = \frac{EO}{a\sqrt{2}} \quad \text{or,} \quad EO = a\sqrt{2} \times \cos 30° = a\sqrt{2} \times 0.866.$$

$\therefore$ area of the (111) plane is $= \frac{1}{2} \times (a\sqrt{2}) \times (a\sqrt{2} \times 0.866) = 0.866a^2$.
Now, total number of atoms in this plane is $\left(\frac{1}{360} \times 60\right) \times 3 = \frac{1}{2} = 0.5$.
Hence, number of atoms per unit area $= \frac{0.5}{0.866a^2} = \frac{\mathbf{0.58}}{\mathbf{a^2}}$ **atoms/mm^2**.

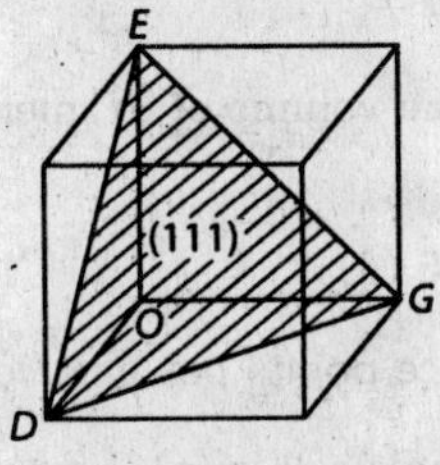

Fig. 12.21(a)

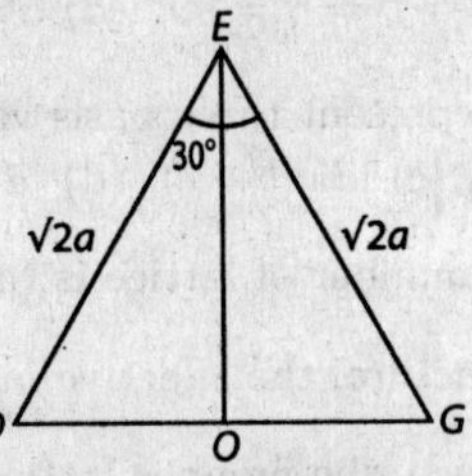

Fig. 12.21(b)

EXERCISES

12(A)–Objective Type Questions

Select the correct answer:

12.1. Solids have definite shape and size because (a) density of solids is larger compared to liquid or gas (b) the atoms or molecules of solid are bound with strong forces of attraction (c) solids are generally of high rigidity (d) solids can be transformed into liquids and gases.

12.2. Most of the solids prefer to be in the crystalline state because (a) volume is generally minimum in crystalline state than liquid or gas state (b) crystalline state is the low energy state (c) crystal has definite shape and size (d) none of these.

12.3. In amorphous solids, the atoms and molecules are (a) regularly arranged in periodic manner (b) randomly arranged (c) partly regularly and partly randomly arranged (d) none of these.

12.4. The physical properties of solids depend on (a) shape and size of the solids (b) the atoms, ions or molecules forming the solid (c) the pressure of impurities (d) the irregularities in the lattice.

12.5. Amorphous solids are isotropic because (a) they don't have directional property (b) they exhibit short range order (c) their colour varies from solid to solid (d) they are brittle.

12.6. Crystals are anisotropic because (a) atoms, ions or molecules are closely packed in a small volume (b) often they contain ions of different elements (c) crystal has different periodic arrangements in different directions (d) the inter-atomic space is void.

12.7. The fundamental building block whose repetition in three dimensions generates a crystal is called (a) unit cell (b) lattice cell (c) primitive cell (d) none of these.

12.8. In a plane lattice the periodic arrangements of atoms occur in (a) three dimensions (b) two dimensions (c) a line (d) both in 2 and 3 dimensions.

12.9. The number of basic crystal system is (a) 2 (b) 3 (c) 4 (d) 7.

12.10. Basis is defined as (a) an atom about a lattice point (b) a group of atoms about a lattice point (c) one atom or a group of atoms about a lattice point (d) none of these.

12.11. Lattice + basis is (a) unit cell (b) primitive cell (c) crystal structure (d) molecular structure.

12.12. The number of Bravais lattices in three dimensions are (a) 7 (b) 9 (c) 14 (d) 19.

12.13. The magnitudes of basis vectors $\vec{a}, \vec{b}, \vec{c}$ are called (a) lattice parameters (b) lattice vectors (c) crystal vectors (d) unit vectors.

12.14. In a primitive unit cell, the number of lattice points is (a) 2 (b) 5 (c) 4 (d) 1.

12.15. For a cubic crystal systems (a) $a = b = c; \alpha = 90°, \beta \neq \gamma$ (b) $a \neq b = c; \alpha = \beta = \gamma = 90°$ (c) $a = b = c; \alpha = \beta = \gamma = 90°$ (d) $a = b \neq c; \alpha = \beta \neq \gamma$.

12.16. If $\vec{a}, \vec{b}$ and $\vec{c}$ represent three basis vectors, then the volume of Wigner-Seitz cell is (a) $|\vec{a} \times \vec{b} \times \vec{c}|$ (b) $|(\vec{a} + \vec{b}) \times \vec{c}|$ (c) $|\vec{a} \cdot (\vec{b} + \vec{c})|$ (d) $|\vec{a} \times \vec{b} \cdot \vec{c}|$

12.17. The effective number of lattice is (a) 4 (b) 2 (c) 5 (d) 1.

12.18. In a BCC structure, the effective number of lattice points per unit cell is (a) 2 (b) 4 (c) 1 (d) 3

12.19. In a FCC lattice, the number lattice points per unit cell is (a) 2 (b) 4 (c) 5 (d) 1.

12.20. NaCl crystal is (a) SC (b) BCC (c) FCC (d) none of these.

12.21. If the number of Bravais lattice is three, then the crystal is (a) monoclinic, (b) hexagonal (c) cubic (d) tetragonal.

12.22. The coordination number of SC structure is (a) 6 (b) 12 (c) 8 (d) 4.

12.23. The FCC structure is (a) primitive (b) non-primitive (c) may be either primitive or non-primitive (d) none of these.

12.24. The unit cell of BCC structure is (a) primitive (b) non-primitive (c) may be either primitive or non-primitive (d) none of these.

12.25. Miller indices (hkl) represent (a) a set of parallel planes (b) a particular plane (c) a set of arbitrarily oriented plane (d) none of these.

12.26. The Miller indices of a plane having intercepts $2, \infty, \infty$ units on the $\vec{a}, \vec{b}$ and $\vec{c}$ axes respectively are (a) 101 (b) 001 (c) 0 (d) 020.

12.27. Miller indices of a plane having intercepts 2, 3, 4 units along X-, Y- and Z-axes are (a) (643) (b) (436) (c) (231). **[WBUT 2006]**

12.28. The inter-planar distance d_{100} for a (100) plane of a SC lattice is (a) $\frac{a}{\sqrt{3}}$ (b) $\frac{a}{\sqrt{2}}$ (c) a (d) $\frac{a}{\sqrt{5}}$

12.29. The inter-planar distance d_{111} for a (111) plane of a SC cubic lattice is (a) a (b) $\frac{a}{\sqrt{3}}$ (c) $\frac{a}{\sqrt{2}}$ (d) $\frac{a}{\sqrt{0.5}}$.

12.30. The inter-planar distance (d_{100}) for (100) plane of a BCC lattice is (a) $\frac{a}{\sqrt{2}}$ (b) $\frac{a}{\sqrt{3}}$ (c) a (d) $\frac{a}{2}$.

12.31. The inter-planar spacing d_{111} for (111) plane of a FCC lattice is (a) $\frac{a}{\sqrt{2}}$ (b) a (c) $\frac{a}{\sqrt{3}}$ (d) $\frac{a}{2}$.

12.32. If a and r be respectively the lattice constant and radius of an atom in a SC structure, then (a) $a = \frac{4}{\sqrt{3}}r$ (b) $a = 2r$ (c) $a = \frac{2}{\sqrt{2}}r$ (d) $a = \frac{r}{2}$.

12.33. If a and r be respectively the lattice constant and radius of an atom in a BCC structure, then (a) $r = \frac{\sqrt{3}}{4}a$ (b) $\frac{\sqrt{2}}{4}a$ (c) a. **[WBUT 2007]**

12.34. For an FCC structure the relation between lattice constant (a) and radius (r) of the atom is (a) $r = \frac{\sqrt{3}}{4}a$ (b) $r = \frac{a}{2}$ (c) $r = a$ (d) $r = \frac{\sqrt{2}}{4}a$.

12.35. The coordination number of SC structure is (a) 6 (b) 3 (c) 2 (d) 4.

12.36. The coordination number of BC structure is (a) 1 (b) 8 (c) 6 (d) 2.

12.37. For FCC cubic structure the coordination number is (a) 6 (b) 8 (c) 12 (d) 10.

12.38. For SC structure atomic packing factor is (a) 74% (b) 52% (c) 68% (d) 47%.

12.39. If the atomic packing factor a crystal is 68% of them, the structure of the crystal is (a) SC (b) FCC (c) BCC (d) none of these.

12.40. The atomic packing factor for FCC structure is (a) 74% (b) 52% (c) 68% (d) 49%.

12.41. For NaCl crystal having lattice constant 5.6 Å the inter-atomic distance is (a) 2.8 Å (b) 1.4 Å (c) 56 Å (d) 7.2 Å.

12.42. If for a crystal $d_{hkl} = 1.414$ nm and $a = 1$ nm, then the Miller indices of the planes are (a) (111) (b) (101) (c) (110) (d) (100).

12.43. (111) plane of a cubic crystal is (a) perpendicular to X-axis (b) perpendicular to Y-axis (c) perpendicular to Z-axis (d) none of these. **[WBUT 2005]**

12.44. $d_{100} : d_{110} : d_{111}$ for a SC structure is (a) $1 : \sqrt{3} : \sqrt{6}$ (b) $\sqrt{2} : \sqrt{5} : \sqrt{6}$ (c) $\sqrt{6} : \sqrt{3} : \sqrt{2}$ (d) $\sqrt{2} : \sqrt{6} : \sqrt{5}$.

12.45. The (111) and (222) planes in a lattice are (a) perpendicular to each other (b) parallel to each other (c) inclined at 60° to each other (d) none of these.

12.46. The element, which crystalizes in BCC structure is (a) Sodium (b) Zinc (c) Copper (d) Iron.

12.47. If the inter-atomic distance of NaCl crystal is 0.28 nm, then the lattice parameter is (a) 0.14 nm (b) 0.28 nm (c) 0.56 nm (d) none of these.

12.48. The crystal structure of aluminum is (a) SC (b) BCC (c) FCC (d) HCP.

12.49. If the primitives $a \neq b \neq c$ and interfacial angles $\alpha \neq \beta \neq \gamma \neq 90°$, then it belongs to the system (a) monoclinic (b) orthorhombic (c) tetragonal (d) cubic.

12.50. In a cubic unit cell, the angles between the normals to the planes (111) and (101) is (a) 50° (b) 64.5° (c) 54.69°(d) 70°.

12.51. Volume of a unit cell in FCC crystal is (a) $16\sqrt{2}r^3$ (b) $4\sqrt{2}r^3$ (c) $\frac{16\sqrt{2}}{r^3}$ (d) $2\sqrt{2}r^3$.

12.52. The density of an FCC structure is (a) $\rho = \frac{M}{Na^3}$ (b) $\rho = \frac{MN}{a^3}$ (c) $\rho = \frac{4M}{Na^3}$ (d) $\rho = \frac{Ma^3}{4N}$, the different letters have their usual meanings.

12.53. The nearest neighbour distance in case of BCC structure is (a) $\frac{a\sqrt{3}}{2}$ (b) $\frac{a\sqrt{2}}{\sqrt{2}}$ (c) $\frac{2a}{\sqrt{3}}$ (d) $\frac{2a}{\sqrt{3}}$.

[WBUT 2008]

12.54. The primitive of a crystal are 1.2 Å, 1.8 Å, 2 Å, 2 Å along the three axes, where a plane with Miller indices (231) cut intercept 1.2 Å along X-axis. What will be the lengths of intercept along Y- and Z-axes? (a) 1.5 Å and 5 Å (b) 1.9 Å and 2.5 Å (c) 1.2 Å and 4 Å (d) 1.5 and 4 Å.

[WBUT 2008]

12.55. In case of simple cubic crystal, the effective number of atoms per unit cell is (a) 2 (b) 1 (c) 3 (d) none of these.

ANSWERS

12.1. (b) **12.2.** (b) **12.3.** (b) **12.4.** (b) **12.5.** (a) **12.6.** (c) **12.7.** (a) **12.8.** (b) **12.9.** (d) **12.10.** (c) **12.11.** (c) **12.12.** (c) **12.13.** (b) **12.14.** (d) **12.15.** (c) **12.16.** (d) **12.17.** (d) **12.18.** (a) **12.19.** (b) **12.20.** (c) **12.21.** (c) **12.22.** (d) **12.23.** (b) **12.24.** (b) **12.25.** (a) **12.26.** (c) **12.27.** (a) **12.28.** (c) **12.29.** (b) **12.30.** (d) **12.31.** (c) **12.32.** (b) **12.33.** (a) **12.34.** (d) **12.35.** (a) **12.36.** (b) **12.37.** (c) **12.38.** (b) **12.39.** (c) **12.40.** (a) **12.41.** (a) **12.42.** (c) **12.43.** (d) **12.44.** (c) **12.45.** (b) **12.46.** (a) **12.47.** (b) **12.48.** (c) **12.49.** (a) **12.50.** (c) **12.51.** (a) **12.52.** (c) **12.53.** (a) **12.54.** (c) **12.55.** (b).

12(B)–Short Answer Type Questions

12.1. Why most of the solids prefer to be in the crystalline state? Why the solids have definite shape and size?

12.2. Why crystalline solids are isotropic in nature but the amorphous are not?

12.3. Distinguish between crystalline and amorphous solids. **[CUH 1996]**

12.4. What do you mean by crystal lattice? What are plane lattice and space lattice?

12.5. Explain the terms lattice *translation operation* and *fundamental translation vectors.* When are the translation vectors are called primitive? **[BUH 1999]**

12.6. What is unit cell? **[WBUT (Old) 2005]**

12.7. If a unit cell has the following characteristics: $a = b = 10.5$ Å, $c = 6$ Å, $\alpha = \beta = \gamma = 90°$, identity to which crystal system does the unit cell belong? **[BUH 2003]**

12.8. What is a Wigner-Seitz primitive cell? How can it be constructed?

12.9. What is the difference between a primitive cell and a unit cell? Show that all primitive cells are unit cells but all unit cells may or may not be primitive cell.

12.10. What are Bravais lattice? What is the number of Bravais lattice in three dimensions?

12.11. What is the basis of dividing the crystal systems into different types?

12.12. Draw the diagrams of SC, FCC and BCC structures.

12.13. Find the number of atoms per unit cell of SC, FCC and BCC types of crystals.

12.14. What is meant by Miller indices. Explain with examples. **[WBUT 2003, '07]**

12.15. What is the utility of Miller indices? **[CUH 1990]**

12.16. Define a crystal lattice. What is a primitive cell? Is it unique? **[BUH 1997]**

12.17. What are basis vectors? How many different *Bravais Lattices* are shown in crystal structure? **[BUH 2006]**

12.18. For an FCC lattice what are (a) the volume of the primitive cell, (b) lattice points per unit volume, (c) nearest neighbour distance and (iv) the number of nearest neighbours? **[BUH 1999]**

12.19. Deduce the inter-planar spacing of a simple cubic lattice of side a. **[WBUT 2007]**

12.20. Show that in a cubic crystal of side a the inter-planar spacing between consecutive parallel planes of Miller indices (hkl) is

$$d_{hkl} = \frac{a}{\sqrt{h^2+k^2+l^2}}.$$

[WBUT 2004(S), '05]

12.21. Calculate atomic packing fraction and atoms per unit cell in crystal having (a) simple cubic (b) body centred cubic and (c) face centred cubic structures considering the atoms as hard spheres. **[BUH 2003; WBUT 2003]**

12.22. Calculate atomic radius for SC, FCC and BCC structures.

12.23. What do you mean by coordination number of atoms. Calculate the coordination number of SC, FCC and BCC structures.

12.24. Prove that in a cubic crystal, the direction $[hkl]$ is perpendicular to the plane (hkl).

12.25. Write down short note on Bravais lattice and Miller indices **[WBUT 2003]**

12(C)–Numerical Problems

12.1. Load is face centred cubic with an atomic radius $r = 1.746$ Å. Find the inter-planar spacing for (220) plane. **[WBUT 2005]**

12.2. Aluminum has FCC structure. If the unit cell dimension and atomic weight of aluminum is 4.048 Å and 26.98 Å, then calculate its density. [*Ans.* 2700 kg/m^3]

12.3. In a cubic unit cell, find the angle between the normals to the planes (111) and (101).

12.4. Copper has an FCC structure and the atomic radius is 0.1278 nm. Calculate the inter-planar spacing for (111) planes. [*Ans.* 0.209 nm]

[BUH 1998]

12.5. Silicon crystallizes in the diamond cubic structure. The radius of silicon atom is 0.1176 nm. The atomic weight and density of silicon are 28.09 and 2.3×10^3 kg/m^3. What is the number of atoms present in a unit cell? [*Ans.* 8]

12.6. Compare the densities of lattice points in (111) and (110) planes in a SC lattice.

[*Ans.* $\rho_{111} : \rho_{110} = \sqrt{2} : \sqrt{3}$]

12.7. Aluminum is FCC crystal with lattice constant $a = 0.405$ nm.

(a) How many unit cells are there in an aluminum foil 0.005 cm thick and a square of each side 25 cm?

(b) If its weight is 0.0085 kg, then how many atoms are present?

(c) What is the number of atoms per unit cell? [*Ans.* $4.7 \times 10^{22}; 1.8 \times 10^{23}; 4$]

12.8. Copper has FCC structure having atomic radius 1.278 Å. Find its density [*Ans.* 8.94 gm/cc]

12.9. Calculate the inter-planar spacing for a (321) plane in a SC lattice whose lattice constant is 4.2×10^{-8} cm. [*Ans.* 1.1×10^{-8} cm]

12.10. In a general lattice $a = b = 2.5$ Å, $c = 1.8$ Å, what is the lattice spacing between (111) planes.

[*Ans.* 3.96 Å]

12.11. A lattice plane in a SC structure cuts the $\vec{a}, \vec{b}$ and $\vec{c}$ axes at distances $6a, 4b$ and $3c$. Find the Miller indices of the plane. [*Ans.* 2,3,4]

12.12. C_s metal (atomic weight 130) has a cubic unit cell of lattice constant 0.6 nm. If the density of C_s is 2 gm/cc, determine whether the unit cell is simple, body centred or face centred.

[WBUT 2010]

12(D)–Broad Answer Type Questions

12.1. (a) Define unit cell.

(b) Show that the spacing d of planes (hkl) in a SC lattice of side a is

$$d = \frac{a}{\sqrt{h^2 + k^2 + l^2}}.$$

(c) Lead is face-centred cubic with atomic radius of $r = 1.746$ Å. Find the inter-planar spacing for (220) planar. **[WBUT 2005]**

12.2. (a) Show that in a cubic crystal of side a the spacing between consecutive parallel planes of Miller indices (hkl) is

$$d_{hkl} = \frac{a}{\sqrt{h^2 + k^2 + l^2}}.$$

(b) Copper has FCC structure and atomic radius 0.1278 nm. Calculate the density and the inter-planar spacing for (321) planes. Take atomic weight of copper as 63.5.

12.3. (a) What are Miller indices? **[WBUT 2005]**

(b) Find the packing factor for SC structure and BCC lattice. **[WBUT 2004]**

(c) Calculate the inter-planar spacing d of planes (111) in a SC lattice of side a.

[WBUT 2003, '04, '07]

12.4. (a) What are the differences between crystalline and amorphous solids? **[CUH 1996]**

(b) Show that FCC structure is more closely packed than BCC structure.

(c) Illustrate the SC, FCC and BCC structures with diagrams.

12.5. (a) Find the atomic radius of SC, FCC and BCC structures of lattice constant a.

(b) What are Miller indices? How can they be determined?

(c) What are the characteristics of Miller indices?

12.6. (a) Explain the terms *lattice translation operation* and *fundamental translation vectors*? When are the translation vectors called primitive?

(b) For an FCC lattice what are (i) volume of the primitive cell (ii) lattice points per unit volume?

(c) What fraction of total volume is occupied by atoms in an FCC structure? **[BUH 1999]**

12.7. (a) Find out expression for the atomic radius and coordination number of SC, FCC and BCC structures of lattice constant a.

(b) Illustrate the SC, FCC and BCC structures with diagrams.

(c) Find out the packing fraction for FCC structure. Find the expression for the separation between the planes with Miller indices h, k, l and lattice constant a for a simple cubic structure.

[WBUT 2008]

12.8. What is coordination number for SC and BCC structures. Draw necessary sketch. *Atomic packing factor* increases with coordination number. Justify the statement. **[WBUT 2010]**

Chapter 13 | Module 5.2

X-rays

13.1 Introduction

While studying the phenomenon of discharge of electricity through rarefied gases, Prof. W.K. Röntgen at Wurzburg (Germany) in 1895 observed *some invisible unknown radiation.* This radiation was found to effect fluorescent screen and possesses high degree of penetrating power. Since the nature of the radiation was not known, Prof. Röntgen called these radiation as *X-rays,* although many call it *Röntgen ray* after the name of the discoverer.

What are X-rays?

X-rays are actually *electromagnetic radiation of very short wavelengths* (0.01 to 10 Å). These rays are produced whenever fast electrons are suddenly stopped by solid target of high atomic weight kept at low pressure ($\sim$ 0.001 mm Hg). The KE of the fast electrons is converted into X-rays.

13.2 Production of X-rays

Since X-rays are produced due to the impact of fast moving electrons on a suitable target, the following are essentially needed for their production: (i) *an electron source* (ii) *high potential difference* (PD) *to accelerate the electrons* (iii) *a target of high atomic weight.*

In the laboratory X-rays are produced by any of the following X-ray tubes: (a) *Röntgen* (or, *gas-filled*) *X-ray tube* (b) *the modern Coolidge tube.*

13.3 Production of X-rays by Coolidge Tube

The modern X-ray tube designed by W.D. Coolidge is an evacuated glass tube (C) consisting of *two electrodes, viz., a cathode and an anode as shown in* Fig. 13.1.

The electrons emitted from the filament are focussed into a narrow beam through a focussing electrode and is incident on the anode (T). The anode consists of a target material of tungsten, or *molybdenum* embedded in a copper bar surrounded by a cooling arrangement. The face of the target is kept at an angle ($\sim 45°$) relative to the incident electron beam.

A potential difference (PD) $\sim$ 150 kV is applied between the anode and the cathode, which accelerates the electron beam to energy $\sim 10^5$ eV. When these electrons are abruptly stopped by the target T, X-rays are emitted. The greater the electron speed the shorter is the X-rays wavelength.

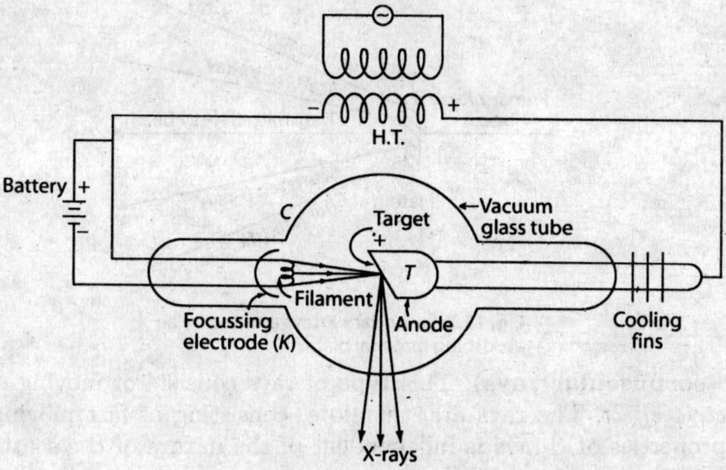

Fig. 13.1: Schematic of coolidge X-ray tube.

13.4 Quality of X-rays

The quality of X-rays represents its penetrating power. X-rays with high penetrating power are *hard X-rays* while those with low penetrating power are *soft X-rays.* If the PD between the anode and the cathode is large the energy of the electrons is large and consequently the X-rays produced are hard X-rays. The penetrating power also increases with increase in atomic weight of target material. So the quality of X-rays can be controlled by properly selecting the (i) *applied PD between the anode and the cathode* (ii) *atomic weight of the material used.*

13.5 Intensity of X-rays

The X-ray intensity depends on the intensity of the incident electron beam on the target. Again, the intensity of the electron beam depends on the filament temperature being regulated by current in the filament. Therefore, *by regulating the current flowing through the filament, the intensity of X-rays can be controlled.*

13.6 Interaction of X-rays with Matter

When primary X-rays are incident on a plate of some material, a complex phenomenon of secondary radiations is observed [Fig. 13.2].

(i) **Characteristic X-rays**: Characteristic X-rays originate due to the *inelastic collision* between X-rays and matter. The nature and frequency of characteristic X-rays is different from those of primary X-rays. They are characteristic to the target material.

(ii) **Scattered X-rays**: Scattered X-rays originates due to *elastic collision* between X-rays and the target material. The nature and frequency of these X-rays are the same as those of primary X-rays. The properties of scattered X-rays do not depend on the scattering material.

This type of scattering is called *coherent scattering.*

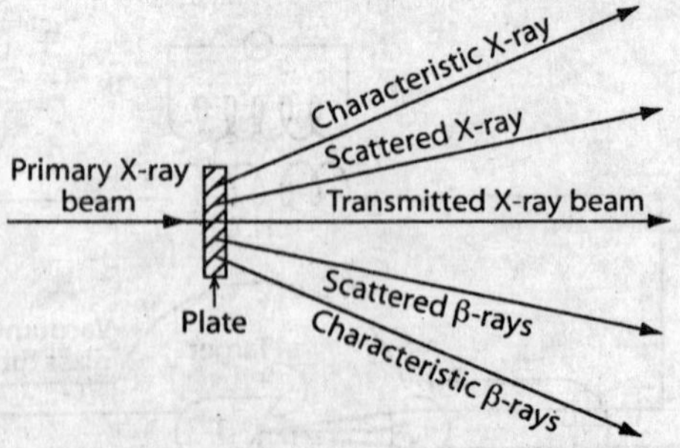

Fig. 13.2: Secondary X-rays emitted by an absorbing material plate.

(iii) **β-rays (corpuscular rays)**: This type of rays consists of moving electrons produced by the *photoelectric effect*. The rays are, therefore, consisting of fast moving electrons (or, β-particles). The properties of β-rays is independent of the nature of the scattering material but depend on the quality of the primary X-rays.

13.7 Origin of X-rays

The origin of X-rays lies in the well-known *principle of classical electrodynamics* according to which *whenever a charged particle undergoes acceleration or deceleration, electromagnetic energy is radiated.* When high energy electrons strike target material of large atomic weight, they suffer loss of KE. A small part of this lost energy is converted into X-rays (electromagnetic energy) and the rest is converted into heat.

Depending on the magnitude of the lost energy of the fast electron two types of X-rays originate, viz., (a) *Continuous X-rays* in which X-rays' intensity varies with continuous variation of wavelength (λ) (b) *Characteristic X-rays*, which arise at particular wavelength (λ) characteristic of the target element.

13.8 X-rays Spectrum

A thorough analysis of the spectrum of X-ray reveals that, it consists of two distinct types of radiations: (i) *continuous X-rays* and (ii) *characteristics X-rays.*

The continuous X-rays spectrum depends on the PD between the anode and the cathode, whereas the characteristic X-rays spectrum depends on the target material and independent on the PD

13.8.1 Continuous X-rays

The continuous X-rays spectrum (also called *Bremsstrahlung* meaning breaking radiation) depends on the applied voltage as shown in Fig. 13.3.

Important features of X-ray spectrum:

The above two curves reveal important features of X-ray spectrum as follows:

(i) *If the accelerating potentials (V) is increased the maximum intensity occurs at lower wavelength (λ).*

(ii) *At higher value of V, total X-ray intensity increases.*

(iii) *The spectral distribution of the emitted radiation is continuous.*

(iv) *For a given value of V, there is a sharply defined minimum wavelength* (λ_m), *which is known as the short wavelength limit.*

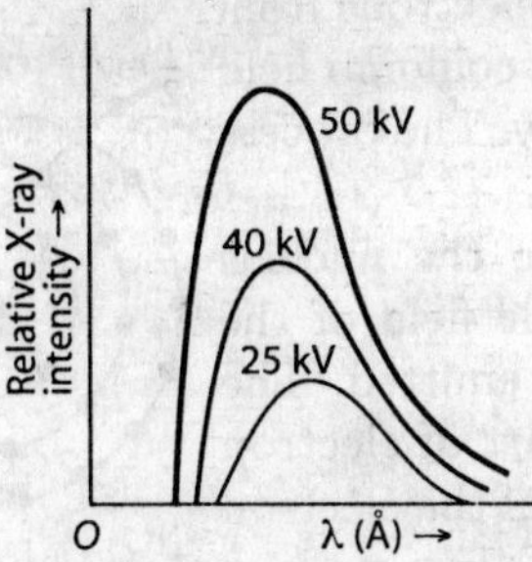

Fig. 13.3: Continuous X-ray spectrum emitted by tungsten.

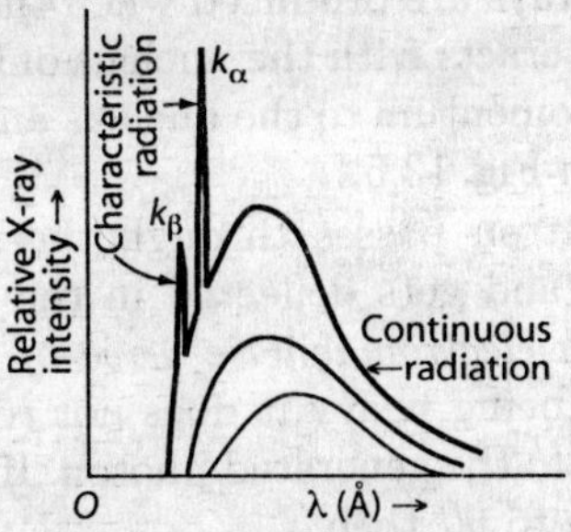

Fig. 13.4: Characteristic X-ray spectrum emitted by molybdenum target.

(v) *With the increase in V,* λ_m *shifts to lower value in accordance with Duane-Hunt law:*

$$\lambda_{min} = \frac{12400}{V} \text{ Å}.$$

Classical electromagnetic theory could provide a satisfactory explanation of the features (i), (ii) and (iii) but the short wavelength limit and its dependence with V [features (iv) and (v)] could not be accounted for by classical theory. *Quantum theory,* however, provides a satisfactory explanation of features (iv) and (v).

13.8.2 Quantum Theory and Duane-Hunt Law

Since the electrons emitted from the cathode in an X-ray tube are accelerated through the same PD(V), they acquire the same energy eV, where e is the charge on the electron. When these electrons are stopped by the target material, this energy eV produces X-rays.

According to quantum theory, the emission of electromagnetic radiation (i.e., X-rays) takes place in the form of photons of energy $h\nu$, where h is the Planck's constant and ν is the frequency of the emitted radiation. If ν_{max} be the frequency corresponding to the maximum energy of the photon, then

$$h\nu_{max} = eV$$

$$\text{or, } \nu_{max} = \frac{eV}{h}$$

$$\text{or, } \frac{c}{\lambda_{min}} = \frac{eV}{h} \quad [\because c = \nu\lambda]$$

$$\text{or, } \lambda_{min} = \frac{hc}{eV}. \tag{13.1}$$

Taking the values of $h = 6.626 \times 10^{-34}$ J-s, $c = 3 \times 10^8$ m/s and $e = 1.602 \times 10^{-19}$ C, we have from eqn (13.1),

$$\lambda_{min} = \frac{1.24}{V} \times 10^{-6} \text{ m} = \frac{12400}{V} \text{ Å}. \tag{13.2}$$

Eqn (13.2) is called the *Duane-Hunt law.* Thus, the quantum theory explains the existence of short wavelength and its dependence on the accelerating PD (V).

Eqn (13.1) is similar to photoelectric equation $h\nu = eV_s + W_0$, if $eV \gg W_0$. In photoelectric effect a photon liberates an electron from the metal, whereas in continuous X-ray generation a high energy electron produces X-ray photon. That is why production of continuous X-rays is termed as *inverse photoelectric effect.*

13.8.3 Origin of Continuous X-rays

Continuous X-rays are produced when the high speed electrons from the cathode interacts with the nucleus of the target via coulomb field and transfer momentum to the nucleus and loose energy. The process is illustrated in Fig. 13.5.

As the electron passes through the atom close to the nucleus it is attracted and gets deflected in the strong electric field of the nucleus and an *electromagnetic wave of energy* $h\nu$ is emitted. The target nucleus being heavy it does not recoil and a part of electron energy is given to the generated photon. If the velocity of the electron changes from v to v', then

$$\tfrac{1}{2}mv^2 - \tfrac{1}{2}mv'^2 = h\nu. \tag{13.3}$$

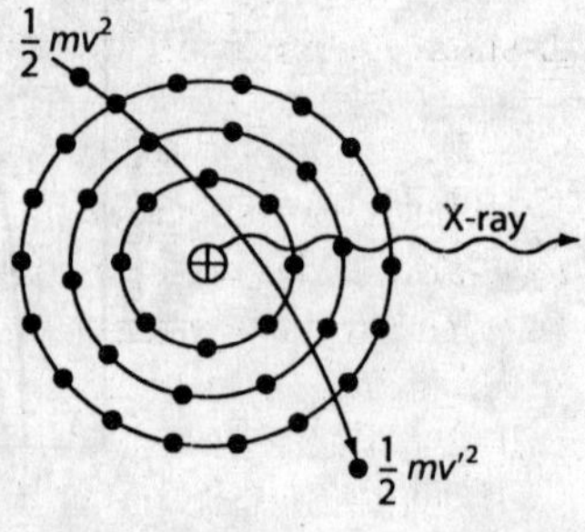

Fig. 13.5: Emission of continuous X-rays.

If the electron is completely stopped, then $v' = 0$ and the frequency of the emitted photon is the highest (ν_{max}) given by

$$\tfrac{1}{2}mv^2 = h\nu_{\text{max}} \quad \text{or,} \quad Ve = h\nu_{\text{max}}, \quad [\text{where } \tfrac{1}{2}mv^2 = eV]. \tag{13.4}$$

The emission of X-rays from decelerated electron is called *Bremsstrahlung process.*

13.9 Characteristic X-rays

Fig. 13.4 shows that the X-radiation radiated by a target consists of a continuous spectrum superimposed with high intensity peaks if the accelerating potential V be greater than a certain value. The peaks occur at specific wavelength and are the characteristic of the target element and hence called *characteristic X-rays.*

The important features of characteristic X-rays spectrum are as follows:

(i) There is a striking similarity in X-ray line spectra of all elements, the frequency being $\nu = a(z - b)$, Z = atomic number.

(ii) The radiation is highly monochromatic and independent of applied voltage (V).

(iii) The line spectrum shifts progressively as the target element changes from lower atomic weight to higher atomic weight.

(iv) The energy of the K-series [Fig. 13.6] is the maximum and forms hard X-rays, whereas that of M-series is much less and forms the soft X-rays. *The origin of X-ray line (characteristic) spectra could not be understood in view of classical electromagnetic theory.*

13.9.1 Origin of Characteristic X-rays

In view of *Bohr's atom model,* an atom of atomic number Z contains Z electrons in normal state. These Z-electrons are distributed in different quantum orbits like K-shell, L-shell, M-shell, N-shell—with quantum numbers $n = 1, 2, 3, 4, \cdots$ respectively [Fig. 13.6]. For the removal of these electrons, they require decreasing excitation energy from lower to higher shells.

If a cathode ray electron of energy eV strikes a K-shell electron having binding energy E_K, then the KE of the knocked out K-electron is

$$\tfrac{1}{2}mv^2 = eV - E_K. \tag{13.5}$$

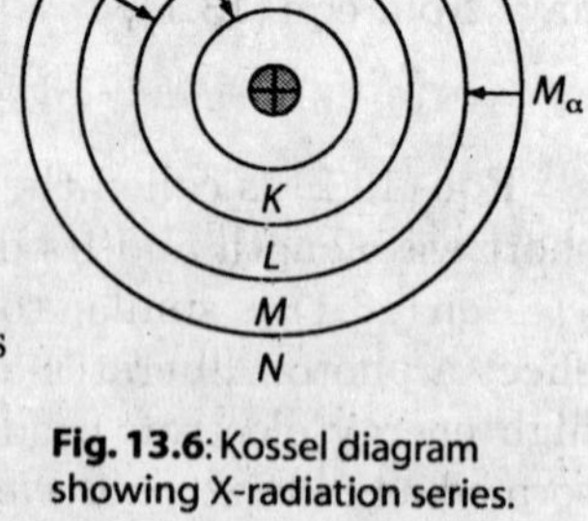

Fig. 13.6: Kossel diagram showing X-radiation series.

Thus, the minimum energy of the incident electron to knock out a K-electron is E_K, when $v = 0$.

The vacancy created by the knocked out electron is filled up by an electron from the higher L-shell. During this transition, a photon of frequency ν_{KL} is emitted, where

$$\nu_{KL} = E_K - E_L. \tag{13.6}$$

The X-ray spectral line due to transition from L- to K-shell is called K_α *line* of frequency ν_{KL} and is the characteristic of the target element. The emitted radiation is not in *visible region* as with target elements of high Z-values, the lower levels have binding energy ~ 1 keV.

If an electron from M-shell fills the vacancy in the K-shell, then resulting spectral line is called K_β *line* of frequency ν_{KM} and so on.

If electrons from M-, N-shell fill the vacancy created in the L-shell, then L-series lines are emitted. Series lines likes $M_\alpha, M_\beta, M_\gamma$, etc. have a similar origin. In each series, the short wavelength limit is known as corresponding *absorption edge*.

Thus, the following table represents the different transition leading to different series lines with their frequency.

Table 10.1: Origin of different series lines and their frequencies

Sl No.	*Transition form*	*Characteristic line*	*Frequency*
1	$K \longrightarrow L$ level	K_α-line	$\nu_{K_\alpha} = \frac{E_K - E_L}{h}$
2	$K \longrightarrow M$ level	K_β-line	$\nu_{K_\beta} = \frac{E_K - E_M}{h}$
3	$L \longrightarrow M$ level	L_α-line	$\nu_{K_\alpha} = \frac{E_L - E_M}{h}$
4	$L \longrightarrow N$ level	L_β-line	$\nu_{K_\beta} = \frac{E_L - E_N}{h}$

The energy level diagram of characteristic X-ray spectra is shown in Fig. 13.7.

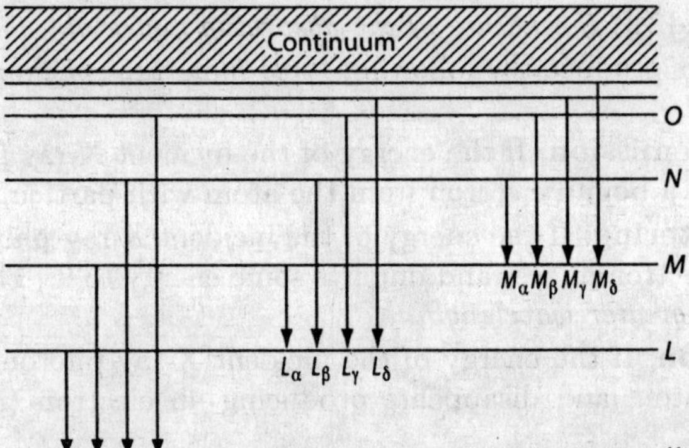

Fig. 13.7: Electric transitions producing characteristic X-ray line.

13.10 Applications of X-rays

X-rays have found important applications in the following fields:

1. **Industrial applications**: (i) X-rays are used to locate internal defects of a body like *flaws* and *blowholes* in metal castings and welding joins. (ii) X-rays are used to study structure

of materials like alloys by determining the structure of ingots with the help of X-ray diffraction study. (iii) The molecular structure of rubber, cellulose, plastic fibres, etc. can be analysed by studying their X-ray diffraction. (iv) X-rays are used to detect any crack in the body of the aeroplane and parts of other fast moving machines.

2. **Medical applications**: (i) *Radiography*: In radiography, X-ray photographs of the affect part of the body are taken using X-rays as source. By analysing the photograph so obtained, fractures, tumours, presence of foreign bodies like bullet, etc. in human body can be detected.

(ii) *X-ray therapy*: Soft X-rays are used to cure certain types of skin diseases like *cancer, tumour*, etc. To destroy tumours formed in the deep of the body, hard X-rays are used.

For bringing about genetic mutation and identification of different types of cells and tissues X-rays are used.

3. **Forensic science**: X-rays have got important applications in the detection of crime. (i) Grenz-rays (very soft X-rays) are used for the detection of *forgery in documents*. Using X-ray fluorescent technique, finger print can be analysed and the criminal can be detected. (ii) *Approximate age* and the *sex* of a highly decomposed or charred body can be determined using X-rays. (iii) Swallowed metals or jewels in human body can be detected through radiography. (iv) The nature of internal injury caused by any quarrel and the cause of death of an animal can be determined by using X-rays.

4. **Scientific research**: X-rays are used to (i) study the structure of complex organic compounds (ii) determine the atomic number of elements in atomic-physics (iii) study the structure of crystalline solids and alloys (iv) analyse the structure of crystals and structure of atoms.

13.11 Absorption of X-rays

An X-ray beam when passes through a material the intensity of the transmitted beam gets reduced. The reduction in intensity for a particular thickness depends on (i) *the frequency of the incident X-ray beam* and (ii) *the nature of the absorbing material.*

The absorbed X-rays produce the following *three important phenomena* through the interaction with the material:

(1) **Photoelectric emission**: If the energy of the incident X-ray photon be from 100 to 200 eV, then it can liberate a bound electron from the atom with particular KE.

(2) **Compton scattering**: If the energy of the incident X-ray photon be $\sim$ 200 keV, then it is scattered by a free electron at rest and imparts some energy to it. The scattered photon moves with *reduced energy* or *greater wavelength.*

(3) **Pair production**: If the energy of the incident X-ray photon be $\sim$ 6 MeV, it interacts with a nucleus of an atom and disappears producing an electron (e^-) positron (e^+) pair as $h\nu \longrightarrow e^- + e^+$.

13.12 Variation of Intensity with Thickness

Let a beam of X-rays of intensity I_0 be incident normally on a material [Fig. 13.8] and is reduced to I after penetrating a distance x of the material.

If the intensity is further reduced from I to $I - dI$ after penetrating a distance from x to $x + dx$, then the rate of reduction in intensity with distance is $-\frac{dI}{dx}$.

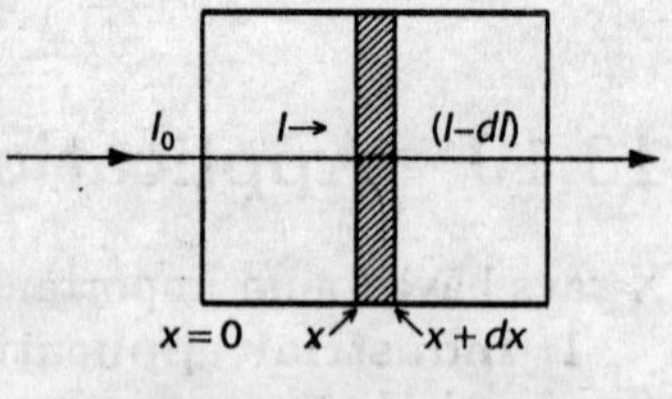

Fig. 13.8: Absorption of X-rays.

But

$$-\frac{dI}{dx} \propto I \quad \text{or,} \quad \frac{dI}{I} = -\mu x,$$

where μ is a constant of proportionality and is called the *linear absorption coefficient.*

Integrating the above equation on both sides, we get

$$\int_{I_0}^{I} \frac{dI}{I} = -\mu \int_0^x dx \quad \text{or,} \quad \ln\left(\frac{I}{I_0}\right) = -\mu x \quad \text{or,} \quad I = I_0 e^{-\mu x}. \tag{13.7}$$

Eqn (13.7) shows that the intensity of X-rays decreases exponentially with distance (x). The exponential variation of I with x is shown in Fig. 13.9.

From eqn (13.7) we note that if $\mu = 1/x$, then $I = I_0/e$. Hence, linear absorption coefficient is defined as *the reciprocal of the thickness* (x) *at which the intensity I falls to* $1/e$ *times its initial value* (I_0).

Again, if $I = I_0/2$ at $x = x'$, then

$$\frac{I}{I_0} = e^{-\mu x_1'} \text{ or, } \frac{1}{2} = e^{-\mu x'} \text{ or, } e^{\mu x'} \text{ or, } \mu = \frac{\ln 2}{x'} = \frac{0.693}{x'}. \tag{13.8}$$

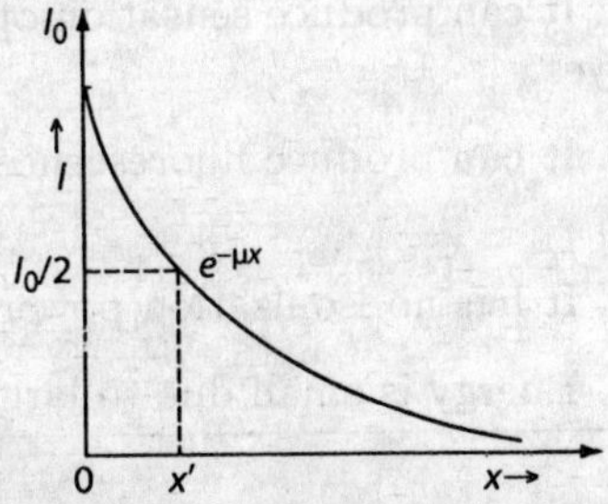

Fig. 13.9: Exponential decay of *I* with *x*.

Thus, measuring the distance x' at which the intensity of the X-ray beam is reduced to 1/2, we can find out the value of μ.

13.13 Properties of X-rays

X-rays have the following important properties:

1. X-rays are electromagnetic waves of very short wavelengths ($\lambda = 0.1$ Å to 100 Å) and show dual nature.

2. They are transverse electromagnetic waves and of non-electrical nature and hence do not get deflected in electric and magnetic fields.

3. X-rays are invisible and travel in straight lines with the velocity of light.

4. They are reflected, refracted, diffracted and polarised like ordinary light.

5. They are differentially absorbed by matter.

6. X-rays can ionize gases and affect the electrical properties of solids and liquids.

7. X-rays can damage living cells and produce genetic mutations.

8. They can affect photographic plate by causing photo-ionization.

9. X-rays can produce fluorescence in substances like barium, cadmium, zinc sulphide, etc.

10. They can cause photoelectric effect in certain metal if the energy of X-ray photon is 100 to 200 eV.

11. If the energy of incident X-ray photon be ~ 200 eV, then Compton effect occurs.

12. X-rays can cause pair production if the energy is ~ 1 MeV.

13. X-rays are diffracted by crystals following Bragg's law.

14. X-rays can produce secondary X-rays when they fall on heavy metals.

13.14 Differences between Visible Light and X-rays

Visible light	*X-rays*
1. Wavelength ranges from $\lambda = 4000$ Å (violet) to 800 Å (red).	1. Wavelength is very small, $\lambda = 0.1$ Å to 100 Å.
2. Penetrating power is very small. It can pass through air, water, glass, etc. transparent medium.	2. Penetrating power is high. It can penetrate through wood, paper, skin, flesh, thin metal plate, etc.
3. It can produce sensation of vision to human eyes.	3. X-ray is invisible to human eye.
4. It can produce fluorescence.	4. X-ray produces fluorescence in many substances.
5. It has no isonization power.	5. X-ray can ionize a gas.
6. Energy is small due to large wavelength.	6. Energy is high as it is of small wavelength.

13.15 Comparison between Cathode-rays and X-rays

Cathode-ray	*X-ray*
Similarities	
1. It travels in a straight line.	1. It travels in a straight line.
2. It can produce fluorescence in many substances.	2. It can produce fluorescence in many substances.
3. Affects photographic plate.	3. Affects photographic plate.
4. It can penetrate thin metal plates.	4. It can penetrate thin metal plates.
Dissimilarities	
1. It is a stream of negatively charged particles.	1. It is an electromagnetic wave having no charge.
2. It is deflected by electric or magnetic field.	2. It does not deflect in electric and magnetic fields.
3. Its speed is always less than the velocity of light.	3. It travels with the speed of light.
4. Velocity of cathode rays increases with the increase of PD between the anode and the cathode.	4. The wavelength decreases and penetrating power increases with the increase of PD between the anode and the cathode.
5. Penetrating power is relatively small.	5. Penetrating power is relatively large.
6. It has no ionizing power.	6. It can ionize a gas.

13.16 X-ray Diffraction: Laue Method

Why a crystal should diffract X-rays?

We know that for the diffraction to occur the size of diffracting obstacle must be comparable with the wavelength of the radiation to be diffracted. *Max von Laue,* a German physicist realized that the spacing between the atoms in a crystal is of the order of 10^{-10} m, which is the order of wavelength of X-rays. Therefore, the closed ordered array of atom in a natural crystal would act as a diffraction grating for X-rays, which are electromagnetic waves. The atoms in the array of the crystal correspond to the *grating lines* and the distance between the consecutive atoms to *grating element.*

A crystal grating, however, behaves as a *three-dimensional space grating.*

13.17 The Laue Photographs

Laue in collaboration with Friedrich and Knipping carried out the following experiment in 1913.

A single crystal C is held fixed and irradiated with white X-rays (X-rays with continuous wavelength). The rays diffracted through the crystal is made to fall on a photographic film M as shown in Fig. 13.10.

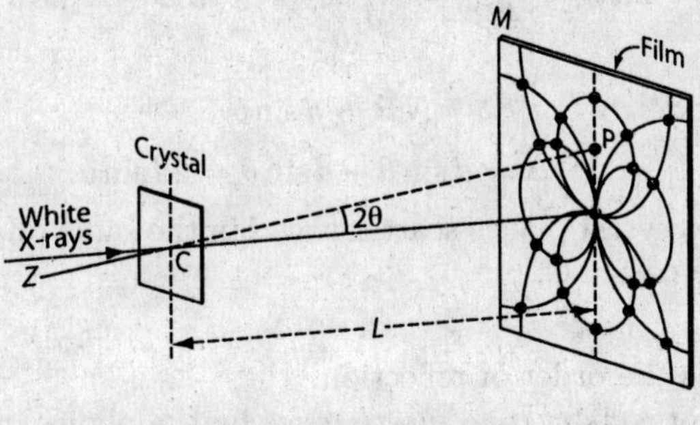

Fig. 13.10: A set of planes lying parallel to a common zone axis ZCP gives rise to a series of spots on a given ellipse.

The crystal C being fixed, the angle of diffraction θ is also fixed. The distance of various reflection spot will be r from the direct beam, where $r = L \tan 2\theta$.

Each spot on the photo-plate is due to the superposition of reflected X-rays of all orders 1, 2, 3, $\cdots$ from a single plane. The structure of the crystal C can be known by the interpretation of the photograph. If the wavelength λ of X-rays be supplied, then the dimensions of the unit cell of the crystal can be determined.

Note: *For obtaining prominent Laue photograph following points are to be kept in mind*: 1. The crystal must be thin. 2. The X-rays should be white if the crystal is fixed. For monochromatic X-rays the crystal is to be rotated. 3. The incident X-ray beam must be narrow.

Conclusion: Laue spot experiment provides experimental proof that (i) X-rays are electromagnetic waves of very short wavelengths and (ii) in a crystal, atoms are arranged in a regular three-dimensional lattice.

For this important contribution Laue was honoured with Nobel prize in 1914.

13.18 Bragg's Laws

We consider a crystal made up of equidistant parallel planes of atoms with inter-planar spacing d as shown in Fig. 13.11. Let a ray of monochromatic X-rays PA of wavelength λ be incident at an angle θ with atomic planes. The corresponding reflected ray AQ must also be inclined at angle θ to the plane. RB is another ray incident on the second plane at the same angle θ and the corresponding reflected ray be BS inclined also at θ with the plane.

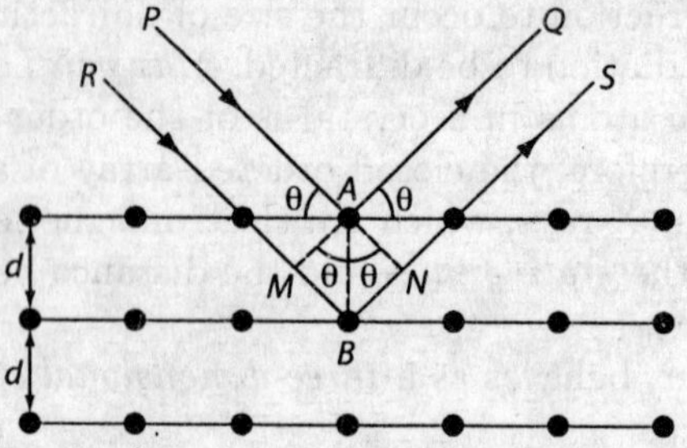

Fig. 13.11: Diffraction of X-rays by crystal planes.

The path difference between the two reflected rays AQ and BS is $\Delta = MB + NB$, where AM and AN are perpendiculars from A on RB and BS respectively.

In ΔMAB,

$$\sin\theta = \frac{MB}{AB} = \frac{MB}{d} \quad \text{or,} \quad MB = d\sin\theta.$$

Similarly, in ΔNAB,

$$NB = d\sin\theta.$$

$$\therefore \Delta = d\sin\theta + d\sin\theta = 2d\sin\theta.$$

If the two consecutive crystal planes scatter waves in the same phase, then the path difference must be integral multiple of λ, i.e.,

$$\Delta = n\lambda,$$

where $n = 0, 1, 2, 3$, etc. is the order of reflection.

Hence, the condition of *constructive interference* by the planes in a crystal is

$$2d\sin\theta = n\lambda. \tag{13.9}$$

Since the maximum possible value of $\sin\theta = 1$,

$$\frac{n\lambda}{2d} \leq 1. \tag{13.10}$$

Hence, the wavelength λ must not exceed twice the interplanar spacing d for Bragg diffraction to occur.

Definition: Bragg's law is defined as the X-rays reflected from different parallel planes of a crystal interfere constructively when the path difference is integral multiple of the wavelength of X-rays.

N.B. Bragg's law of X-ray reflection eqn (13.10) has the following utilities: 1. If λ is known, then measuring θ experimentally the value of inter-planar spacing and hence lattice constant can be determined. 2. If d is supplied, then measuring θ we can find out the wavelength of the unknown X-rays.

13.19 Determination of Lattice Constant

Using *Bragg's equation* the lattice constant of cubic crystals can be determined by measuring the angle of diffraction by Bragg X-ray crystal spectrometer as shown in Fig. 13.12.

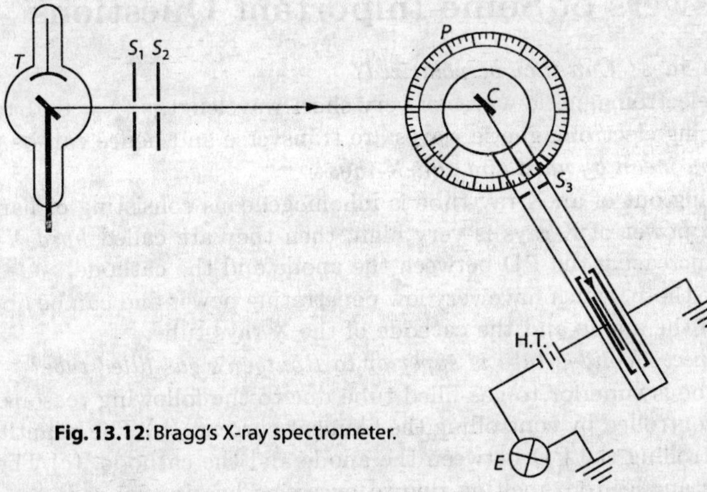

Fig. 13.12: Bragg's X-ray spectrometer.

X-rays from the X-ray tube T pass through two narrow slits S_1 and S_2 and then fall on the crystal C mounted on a turntable P. Using the ionization chamber I, the intensity of the reflected rays is measured at different angles. The turntable P and the ionization chamber are capable of rotation about a central vertical axis. In fact, P and I are geared together such that the detector turns through 2θ, when the crystal rotates through θ. Thus, the detector is always self-set to receive the reflected beam from C.

According to Bragg's law $2d\sin\theta = n\lambda$, where d is the inter-planar spacing, θ the Bragg angle, λ the wavelength of the incident X-rays and n, the order number of the spectrum.

For the first order, $n = 1$,

$$\therefore \frac{1}{d} = \frac{2\sin\theta}{\lambda}.$$

Again, $\lambda = 2d_{100}\sin\theta_1 = 2d_{110}\sin\theta_2 = 2d_{111}\sin\theta_3$, where d_{100}, d_{110} and d_{111} are the inter-planar spacing for (100), (110) and (111) planes respectively and θ_1, θ_2 and θ_3 the corresponding glancing angles.

$\therefore \frac{1}{d_{100}} : \frac{1}{d_{110}} : \frac{1}{d_{111}} = \sin\theta_1 : \sin\theta_2 : \sin\theta_3.$

Now, the ratios $\frac{1}{d_{100}} : \frac{1}{d_{110}} : \frac{1}{d_{111}}$ for the three types of cubic crystal are:

Crystal $\frac{1}{d_{100}} : \frac{1}{d_{110}} : \frac{1}{d_{111}}$

Simple cubic $\sqrt{1} : \sqrt{2} : \sqrt{3}$

Body centred cubic $1 : \frac{1}{\sqrt{2}} : \sqrt{3}$

Face centred cubic $1 : \sqrt{2} : \frac{\sqrt{3}}{2}$.

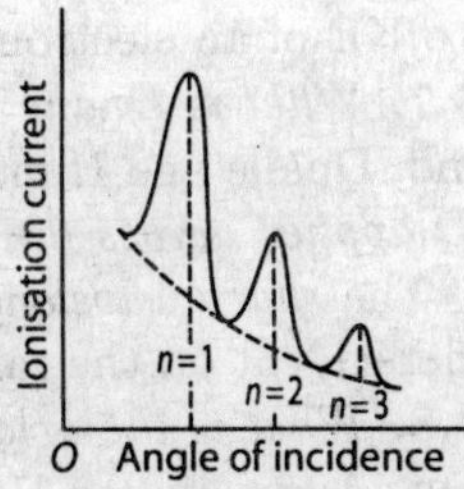

Fig. 13.13: Variation of ionisation current with angle of incidence.

Knowing the values of θ_1, θ_2 and θ_3 from the angle of incidence verses ionization current curve [Fig. 13.13] we can find out the values of $\sin\theta_1 : \sin\theta_2 : \sin\theta_3$ and then the inter-planar spacing d_{100}, d_{110} and d_{111}. From these d values the lattice constant a can be found out and the type of the lattice can be identified.

Example: In case of NaCl crystal, the first order reflection maxima from (100), (110) and (111) planes occur at $5.9°, 8.4°$ and $5.2°$.

Hence, $\frac{1}{d_{100}} : \frac{1}{d_{110}} : \frac{1}{d_{111}} = \sin 5.9° : \sin 8.4° : \sin 5.2° = 1 : \sqrt{2} : \frac{\sqrt{3}}{2}$.

$\therefore$ NaCl crystal is face centred cubic.

13.20 Answers of Some Important Questions

13.1. *What are X-rays? Can they be polarized?*

Ans. X-rays are electromagnetic waves of very short wavelengths ($\lambda = 0.01$ to $10\,\text{Å}$).

Yes. X-rays being electromagnetic waves are transverse and hence can be polarized.

13.2. *What do you mean by hard and soft X-rays?*

Ans. X-rays coming out of an X-ray tube is inhomogeneous consisting of hard and soft X-rays. If the penetrating power of X-rays is very high, then they are called *hard X-rays.* Hard X-rays are produced by increasing the PD between the anode and the cathode.

Soft X-rays are those, which have very low penetrating power and can be obtained by applying small PD between the anode and the cathode of the X-rays tube.

13.3. *In what respect Coolidge tube is superior to Röntgen's gas-filled tube?*

Ans. Coolidge tube is superior to gas-filled tube due to the following reasons: (a) The intensity of X-rays can be controlled by controlling the filament current. (b) The quantity of X-rays can be controlled by controlling the PD between the anode and the cathode. (c) There is arrangement for protecting the anode from melting due to excessive heating by radiator fins or circulating cold water through small tubes. (d) Coolidge tube is more *stable* than gas-filled tubes during continued operation.

13.4. *What are continuous X-rays? What is their origin?*

Ans. X-rays in which X-ray intensity varies with the continuous variation of wavelength are called *continuous X-rays.*

Continuous X-rays originate when high energy electrons strike a target material of large atomic weight. The energetic electrons suffer loss of KE, a small part of which is converted into X-rays and the rest is converted into heat.

13.5. *Continuous X-rays are called white radiation—why?*

Ans. Continuous X-rays originate due to energy loss of fast electrons at the target. The intensity of the emitted X-rays appears in the form of continuously varying wavelength of X-rays with certain maximum and minimum values. In analogy with visible white light consisting of wavelength from ultraviolet to infrared, continuous X-rays are termed as *white radiation* due to the presence of continuously varying wavelength in it, though it is invisible.

13.6. *What do you mean by inverse photoelectric effect?*

Ans. The phenomenon of *short wavelength limit* of continuous X-ray spectra is called *inverse photoelectric effect.* This is because in the emission of X-rays, KE of fast electron is converted into radiant energy, but in photoelectric effect radiant energy (at least partially) is converted into KE of an electron.

13.7. *What is Duane and Hunt law?*

Ans. Duane and Hunt law states that *the short wavelength limit is inversely proportional to the PD applied across the anode and the cathode of the X-ray tube* (*i.e.*, $\lambda_{\text{min}} \propto \frac{1}{V}$)

The *sharp wavelength limit,* an outstanding property of continuous X-ray spectrum, which is independent on the target material, could be understood in view of this law.

13.8. *What is the origin of characteristic X-rays? Why are they so-called?*

Ans. According to Bohr the electrons in an atom revolve in definite orbits like K, L, M, etc. around a heavy positively charged nucleus. The electrons in the innermost orbit are strongly bound to the nucleus by electrostatic force of attraction and as such considerable external energy is required to knock out an electron from the innermost orbit.

When a fast moving electron from the cathode is incident on a target, it penetrates into the interior of the atom and may knock out an electron from the innermost orbit (i.e., K-shell). The vacancy in the innermost, say, K-shell is filled up by electrons from say L-shell and the different in energy $(E_K - E_L)$ between the orbits is given out as X-rays of definite wavelength.

The wavelength of the X-radiation in this process depends upon the character of the target element (i.e., Z-value) and hence called *characteristic X-rays.*

13.9. *Arrange in increasing order the frequencies* ν_K, ν_L, ν_M *of characteristic* K_α, L_α *and* M_α *X-rays lines. Which of these correspond to hard and soft X-rays and why?*

Ans. The required arrangement according to increasing value is $\nu_K > \nu_L > \nu_M$.

The energy of the X-rays photon $h\nu_K$ emitted from K-shell is more due to greater binding energy of the electron and hence X-rays emitted with energy $h\nu_K$ are called *hard X-rays.* On the other hand, X-ray photon of energy $h\nu_M$ emitted from M-shell is low and are called *soft X-rays.*

13.10. *Fill up the blanks: The frequency of characteristic X-rays depends on* (1) *but independent on* (2).

Ans. (1) the atomic number (Z) of the target element (2) PD applied between the anode and the cathode of the X-ray tube.

13.11. *Light elements are unable to emit characteristic X-rays—why?*

Ans. It is observed that the shortest wavelength limit of X-rays emitted by light elements in far more greater than X-ray wavelength range, i.e., 0.1 to 100 Å. So the light elements having low value of Z are unable to emit characteristic X-rays.

13.12. *Explain why an optical grating is unsuitable but a natural crystal is suitable for X-ray diffraction.*

Ans. We know that for the diffraction to occur, the sharpness of the diffraction obstacle must be comparable with the wavelength of the radiation to be diffracted.

For optical grating the size of the diffraction obstacle is $\sim 1.6 \times 10^{-6}$ m but the wavelength of X-rays is $\sim 10^{11}$ m, which is much smaller than the size of the obstacle. Hence, optical gratings are unsuitable for diffraction of X-rays.

But the spacing between the atoms in a crystal is of the order of 10^{-10} m, which is almost of the order X-ray wavelength. That is why closed ordered array of atoms in a natural crystal would act as a diffraction grating for X-rays. The atoms in the array of the crystal correspond to grating lines and the distance between the consecutive atoms to grating element.

A crystal grating, however, behaves as a three-dimensional space grating.

13.13. *What is fundamental difference between reflection of light of plane mirror and reflection of X-rays by crystal planes?*

Ans. Reflection of light by plane mirrors is secular, i.e., for any angle of incidence there is a reflected ray. But X-ray reflection from crystal planes is secular, i.e., there is a reflected ray only for those angles of incidence θ for which Bragg's condition, $2d \sin\theta = n\lambda$ is satisfied.

13.14. *Why is Bragg reflection sometimes called diffraction?*

Ans. Bragg assumed that in crystal grating the Bragg diffraction pattern is obtained by actual reflection of X-rays from crystal atomic planes. Moreover, the Bragg condition of X-ray reflection, viz., $2d \sin\theta = n\lambda$ is the same as the condition of optical diffraction. That is why the Bragg reflection is sometimes called *Bragg diffraction.*

13.15. *What is the difference between the origin of visible light and X-rays?*

Ans. When an electron in the outermost orbit is excited to higher energy state it reverts back to lower energy state due to small life-time ($\sim 10^{-8}$ s) of the excited state. The difference of energy between the initial and final states gives rise to a photon in the visible region.

But characteristic X-rays originate due to the transition of electron in the innermost shell (say K-shell, etc.) having higher energy. The innermost electron is knocked out by the incident fast electron and the vacancy thus created is filled up by an electron in the higher shell (like L-shell, etc.) and the difference of energy between the two shells leads to the creation of a photon in the X-ray region.

13.16. *Indicate the areas in which X-rays behave as wave and particle.*

Ans. X-rays play dual role, viz., wave and particle.

In reflection, refraction, interference, polarization, classical scattering, etc. X-rays behave as wave.

But in Compton effect X-rays behave as particle.

13.17. *If the energy levels of K, L and M states of* Pt *are* 78 keV, 12 keV *and* 3 keV *respectively, then calculate the wavelengths of K_α and K_β-lines emitted from platinum.* [$h = 6.626 \times 10^{-34}$ J-s, $c = 3 \times 10^8$ m/s] **[WBUT 2008]**

Soln. If ν_{K_α} be the frequency of K_α-line and E_K and E_L, the energy values of K and L orbits, then

$$h\nu_{K_\alpha} = E_K - E_L$$

$$\text{or, } \nu_{K_\alpha} = \frac{E_K - E_L}{h} \quad \text{or, } \frac{c}{\lambda_{K_\alpha}} = \frac{E_K - E_L}{h}$$

$$\text{or, } \lambda_{K_\alpha} = \frac{ch}{E_K - E_L} = \frac{3 \times 10^8 \times 6.626 \times 10^{-34}}{(78 - 12) \times 10^3 \times 1.6 \times 10^{-19}}$$

$$= \mathbf{0.1882\ Å}.$$

Similarly, the wavelength of K_β-line of Pt is

$$\lambda_{K_\beta} = \frac{ch}{E_K - E_M} = \frac{3 \times 10^8 \times 6.626 \times 10^{-34}}{(78 - 3) \times 10^3 \times 1.6 \times 10^{-19}} = \mathbf{0.1656\ Å}.$$

13.21 Solved Problems

Ex. 1 *The operating voltage of an X-ray tube is* 50 kV. *Find the maximum speed of the electrons striking the surface of the anode and the shortest wavelength of the X-ray produced.*

Soln. If V be the PD between the anode and the cathode of an X-ray tube and e the electronic charge, then the energy of the electrons generating X-rays is eV. If v_m be the maximum velocity of the electron, then the maximum energy gained is $\frac{1}{2}mv_m^2$.

$$\therefore \frac{1}{2}mv_m^2 = eV \quad \text{or,} \quad v_m = \sqrt{\frac{2eV}{m}} = \sqrt{V}\sqrt{\frac{2 \times 1.6 \times 10^{-19}}{9 \times 10^{-31}}} = \sqrt{V} \times 0.596 \times 10^6 \text{ m/s}.$$

[$\because e = 1.6 \times 10^{-19}$ C, $m = 9 \times 10^{-31}$ kg]

Since $V = 50\,\text{kV} = 5000\,\text{V}$.

$\therefore v_m = \sqrt{50000} \times 0.596 \times 10^6 = \mathbf{133.27 \times 10^6}$ **m/s**

The shortest wavelength λ_0 is given by

$$\lambda_0 = \frac{ch}{eV} = \frac{12420 \times 10^{-10}}{V} = \frac{12420 \times 10^{-10}}{50000} = \mathbf{0.2484\ Å}.$$

$$\left[\because \frac{ch}{e} = \frac{(3 \times 10^8) \times (6.624 \times 10^{-34})}{1.6 \times 10^{-19}} = 12420 \times 10^{-10}\right].$$

Ex. 2 *An X-ray tube gives an X-ray of wavelength* 1 Å. *What is the voltage between the anode and the cathode?* [$e = 1.6 \times 10^{-19}$ C, $c = 3 \times 10^8$ m/s *and* $h = 6.6 \times 10^{-34}$ J-s]

Soln. If ν be the frequency and h, the Planck's constant, then the energy of the electron = energy of the photon

$$\text{or,} \quad E = h\nu = \frac{hc}{\lambda} = \frac{6.6 \times 10^{-34} \times 3 \times 10^8}{1 \times 10^{-10}} = 19.8 \times 10^{-16} \text{ J}.$$

If V be the required voltage, then

$$E = eV = 19.8 \times 10^{-16} \quad \text{or,} \quad 1.6 \times 10^{-19} V = 19.8 \times 10^{-16} \quad \text{or,} \quad V = \frac{19.8 \times 10^{-16}}{1.6 \times 10^{-19}} = \mathbf{12375\ V}.$$

Ex. 3 *If the PD between the anode and the cathode is* 25 kV, *what is the cut-off wavelength and the cut-off frequency of the emitted X-rays?*

Soln. The cut-off wavelength

$$\lambda_0 = \frac{hc}{eV} = \frac{12400}{V} = \frac{12400}{25000} = \mathbf{0.496\,\text{Å}}.$$

The maximum frequency of the emitted X-ray is

$$\nu_0 = \frac{c}{\lambda_0} = \frac{3 \times 10^8}{0.496} = \mathbf{6.05 \times 10^{18}\,Hz}.$$

$$[\because \lambda_0 = 0.496\,\text{Å} = 0.496 \times 10^{-10}\,\text{m}]$$

Ex. 4 *Find the wavelength of short wave limit of X-ray continuous spectrum, if it shifts by* 0.5 Å, *when the voltage applied to X-ray tube is doubled.* **[CUH 2002]**

Soln. If V_1 and V_2 be the PD between the anode and the cathode of the X-ray tube in the two cases, then

$$\frac{V_2}{V_1} = 2.$$

If λ_{01} and λ_{02} be the minimum wavelength of the emitted X-rays, then

$$eV_1 = \frac{hc}{\lambda_{01}} \quad \text{or,} \quad \lambda_{01} = \frac{hc}{eV_1}.$$

Similarly, $\lambda_{02} = \frac{hc}{eV_2}$.

$$\therefore \frac{\lambda_{01}}{\lambda_{02}} = \frac{V_2}{V_1} = 2 \quad \text{or,} \quad \lambda_{01} = 2\lambda_{02}.$$

Again,

$$\lambda_{01} - \lambda_{02} = 0.5\,\text{Å}$$

$$\text{or,} \quad 2\lambda_{02} - \lambda_{02} = 0.5\,\text{Å}$$

$$\text{or,} \quad \lambda_{02} = \mathbf{0.5\,\text{Å}}$$

$$\therefore \quad \lambda_{01} = 2\lambda_{02} = 2 \times 0.5\,\text{Å} = \mathbf{1\,\text{Å}}.$$

Ex. 5 *An X-ray tube is operating at* 15 kV. *Calculate the upper limit of the speed of the electrons striking the anti-cathode and the lower limit of the wavelength of the X-rays produced.* $e = 1.6 \times 10^{-19}$ C, $m = 9.1 \times 10^{-31}$ kg, $h = 6.6 \times 10^{-34}$ J-s. **[DUH 1963]**

Soln. If V be the applied PD and v_m be the maximum velocity of the ejected electron, then

$$Ve = \frac{1}{2}mv_m^2 \quad \text{or,} \quad v_m = \sqrt{\frac{2Ve}{m}} = \left[\frac{2 \times (15000) \times (1.6 \times 10^{-19})}{9.1 \times 10^{-31}}\right]^{1/2} = \mathbf{7.261 \times 10^7\,m/s}.$$

Again, if λ_0 be the minimum wavelength limit of X-rays, then

$$h\nu = \frac{hc}{\lambda_0} = \frac{1}{2}mv_m^2 = Ve.$$

$$\therefore \lambda_0 = \frac{hc}{Ve} = \frac{6.6 \times 10^{-34} \times 3 \times 10^8}{15000 \times 1.6 \times 10^{-19}} = \mathbf{0.8249\,\text{Å}}.$$

Ex. 6 *Electrons bombarding the anode of a Coolidge tube produce X-rays of wavelength* 1 Å. *Find the energy of each electron at the moment of impact.* **[CUH 1961]**

Soln. If v be the velocity of the electron at the time of impact with the anode, then

$$\text{Energy of the electron } E = \frac{1}{2}mv^2 = Ve = \frac{hc}{\lambda},$$

where λ is the wavelength of the X-rays and V is the applied PD to the tube.

$$\therefore \text{Energy } E = \frac{hc}{\lambda} = \frac{(6.62 \times 10^{-34}) \times (3 \times 10^8)}{1 \times 10^{-10}} = 19.86 \times 10^{-16}\,\text{J} = \mathbf{1.24 \times 10^4\,eV}.$$

Ex. 7 *The energy levels of K, L and M orbits of uranium are respectively* 68 keV, 12 keV *and* 2.5 keV. *What are the wavelengths of K_α and K_β characteristic lines emitted by uranium.*

Soln. If ν_{K_α} be the frequency of K_α-line and E_K and E_L, the energy values of K and L orbits respectively, then

$$h\nu_{K_\alpha} = E_K - E_L \quad \text{or,} \quad \nu_{K_\alpha} = \frac{c}{\lambda_{K_\alpha}} = \frac{E_K - E_L}{h}.$$

$$\therefore \lambda_{K_\alpha} = \frac{ch}{E_K - E_L} = \frac{(3 \times 10^8) \times (6.62 \times 10^{-34})}{(68 - 12) \times 10^3 \times 1.6 \times 10^{-19}} = \mathbf{0.221\,Å}.$$

Similarly, the wavelength of K_β-line of uranium is

$$\lambda_{K_\beta} = \frac{ch}{E_K - E_M} = \frac{(3 \times 10^8) \times (6.62 \times 10^{-34})}{(68 - 2.5) \times 10^3 \times 1.6 \times 10^{-19}} = \mathbf{0.189\,Å}.$$

Ex. 8 *A beam of X-rays of wavelength* 0.842 Å *is incident on a crystal at a glancing angle of* 8°35′, *when the first order Bragg's reflection occur. Calculate the glancing angle of the third order reflection.* **[WBUT 2002, '07]**

Soln. From Bragg's law of X-ray reflection

$$2d \sin\theta_n = n\lambda.$$

For the first order spectrum, $n = 1, \lambda = 0.842\,Å, \theta_1 = 8°35' = 8.6°$.

$$\therefore d = \frac{\lambda}{2\sin\theta_1} = \frac{0.842}{2 \times 0.1347} = 3.125\,Å.$$

If θ_3 be the glancing angle for the third order ($n = 3$) reflection, then

$$\sin\theta_3 = \frac{3\lambda}{2d} = \frac{3 \times 0.842}{2 \times 3.125} = 0.40416.$$

$$\therefore \theta_3 = \sin^{-1} 0.40416 = \mathbf{26°29'}.$$

Ex. 9 *The spacing of planes in a crystal is* 1.2 Å *and the angle for the first order Bragg's reflection is* 30°. *Determine the energy of the X-ray beam in* eV. **[WBUT 2005, '08]**

Soln. We have Bragg's equation,

$$2d \sin\theta_n = n\lambda.$$

For the 1st order $n = 1$, glancing angle $\theta_1 = 30°$, and inter-planar spacing $d = 1.2\,Å$.
$\therefore$ wavelength of the X-rays is

$$\lambda = \frac{2d\sin\theta_1}{n} = \frac{2 \times 1.2 \times \sin 30°}{1} = 2 \times 1.2 \times \frac{1}{2} = 1.2\,Å.$$

Now, the energy of the incident X-ray beam is

$$E = \frac{hc}{\lambda} = \frac{(6.623 \times 10^{-34}) \times (3 \times 10^8)}{1.2 \times 10^{-10} \times 1.6 \times 10^{-19}}\,\text{eV} = \mathbf{10.35 \times 10^3\,eV}.$$

Ex. 10 *The spacing between the principal planes of* NaCl *crystal is* 2.82 Å. *It is found that the first order reflection of a beam of monochromatic X-rays occurs at an angle of* 10°. *What is the wavelength of X-rays?*

Soln. According to Bragg's equation,

$$2d\sin\theta_n = n\lambda.$$

Here $d = 2.82\,\text{Å}, n = 1, \theta_1 = 10°, \lambda = ?$

$\therefore 2 \times 2.82 \times \sin 10° = 1 \times \lambda$ or, $\lambda = 2 \times 2.82 \times 0.1736\,\text{Å} = \mathbf{0.98\,Å}$.

Ex. 11 *Monochromatic X-rays* ($\lambda = 1.54\,\text{Å}$) *are incident on a cleavage face of rock salt crystal and the first order Bragg reflection is observed at* 16°. *Calculate the Avogadro's number.* [*Given: molecular weight of* NaCl = 58.5, *density* = 2.20 gm/cc.] **[CUH 1989]**

Soln. Bragg's equation,

$$2d\sin\theta = \lambda \quad \text{or,} \quad d = \frac{\lambda}{2\sin\theta} = \frac{1.54 \times 10^{-8}}{2 \times \sin 16°} = 2.794 \times 10^{-8} \text{ cm.}$$

$\therefore$ the number of atoms in 1 cc $= \frac{1}{d^3}$.

$\therefore \frac{1}{d^3} \times$ mass of an atom of NaCl = 2.20.

$\therefore$ mass of each atom of NaCl $= 2.20d^3$ or, mass of molecule of NaCl $= 2 \times 2.20d^3$.

$$\therefore N \times 2 \times 2.20d^3 = 58.5 \quad \text{or,} \quad N = \frac{58.5}{2 \times 2.20 \times (2.794)^3 \times 10^{-24}} = \mathbf{6.1 \times 10^{23}}.$$

Ex. 12 *X-rays of wavelength* 0.71 Å *are reflected from the* (110) *plane of a rock salt crystal of lattice constant* $a = 2.82\,\text{Å}$. *Calculate the corresponding glancing angle for second order reflection.*

Soln. The inter-planar distance between (110) planes is

$$d = \frac{a}{\sqrt{h^2 + k^2 + l^2}}.$$

Here, $a = 2.82\,\text{Å}, h = 1, k = 1, l = 0$.

$$\therefore d = \frac{2.82}{\sqrt{1^2 + 1^2 + 0^2}} = \frac{2.82}{\sqrt{2}} = \frac{2.82}{1.41} = 2\,\text{Å}.$$

Now, from Bragg's law,

$$2d\sin\theta = n\lambda.$$

Here, $d = 2\,\text{Å}, n = 2$, and $\lambda = 0.71\,\text{Å}$

$$\therefore 2 \times 2\sin\theta = 2 \times 0.71 \quad \text{or,} \quad \sin\theta = \frac{2 \times 0.71}{2 \times 2} = 0.355 \quad \text{or,} \quad \theta = \mathbf{20°42'}.$$

Ex. 13 *A crystal of* NaCl *mounted on a Bragg spectrometer reflects the* K_α*-line of X-rays at an angle of* 60° *in the first order. Calculate the wavelength of the X-ray* K_α*-line.* [*Given: Density of* NaCl = 2.17 g/cc; *molecular weight of* NaCl = 58.46]

Soln. The inter-planar distance (d) of NaCl is

$$d = \left(\frac{M}{2N\rho}\right)^{\frac{1}{3}},$$

where $M = 58.46$, the mol. wt. of NaCl, $N = 6.02 \times 10^{23}$, the Avogadro's number and $\rho = 2.17$, the density of NaCl.

$$\therefore d = \left[\frac{58.5}{2 \times (6.02 \times 10^{23}) \times (2.17)}\right]^{1/3} = 2.82 \times 10^{-8} \text{ cm.}$$

Glancing angle = 90° − angle of reflection = 90° − 60° = 30°.

Now, from Bragg's law,

$$2d\sin\theta = \lambda \quad [\because n = 1]$$

$$\text{or,} \quad \lambda = 2 \times 2.82 \times 10^{-8} \times \frac{1}{2} = \mathbf{2.82\,\text{Å}}.$$

Ex. 14 *The spacing between successive* (100) *planes of* NaCl *is* 2.82 Å. *X-rays incident on the surface of the crystal is found to give rise to first order Bragg reflection at glancing angle* 8.8°. *Calculate the wavelength of X-rays.* **[BUH 2003]**

Soln. The Bragg's law of X-rays reflection

$$2d\sin\theta = n\lambda.$$

Here, $d = 2.82$ Å, $\theta = 8.8°$ and $n = 1, \lambda = ?$

$$\therefore \lambda = 2 \times 2.82 \times \sin 8.8° = 2 \times 2.82 \times 0.1530 = \mathbf{0.8629\,\text{Å}}.$$

Ex. 15 *Using a crystal of* KCl, *Bragg found the following glancing angles for intense reflection from the lattice planes* (100), (110) *and* (111): $\theta_1 = 5.22°$, $\theta_2 = 7.30°$ *and* $\theta_3 = 9.05°$ *respectively. Show that* KCl *has SC structure.* **[cf. CUH 1996]**

Soln. We have Bragg's law,

$$2d\sin\theta = n\lambda.$$

$$\therefore \frac{1}{d} = \frac{2\sin\theta}{\lambda} \quad [\because \text{for the 1st order spectrum } n = 1].$$

$$\therefore \frac{1}{d_{100}} : \frac{1}{d_{110}} : \frac{1}{d_{111}} = \sin 5.22° : \sin 7.30° : \sin 9.05°$$

$$= 0.0910 : 0.1272 : 0.1570$$

$$= 1 : 1.40 : 1.73$$

$$\simeq 1 : \sqrt{2} : \sqrt{3}.$$

$\therefore$ KCl has a SC structure.

Ex. 16 *Copper has FCC structure and the atomic radius is* 0.1278 nm. *Calculate its density and the inter-planar spacing for* (321) *planes. The atomic weight of copper is* 63.5. **[WBUT 2005]**

Soln. The lattice constant for a SC structure is

$$a = \left(\frac{nM}{\rho N}\right)^{1/3}$$

where n = no. of atoms/cell = 4, M = atomic weight = 63.5, N = Avogadro's number = 6.023×10^{23} and ρ = density of copper.

Now the lattice constant for an FCC lattice is

$$a = \frac{4r}{\sqrt{2}} = \frac{4 \times (0.1278 \times 10^{-9})}{\sqrt{2}},$$

where r = atomic radius of the crystal = 0.1278 nm = = 0.1278×10^{-9} m.

$$\therefore \rho = \frac{nM}{Na^3} = \frac{4 \times 63.5}{6.023 \times 10^{23} \times \left[\dfrac{4 \times 0.1278 \times 10^{-9}}{\sqrt{2}}\right]^3} = \mathbf{8.9726\ gm/cc.}$$

Again, inter-planar spacing for (321) planes is

$$d_{hkl} = \frac{a}{(h^2 + k^2 + l^2)^{\frac{1}{2}}} = \frac{\left(\frac{4 \times 0.1278 \times 10^{-9}}{\sqrt{2}}\right)}{(3^2 + 2^2 + 1^2)^{\frac{1}{2}}} = \mathbf{0.0966\ nm.}$$

EXERCISES

13(A)–Objective Type Questions

Select the correct answer:

13.1. X-rays are (a) elastic waves (b) electromagnetic waves (c) stationary waves (d) radio waves.

13.2. The wavelength of X-rays is of the order of (a) metre (b) micron (c) nm (d) angstrom.

13.3. The wavelength, which falls in the X-rays region is (a) 1 Å (b) 10^{-6} Å (c) 10^{6} Å (d) 10^{-5} nm.

13.4. X-rays suffer deflection when subjected to (a) electric field (b) magnetic field (c) cross-electric and magnetic fields (d) neither of these.

13.5. If λ_x and λ_γ be the wavelengths of X-rays and γ-rays, then (a) $\lambda_x = \lambda_\gamma$ (b) $\lambda_x > \lambda_\gamma$ (c) $\lambda_x < \lambda_\gamma$ (d) $\lambda_x = \frac{1}{\lambda_\gamma}$.

13.6. According to the principle of classical electrodynamics when a charged particle undergoes deceleration it emits (a) elastic waves (b) electromagnetic waves (c) heat waves (d) none of these.

13.7. Which experiment established that X-rays are transverse in nature? (a) interference (b) diffraction (c) polarization (d) scattering.

13.8. If the number of electrons striking the target be increased, then the (a) frequency of X-rays will decrease (b) wavelength of X-rays will decrease (c) intensity of X-rays will increase (d) power of X-rays will increase.

13.9. The quality and intensity of X-rays can be independently controlled in (a) modern Coolidge tube (b) Röntgen gas-filled tube (c) both (a) and (b) (d) none of these.

13.10. The thermal conductivity of the target material in an X-ray tube should be (a) low (b) high (c) medium (d) zero.

13.11. The intensity of X-rays in a Coolidge tube can be controlled by changing the (a) PD between the anode and the cathode (b) the current in the tube (c) pressure in the tube (d) atomic number of the target material.

13.12. The quality of X-rays in a Coolidge tube can be increased by (a) increasing the current in the tube (b) increasing the PD between the anode and the cathode (c) reducing the PD between the anode and the cathode (d) reducing the pressure in the tube.

13.13. X-rays suffer Compton scattering by elements, which are (a) heavy (b) light (c) both (a) and (b) (d) none of these.

13.14. X-rays show (a) wave property (b) particle property (c) dual property (d) none of these.

13.15. 1 MeV X-ray photon can produce (a) wave property (b) particle property (c) dual property (d) none of these.

13.16. Origin of X-rays is de to the process of (a) ionization (b) inner orbital transition (c) bremsstrahlung.

[WBUT 2007]

13.17. If V be the accelerating PD and λ_m the minimum wavelength of continuous X-rays spectra, then (a) $\lambda_m \propto V$ (b) $\lambda_m \propto \frac{1}{V}$ (c) $\lambda_m = aV + c$ (d) $\lambda_m \propto \frac{1}{V^3}$.

13.18. If $\lambda = \frac{12400}{\sqrt{V}}$, then the wavelength ($\lambda$) of the continuous X-rays is the (a) maximum (b) minimum (c) medium (d) infinity.

13.19. Production of continuous X-ray is termed as (a) Compton effect (b) photoelectric effect (c) inverse photoelectric effect (d) scattering.

13.20. Characteristic X-rays originate due to transition of electrons in (a) outermost orbit (b) innermost orbit (c) retardation of fast electrons (d) acceleration of slow electrons.

13.21. The wavelength of characteristic X-rays depends on (a) the atomic number of the target element (b) PD applied between the electrodes (c) pressure in the tube (d) atomic weight of the target element

13.22. Characteristic X-ray spectrum has (a) continuous distribution of intensity over a spread of wavelength (b) large intensity at a particular wavelength (c) intensity peaks superimposed on continuous intensity distribution (d) small intensity peaks enclosed by a continuous intensity distribution envelope.

13.23. If ν_{K_α} and ν_{L_α} be the frequencies of K_α ad L_α characteristic X-ray lines, then (a) $\nu_{K_\alpha} = \nu_{L_\alpha}$ (b) $\nu_{K_\alpha} < \nu_{L_\alpha}$ (c) $\nu_{K_\alpha} > \nu_{L_\alpha}$ (d) $\nu_{K_\alpha} = \frac{1}{\nu_{L_\alpha}}$.

13.24. If λ_L and λ_K are the wavelengths L and K series X-ray lines respectively, then (a) $\lambda_L > \lambda_K$ (b) $\lambda_L < \lambda_K$ (c) $\lambda_L = \lambda_K$. **[WBUT 2005]**

13.25. X-rays can be diffracted by a (a) plane transmission grating (b) crystal grating (c) sharp edge of a blade (d) grid of a window.

13.26. The natural crystal acts as three-dimensional space grating because (a) the wavelength (λ) of X-rays is less than the lattice constant (a) (b) $\lambda = a$ (c) $\lambda < a$ (d) $\lambda = \frac{1}{a}$.

13.27. Which of the following wavelengths falls in the X-ray range? (a) 1 nm (b) 100 nm (c) 0.001 nm (d) 1000 nm. **[WBUT 2006]**

13.28. When two parallel rays of X-rays of wavelength λ are incident at an angle θ on a crystal with lattice separation d, constructive interference would be observed when (n is an integer) (a) $n\lambda = 2d\sin\theta$ (b) $n\lambda = d\sin\theta$ (c) $n\lambda = d\sin 2\theta$ (d) $n\lambda = 2d\sin 2\theta$. **[WBUT 2006]**

13.29. [111] plane of a cubic crystal is (a) perpendicular to the X-axis (b) perpendicular to the Y-axis (c) perpendicular to the Z-axis (d) none of these. **[WBUT 2005]**

13.30. X-ray reflection is (a) regular reflection (b) secular reflection (c) internal reflection (d) none of these.

13.31. If the X-ray intensity I_0 reduces to $I_0/2$ after traversing a distance $x_{1/2}$ of a substance, then the linear absorption coefficient is given by (a) $\mu = 0.693x_{1/2}$ (b) $\mu = \frac{x_{1/2}}{0.693}$ (c) $\mu = \frac{0.693}{x_{1/2}}$ (d) $\mu = x_{1/2}$

13.32. Laue spot experiment established that (a) X-rays are electromagnetic waves (b) X-rays are elastic waves (c) The atoms in a crystal are arranged in a regular three-dimensional lattice (d) X-rays can be diffracted by a plane grating.

13.33. In Bragg spectrometer experiment it is observed that the diffracted X-ray intensity (a) increases with order number (n) (b) decreases with n (c) remains constant with n (d) neither of these.

13.34. In Bragg's spectrometer experiment, the ionization current never falls to zero because (a) the continuous X-rays exist (b) the characteristic X-rays exist (c) both (a) and (b) exist (d) characteristic X-rays do not exist

13.35. In case of a SC structure $1/d_{100} : 1/d_{110} : 1/d_{111}$ is (a) $\sqrt{1} : \sqrt{2} : \sqrt{3}$ (b) $1 : \frac{1}{\sqrt{2}} : \sqrt{3}$ (c) $1 : \sqrt{2} : \frac{\sqrt{3}}{2}$ (d) $\frac{1}{\sqrt{3}} : \sqrt{2} : 1$.

13.36. In order to produce X-ray by striking electron beam on a metal target the KE of the electron should be (a) less than one eV (b) of the order of eV (c) of the order of keV (d) of the order of GeV. **[WBUT 2010]**

ANSWERS

13.1. (b) **13.2.** (d) **13.3.** (a) **13.4.** (d) **13.5.** (b) **13.6.** (b) **13.7.** (c) **13.8.** (c) **13.9.** (b) **13.10.** (b) **13.11.** (b) **13.12.** (b) **13.13.** (b) **13.14.** (c) **13.15.** (d) **13.16.** (c) **13.17.** (b) **13.18.** (b) **13.19.** (c) **13.20.** (b) **13.21.** (a) **13.22.** (b) **13.23.** (c) **13.24.** (a) **13.25.** (b) **13.26.** (b) **13.27.** (a) **13.28.** (a) **13.29.** (d) **13.30.** (b) **13.31.** (c) **13.32.** (a), (c) **13.33.** (b) **13.34.** (a) **13.35.** (a) **13.36.** (c).

13(B)–Short Answer Type Questions

13.1. What are X-rays? How can they be produced?

13.2. How can you prove that X-rays are transverse electromagnetic waves?

13.3. In what respect modern Coolidge tube is superior to Röngen gas-filled X-ray tube?

13.4. What do you mean by intensity and quality of X-rays. How can they be controlled?

13.5. What are the important properties of X-rays? **[WBUT 2004]**

13.6. Can X-rays be deflected from a single slit of width 0.1 mm? Explain. **[WBUT 2005]**

13.7. Explain the origin of continuous and characteristics X-rays. **[BUH 1993, 2006, CUH 2002]**

13.8. Draw continuous X-ray spectra for two different applied voltages. **[CUH 2002]**

13.9. The energy gap in Ge is 0.72 eV. Find the maximum wavelength of a radiation, which can cause transition from the valence band to conduction band. **[WBUT 2007]**

[Use the relation: $\lambda_m = \frac{ch}{E_g}$]

13.10. Justify the statement that, *X-ray production is the inverse process of photoelectric effect.*

13.11. What is the basic difference between optical spectra and X-ray spectra?

13.12. Point out the fundamental difference between continuous and characteristic X-rays.

13.13. On the basis of Bohr's theory deduce an expression for the frequency of the K_α X-ray line of an element of atomic number Z. **[BUH 1998]**

[**Derivation of frequency of K_α-line using Bohr's theory**

According to Bohr's theory, the energy of an electron in the n_1th orbit of an element of atomic number Z is

$$E_{n_1} = -\frac{mZ^2e^4}{8\epsilon_0 n_1^2 h^2}, \tag{1}$$

where m and e are respectively the mass and charge of the electron and ϵ_0, h are the electrical permittivity of the medium and *Planck's constant* respectively.

Similarly, energy of an electron in the n_2th orbit is

$$E_{n_2} = -\frac{mZ^2e^4}{8\epsilon_0 n_2^2 h^2}. \tag{2}$$

If the nuclear charge Z be reduced to $(Z-a)$ and $(Z-b)$ for the electrons in the n_1th and n_2th orbits respectively, then

$$E_{n_1} = -\frac{m(Z-a)^2e^4}{8\epsilon_0 n_1^2 h^2} \quad \text{and} \quad E_{n_2} = -\frac{m(Z-b)^2e^4}{8\epsilon_0 n_2^2 h^2} \tag{3}$$

Now,

$$E_{n_1} - E_{n_2} = \frac{me^4}{8\epsilon_0^2 h^3}\left[\frac{(Z-b)^2}{n_2^2} - \frac{(Z-a)^2}{n_1^2}\right].$$

Now, $a = b$, when Z is very large, then

$$\Delta E = E_{n_1} - E_{n_2} = \frac{me^4}{8\epsilon_0^2 h^2}(Z-a)^2\left[\frac{1}{n_2^2} - \frac{1}{n_1^2}\right].$$

Now, the frequency of the K_α X-ray line is

$$\nu_{K_\alpha} = \frac{\Delta E}{h} = \frac{me^4}{8\epsilon_0^2 h^3}(Z-a)^2\left(\frac{1}{1^2} - \frac{1}{2^2}\right).$$

Since K_α-radiation is emitted when an electron in $n_1 = 2$ orbit makes a transition to a vacancy created in $n_2 = 1$ orbit

$$\text{or,}\quad \nu_{K\alpha} = c(Z-a)^2, \qquad (4)$$

where $c = \frac{me^4}{8\epsilon_0^2 h^3}\left(\frac{1}{1^2} - \frac{1}{2^2}\right)$ = a constant.

Eqn (4) is known as the *Moseley's law.*]

13.14. Discuss the importance of Laue experiment regarding X-ray diffraction by crystal. Can you obtain X-ray diffraction pattern by plane transmission grating? Explain with reason.

13.15. Derive Bragg's law of X-ray reflection. Can you expect a reflected ray for any angle of incidence?

13.16. Write a note on the application of X-rays.

13.17. Give the theory of crystal structure determination using Bragg spectrometer.

13(C)–Numerical Problems

13.1. Electrons bombarding the anode of a Coolidge tube produce X-rays of wavelength 1 Å. Find the energy of each electron at the moment of impact. [Given: $h = 6.56 \times 10^{-27}$ ergs-sec, $e = 4.77 \times 10^{-10}$ esu, mass of electron $= 9.1 \times 10^{-28}$ gm] **[CUH 1961]**

[*Ans.* 1.24×10^4 eV]

13.2. A TV set operates with 20 kV accelerating potential. What are the maximum-energy X-rays from the TV set? [*Ans.* 0.62]

13.3. An electron is accelerated through a 10^5 V potential. Find the smallest possible wavelength produced when this electron interacts with a heavy target. [*Ans.* 0.124 Å]

13.4. An X-ray tube operates at 50 kV. Calculate the shortest wavelength of the X-rays produced. [$h = 6.62 \times 10^{-27}$ erg-sec and $e = 4.8 \times 10^{-10}$ esu] [*Ans.* 0.248 Å]

13.5. NaCl crystal has a density of 2.16 gm/cm³. The molecular weight of NaCl is 58.5. If the glancing angle for the first order (i.e., $n = 1$) is $11°8'$, calculate the wavelength of X-rays. [Avo. No. $= 6.06 \times 10^{23}$] **[RUH 1970]**

[*Ans.* 1.085 Å]

13.6. Calculate the wavelength of K_β and K_γ-lines of molybdenum if the wavelength of K_α-line is 0.71 Å. [*Ans.* 0.599 Å, 0.568 Å]

13.7. Calculate the lattice constant of rock salt from the following data: Atomic weight of Na = 23. Atomic weight of Cl = 35.45. Density of rock salt = 2.17 gm/cc. Avogadro's number $= 6.019 \times 10^{23}$

[IAS 1971]

[*Ans.* 5.64 Å]

13.8. Calculate the wavelength of an X-ray beam incident at 12° for the first order reflection from a calcite crystal if the grating constant of the crystal is 3.035 Å. **[Osmania U. 1964]**

[*Ans.* 1.26 Å]

13.9. The wavelength of X-ray radiation from copper target is 0.1537 nm. This radiation, when diffracted from (111) planes of a crystal with FCC structure, corresponds to Bragg angle of 19.2°. If the density of the crystal is 2698 kg/m^3 and atomic weight 26.98 kg/k mol, calculate the Avogadro's number.

[BUH 2008]

13.10. The energy gap in Ge is 0.72 eV. Find the maximum wavelength of electromagnetic radiation, which can cause transition from the valence band to conduction band. **[BUH 1997, WBUT 2007]**

13.11. If a narrow beam of K_α radiation ($\lambda = 1.541$ Å) from Cu is reflected from (100) planes of such a crystal, calculate the glancing angle corresponding to the first order spectrum. **[CUH 1991, '92]**

13.12. A beam of X-rays in an X-ray spectrometer is incident on a crystal of inter-planar spacing 2.82 Å at an angle of 14°. The operating voltage is gradually increased and an intense radiation appears when the operating voltage is 9.1 kV. Find the value of the Planck's constant.

[*Ans.* 6.62×10^{-34} J-s]

13.13. The linear absorption coefficient of an element of X-rays having $\lambda = 0.3$ Å is 135 m^{-1}. Find the half-value thickness for that material. **[WBUT 2010]**

[*Hint.* Use the relation,

$$\mu = \frac{0.693}{x'} \quad \text{or,} \quad x' = \frac{0.693}{\mu}.$$

Here x' = half value thickness of the material and μ = the linear absorption coefficient.

$$\therefore\ x' = \frac{0.693}{135} \quad [\because\ \mu = 135\,\text{m}^{-1}]$$

$$= \mathbf{5.13 \times 10^{-3}\ m}].$$

13.14. Find the short wave limit of continuous X-ray spectrum. Does it shift by 0.5 nm when the voltage applied to X-ray tube is doubled? **[WBUT 2010]**

13(D)–Broad Answer Type Questions

13.1. (a) What are X-rays?

(b) Describe an experimental outfit for producing them.

(c) What are the chief properties of the X-rays? Mention some of their uses.

13.2. (a) Describe the method of production and properties of X-rays. **[CUH 1967, '69]**

(b) Describe the essential feature of an X-ray tube and explain its mode of action. **[CUH 1961]**

13.3. (a) Describe the production of X-rays by a modern Coolidge tube.

(b) How the intensity and quality of X-rays can be controlled in a Coolidge tube?

(c) Discuss some of the important applications of X-rays.

13.4. (a) State the basic principle of production of X-rays?

(b) Explain the origin of continuous and characteristic X-rays? **[CUH 2002]**

(c) Draw the continuous X-rays spectra for two different voltages. **[CUH 2002]**

13.5. (a) Explain the origin of continuous X-rays. Why are they so-called?

(b) What is Duane and Hunt law relating to continuous X-rays?.

(c) If V be the applied PD between the anode and the cathode of an X-ray tube, then show that the minimum wavelength of continuous X-rays produced is $\lambda_{\text{min}} = \frac{12450}{V}$ Å.

13.6. (a) Explain the origin of characteristic X-rays? **[BUH 1997, 2002]**

(b) Describe the applied voltage dependence of characteristic X-rays.

(c) What are the chief characteristics of the above.

13.7. (a) Distinguish between continuous and characteristic X-ray spectra.

(b) Explain and derive Bragg's law of X-ray diffraction from a crystal.

(c) The spacing of the planes in a crystal is 1.2 Å and the angle for the first order Bragg reflection is 30°. Determine the energy of the X-rays in eV. **[WBUT 2005]**

13.8. (a) Discuss how crystals offer a natural grating for X-ray diffraction.

(b) Work out the theory of diffraction of X-rays by a crystal.

(c) Describe how an experimental study of such diffraction throws light on the structure of a crystal. **[CUH 1995]**

13.9. (a) What are Laue spots? What are the utility of Laue photograph?

(b) Distinguish between optical spectra and X-ray spectra.

(c) What are *soft* and *hard X-rays*? How can there be produced?

13.10. (a) What are continuous and characteristic X-ray spectra?

(b) State Moseley's law and explain its importance. **[BUH 1993, 2001, '06]**

13.11. (a) What are the products of interaction of X-rays with matter?

(b) Show that intensity of X-rays falls exponentially when it passes through an observer.

(c) What is linear absorption coefficient?

If x be the thickness of an absorber to reduce the intensity to half the initial value, then show that the linear absorption coefficient is $\mu = \frac{0.693}{x}$.

Chapter 14 | Module 6

Fibre Optics

14.1 Introduction

J. Tyndall in 1870 demonstrated to *Royal Society* that *light can be guided along a curved stream of water.* This luminous stream of water was the precursor of the present day *optical fibre.*

Fibre optics technology, due to large frequency of light waves ($\sim 10^{15}$ Hz), fulfills the present day demand of easy and rapid transmission of data over long distances by offering *enormous number of channels.* Electronic communication systems use radio waves (frequency $\sim 10^6$ Hz) which suffers from the limitation of the number of channels. Since light waves have the capacity of transmitting larger number of signals simultaneously compared to electric communication system, it is becoming more and more popular day by day.

Apart from optical communication, optical fibres find important applications in *sensors* for detecting electrical, thermal and mechanical energies, copying machines, fibroscopes for medical diagnostics, simple display system.

14.2 What is an Optical Fibre?

An optical fibre is a flexible cylindrical glass or plastic conduit as thin as human hair, designed to guide light waves over a long distance.

Principle. It works on the principle of *total interval reflection.* Light entering at one end of the fibre, undergoes successive total interiial reflections at side walls and travels down the length along a zigzag path as shown in Fig. 14.1.

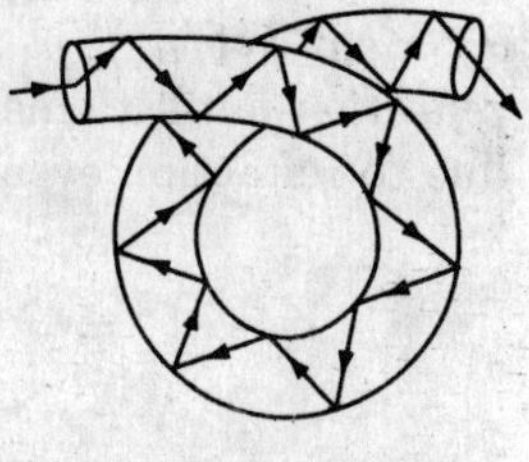

Fig. 14.1: Fibre guiding light along the length.

14.3 Structure of an Optical Fibre

An optical fibre consists of the following three coaxial sections, viz., (i) *the core* (ii) *the cladding* and (iii) *the sheath as shown in* Fig. 14.2(a). Fig. 14.2(b) shows that the dimensions of different sections in transverse view.

(i) **The core**: It is the innermost region which guides the light through total internal reflection. Its diameter varies from 50 to 100 μm and refractive index $n_1 > n_2$, where n_2 is the refractive index of the cladding.

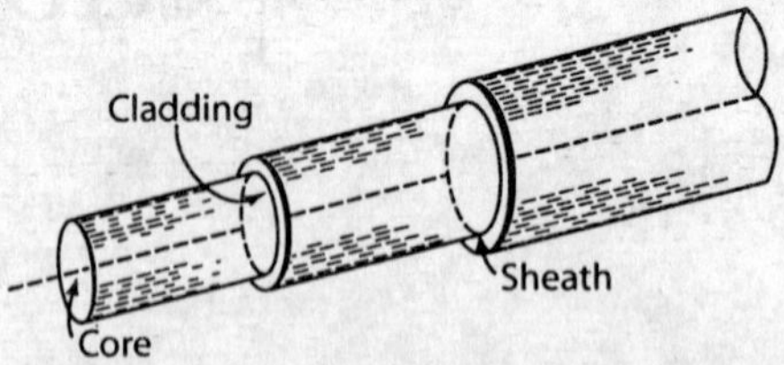

Fig. 14.2(a): Longitudinal view of an optical fibre showing the different sections.

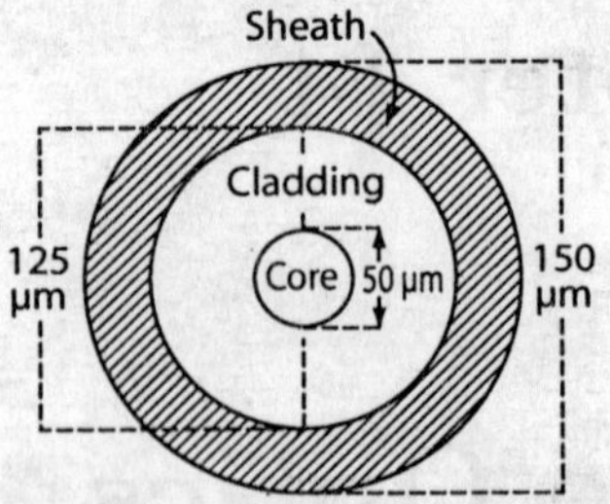

Fig. 14.2(b): Transverse view of an optical fibre.

(ii) **The cladding**: The cladding surrounding the core is coaxial with it and is of diameter $\sim 125\,\mu$m. The refractive index of cladding is always lower than that of the core, i.e., $n_2 < n_1$. The purpose of the cladding is to help the light to be confined within the core through total internal reflection.

(iii) **The sheath (or Jacket)**: The outermost part is made of *plastic* or *polymer* and is called the *sheath.* It protects the cladding and the core from the harmful influences of moisture, contamination, etc. and provides mechanical strength to the fibre. Its diameter is $\sim 150\,\mu$m.

14.4 Working Principle of Optical Fibres

An optical fibre works on the principle of total internal reflection. The light launched inside the core through its one end propagates to the other end due to multiple total internal reflection at the core and cladding interface.

Let a ray of light incident at one end of the fibre make a nearly glancing angle i (due to small radius of the fibre) with the axis [Fig. 14.3] and r the angle of refraction. If the ray now be incident at an angle θ with the core-cladding interface, then $\theta = (90° - r)$. If $\theta > \theta_c$, where $\theta_c = \sin^{-1}(1/n_1)$, n_1 being the refractive index of the core, then the ray suffers *total internal reflection* until it emerges out at the other end. The ray does not suffer from any loss of energy due to refraction even if it is bent.

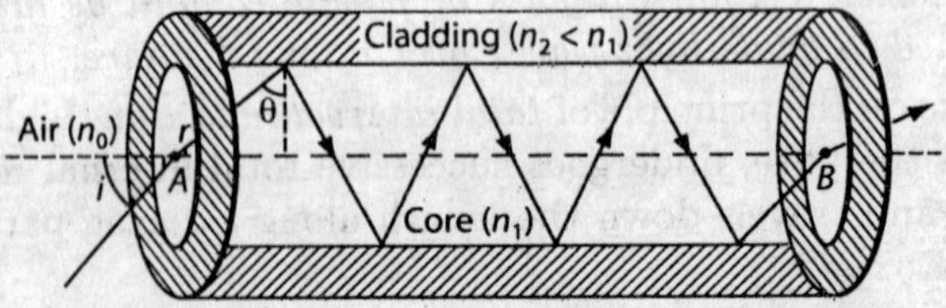

Fig. 14.3: Path of a typical light ray launched into a fibre.

N.B. (i) The loss due to absorption of light during refraction is minimized by *purification* and *special preparation* of the material. (ii) The efficiency of the fibre is increased by *selecting suitable value of numerical aperture and core size.*

14.5 Classification of Optical Fibres

On the basis of refractive index profile, optical fibres are classified into two types, viz., (i) *step-index fibre* and (ii) *graded index fibre.*

1. **Step-index (SI) fibre**: A step-index fibre (of radius a) has a *uniform refractive index* $n_1$of slightly greater than the refracting index of the cladding (of radius b) n_2 as shown in [Fig. 14.4].

Since the refractive index profile is in the form of a step, these fibres are called *step-index fibres* (SI).

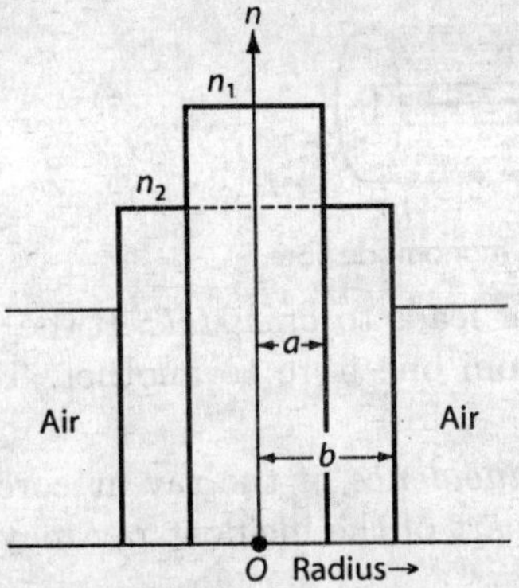

Fig. 14.4: Refractive index profile of a step-index fibre.

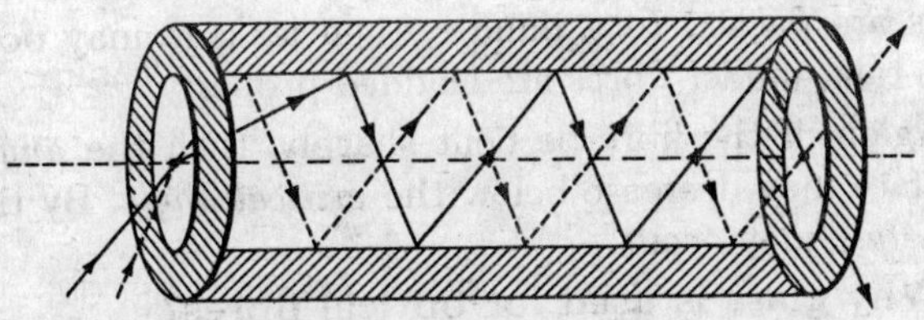

Fig. 14.5: Signal transmission in step-index multi-mode fibre.

The signal in the digital (pulsed) form travels in different zigzag paths (or modes) depending on the *launch angle* of the signal into the fibre. Fig. 14.5 shows two rays entering at two different angles of incidence with the fibre axis. Since the time taken by these two rays to emerge out of the fibre is different, there is a *widening of the input signal pulse as it traverses the fibre.* This is known as the *inter-modal dispersion* whose effect is to *distort* the transmitting signal. Since the fibre supports more than one mode, it is called a *multimode step-index fibre.*

2. **Graded Index (GI) fibre**: In a graded index multimode fibre the refractive index of the core varies radially as shown in Fig. 14.6. *The refractive index is the maximum at the centre and gradually decreases with the increase of radius of the fibre and becomes equal to the refractive index of the cladding at the core-cladding interface.*

The nature of variation of the refractive index (n) of the core with radial distance (x) from the centre of the graded index is given by

$$n(x) = n_1 \left[1 - 2\Delta \left(\frac{x}{a}\right)^p\right]^2, \qquad (14.1)$$

where n_1 = refractive index at the centre of the core; a = core radius; $\Delta = \frac{n_1 - n_2}{n_1}$, n = uniform refractive index of the cladding and p = grading profile index number.

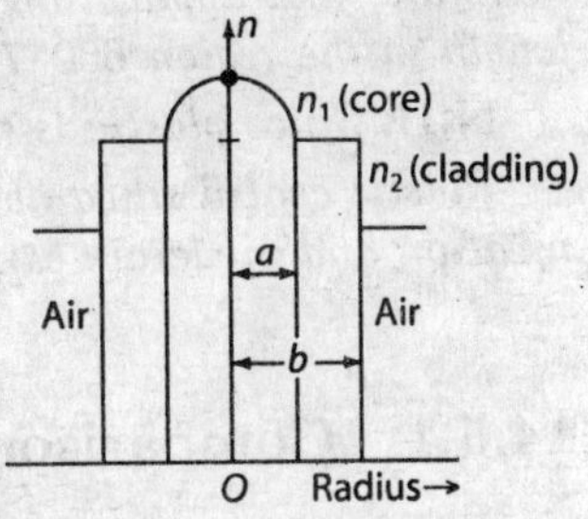

Fig. 14.6: Multi-mode graded index fibre of a graded index.

In a multimode graded index fibre, the rays are refracted gradually. Due to higher refractive index at the central region, the velocity of light is *smaller*, and the velocity of light is *greater* near the cladding due to smaller refractive index of the core. As a result the rays *spiral around* near the axis and travel shorter paths and travel slowly. In the vicinity of cladding the rays spiral and travel faster in longer paths [Fig. 14.7].

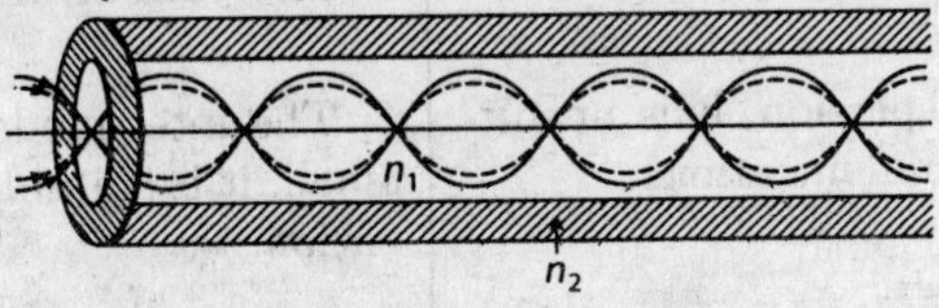

Fig. 14.7: Propagation of different modal rays in a graded index fibre.

As they travel along the core the rays are continuously *refocussed* within it and the signal distortion is reduced.

Mono-mode fibre. Mono-mode fibres due to narrow cores ($\sim 10\,\mu$m) allows only one mode to propagate along the length [Fig. 14.8]. In this fibre *inter-modal dispersion* is almost absent as the rays travel only parallel to the axis of the fibre.

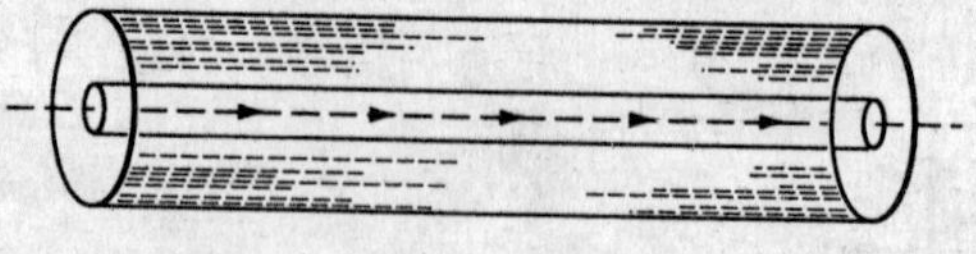

Fig. 14.8: Propagation of signal through a monomode fibre.

Cross-talk. Leakage of signal from one fibre to another leads to *cross-talk*. If the *uncladded* fibres are packed to a bundle, then leakage may occur from one fibre to another. To prevent cross-talk, moder fibres are cladded properly.

Again, if the fibre be bent sharply, then the *angle of incidence* of the ray at core-cladding interface may decrease below the *critical angle*. By this a part of the incident ray may leak and *cross-talk* may occur.

Why glass is used for optical fibres?

Due to the following reasons glass is used for the optical fibres: (i) *Glass does not solidify at a fixed freezing point, instead gradually becomes harder and harder in the transition region from liquid to solid and hence can be drawn into a thin fibre.* (ii) *The dopant pure silica is characterized by a low loss* ($\sim 4\%$/km). (iii) *Its strength is very high* ($\sim 2 \times 10^{16}$ lb/inch^2). (iv) *Silica is very cheap.* (v) *Glass fibre can be easily prepared with a relatively wide range of refractive indices.*

Why plastic fibres are suitable only for short distance communications?

Plastic fibres are made either entirely from plastics or with silica cores and plastic cladding having the following limitations: (i) *Full plastic fibres are characterized by high attenuation which, therefore, find application in short distance communications.* (ii) *Plastic fibres use light of wavelength in the region* 600–700 nm, *in low bandwidth system.*

N.B. *Since plastic is more flexible than glass, these fibres can be made with large diameters.*

Plastic coated silica fibres are easily manufactured and are used for medium distance communication and moderate bandwidth communication systems.

14.5.1 Comparison of Step-Index and Graded-Index Optical Fibres

Table 14.1

Step-index fibre	*Graded-index fibre*
1. The refractive index (n_1) of the core is constant over the entire cross section with $n_1 > n_2$, n_2 being the refractive index of the cladding.	1. The refractive index of the core decreases with increase of radius and $n_1 = n_2$ at the core-cladding interface.
2. The path of the ray is zigzag combination of straight line segments.	2. The path of the ray is essentially *sinusoidal.*
3. Due to inter-modal dispersion, it is unsuitable for long distance communications.	3. The inter-modal dispersion being relatively small, it is suitable for long distance communication.
4. It is least expensive.	4. It is medium priced.
5. It is suitable for high bandwidth applications.	5. For high bandwidth applications it is more suitable than step-index fibres.

14.6 Acceptance Angle and Acceptance Cone

When a light beam is launched into a fibre at its one end (called the *launching end*) using a focussing lens, the entire light does not pass through the core. Only those rays which are incident at an angle greater than the critical angle at the core-cladding interface undergo total internal reflection and propagate through the core. The rest of the incident rays are refracted at the interface and lost. So, *there is a maximum value of the launching angle for which entire light passes through the core. This angle is called the acceptance angle.*

To find out an expression for *the acceptance angle* the light is launched from a medium of refractive index n_0 into the core of an optical fibre of refractive index n_1. Let the angle of incidence and refraction of a particular ray at A be θ_i and θ_r, respectively. The ray is, then incident at B at an angle ϕ. We shall find out the value of ϕ for which total internal reflection is possible at B [Fig. 14.9].

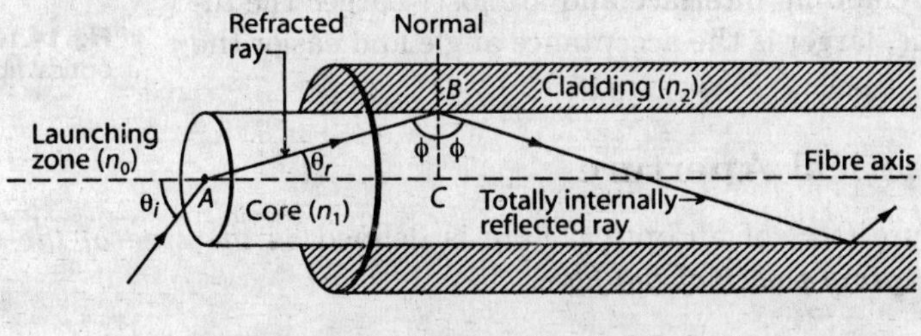

Fig. 14.9: Path of a light ray launched into a fibre.

Applying *Snell's law* at A, we get

$$\frac{\sin\theta_i}{\sin\theta_r} = \frac{n_1}{n_0}. \tag{14.2}$$

Let θ_{im} be the maximum value of angle of incidence at A for which $\phi = \phi_c$, ϕ_c = the critical angle at B.

Now, from ΔABC,

$$\sin\theta_r = \sin(90^\circ - \phi) = \cos\phi. \tag{14.3}$$

From eqns. (14.2) and (14.3), we get

$$\sin\theta_i = \frac{n_1}{n_0}\cos\phi.$$

But at $\theta_i = \theta_{im}$, $\phi = \phi_c$,

$$\therefore \sin\theta_{im} = \frac{n_1}{n_0}\cos\phi_c. \tag{14.4}$$

Again, $\sin\phi_c = \frac{n_2}{n_1}$, n_2 = refractive index of the cladding.

$$\therefore \cos\phi_c = \sqrt{1-\sin^2\phi_c} = \sqrt{1-\frac{n_2^2}{n_1^2}} = \frac{\sqrt{n_1^2-n_2^2}}{n_1}. \tag{14.5}$$

Using this value of $\cos\phi_c$ in eqn. (14.4), we get

$$\sin\theta_m = \frac{n_1}{n_0}\times\frac{\sqrt{n_1^2-n_2^2}}{n_1}$$

$$\text{or,}\quad \sin\theta_{im} = \frac{\sqrt{n_1^2-n_2^2}}{n_0} \tag{14.6}$$

$$\text{or,}\quad \boxed{\theta_{im} = \sin^{-1}\left[\frac{\sqrt{n_1^2-n_2^2}}{n_0}\right]} \tag{14.7}$$

Generally, the incident ray is launched from air ($n_0 = 1$), and denoting θ_{im} by θ_0,

$$\boxed{\theta_0 = \sin^{-1}\left[\sqrt{n_1^2 - n_2^2}\right]}, \tag{14.8}$$

θ_0 is called the *acceptance angle* of the fibre and may be defined as follows:

The acceptance angle is the maximum angle of incidence of a ray at the launching end of the fibre, so that it can propagate down the fibre.

Acceptance cone. *The cone of semi-vertical angle θ_0 within which the enclosed rays are accepted and transmitted through the fibre is called he acceptance cone* [Fig. 14.10].

Light rays which are incident at an angle $\theta > \theta_0$, refracts through the core-cladding interface and are lost. Larger the diameter of the cone, larger is the acceptance angle and easier the launching.

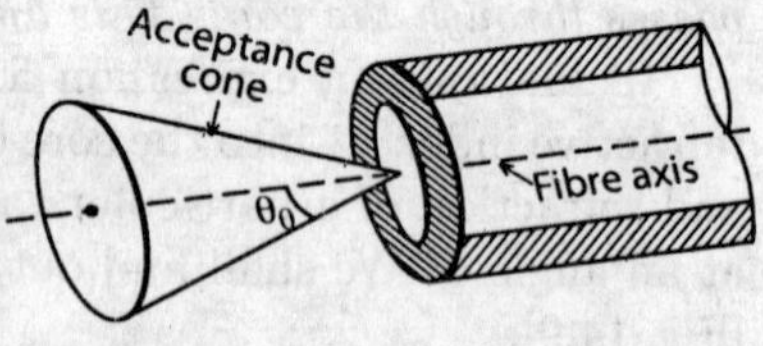

Fig. 14.10: Acceptance cone of an optical fibre.

14.6.1 Numerical Aperture

Numerical aperture (NA) of an optical fibre is defined as *the sine of the acceptance angle and measures the accepting power of the fibre.*

So,

$$\text{NA} = \sin\theta_0 \quad \text{or,} \quad \text{NA} = \sqrt{n_1^2 - n_2^2}. \tag{14.9}$$

Now

$$n_1^2 - n_2^2 = (n_1 + n_2)(n_1 - n_2) = \left(\frac{n_1 + n_2}{2}\right)\left(\frac{n_1 - n_2}{n_1}\right) 2n_1.$$

Putting

$$\frac{n_1 + n_2}{2} \simeq n_1 \quad \text{and} \quad \frac{n_1 - n_2}{n_1} = \Delta,$$

we get the *fractional change of refractive index*,

$$\boxed{\text{NA} = n_1\sqrt{2\Delta}}. \tag{14.10}$$

NA provides a measure of *light gathering capacity* of the fibre. The value of NA ranges from 0.13 to 0.50.

It is to note that,

(i) NA *depends only on the refractive index of the core* (n_1) *and cladding* (n_2) *materials.*

(ii) *The larger the value of* NA, *larger is the energy gathered by the fibre from the source.*

14.7 Attenuation or Transmission Losses

During the propagation of an optical signal along the fibre the signal gets progressively attenuated causing a *loss of energy with distance.* The attenuation is found to depend on *quality of fibre material, wavelength of light and length of the fibre.*

The different sources of losses are as follows:

(i) **Scattering losses.** The inhomogeneities due to the fluctuation in refractive index (due to submicroscopic variation in density, dopants added to silica to modify the refractive index, etc.) act as the reflecting and refracting facets to scatter a small portion of the light passing through the glass, contribute for the *energy loss of the signal.*

(ii) **Absorption losses.** The different mechanisms causing absorption losses in glass fibre are (a) *ultraviolet absorption* (b) *infrared absorption* and (c) *ion resonance absorption.*

Apart from that, the impurities like *iron* and *chromium* cause energy loss of the signal.

(iii) **Bending losses.** The distortion of the fibre from the ideal straight line may cause fibre losses. The loss can be expressed by

$$\alpha = Ce^{-R/R_c}, \tag{14.11}$$

where α is the *absorption coefficient* and C is a constant. R is the radius of curvature of the bend and $R_c = a/(\text{NA})^2$, a being the radius of the fibre.

Loss in Decibel

An optical signal during propagation along the fibre will get *progressively attenuated.*

The signal attenuation is defined as *the logarithmic ratio of the input power (P_i) of the signal launched at one end to the power (P_0) emerging from the fibre at a distance L per unit length.* It is expressed in dB/km, i.e.,

$$\alpha = \frac{10}{L} \ln\left(\frac{P_i}{P_0}\right). \tag{14.12}$$

For ideal fibre $P_i = P_0$, the attenuation coefficient $\alpha = 0\,\text{dB/km}$.

The value of α is generally 3 dB/km.

It is to note that the attenuation depends on the wavelength (λ) of the light used.

14.8 Advantages of the Fibre Optic Communication

Optical fibre communication has many advantages over the conventional microwave communication. Some of them are as follows:

1. **Enormous bandwidth.** The bandwidth in fibre optical communication is $\sim 10^5$ GHz as against 500 MHz in case of coaxial cable transmission. Hence the information carrying capacity of optical fibre system is much more than radio or microwave carriers due to the availability of a large number of channels. Energy loss in optical fibre is very low and clarity of signal transmission is much better.

2. **Less weight and small size.** Optical fibres are highly suitable for *aeronautical* and space applications because of their *light weight* and *smaller size.*

3. **Electrical isolation.** Since optical fibres are *insulators,* there is no danger of spark hazards in electrical hazardous environment.

4. **Low cost.** Optical fibres are made of glass which is too cheap and available in plenty in nature.

5. **No interference and cross-talk.** Optical fibre is well protected from *interference* and *coupling* with signals in other fibres or communication channel. There is no possibility of *cross-talk.*

6. **Security of the signal.** The signal is well protected and privacy is maintained in optical fibre communication.

7. **Ruggedness and flexibility.** Glass fibres can withstand high temperature $\sim 800\,°\text{C}$, corrosive effect due to water or chemicals is negligible. These are flexible and extremely rugged.

8. **Safety against voltage problems.** Optical fibres do not need additional equipment for protection against grounding and voltage problems.

14.9 Applications of Fibre Optics

Following are some important applications of fibre optics:

1. Fibre optics due to large bandwidth is suitable for handling a large number of channels. Due to this property it has wide application in the field of communication.
2. Fibre optics is suitable for transmission of digital data generated by computers, etc.
3. In hazardous areas, for example in mines, on oil rings, wells, etc. it is used for communications and control.
4. The areas, where high privacy is required as in defense services, optical fibre system is used.
5. Optical fibre system is used for signalling purposes.
6. Fibre optic sensors are used to measure temperature, pressure, liquid levels, rotary and linear positions, etc.
7. It has important applications in cable television, space vehicles, ships and submarine cables.
8. It is used to view interior of the lungs and the other parts of the body that cannot be viewed directly. The process is called *endoscopy* (meaning to see within).
9. Optical fibre system has applications in military services like communications, command and control links on ships and air-craft, data links for satellite and earth stations.

14.10 Answers of Some Important Questions

14.1. *What is an optical fibre?*

Ans. An optical fibre is a flexible cylindrical glass or plastic conduit as thin as human hair, designed to guide light waves over a long distance.

It works on the principle of total internal reflection.

14.2. *What are the different parts of an optical fibre?*

Ans. The different structural parts of an optical fibre are (i) the core of diameter $\sim 50\,\mu$m (ii) the cladding of diameter $\sim 120\,\mu$m and (iii) the sheath (or jacket) of diameter $\sim 150\,\mu$m.

14.3. *What are the different types of optical fibres?*

Ans. Depending on the refractive index profile optical fibres are of two types:

(i) *Step-index fibre*: In this fibre the refractive index of the core (n_1) has a constant value and that of the cladding (n_2) has another constant such that ($n_1 > n_2$). This is essential for total internal reflection of the propagating light ray at the core-cladding interface.

(ii) *Graded-index fibre*: In this type of fibre the refractive index is the maximum at the centre and gradually decreases with the increase of radius of the fibre and becomes equal to the refractive index of the cladding at the core-cladding interface.

Again, depending on the number of modes that a fibre can support, optical fibres are of the following two types:

(i) *Single-mode fibre.* The rays which travelling in a particular direction in the fibre satisfying the conditions for constructive interference are called *modes.*

If the thickness of the fibre is so small that it can support only one mode, then the fibre is called *single mode* (or, mono-mode).

(ii) *Multi-mode fibre.* If the dimension of the fibre is such that it can support more than one mode, then it is a *multi-mode fibre.*

14.4. *What is the working principle of an optical fibre?*

Ans. Total internal reflection is the basic principle of an optical fibre. The light launched at one end of the core propagates to the other end due to multiple total internal reflection at the core-cladding interface.

14.5. *State the mathematical relation representing the variation of refractive index with radius of the core in case of a graded-index fibre.*

Ans. The mathematical relation representing the variation of refractive index n with radius x from the centre of the core of a graded-index fibre is given by

$$n(x) = n_1 \left[1 - 2\Delta\left(\frac{x}{a}\right)^p\right]^2,$$

where n_1 = refractive index at the centre of the core; $\Delta = \frac{n_1 - n_2}{n_1}$, the fractional change in refractive index; a = radius of the core and p = grading profile index number.

14.6. *What are the different losses in optical fibre?*

Ans. The different losses in optical fibre are as follows:

(i) Scattering losses (due to fluctuation in refractive index)

(ii) Absorption losses (due to UV, IR and ion resonance absorption)

(iii) Bending losses (due to bending of the fibre).

14.7. *What is optical window?*

Ans. In some region of the electromagnetic spectrum, attenuation of the transmitted signal has a minimum. The signal having wavelength lying in this region suffers minimum attenuation.

The *optical window* (or *transmission window*) *is the band of wavelength at which attenuation is a minimum.*

14.8. *What do you mean by transit time distortion?*

Ans. The time taken by different light rays while propagating along paths of different lengths up to the receiving end of a step-index multi-mode fibre are different. As a result the variation of the successive pulses of light may overlap leading to distortion. This type of distortion are called *transit-time distortion.*

14.9. *Express energy loss due to attenuation in an optical fibre.*

Ans. The signal attenuation is defined as the logarithmic ratio of the input power (P_i) emerging from the fibre at a distance L per unit length and is represented by α, i.e.,

$$\alpha = \frac{10}{L} \ln\left(\frac{P_i}{P_0}\right) \text{ dB/km}.$$

14.10. *What do you mean by acceptance angle and acceptance cone of an optical fibre?*

Ans. *Acceptance angle*: The maximum value of the launching angle for which entire light passes through the core of the fibre is called the acceptance angle.

If n_0, n_1 and n_2 be the refractive indices of the medium from which light is launched, the core and the cladding respectively, then the acceptance angle is given by

$$\theta_{im} = \sin^{-1}\left[\frac{\sqrt{n_1^2 - n_2^2}}{n_0}\right].$$

If the outside medium be air, then $n_0 = 1$ and the corresponding acceptance angle is

$$\theta_0 = \sin\sqrt{n_1^2 - n_2^2}.$$

Acceptance cone: The cone of semi-vertical angle θ_0 within which the enclosed rays are accepted and transmitted through the fibre is called the acceptance cone.

14.11. *In what respect fibre communication system* is superior to radio or microwave communication system?

Ans. The bandwidth of fibre optical communication is $\sim 10^5$ GHz as against 50 MHz in case of microwave communication. So the information carrying capacity of optical fibre system is much more than radio or microwave communication system due to the availability of large number of channels.

Moreover, energy loss in optical fibre is very low ($\sim$ 4%/km) and clarity of signal transmission is much better.

14.12. *What is cross-talk?*

Ans. When many uncladded fibres are packed to form a bundle, then light may leak through from one fibre to the other. This is *cross-talk.*

Cross-talk may also occur in case of *sharply bent fibre,* since then the angle of incidence at one core-cladding interface may be less than the critical angle and a part of light may escape from the fibre.

14.13. *What is numerical aperture of an optical fibre?*

Ans. Numerical aperture of an optical fibre is defined as *the sine of the acceptance angle and measures the accepting power of the fibre.*

Mathematically, numerical aperture is given by $\text{NA} = n_1\sqrt{2\Delta}$, where $\Delta = \frac{n_1 - n_2}{n_1}$ is the fractional change of refracting index.

14.14. *What is the significance of* NA *of a fibre?*

Ans. NA aperture provides a measure of light gathering capacity of the fibre. This means the larger the value of NA, larger is the energy gathered by the fibre from the source and greater is the clarity of the signal. The value of NA of an optical fibre depends on the refractive index of the core (n_1) and that of the cladding. The value of NA ranges from 0.13 to 0.50.

14.15. *Why glass is preferred for the construction of optical fibre?*

Ans. Glass has certain advantages in the construction of optical fibre, e.g., (i) *glass can be drawn into a thin fibre as it does not solidify at a certain fixed temperature;* (ii) *the dopant pure silica is characterized by a low loss* ($\sim 4\%$/km). *Moreover, its strength is high* ($\sim 2 \times 10^{16}\,\text{lb/inch}^2$); (iii) *silica is very cheap;* (iv) *glass fibre can be easily prepared with a relatively wide range of refractive indices.*

14.11 Solved Problems

Ex. 1 *An optical fibre made of silica glass has a core of refractive index* 1.50 *and the cladding of refractive index* 1.45. *Calculate the critical angle for the core-cladding interface.*

Soln. The critical angle ϕ_c is given by $\sin\phi_c = n_2/n_1$, where $n_1 = 1.50$ and $n_2 = 1.45$.

$$\therefore\ \sin\phi_c = \frac{1.450}{1.50} = 0.966 \quad \text{or,} \quad \phi_c = \sin^{-1}(0.966) = \mathbf{75^\circ}.$$

Ex. 2 *Calculate the angle of acceptance of a given optical fibre, if the refractive indices of the core and the cladding are* 1.563 *and* 1.498, *respectively.*

Soln. Numerical aperture of the optical fibre is given by

$$\text{NA} = \sin\theta_i = \sqrt{n_1^2 - n_2^2} = \sqrt{(1.563)^2 - (1.498)^2} = \sqrt{3.061 \times 0.065} = 0.4461.$$

The angle of acceptance $\theta_0 = \sin^{-1}(\text{NA}) = \sin^{-1}(0.4461) \simeq \mathbf{26^\circ 30'}$.

Ex. 3 *What is the value of fractional change of refractive index of a given optical fibre if the refractive indices of the core and cladding are* 1.563 *and* 1.498, *respectively?*

Soln. The fractional change in refractive index of the given optical fibre is

$$\Delta = \frac{n_1 - n_2}{n_1} = \frac{1.563 - 1.498}{1.563} \simeq \mathbf{0.0416}.$$

Ex. 4 *The numerical aperture of an optical fibre is* 0.5 *and the refractive index of the core is* 1.54. *Find the refractive index of the cladding.*

Soln. The refractive index of the cladding is given by $n_2 = \sqrt{n_1^2 - (\text{NA})^2} = \sqrt{(1.54)^2 - (0.5)^2}$ $= \mathbf{1.46}$.

Ex. 5 *Compute the numerical aperture and the acceptance angle of an optical fibre having refractive indices* 1.54 (*core*) *and* 1.50 (*cladding*) *surrounding medium being air.* **[WBUT 2005]**

Soln. The numerical aperture of the fibre is $\text{NA} = \sin\theta_i = \sqrt{n_1^2 - n_2^2}$.

Here $n_1 = 1.54, n_2 = 1.50$.

$\therefore \text{NA} = \sqrt{(1.54)^2 - (1.50)^2} = 0.3487 = 0.35$.

Acceptance angle $\theta_0 = \sin^{-1}\sqrt{n_1^2 - n_2^2} = \sin^{-1}(0.35) \simeq \mathbf{20°48'}$.

Ex. 6 *An optical fibre has a core diameter of* 50 μm. *If the refractive index of the core is* 1.5, *what is the axial distance travelled by a ray incident at* 30° *between two successive internal reflections?*

Soln. In Fig. 14.11 θ_i = angle of refraction; D = diameter of the fibre and x = the axial distance travelled by the ray between two successive internal reflections.

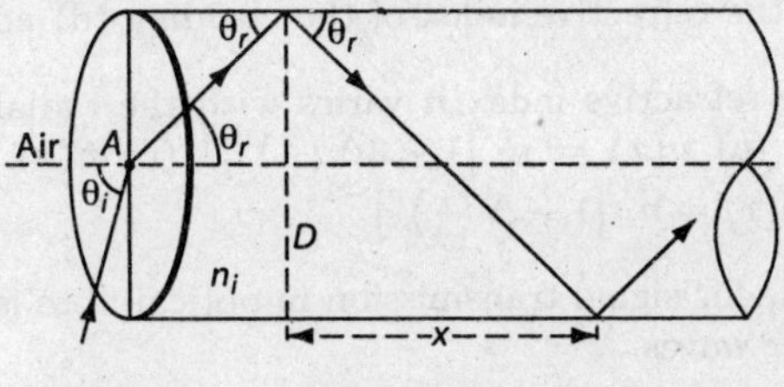

Fig. 14.11: Axial distance (x) between two successive internal reflections.

From the Fig. 14.11,

$$\frac{D}{x} = \tan\theta_r \quad \text{or,} \quad x = \frac{D}{\tan\theta_r},$$

where θ_r = the angle of refraction.

From *Snell's law*, we have

$$n_1 = \frac{\sin\theta_i}{\sin\theta_r} \quad \text{or,} \quad \sin\theta_r = \frac{\sin\theta_i}{n_1}.$$

$$\therefore x = \frac{D\cos\theta_r}{\sin\theta_r} = \frac{D\sqrt{1-\sin^2\theta_r}}{\sin\theta_r} = \frac{D\sqrt{1-(\sin^2\theta_i)/n_1^2}}{(\sin\theta_i)/n_1} = \frac{D\sqrt{n_1^2 - \sin^2\theta_i}}{\sin\theta_i}.$$

Putting $n_1 = 1.5$, $D = 50 \times 10^{-6}$ m and $\theta_i = 30°$, we get

$$x = \frac{(50 \times 10^{-6}) \times \left(\sqrt{1.5^2 - \frac{1}{4}}\right)}{\frac{1}{2}} = 1.41 \times 10^{-4}\,\text{m} = \mathbf{1.41 \times 10^{-4}\,m}.$$

Ex. 7 *A signal of* 100 mW *is injected into a fibre. The outcoming signal from the other end is* 40 mW. *Calculate the power loss in* dB.

Soln. Power loss $= -10\ln\left(\frac{P_0}{P_i}\right)$ dB $= -10\ln\left(\frac{40}{100}\right)$ dB $= \mathbf{3.98\,dB}$.

Ex. 8 *Find the numerical aperture and angle of acceptance of a given optical fibre if the refractive indices of the core and cladding are* 1.562 *and* 1.497. **[WBUT 2004]**

Soln. The numerical aperture of the fibre is $\text{NA} = \sin\theta_0 = \sqrt{(1.562)^2 - (1.497)^2} = 0.446$.

Angle of acceptance $\theta_0 = \sin^{-1}(0.446) = \mathbf{26°5'}$.

EXERCISES

14(A)–Objective-Type Questions

Select the correct answer:

14.1. Optical fibre communication system is superior to radio or microwave communication because (a) Optical fibres are cheaper than metal wires (b) Optical fibres have very high bandwidth offering enormous number of channels (c) The fibres can be bent like tubes (d) Instead of radio or microwaves, optical fibres use amplified light waves.

14.2. Propagation of light through optical fibre is due to (a) refraction (b) reflection (c) total internal reflection (d) interference.

14.3. If n_1 and n_2 be the refractive indices of the core and cladding respectively, then (a) $n_1 = n_2$ (b) $n_1 > n_2$ (c) $n_1 < n_2$ (d) $n_1 \geq n_2$.

14.4. In a step-index fibre the refractive index of the core (a) varies with radius (b) remains constant w.r.t. radius (c) equals the refractive index of the cladding (d) none of these.

14.5. In a graded-index fibre refractive index n varies with the radial distance x from the centre according to the relation (a) $n(x) = n_1\left[1+2\Delta\left(\frac{x}{a}\right)^p\right]^2$ (b) $n(x) = n_1\left[1+\Delta\left(\frac{x}{a}\right)^p\right]^2$ (c) $n(x) = n_1\left[1-2\Delta\left(\frac{x}{a}\right)^p\right]^2$ (d) $n(x) = n_1\left[1+\Delta\left(\frac{x}{a}\right)^p\right]^{-2}$.

14.6. The medium responsible for signal transmission in optical fibre is (a) radio waves (b) microwaves (c) light waves (d) water waves.

14.7. In optical fibre attenuation of light waves takes place (a) linearly (b) inversely (c) logarithmically (d) exponentially with the length of the fibre.

14.8. A step-index fibre can be (a) multimode fibre only (b) mono-mode fibre only (c) either mono-mode or multimode (d) mono-mode as well as multimode.

14.9. The numerical aperture (NA) of an optical fibre of core r.i. n_1 and cladding r.i. n_2 and length l (a) depends on n_1 and n_2 (b) independent of n_1 and n_2 (c) depends on l (d) depends on its diameter (*d*).

14.10. If n_0, n_1 and n_2 be the refractive indices of the medium from which light is launched, the core and cladding respectively, then (a) NA $= \frac{\sqrt{n_2^2-n_1^2}}{n_0}$ (b) NA $= \frac{\sqrt{n_1^2-n_2^2}}{n_1}$ (c) NA $= \frac{\sqrt{n_1^2-n_2^2}}{n_1}$ (d) NA $= \frac{\sqrt{n_1^2-n_2^2}}{n_0}$.

14.11. The loss in dB per unit length of an optical fibre of input and output signals respectively, 100 mW and 50 mW is (a) 50 (b) 3 (c) 2 (d) 0.5.

14.12. Plastic fibres are suitable for communication if the distance is (a) short (b) long (c) medium (d) both (a) and (b).

14.13. Pure silica doped glass fibres are suitable for communication if the distance is (a) long (b) medium (c) short (d) infinity.

14.14. The spreading of pulse width in optical fibre is due to (a) scattering (b) total internal reflection (c) dispersion (d) diffraction.

14.15. The quality factor of an optical fibre is given by (a) bandwidth (b) dispersion (c) scattering (d) product of (a) and (b).

14.16. The optical fibre is protected from the harmful effect by the (a) core (b) cladding (c) sheath (jacket) (d) none of the these.

14.17. If n_1 be the refractive index of the core and $\Delta = \frac{n_1 - n_2}{n_1}$, n_2 being the refractive index of the fibre, then (a) NA $= n_2\sqrt{2\Delta}$ (b) NA $= n_2(2\Delta)$ (c) NA $= n_1\sqrt{2\Delta}$ (d) NA $= n_1\sqrt{4}$.

14.18. Glass is used for the construction of optical fibre because it (a) can be converted into liquid (b) does not have a fixed freezing point (c) has a high rigidity (d) can be bent like a pipe.

14.19. In a step-index fibre the path of the ray is (a) a straight line (b) zigzag line segments (c) a wave (d) none of these.

14.20. Graded-index fibres are preferred to step-index fibre for long distance communication (a) due to less energy loss (b) for variation of refractive index (c) since the light is periodically focussed in it (d) none of these.

14.21. In a graded-index fibre the NA and acceptance angle decrease with (a) length (b) radial distance from the axis (c) bending of the fibre (d) impurity present in it.

ANSWERS

14.1. (b) **14.2.** (c) **14.3.** (b) **14.4.** (b) **14.5.** (c) **14.6.** (c) **14.7.** (a) **14.8.** (c) **14.9.** (b) **14.10.** (d) **14.11.** (b) **14.12.** (a) **14.13.** (b) **14.14.** (c) **14.15.** (d) **14.16.** (c) **14.17.** (c) **14.18.** (b) **14.19.** (b) **14.20.** (a) **14.21.** (b).

14(B)–Short Answer-Type Questions

14.1. What is an optical fibre?

14.2. Discuss the advantages of optical fibre communication system over that of radio or microwave system.

14.3. What is the working principle of an optical fibre?

14.4. What do you mean by step-index and graded-index optical fibres?

14.5. Give a comparative study of step-index and graded-index fibres.

14.6. What do you mean by core, cladding and sheath of an optical fibre? Draw a block diagram showing the above.

14.7. What do you mean by numerical aperture and acceptance angle of an optical fibre? Write down their expressions.

14.8. What are the causes of energy loss of an optical fibre?

14.9. What are cross-talk and inter-modal dispersion?

14.10. What do you mean by power loss in optical fibre?

14.11. What do you mean by V-number?

[**Ans.** The normalised frequency V, associated with the cut-off condition is called the *V-number* or *V-parameter* given by

$$V = \frac{2\pi a}{\lambda}\sqrt{n_1^2 - n_2^2} = \frac{2\pi a}{\lambda}\,(\text{NA}),$$

where a is the radius of the fibre, λ, the wavelength of light and NA is the numerical aperture. V is an important parameter which determines the number of modes a fibre can support.]

14(C)–Numerical Problems

14.1. The refractive indices of the core and cladding of an optical fibre are 1.523 and 1.472, respectively. Find the numerical aperture and the angle of acceptance of the optical fibre. **[WBUT 2005, '06]**

[*Hint.* NA $= \sin\theta_0 = \sqrt{n_1^2 - n_2^2}$. Here $n_1 = 1.523$ and $n_2 = 1.572$.

$\therefore$ NA $= \sin\theta_0 = \sqrt{(1.523)^2 - (1.482)^2} = \sqrt{2.995 \times 0.051}$.

$\therefore$ NA $= 0.3907$ and $\theta_0 = \sin^{-1} 0.3907 = \mathbf{22.998°}$.]

14.2. In a graded index fibre the radial distribution of index is given by $n(r) = 1.52 - 2r^2$, where r in mm up to a radius $r_0 = 0.2$ mm. Calculate the acceptance angle of the fibre. [*Ans.* 29.12°]

14.3. For a step-index fibre $n_1 = 1.5$, and $\Delta n = 0.015$. Calculate n_2, NA and acceptance angle.

[*Ans.* 1.485, 0.173, 10°]

14.4. If the NA of an optical fibre is 0.39 and $n_1 - n_2 = 0.05$, then find the refractive index n_1 of the core.

[*Hint.* NA $= n_1\sqrt{2\Delta}$. $\therefore n_1 = \frac{\text{NA}}{\sqrt{24}} = \frac{0.39}{\sqrt{2\times 0.05}} \simeq \mathbf{1.2334}$]

14.5. A step-index fibre in air has NA of 0.16, refractive index of core 1.45 and diameter of the core is 60 cm. Determine the normalized frequency for the fibre when light at a wavelength of 0.9 μm is transmitted.

[*Hint.* Here NA $= 0.16$; $n_1 = 1.45$; $d = 2a = 60$ cm $= 0.6$ m; $\lambda = 0.9\,\mu\text{m} = 9 \times 10^{-7}$ m.

$\therefore$ the normalized frequency (or V-number) is $V = \frac{\pi d}{\lambda}\sqrt{n_1^2 - n_2^2} = \frac{3.143\times 0.60}{9\times 10^{-7}} \times 0.16 = \mathbf{3.35 \times 10^5}$.]

14.6. An optical signal has lost 85% of its power after traversing 500 m of the fibre. What is the loss in dB/km of the fibre?

[Loss/km of the power $= \frac{10}{L}\log\left(\frac{P_i}{P_0}\right) = \frac{10}{0.5} \times \log\left(\frac{1}{0.85}\right) = \mathbf{1.41\ dB/km}$.]

14.7. An optical fibre has a NA of 0.20 and a cladding refractive index of 1.59. Determine the acceptance angle for the fibre in water which has a refractive index of 1.33. [*Ans.* $\theta_0 = 8.6°$]

14.8. A step-index optical fibre has refractive indices of the core and cladding 1.44 and 1.40, respectively, and is surrounded by air. Calculate the numerical aperture. **[WBUT 2008]**

Soln. Numerical aperture of the fibre is

$$\text{NA} = \sqrt{n_1^2 - n_2^2}, \quad n_1 \text{ and } n_2 \text{ are the refractive indices}$$

$$= \sqrt{1.44^2 - 1.40^2} = \sqrt{2.84 \times 0.04} = \sqrt{0.1136} = \mathbf{0.3370}.$$

14(D)–Broad Answer Type Questions

14.1. (a) What is a fibre guide? Discuss the working principle of a fibre guide.

(b) What do you mean by graded-index fibre and step-index fibre?

(c) Define numerical aperture and find out an expression for the same in a step-index fibre.

[BUH 2000, '05]

14.2. (a) What is an optical fibre?

(b) Discuss briefly the structure and the principle of optical fibre. **[WBUT 2004]**

(c) What do you mean by acceptance angle and numerical aperture of a fibre? **[WBUT 2004]**

(d) Find the numerical aperture and angle of acceptance of a given optical fibre if the refractive indices of the core and cladding are 1.562 and 1.497, respectively. [*Ans.* 0.446, 26.5°]

14.3. (a) What are step-index fibre and graded-index fibre?

(b) Give a comparative study of step-index and graded-index fibre.

(c) What are the different losses in an optical fibre? Find the expression for power loss in dB/km of a fibre.

14.4. (a) Discuss the advantages of fibre communication over radio wave or microwave communication.

(b) Discuss the advantages of glass doped with pure silica for the construction of optical fibre.

(c) Discuss some important uses of optical fibre.

14.5. (a) Derive expressions for numerical aperture and acceptance angle of an optical fibre.

(b) What are the significances of the above two quantities?

(c) What are mono-mode and multi-mode fibres? What is *V-number* and what is its significance?

Appendix A

Some Fundamental Constants

Sl. No.	*Constants*	*Symbol*	*Value*
1.	Gravitational constant	G	6.67×10^{-11} N-m^2/kg^2
2.	Acceleration due to gravity	g	9.81 m/s^2
3.	Velocity of sound in air at NTP	V	~ 332.5 m/s
4.	Ratio of C_ρ/C_V of a diatomic molecule	γ	1.41
5.	Density of air at NTP	ρ	1.293 kg/m^3
6.	Velocity of light in vacuum	c	3×10^8 m/s
7.	Planck's constant	h	6.63×10^{-34} J-s
			$= 4.14 \times 10^{-21}$ MeVs
8.	Charge of an electron	e	1.6×10^{-19} C
9.	Rest mass of an electron	m_e	9.11×10^{-31} kg
10.	Specific charge of the electron	e/m_0	1.76×10^{11} C/kg
11.	Boltzmann's constant	k	1.38×10^{-23} J/K
			$= 8.62 \times 10^{-5}$ eV/K
12.	Permeability of vacuum	μ_0	$4\pi \times 10^{-7} = 1.26 \times 10^{-6}$ H/m
13.	Permittivity of vacuum	$\epsilon_0 = \frac{1}{\mu_0 c^2}$	8.85×10^{-12} F/m
			$\frac{1}{4\pi\epsilon_0} = 8.99 \times 10^9 \frac{\text{m}}{\text{F}}$
14.	Faraday's constant	$F = N_A e$	9.65×10^4 C/mole
15.	Avogadro's number	N_A	6.602×10^{23} / mole
16.	Universal gas constant	R	8.314 J/mol-K
17.	Standard atmospheric pressure	P_s	1.013×10^5 N/m^2
18.	Stefan's constant	σ	5.67×10^{-8} W/m^{2}K^4

Appendix B

Miscellaneous Conversion Factors

1.	1 erg	=	10^{-7} J
2.	1 light year	=	9.46×10^{15} m
3.	1 Å(angstrom)	=	10^{-10} m
4.	1 fermi	=	10^{-15} m
5.	1 barn	=	10^{-28} m^2
6.	1 gauss	=	10^{-4} T
7.	1 eV	=	1.6×10^{-12} erg
		=	1.6×10^{-19} J
8.	1 calorie	=	4.18 J
9.	1 radian	=	$57.3°,\ 1° = 0.0174$ radian
10.	1 kWH	=	3.6×10^{6} J

Appendix C

SI Prefixes

Sl. No.	*Prefix*	*Symbol*	*Factors by which base unit is to be multiplied*
1.	atto	a	10^{-18}
2.	femto	f	10^{-15}
3.	pico	p	10^{-12}
4.	nano	n	10^{-9}
5.	micro	μ	10^{-6}
6.	milli	m	10^{-3}
7.	centi	c	10^{-2}
8.	deci	d	10^{-1}
9.	deca	da	10
10.	hecto	h	10^{2}
11.	kilo	k	10^{3}
12.	mega	M	10^{6}
13.	giga	G	10^{9}
14.	tera	T	10^{12}
15.	peta	P	10^{15}
16.	exa	E	10^{18}

Appendix D

Some Derived Units

Sl. No.	*Physical quantity*	*Unit*	*Symbol*	*Dimension*
1.	Area	square metre or hectare	m^2 or ha	$[L^2]$
2.	Volume	cubic metre or litre	m^3 or l	$[L^3]$
3.	Force	newton	N	$[MLT^{-2}]$
4.	Pressure and Stress	pascal	Pa or N/m^2	$[ML^{-1}T^2]$
5.	Work, energy, heat	joule	J	[Nm]
6.	Power	watt	W	$[Js^{-1}]$
7.	Moment of inertia	kilogram-metre^2	kg-m^2	$[Mm^2]$
8.	Torque	newton-metre	N-m	$[ML^2T^{-2}]$
9.	Specific heat capacity	joule per kilogram-kelvin	J/kg-K	—
10.	Thermal conductivity	watt per metre-kelvin	W/m-K	$[MLT^{-3}\theta^{-1}]$
11.	Radiant energy intensity	watt per steradian	WSr^{-1}	
12.	Induction	henry	H	$[VAs^{-1}]$
13.	Magnetic flux	weber	Wb	[Vs]
14.	Magnetic field strength	ampere per metre	Am^{-1}	
15.	Magnetic flux density	tesla	T	$[Wbm^{-2}]$
16.	Magnetomotive force	ampere	A	
17.	Entropy	joule per kelvin	JK^{-1}	

Appendix E

Some Useful Mathematical Formulae

1. $1 + 2 + 3 + \cdots + n = \frac{n(n+1)}{2}$.

2. $1^2 + 2^2 + 3^2 + \cdots + n^2 = \frac{1}{6}n(n+1)(2n+1)$.

3. $1^3 + 2^3 + 3^3 + \cdots + n^3 = \left[\frac{n(n+1)}{2}\right]^2$.

4. If an arithmetic progression (A.P.) is expressed as $a, a+d, a+2d, \cdots$, then the sum of its first n terms is $S = \frac{n}{2}[2a + (n-1)d]$.

5. If a geometrical progression is expressed as $a, ar, ar^2, \cdots$, then the sum of its first n terms is

$$S = a\frac{1-r^n}{1-r}, \text{ when } -1 < r < 1$$
$$S = a\frac{r^n-1}{r-1}, \text{ when } -1 > r > 1.$$

6. (a) $(1-x)^{-1} = 1 + x + x^2 + x^3 + \cdots \infty$

(b) $(1+x)^{-1} = 1 - x + x^2 - x^3 + \cdots \infty$

(c) $(1-x)^{-2} = 1 + 2x + 3x^2 + 4x^3 + \cdots \infty$

(d) $(1+x)^{-2} = 1 - 2x + 3x^2 - 4x^3 + \cdots \infty$.

7. (a) $e^x = 1 + \frac{x}{1!} + \frac{x^2}{2!} + \frac{x^3}{3!} + \cdots + \frac{x^r}{r!} + \cdots \infty$

(b) $e = 1 + \frac{1}{1!} + \frac{1}{2!} + \frac{1}{3!} + \cdots \infty$

(c) $a^x = 1 + (\log_e a) \cdot x + \frac{(\log_e a)^2}{2!} \cdot x^2 + \frac{(\log_e a)^3}{3!} + \cdots + \infty$.

8. (a) $\log_e(1+x) = x - \frac{x^2}{2} + \frac{x^3}{3} - \frac{x^4}{4} + \cdots \infty, \ (-1 < x \le 1)$

(b) $\log_e(1-x) = -x - \frac{x^2}{2} - \frac{x^3}{3} - \frac{x^4}{4} + \cdots \infty, \ (-1 \le x \le 1)$

(c) $\frac{1}{2}\log_e \frac{1+x}{1-x} = x + \frac{x^3}{3} + \frac{x^5}{5} + \cdots \infty, \ (-1 < x < 1)$.

Appendix F

Some Useful Trigonometric Formulae

1. (a) $\sin(A \pm B) = \sin A \cos B \pm \cos A \sin B$

(b) $\cos(A \pm B) = \cos A \cos B \mp \sin A \sin B$.

2. (a) $\tan(A \pm B) = \frac{\tan A \pm \tan B}{1 \mp \tan A \tan B}$

(b) $\cot(A \pm B) = \frac{\cot A \cot B \mp 1}{\cot B \pm \cot A}$.

3. $\tan(A + B + C) = \frac{\tan A + \tan B + \tan C - \tan A \tan B \tan C}{1 - \tan A \tan B - \tan B \tan C - \tan C \tan A}$.

4. (a) $\sin C + \sin D = 2 \sin \frac{C+D}{2} \cos \frac{C-D}{2}$

(b) $\sin C - \sin D = 2 \cos \frac{C+D}{2} \sin \frac{C-D}{2}$.

5. (a) $\cos C + \cos D = 2 \cos \frac{C+D}{2} \cos \frac{C-D}{2}$

(b) $\cos C - \cos D = 2 \sin \frac{C+D}{2} \sin \frac{D-C}{2}$.

6. (a) $\sin 3\theta = 3 \sin\theta - 4 \sin^3\theta$

(b) $\cos 3\theta = 4 \cos^3\theta - 3 \cos\theta$

(c) $\tan 3\theta = \frac{3 \tan\theta - \tan^3\theta}{1 - 3\tan^2\theta}$.

7. (a)
   i. If $\sin\theta = 0$, then $\theta = n\pi$
   ii. If $\sin\theta = 1$, then $\theta = (4n+1)\frac{\pi}{2}$
   iii. If $\sin\theta = -1$, then $\theta = (4n-1)\frac{\pi}{2}$
   iv. If $\sin\theta = \sin\alpha$, then $\theta = n\pi + (-1)^n\alpha$

(b)
   i. If $\cos\theta = 0$, then $\theta = (2n+1)\frac{\pi}{2}$
   ii. If $\cos\theta = 1$, then $\theta = 2n\pi$
   iii. If $\cos\theta = -1$, then $\theta = (2n+1)\pi$
   iv. If $\cos\theta = \cos\alpha$, then $\theta = 2n\pi \pm \alpha$

(c)
   i. If $\tan\theta = 0$, then $\theta = n\pi$
   ii. If $\tan\theta = \tan\alpha$, then $\theta = n\pi + \alpha$, where $\theta = 0$ or any integer.

8. (a) $\cos(A \pm 90°) = \mp \sin A$

(b) $\sin(A \pm 90°) = \pm \cos A$

(c) $\tan(A \pm 90°) = -\cot A$

(d) $\cos(A \pm 180°) = -\cos A$

(e) $\sin(A \pm 180°) = -\sin A$

(f) $\cot(A \pm 180°) = \tan A$.

Inverse Circular Functions

1. $\cos^{-1} x \pm \cos^{-1} y = \cos^{-1}\left[xy \mp \sqrt{(1-x^2)(1-y^2)}\right]$

2. $2\tan^{-1} x = \sin^{-1} \frac{2x}{1+x^2} = \cos^{-1} \frac{1-x^2}{1+x^2} = \tan^{-1} \frac{2x}{1-x^2}$

3. $\tan^{-1} x + \tan^{-1} y + \tan^{-1} z = \tan^{-1}\left[\frac{x+y+z-xyz}{1-xy-yz-zx}\right]$.

Appendix G

(a) Some Indefinite Integrals

1. $\int \tan x \, dx = \log |\sec x|.$

2. $\int \cot x \, dx = \log |\sin x|.$

3. $\int \frac{dx}{x^2 + a^2} = \frac{1}{a} \tan^{-1} \frac{x}{a}.$

4. $\int \frac{dx}{x^2 - a^2} = \frac{1}{2a} \log \left| \frac{x-a}{x+a} \right|, \; x > a.$

5. $\int \frac{dx}{a^2 - x^2} = \frac{1}{2a} \log \left| \frac{a+x}{a-x} \right|, \; (x < a).$

6. $\int \frac{dx}{\sqrt{x^2 \pm a^2}} = \log \left| x + \sqrt{x^2 \pm a^2} \right|.$

7. $\int \frac{dx}{\sqrt{a^2 - x^2}} = \sin^{-1} \frac{x}{a}.$

8. $\int \frac{dx}{x\sqrt{x^2 - a^2}} = \frac{1}{a} \sec^{-1} \frac{x}{a}.$

9. $\int e^{ax} \sin bx \, dx = e^{ax} \left(\frac{a \sin bx - b \cos bx}{a^2 + b^2} \right) = \frac{e^{ax}}{a^2 + b^2} \sin \left(bx - \tan^{-1} \frac{b}{a} \right).$

10. $\int e^{ax} \cos bx \, dx = e^{ax} \left(\frac{a \cos bx + b \sin bx)}{a^2 + b^2} \right) = \frac{e^{ax}}{\sqrt{a^2 + b^2}} \cos \left(bx - \tan^{-1} \frac{b}{a} \right).$

11. $\int \sqrt{x^2 \pm a^2} dx = \frac{x\sqrt{x^2 \pm a^2}}{2} \pm \frac{a^2}{2} \log \left| x + \sqrt{x^2 \pm a^2} \right|.$

12. $\int \sqrt{a^2 - x^2} dx = \frac{x\sqrt{a^2 - x^2}}{2} + \frac{a^2}{2} \sin^{-1} \frac{x}{a}.$

(b) Some Definite Integrals

1. $$\int_0^{\pi} \sin mx \sin nx\, dx = \int_0^{\pi} \cos mx \cos nx\, dx = \begin{cases} 0, & m \neq n \\ \frac{\pi}{2}, & m = n. \end{cases}$$

2. $$\int_0^{\pi} \sin mx \cos nx\, dx = \begin{cases} 0, & m+n = \text{even} \\ \frac{2m}{m^2-n^2}, & m+n = \text{odd}. \end{cases}$$

3. $$\int_0^{2\pi} \sin mx \sin nx\, dx = \int_{-\pi}^{+\pi} \sin mx \sin nx\, dx = \begin{cases} 0, & m \neq n \\ \pi, & m = n. \end{cases}$$

4. $$\int_0^{\infty} \frac{\sin ax}{x} dx = \begin{cases} \frac{\pi}{2}, & a > 0 \\ 0, & a = 0 \\ -\frac{\pi}{2}, & a < 0. \end{cases}$$

5. $$\int_0^{\infty} \frac{\sin^2 ax}{2} dx = \frac{\pi}{2}.$$

6. $$\int_0^{\infty} \frac{\sin^2 ax}{x^2} dx = |a| \frac{\pi}{2}.$$

7. $$\int_0^{\pi} x^n e^{-ax} dx = \frac{n!}{a^{n+1}}.$$

8. $$\int_0^{\infty} e^{-ax^2} dx = \frac{1}{2}\sqrt{\frac{\pi}{a}}.$$

9. $$\int_{-\infty}^{+\infty} e^{-(ax^2+bx+c)} dx = \sqrt{\frac{\pi}{a}} e^{(b^2-4ac)/4a}.$$

10. $$\int_0^{\infty} e^{-ax} \cos bx\, dx = \frac{a}{a^2+b^2}.$$

11. $$\int_0^{\infty} e^{-ax} \sin bx\, dx = \frac{b}{a^2+b^2}.$$

12. $$\int_0^{\infty} \frac{x^3}{e^x - 1} dx = \frac{\pi^4}{15}.$$

13. $$\int_0^{\infty} \frac{x}{e^x - 1} dx = 1.51.$$

14. $$\int_0^{\infty} e^{-x} dx = \frac{\sqrt{\pi}}{2}.$$

15. $$\int_0^{\infty} e^{-x} x\, dx = \frac{1}{2}.$$

16. $$\int_0^{\infty} e^{-x} x^2 dx = \frac{\sqrt{\pi}}{4}.$$

17. $$\int_0^{\infty} e^{-x} x^3 dx = \frac{1}{2}.$$

18. $$\int_0^{\infty} e^{-x} x^4 dx = \frac{3}{8}\sqrt{\pi}.$$

19. $$\int_0^{\infty} e^{-x} x^5 dx = 1.$$

Appendix H

Some Important Hyperbolic Functions

1. $e^{\pm ix} = \cos x \pm i \sin x.$

2. $\sinh x = \dfrac{e^x - e^{-x}}{2}.$

3. $\cosh x = \dfrac{e^x + e^{-x}}{2}.$

4. $\tanh x = \dfrac{\sinh x}{\cosh x} = \dfrac{e^x - e^{-x}}{e^x + e^{-x}}.$

5. $\coth x = \dfrac{1}{\tanh x} \dfrac{e^x + e^{-x}}{e^x - e^{-x}}.$

6. $\text{sech} x = \dfrac{1}{\cosh x} \dfrac{2}{e^x + e^{-x}}.$

7. $\cosh^2 x - \sinh^2 x = 1.$

8. $\text{sech}^2 x + \tanh^2 x = 1.$

9. $\dfrac{d}{dx}(\sinh u) = \cosh u \dfrac{du}{dx}.$

10. $\dfrac{d}{dx}(\cosh u) = \sinh u \dfrac{du}{dx}.$

11. $\dfrac{d}{dx}(\tanh u) = \text{sech}^2 u \dfrac{du}{dx}.$

Appendix I

Solutions of Some Differential Equations

Sl. No.	*Differential equation*	*Solution*
1.	$\frac{d^2y}{dt^2} + \omega^2 y = 0$	(i) $y(t) = A\sin\omega t + B\cos\omega t$ (ii) $y(t) = C_1 e^{i\omega t} + C_2 e^{-i\omega t}$ (iii) $y(t) = A\sin(\omega t + \delta)$ (iv) $y(t) = A\cos(\omega t + \phi)$, where A, B, C_1 and C_2 are constants.
2.	$\frac{d^2y}{dt^2} + \gamma\frac{dy}{dt} + \omega_0^2 y = 0$, where $\gamma = \frac{p}{m}$ and $\omega_0^2 = \frac{k}{m}$	$y(t) = A_1 \exp\left[-\frac{\gamma}{2} + \left(\frac{\gamma^2}{4} - \omega_0^2\right)^{1/2}\right] t$ $+A_2 \exp\left[-\frac{\gamma}{2} + \left(\frac{\gamma^2}{4} - \omega_0^2\right)^{1/2}\right] t$, where A_1 and A_2 are constants.
3.	$\frac{d^2y}{dt^2} + \gamma\frac{dy}{dt} + \omega_0^2 y = f_0\cos\omega t$, where $\gamma = \frac{p}{m}, \omega_0 = \sqrt{\frac{k}{m}}$ and $f_0 = \frac{F_0}{m}$	$PI = B\cos(\omega t - \phi)$, $CF = Ae^{-\frac{\gamma t}{2}}\cos(\omega^* t - \delta)$ $\therefore\ CS = PI + CF$ i.e., $y(t) = Ae^{-\frac{\gamma t}{2}}\cos(\omega^* t - \delta) + B\cos(\omega t - \phi)$, where A and B are constants. $\omega^* = \left(\omega_0^2 - \frac{\gamma^2}{4}\right)^{1/2}$ and δ, ϕ represent phase terms.
4.	$\frac{\partial^2 y}{\partial t^2} = c^2\frac{\partial^2 y}{\partial x^2}$	$y(t) = \sum_{s=1}^{\infty}\left[a_s\cos\frac{s\pi ct}{l} + b_s\sin\frac{s\pi ct}{l}\right]\sin\frac{s\pi x}{l}$ $= \sum_{s=1}^{\infty} A_s\sin\frac{s\pi x}{l}\cos\left(\frac{s\pi ct}{l} - \phi_s\right)$

Appendix J

Gamma Functions

Gamma function is defined as $\Gamma(n) = \int_0^\infty e^{-x} x^{n-1} dx$, then

(i) $\Gamma(n) = (n-1)\Gamma(n-1)$

(ii) $\Gamma(n+1) = n\Gamma(n)$.

Value of some useful Gamma functions:

1. $\Gamma(0) = \infty$

2. $\Gamma(-n) = \infty$ (where n is a +ve integer)

3. $\Gamma\left(\frac{1}{2}\right) = \sqrt{\pi}$

4. $\Gamma(1) = 1$

5. $\Gamma\left(\frac{3}{2}\right) = \frac{1}{2}\Gamma\left(\frac{1}{2}\right) = \frac{1}{2}\sqrt{\pi}$

6. $\Gamma\left(\frac{5}{2}\right) = \frac{3}{2}\Gamma\left(\frac{3}{2}\right) = \frac{3}{2} \cdot \frac{\sqrt{\pi}}{2} = \frac{3}{4}\sqrt{\pi}$

7. $\Gamma\left(\frac{7}{2}\right) = \frac{5}{2}\Gamma\left(\frac{5}{2}\right) = \frac{5}{2} \cdot \frac{3}{4}\sqrt{\pi} = \frac{15}{8}\sqrt{\pi}$

8. $\Gamma\left(n+\frac{1}{2}\right) = \frac{\sqrt{n}}{2^{2n}-1}\Gamma(2n)$

9. $\Gamma(n)\Gamma\left(\frac{1-n}{2}\right) = \frac{\sqrt{\pi}\Gamma\left(\frac{n}{2}\right)}{2^{1-n}\cos\left(\frac{n\pi}{2}\right)}$, $0 < n < l$.

Bibliography

1. Ali Omar M., *Elementary Solid State Physics,* Addison-Wesley, Reading, 1975.
2. Anderson R.A., *Fundamentals of Vibrations,* Amerind Publishcing Co. Pvt. Ltd., 1967.
3. Baggot J., *The Meaning of Quantum Theory,* Oxford Univeristy Press, N.Y.
4. Bajaj N.K., *The Physics of Waves and Oscillations,* Tata McGraw-Hill, Publishing Company Limited, New Delhi, 1989.
5. Beiser A., *Perspective of Modern Physics,* McGraw-Hill, N.Y., 1969.
6. Bishop R.E.D., *Vibration, Cambridge University Press,* N.Y., 1965.
7. Bohm D., *Quantum Theory,* N.J., Prentice Hall.
8. Born M. and Wolf E., *Principles of Optics,* Pergamon Press, Oxford, 1975.
9. Brown J.G., *X-rays and their Applications,* Plenum, 1975.
10. Coulson C.A., *Waves,* Oliver and Boyd, Edinburgh, 1941.
11. Crawford (Jr.) F.S., *Waves,* Barkley Physics Course, Vol. 3, McGraw-Hill Book Company, 1968.
12. Ditchburn R.W., *Light,* Academic Press, London, 1976.
13. Fano U. and Fano L., *Physics of Atoms and Molecules,* University of Chicago Press, Chicago, 1970.
14. Fermi E., *Notes on Quantum Mechanics,* Univ. of Chicago Press, Chicago, 1961.
15. Francon M., *Diffraction: Coherence in Optics,* Pergamon Press, Oxford, 1966.
16. French A.P., *Special Relativity,* 1968.
17. Ghatak A., *Optics,* Tata McGraw Hill Publishing Company Ltd., New Delhi, 1997.
18. Ghosal S.N., *Atomic and Nuclear Physics,* Vol. I, S. Chand and Company Ltd., New Delhi, 1997.
19. Gray P., *Crystalline State,* Oliver and Boyd, 1972.
20. Hary J.E., *Industrial Laser and their Applications,* McGraw-Hill, London, 1974.
21. Hect J.E. and Zajac A., *Optics,* Addison-Wesley, Reading, Mass, 1974.
22. Jenkins F.A. and White H.E., *Fundamental of Optics,* 3rd Edn., McGraw-Hill Book Co. N.Y., 1957.
23. Kittel C., *Introduction to Solid State Physics,* John Wiley, N.Y. 1976.

24. Kuila S.P., *Perspective of Quantum Mechanics,* New Central Book Agency, Kolkata, India, 2008.

25. Maitland A. and Dunn M.H., *Laser Physics,* North Holland Publishing Co., Amsterdam, 1969.

26. Mani H.S. and Mehta G.K., *Introduction to Modern Physics,* Affiliated East-West Press Pvt. Ltd., New Delhi, 2006.

27. Merzabacher E., *Quantum Mechanics,* John Wiley, N.Y., 1979.

28. Morrison M., *Understanding of Quantum Mechanics,* N.J., Prentice Hall, 1990.

29. Pain A.P., *Vibrations and Waves,* M.I.T. Introductory Physics Series, Arnold-Heinemann, India, 1973.

30. Peaslec D.C., *Elements of Atomic Physics,* Prentice Hall, Englewood Cliffs, 1955.

31. Personic S.D., *Fundamentals of Optical Fibre Communication,* Academic Press, N.Y., 1976.

32. Powel J.L. and Crasemann B., *Quantum Mechanics,* Addison-Wesley Publishing Co., Inc., Reading, USA, 1971.

33. Resnick R., *Introduction to Special Theory of Relativity,* John Wiley, N.Y., 1968.

34. Schurchliff W.A. and Ballard S.S., *Polarised Light,* Van Nostrand, Princeton, N.J., USA, 1964.

35. Shiff, L.I., *Quantum Mechanics,* McGraw-Hill Book Company, N.Y. 1968.

36. Sommerfeld A., *Optics,* Academic Press, N.Y., 1964.

37. Stefan R.W.B. and Bate A.E., *Wave Motion and Sound,* Adward Arnold and Co., London, 1950.

38. Thyagarajan K. and Ghatak A.K., *Laser Theory and Applications,* Macmillan India Ltd., 1991.

39. Waldron R.A., *Waves and Oscillations,* Momentum Books, Van Nostrand, Reinhold, New York, 1964.

40. Wheeler J.A. and Taylor E.F., *Space-time Physics,* W.H. Freeman, San Fransciso, California, 1966.

41. Wood A., *Acoustics,* Blackie, London, 1940.

42. Wyckoff R.W.G., *Crystal Structure,* John Wiley, 1963.

Index

Absorption edge, 355
Acceptance cone, 380
Active medium, 201
Amorphous solids, 319
Amplitude, 3
Analyser, 167

Beats, 12
Blackbody, 261
Bremsstrahlung process, 354
Brewster's window, 163

Calculation of packing factor, 325
Characteristics of matter waves, 296
Coherent sources, 88
Compton shift, 272
Compton wavelength, 272
Condition for destructive, Constructive interference, 82
Conservative system, 7
Contraction hypothesis, 230
Corpuscular rays, β-rays, 352
Corpuscular theory, 77
Critical damping, 38
Cross-talk, 378
Crystallographic direction, 333

Damped vibration, 34
Data storage, 219
De Broglie model of atom, 306
Dead beat motion, 37
Decay constant ($\gamma/2$), 56
Decrement of motion, 41
Dichroism, Selective absorption, 162
Diffraction, 122
Dimension of force constant of a vibrating spring, 21
Division of wave front, amplitude, 88
Double refraction, 165, 170

Early stage, 56
Effect of damping force, 41
Einstein's A and B constants, 195
Elliptically polarized light, 172
Epoch, initial phase, 6
Ether drag hypothesis Lorentz-Fitzerald, 230
Existence of zero point energy, 304

Figure of merit, 71
Force of restitution, 35
Foundation of physics, 261
Fraunhofer diffraction, 123
Fresnel's diffraction, 123
Fringes of equal inclination, thickness, chromatic order, 95
Full radiation, 261

Graded index fibre, 377
Grain boundaries, 319

Half-wave plate, 177
Hologram, 215
Huygen's secondary wavelets, 128

Inertial and non-inertial frames, 223
Injection current, 199
Interference by transmitted light, 94
Inverse photoelectric effect, 353

k-polaroid, 170

Large damping, 36
Lissajous figures, 17
Lower and upper half power frequencies, 69

Mass controlled motion, 60
Mass variation relation, 239
Metastable state, 198
Miller indices, 333
Missing order of interference maxima, 130
Modern coolidge tube, 350
Monomode fibre, 378

Natural frequency, 36
Negative and positive crystals, 165
Non-destructive testing, 220
Normal spectrum, 138

Optic axis, Principal section, Principal plane, 166
Optical pumping, 199

Partial polarized light, 161
Phase, 4
Physical significance of uncertainty principle, 300
Plane-, circularly polarized light, 171
Polariser, 162
Polaroid, 170
Poly crystalline, 319
Population inversion, 197
Principal maxima, 126
Principle of complimentarity, 276

Quality factor, 66
Quarter-wave plate, 176

Reconstruction of the image, 216
Relaxation time or modulus of decay, 44
Resistance controlled motion, 60
Resonant frequency, 59

Secondary maxima, 126
Secondary point sources, 124
Secondary waves, 79
Sharpness of resonance, 66
Size of hydrogen atom, 303
Sources of damping, 45
Space lattice and crystal lattice, 332
Spatial coherence, 86
Spectrometer, 138
Spherical, Cylindrical, Plane wave front, 78, 79
Spontaneous emission, 193
Step-index fibre, 376
Stiffness controlled motion, 60
Stimulated emission, –absorption, 193
Stopping potential, 268
Structural damping, 35
Sustained interference, 82

Temporal coherence, 85
The core, the cladding, the sheath (Jacket), 375, 376
Thermal de Broglie wavelength, 307
Threshold frequency, 268
Transient beats, 55
Transit time distortion, 383

Ultraviolet catastrophe, 264

Wave packet, 298
Weak damping, 39
Wien's displacement law, 262
Wien's distribution law, 263

ENGINEERING AND MANAGEMENT EXAMINATIONS

ENGINEERING PHYSICS

June 2005

Semester 2

Old Syllabus

Answer *Q. No.* 11 and any *five* from the rest

1. (a) State and explain de Broglie's hypothesis.

(b) Show that the de Broglie wavelength λ of electrons of energy e is given by $\lambda = \frac{h}{\sqrt{2\,\text{meV}}}$.

(c) If the uncertainty in the position of an electron is 4×10^{-10} m, calculate the uncertainty in its momentum.

2. (a) What are the basic postulates of Quantum Mechanics?

(b) Deduce time dependent Schrödinger's equation for a free particle of mass m.

(c) Prove that any coordinate and its conjugate momentum do not commute.

3. (a) Define unit cell.

(b) Show that the spacing d of planes (hkl) in a SC lattice of side a is $d = \frac{a}{\left[\sqrt{h^2+k^2+l^2}\right]}$.

(c) Lead is face centred cubic with an atomic radius of $r = 1.746$ Å. Find the inter-planar spacing for (2 2 0) plane.

4. (a) Distinguish between Bose-Einstein statistics and Fermi-Dirac statistics.

(b) Plot electron distribution function governed by Fermi-Dirac statistics in metal at $T = 0\,\text{K}$ and $T > 0\,\text{K}$ and discuss it.

5. (a) State Gauss's divergence theorem.

(b) If $\psi(x, y, z) = 3(x^2y - y^2z)$, calculate grad ψ at the point $(1, -2, 1)$.

(c) Prove: $\vec{A} \times (\vec{B} \times \vec{C}) + \vec{B} \times (\vec{C} \times \vec{A}) + \vec{C} \times (\vec{A} \times \vec{B}) = 0$.

(d) Find the work done to move on object along a vector $\vec{r} = 3\hat{i} + 2\hat{j} - 5\hat{k}$, if the applied force is $\hat{F} = 2\hat{i} - \hat{j} - \hat{k}$.

6. (a) What is Seebeck effect?

(b) Draw the emf vs. temperature curve for a thermocouple and find the expressions for neutral and inversion temperatures.

(c) A thermocouple is made up of iron and constantan. Find the emf developed per °C difference of temperatures between the junctions, given that thermo emfs of iron and constantan against platinum are $16\mu\text{V}/°\text{C}$ and $-34\mu\text{V}/°\text{C}$, respectively.

7. (a) Deduce Gauss's law in differential form.

(b) Show that for an electrostatic field curl $\vec{E} = 0$.

(c) In a region of space, the electric field is given by $\vec{E} = 8\hat{i} + 4\hat{j} + 3\hat{k}$. Calculate the electric flux through a surface of 100 units in XY-plane.

8. (a) Write down four Maxwell's equation for electromagnetic wave.

(b) Explain why the introduction of a dielectric between the plates of a capacitor changes its capacitance.

(c) Show that the potential function $V = V_0(x^2 - y^2 + z)$ satisfies Laplace's equation, where V_0 is a constant.

9. (a) Discuss the important characteristics of nuclear forces.

(b) What is nuclear binding energy? Draw binding energy curve. What information do you get from such a curve?

(c) Find the energy released for the following nuclear fusion reaction $4\,{}_1H^1 \rightarrow {}_2He^4 + 2\,{}_{+1}e^0 +$ energy. Given that mass of ${}_1H^1$ atom $= 1.007825$ amu, mass of ${}_2He^4$ atom $= 4.0026$ amu and mass of positron ${}_{+1}e^0 = 0.00055$ amu.

10. Write short notes on any *two* of the following: (a) Nuclear fission (b) One-dimensional square well potential (c) Peltier effect (d) Law of equipartition of energy.

11. (a) Select the correct answer from the given alternatives in each case:

i. If $\hat{n}$ is unit vector in the direction of $\vec{A}$, then (a) $\hat{n} = \vec{A}/|\vec{A}|$ (b) $\hat{n} = \vec{A}|\vec{A}|$ (c) $\hat{n} = |\vec{A}|/\vec{A}$.

ii. The photoelectric current depends on (a) the frequency of incident radiation (b) the intensity of incident radiation (c) none of the above.

iii. Two vectors $\hat{A}$ and $\vec{B}$ are perpendicular when (a) $\vec{A}\vec{B} = 0$ (b) $\vec{A} \times \vec{B} = 0$ (c) $\vec{A}\vec{B} = 1$.

iv. The displacement current arises due to (a) positive charges only (b) time varying electric field (c) negative charges only.

v. The wavelength λ associated with a particle of mass m moving with velocity v is given by (a) $\lambda = h/mv$ (b) $\lambda = h/mv^2$ (c) $\lambda = mv/h$.

vi. The Miller indices of a plane which cuts the intercepts of 2, 3 and 4 units along the three axes is (a) (643) (b) (234) (c) (432).

vii. Electrons are (a) Fermions (b) Bosons (c) none of the above.

viii. The value of $\left[\hat{x}, \frac{\partial}{\partial x}\right]$ is (a) 1 (b) -1 (c) none of the above.

ix. The nuclear forces are (a) charge independent (b) charge dependent (c) none of the above.

x. One milligram of matter converted into energy will give (a) 90 joules (b) 90×10^9 joules (c) 99 joules.

(b) Answer any *five* of the following questions:

i. *The gradient of a scalar field is a vector quantity.*—Explain.

ii. What is Fermi energy?

iii. Why nuclear fission reaction must be done first before nuclear fusion?

iv. What are the differences between conductor and semiconductor on the basis of energy band structure?

v. What is Zeeman effect?

vi. State Ampere's law.

vii. Write down Einstein's photoelectric equation.

New Syllabus

Answer *Q. No.* 11 and any *five* from the rest

1. A. Answer the following questions:

(a) Relative velocity of two particles moving with velocity of light in opposite directions is (i) 0 (ii) $2c$ (iii) c (iv) none of these.

(b) The wavelength of which the spectral energy density of emitted radiation at temperature T from a black body attains maximum value, is proportional to (i) $1/T$ (ii) T (iii) T^4 (iv) $T^{3/2}$.

(c) de Broglie wavelength of a body of mass m and kinetic energy E is (i) $\frac{2mh}{\sqrt{E}}$ (ii) $\frac{h}{\sqrt{2mE}}$ (iii) $\frac{h}{2mE}$ (iv) $\frac{\sqrt{2mh}}{E}$.

(d) Total probability of finding a moving particle represented by a wave function $\psi(\vec{x})$ in the whole space is (i) zero (ii) ∞ (iii) 1 (unity) (iv) $|\psi(x)|^2$.

(e) A coin and six-faced dice are thrown. The probability that the coin shows tail and dice shows five is (i) 7/12 (ii) 1/8 (iii) 1/12 (iv) 1/6.

(f) [1 1 1] plane of cubic crystal is (i) perpendicular to x-axis (ii) perpendicular to y-axis (iii) perpendicular to z-axis (iv) none of these.

(g) If λ_L and λ_K are the wavelengths of L, K X-rays, respectively, then (i) $\lambda_L > \lambda_K$ (ii) $\lambda_L < \lambda_K$ (iii) $\lambda_L = \lambda_K$ (iv) $\lambda_K = 4\lambda_L$.

(h) A nuclear reaction is called *exoergic*, if Q-value is (i) negative (ii) positive (iii) zero (iv) none of these.

(i) With the increases in wavelength of the incident radiation the kinetic energy of the emitted photoelectron (i) increases (ii) decreases (iii) remains constant.

(j) For laser action to occur, the medium used must have at least (i) 4 energy levels (ii) 2 energy levels (iii) 3 energy levels.

B. Answer any *five* from the following questions:

(a) State the characteristics of the laser beam.

(b) What do you understand by the terms Fermi-energy?

(c) What is Heisenberg's uncertainty principle?

(d) What is the distinction between inertial and non-inertial frame?

(e) What is Bravais lattice?

(f) What is meant by charge independence of nuclear force?

(g) Write the time independent Schrödinger's equation for a free particle of mass m and momentum $\vec{p}$.

(h) Write the Lorentz transformation relation for time coordinate in a frame moving with velocity v in the Z-direction relative to a stationary frame.

2. (a) What are basic postulates of special theory of relativity?

(b) Establish Einstein's mass-energy relation.

(c) Calculate the energy equivalent to 1 gm of mass.

3. (a) Derive an expression for the Compton shift in wavelength for a photon scattered from a free electron, at an angle θ.

(b) At which angle the shift will be maximum?

(c) Calculate the Compton wavelength (in Å) for an electron.

4. (a) What is the experimental evidence in favour of de Broglie's hypothesis of matter waves?

(b) Why in case of moving electrons quantum mechanics is used while for moving cars we use Newtonian mechanics? Explain.

(c) Show that there cannot be any existence of free electrons inside the nucleus of an atom.

5. (a) Give the physical interpretation of the wave function $\psi(\vec{x})$.

(b) Obtain the expression for stationary energy levels for a particle of mass m which is free to move in a region of zero potential between two rigid walls of $x = 0$ and $x = L$. Are the energy levels degenerate?

(c) Find the corresponding wave-function pertaining to (b) and normalize.

6. (a) Show that in a cubic crystal of side a the spacing between consecutive parallel planes of Miller indices (hkl) is $d_{hkl} = a/\sqrt{h^2 + k^2 + l^2}$.

(b) Copper has FCC structure and the atomic radius is 0.1278 nm. Calculate its density and the inter-planar spacing for (3 2 1) planes. Take the atomic weight of copper as 63.5.

7. (a) Distinguish between continuous and characteristic X-ray spectra.

(b) Explain and derive Bragg's law of X-ray diffractions from a crystal.

(c) The spacing of the planes in a crystal is 1.2 Å and the angle for the 1st order Bragg reflection is 30°. Determine the energy of the X-rays in eV.

8. (a) What is population inversion? Explain.

(b) How higher probability of stimulated emission compared to that of spontaneous emission is achieved in a laser?

(c) Calculate the numerical aperture of a fibre with refractive indices of the core and cladding 1.54 and 1.50, respectively.

9. (a) Explain with suitable diagram the variation of per nucleon binding energy with mass number.

(b) Hence explain why fusion takes place in the light mass region while fission occurs for heavy nuclei.

(c) Why high temperature is needed to initiate fusion?

10. (a) Compare MB, FD and BE statistics mentioning at least three characteristics.

(b) Draw the Fermi-distribution curve of (i) $T = 0\,°K$ and (ii) $T > 0\,°K$. Explain their significance.

(c) Indicate which statistics will be applicable for (i) $_1H^1$ (ii) e^- (iii) π° (iv) photon.

11. Write short notes on any *two* of the following: (a) Planck's radiation law (b) Macro-states and thermodynamic probability (c) Nuclear chain reaction (d) Energy band diagram of a crystal vis-à-vis metal, insulator and semiconductor.

June 2006

Semester 2

Group A: Multiple Choice Questions

1. Choose the correct alternatives for any *ten* of the following:

(i) 1 milligram of matter after getting completely converted to energy will yield (a) 90 joules (b) 9×10^{10} joules (c) 900 joules (d) 9 joules.

(ii) Lorentz transformation is LT, Galilean transformation is GT, velocity between two frames is v and velocity of light is given by c.

Which of the following statements is correct?

(a) LT can never be reduced to GT under any approximation. (b) LT is reduced to GT when $v \gg c$. (c) LT is reduced to GT when $v = c$. (d) LT is reduced to GT when $v \ll c$.

(iii) Mass of a photon of frequency v is given by (Planck's constant $= h$ and velocity of light $= c$) (a) $h\nu$ (b) $h\nu/c$ (c) $h\nu/c^2$ (d) $h\nu^2/c$.

(iv) An α-particle is 4 times heavier than a proton. If a proton and an α-particle are moving with the same velocity, how do their de Broglie wavelengths compare?

(a) $\lambda_p = \lambda_\alpha$ (b) $\lambda_p = 4\lambda_\alpha$ (c) $\lambda_p = \lambda_\alpha/2$ (d) $\lambda_p = \lambda_\alpha/4$.

(v) Uncertainty principle tells that (a) A particle can have only position but no momentum. (b) A particle can have only momentum but no position. (c) One can determine simultaneously the position and momentum of a particle. (d) One cannot determine simultaneously the position and momentum of a particle.

(vi) Which of the following statements is wrong?

(a) Maxwell-Boltzmann statistics deals with distinguishable particles. (b) Fermi-Dirac statistics deals with indistinguishable particles. (c) Bose-Einstein statistics deals with indistinguishable particles. (d) Bosons are subjected to Pauli's exclusion principle.

(vii) Miller indices of a plane which cut intercepts of 2, 3 and 4 units along the three axes are (a) (6, 4, 3) (b) (2, 3, 4) (c) (3, 2, 1) (d) (2, 3, 2).

(viii) Which of the following wavelengths falls in the X-ray range?
(a) 1 nm (b) 100 nm (c) 0.001 nm (d) 1000 nm.

(ix) The population inversion in preparing laser beam can be achieved (a) when one of the excited states is more populated than the ground state. (b) when one of the excited states is less populated than the ground state. (c) when the population of one excited state and the ground state are equal. (d) irrespective of any relationship between the population in the excited and the ground states.

(x) Which of the following statements is correct?
(a) Optical fibres can bend light using the principle of Poiseuille's law. (b) Optical fibres cannot bend the light beam. (c) Optical fibres bend the light beam using the principles of total internal reflection. (d) Optical fibres use a liquid which helps in bending the light beam.

(xi) Nuclear binding energy of an isotope is $c^2\Delta M$, where c is the velocity of light and ΔM is (a) difference between the mass of the particles forming the nucleus and the mass of the nucleus itself. (b) difference between the mass of the protons and the neutrons present. (c) difference between the mass of the isotope and its most abundant variety. (d) difference between the mass of the protons and the electrons.

(xii) Emissive power of a black body kept at any absolute temperature T is proportional to (a) T^3 (b) T^4 (c) T^5 (d) T^{-1}.

(xiii) When two parallel rays of X-rays, of wavelength λ, are incident at an angle θ on a crystal with lattice separation d, constructive interference would be observed when (n is an integer) (a) $n\lambda = 2d\sin\theta$ (b) $n\lambda = d\sin\theta$ (c) $n\lambda = d\sin 2\theta$ (d) $n\lambda = 2d\sin 2\theta$.

(xiv) Which of the following statements is wrong?
(a) Strong forces are of short range. (b) Electromagnetic forces can be both attractive and repulsive. (c) Weak forces are of larger range than electromagnetic forces. (d) Electromagnetic forces are weaker than the strong force.

Group B: Short Answer Type Questions
Note: (i) Each answer should not exceed 50 words (ii) Answer any *three* questions.

2. Write down the Schrödinger's equation for a particle confined in a one-dimensional box.

3. Consider a two-state system each of which can exist in states ϵ_1, ϵ_2 and ϵ_3. What are the possible states if the particles are (a) bosons (b) fermions?

4. What are the postulates of special theory of relativity?

5. What is the ratio of Compton shift for scattering at 0° and 90°?

6. The distance between (100) planes in a body-centred cubic structure is 0.232 nm. What is the size of the unit cell? What is the radius of the atom?

7. Why are slow neutrons used in nuclear reactor?

8. Describe briefly the principle of operation of a laser.

Group C: Long Answer Type Questions
Note: Answer any *three* of the following questions.

9. Explain the terms packing fraction, mass defect and binding energy of a nucleus. Draw a curve showing the variation of binding energy per nucleon against the mass number. Use t his curve to explain the release of energy in fusion of light nucleus and in fission of heavy nucleus.

10. State clearly, explaining all the terms, the Planck's law, Rayleigh-Jeans law and Wien's displacement law for radiation. Find out the two limits at which the Planck's formula reduces to the other two.

11. Describe briefly the Davisson-Germer experiment. What inferences may you draw from this experiment? Explain what you mean by *wave packet.*

12. Describe briefly the Michelson-Morley experiment and the significance of the experimental results.

13. What is a crystal band? How to differentiate between metal, insulator and semiconductor on the basis of energy band diagram?

14. What is an optical resonator? The r.i.s. of the core and cladding of an optical fibre are 1.523 and 1.472, respectively. Find numerical aperture and the angle of acceptance of the optical fibre.

June 2007

Semester 2

Group A: Multiple Choice Type Questions

1. Choose the correct alternatives for any *ten* of the following:

(i) Time interval between two events measured in inertial frames S1, S2, S3 and S4 are 1s, 1.1s, 1.5s and 2s, respectively. The events occurred in the same spatial location in one of those frames of reference, which is (a) S1 (b) S2 (c) S3 (d) S4.

(ii) How fast a particle must travel so that its mass becomes twice its rest mass?
(a) $0.5c$ (b) $2c$ (c) $\frac{\sqrt{3}}{2}c$ (d) $0.25c$.

(iii) Davisson and Germer studied electron diffraction with Nickel crystal and found a first order peak at 65° with electron beam of 54 eV. If instead a 216 eV beam were used, then the peak would have been at (a) 27° (b) 54° (c) 130° (d) 260°.

(iv) If visible light is used to study the Compton scattering, then the Compton shift will be (a) negative (wavelength of the scattered light will be lesser) (b) more positive than what is observed with X-ray (c) zero (d) positive but not detectable in the visible window.

(v) What is the value of $\left[\hat{x}, \frac{\hat{\partial}}{\partial x}\right]$?
(a) -1 (b) 0 (c) -2 (d) 1.

(vi) Which of the following wave-function is the solution of Schrödinger's equation?
(a) $A\sec(x)$ (b) $A\tan(x)$ (c) $A\exp(-x^2)$ (d) $A\exp(x^2)$.

(vii) Fermi-Dirac distribution approaches Maxwell-Boltzmann distribution at (a) low temperature and high pressure (b) low temperature and low particles mass (c) high temperature and high particle mass (d) high density and low energy range.

(viii) A nucleus is more stable if both atomic number (Z) and the mass number (A) are even. This indicates that (a) inter-nuclear force is charge dependent (b) inter-nuclear force is spin dependent (c) inter-nuclear force saturates over a small distance (d) nucleons are fermions.

(ix) In the nuclear reaction ${}^{23}_{11}\text{Na}+{}^{4}_{2}\text{He} \rightarrow {}^{26}_{13}\text{Al}+\cdots$, the missing particle is (a) electron (b) proton (c) neutron (d) positron.

(x) Assuming that the atoms in a crystal are spheres of equal size and touching each other, it can be shown that atomic radius of BCC is equal to (a) $\frac{a}{2}$ (b) $\sqrt{3}\frac{a}{4}$ (c) $\sqrt{3}\frac{a}{2}$ (d) $a/2\sqrt{2}$.

(xi) Average energy of an electron in a metal at $T = 0\,\text{K}$ is (a) ϵ_F (b) $\frac{1}{2}\epsilon_F$ (c) $\frac{3}{5}\epsilon_F$ (d) $\frac{5}{3}\epsilon_F$.

(xii) Origin of continuous X-ray is due to the process of (a) ionization (b) inner orbital transition (c) bremsstrahlung (d) none of these.

(xiii) In an optical fibre the intensity of light (a) attenuates linearly with the length of the fibre (b) attenuates exponentially with the length of the fibre (c) attenuates logarithmically with the length of the fibre (d) does not attenuate with the length of the fibre.

(xiv) In a ruby laser, population inversion is achieved by (a) optical pumping (b) inelastic atom-atom collisions (c) chemical reactions (d) applying strong electric field.

Group B: Short Answer Type Questions
Answer any *three* of the following.

2. (a) When a particle moves with a velocity much less compared to the velocity of light in free space, then find the relativistic expression of kinetic energy approaches the classical result.

(b) If two photons are approaching each other in free space, then find the relative velocity of one photon w.r.t. the other.

3. (a) What are the features of photoelectric effect which cannot be explained from wave theory of light?

(b) Show graphically how the energy density vs. frequency plot of black body radiation is changed if the temperature is increased.

4. (a) If the wave function $\psi(x)$ of a quantum mechanical particle is given by

$$\psi(x) = a \sin \frac{\pi x}{L}, \quad \text{for } 0 \le x \le L$$
$$= 0, \quad \text{otherwise,}$$

then determine the value of a.

Also determine the value of x, where the probability of finding the particle is maximum.

(b) Define, with an example, the Q-value of a nuclear reaction.

5. What are energy bands? What is the difference between valency band and conduction band? What is the relation between the energy bands and fermi level of a conductor?

6. (a) 3 distinguishable particles, each of which can be in one of the ϵ, 2ϵ, 3ϵ, 4ϵ energy states, have total energy 6ϵ. Find all possible distribution of particles in the energy states.

(b) Find the number of micro-states in each case.

Group C: Long Answer Type Questions
Answer any *three* questions.

7. (a) Two rods of proper length l_0 move length-wise towards each other parallel to their common axis with the same velocity v relative to laboratory frame. Show that the length of each rod in the reference frame fixed to the other rod is $l = l_0 \frac{1-\beta^2}{1+\beta^2}$, where $\beta = \frac{v}{c}$.

(b) Starting from the Lorentz transformation show that if (x_1, y_1, z_1, t_1) and (x_2, y_2, z_2, t_2) represent two events, then the quantity $(x_1 - x_2)^2 + (y_1 - y_2)^2 + (z_1 - z_2)^2 - c^2(t_1 - t_2)^2$ is invariant in all inertial frames.

(c) Show that $\hat{x} = i\hbar \frac{\partial}{\partial p}$ and $\hat{p} = p$ is consistent with the commutation relation $[\hat{x}, \hat{p}] = i\hbar$.

(d) Starting from de-Broglie's hypothesis show that the group velocity associated with a particle is same as the particle velocity.

8. (a) Describe briefly Michelson-Morley's experiment to measure the velocity of light. What was the observation?

(b) A system has two energy eigenstates ϵ_0 and $3\epsilon_0$. ψ_1 and ψ_2 are the corresponding normalized wave functions. At an instant the system is in a superposed state $\psi = c_1\psi_1 + c_2\psi_2$ and $c_1 = \frac{1}{\sqrt{2}}$.

i. Find the value of c_2, if ψ is normalized.

ii. What is the probability that an energy measurement would yield a value $3\epsilon_0$?

iii. Find out the expectation value of energy.

(c) Given that for a particle in a rigid box (spanning from $x = 0$ to a) the eigenfunctions are given by $\psi_n = \sqrt{\frac{2}{a}} \sin \frac{n\pi x}{a}$. What are the energy eigenvalues in these states?

9. (a) Assume that an electron is inside a nucleus of radius 10^{-15} m. Calculate from the uncertainty principle the minimum kinetic energy of the electron.
[Given: $h = 6.63 \times 10^{-34}$ J-s, electron mass $= 9.1 \times 10^{-31}$ kg.]

(b) If a light of wavelength 6000 Å falls on a metal surface and emits photoelectrons with velocity of 4×10^5 m/s, what is the threshold wavelength?
[Given: $c = 3 \times 10^8$ m/sec.]

(c) Explain the working principle of He-Ne laser with energy level diagram.

(d) Determine the numerical aperture of a step-index fibre when r.i. of the core $n_1 = 1.50$ and that of the cladding $n_2 = 1.48$. Find the maximum angle for entrance of light if the fibre is placed at air.

10. (a) Calculate the total number of particles in a Fermionic gas in terms of the Fermi level absolute zero temperature.

(b) What is meant by Miller indices? Explain with example.

(c) Calculate the inter-planar spacing d of planes (111) in a simple cubic lattice of side a. Deduce the formula that you use.

11. (a) Show how binding energy per nucleon ($\bar{B}$) varies with mass number A. From that curve explain the possibility of fusion and fission energy sources.

(b) Can a ${}^{238}_{92}$U nucleus decay into a ${}^{206}_{82}$Pb nucleus? Given $\bar{B}$ of ${}^{238}_{92}$U is 7.57 MeV and $\bar{B}$ for ${}^{206}_{82}$Pb is 7.83 MeV.

(c) A beam of X-rays of wavelength 0.842 Å is incident on a crystal at a glancing angle of $8°35'$, when first order Bragg's reflection occurs. Calculate the glancing angle for third order reflection.

(d) The energy gap in Ge is 0.72 eV. Find the maximum wavelength of e.m. radiation which can cause transition from the valence boud to the conduction band.

June 2008

Semester 2

Group A: Multiple Choice Type Questions

1. Choose the correct alternatives for any *ten* of the following:

(i) Double-slit interference pattern is the limiting case of double-slit diffraction pattern when the (a) distance between the slits tends to zero (b) distance of the source and the slits tends to infinity (c) slit widths tend to zero (d) distance between the slits tends to infinity.

(ii) A light ray is passing through a calcite crystal. If the plane vibration of the light ray is perpendicular to the optical axis, then the ray (a) is an O-ray (b) is an E-ray (c) has both nonzero O and E components (d) cannot propagate through the crystal.

(iii) A source of light emits lights of frequencies between ν to $\nu + \Delta\nu$. Coherence time of the emergent light beam is τ_c, then (a) $\tau_c \propto \Delta\nu$ (b) $\tau_c \propto 1/\Delta\nu$ (c) $\tau_c \propto \nu$ (d) $\tau_c \propto 1/\nu$.

(iv) If the quantum mechanical state of a particle is described by $\psi(x) = \psi_0 e^{ik(x-a)}$, for $|x| \leq L$ and $\psi(x) = 0$, otherwise, then (a) the particle has a definite position but uncertainty in momentum (b) the particle has a definite momentum but uncertainty in position (c) the particle has uncertainties both in position and momentum (d) the particle has no uncertainties either in momentum or in position.

(v) Which of the facts below about optical fibre is not true?
(a) In multimode graded index fibre the distortion due to difference in path length is lesser than in multimode step index fibre. (b) The diameter of the core in single mode fibre is more than 10 times the maximum wavelength to be conducted through. (c) Distortion of the signal due to dispersion is less in single mode fibre than in multimode. (d) Single mode fibre is preferred over multimode in long distance communication.

(vi) The intensity of principal maxima in the spectrum of a grating with N number of lines is proportional to (a) $1/N$ (b) N (c) N^2 (d) $1/N^2$.

(vii) A stone is dropped from the top of a building. What happens to the de Broglie wavelength of the stone as it falls?

(a) It increases. (b) It decreases. (c) It remains constant. (d) de Broglie wavelength cannot be defined.

(viii) Number of oscillation modes for the electromagnetic standing waves of frequency ν to $\nu+\Delta\nu$ or the cavity radiation is proportional to (a) ν (b) ν^2 (c) ν^4 (d) $\frac{h\nu}{e^{\frac{h\nu}{kT}}-1}$.

(ix) If $\psi(x,t)$ is a normalized wave function, we must have (a) $\int \psi^* \psi dx = 1$ (b) $\int \psi^* \psi dx = 0$ (c) $\int \psi^* \psi dx = 1/2$ (d) $\int \psi \psi^* dx = 1$.

(x) Which one of the following functions is an eigenfunction of the operator $\frac{d^2}{dx^2}$?

(a) x (b) x^2 (c) e^{-x^2} (d) $\cos x$.

(xi) The number of possible arrangement of two fermions in 3 cells is (a) 9 (b) 6 (c) 3 (d) 1.

(xii) If n_i is the number of identical and indistinguishable particles in the ith energy state with degeneracy g_i, then classical statistics can be applied, if (a) $\frac{n_i}{g_i} = 1$ (b) $\frac{n_i}{g_i} \ll 1$ (c) $\frac{n_i}{g_i} \gg 1$ (d) $g_i = 0$.

(xiii) The nearest neighbour distance in the case of BCC structure is (a) $\frac{a\sqrt{3}}{2}$ (b) $\frac{a\sqrt{2}}{2}$ (c) $\frac{2a}{\sqrt{3}}$ (d) $\frac{2a}{\sqrt{2}}$.

(xiv) The primitives of a crystal are 1.2 Å, 1.8 Å, 2 Å, 2 Å along the three axes, where a plane with Miller indices (2 3 1) cut intercept 1.2 Å along x-axis. What will be the lengths of intercept along y- and z-axes?

(a) 1.5 Å and 5 Å (b) 1.9 Å and 2.5 Å (c) 1.2 Å and 4 Å (d) 1.5 Å and 4 Å.

(xv) In He-Ne Laser, neon atoms obtain energy (a) on collision with He atoms (b) from chemical reactions (c) from electrical pumping (d) optical pumping.

Group B: Short Answer Type Questions
Answer any *three* of the following.

2. Two polarizers are placed at crossed position (angle between the polarizing planes are 90°), a third polarizer with an angle θ with the first one is placed between them. An unpolarized light of intensity I is incident on the first one and passes through all three polarizers. Find the intensity of the light that comes out.

3. (a) Calculate the polarizing angle for light travelling from water of refractive index 1.33 to glass of refractive index 1.53.

(b) The spacing of the planes in a crystal is 1.2 Å and the angle for the first order Bragg reflection is 30°. Determine the energy of the X-ray in eV.

4. (a) An electron and a proton has same de Broglie wavelength. Prove that the energy of electron is greater.

(b) Why Compton effect cannot be observed with visible light?

(c) Show that the temperature dependence in Stefan's law can be derived from Planck's radiation law.

5. (a) When electromagnetic radiations of wavelength λ is incident on a photosensitive surface, the kinetic energy of the photoelectrons emitted from the surface is 2 eV. When the wavelength of the incident radiation is 2λ the kinetic energy of the photoelectrons emitted from the surface is 0.5 eV. Then find the threshold wavelength (maximum wavelength) for photoelectric emission from the surface in terms of λ.

(b) A He-Ne laser transition from $3s$ to $2p$ level gives a laser emission of wavelength 6328 Å. If the energy of the $2p$ level is 15.2×10^{-19} J, how much pumping energy is required, if there is no loss in He-Neon laser?

6. (a) A step index optical fibre has refractive indices of the core and cladding as 1.44 and 1.40, respectively and is surrounded by air. Calculate the numerical aperture.

(b) A system has non-degenerate single-particle states with 0, 1, 2, 3 energy units. Three particles are to be distributed in these states such that the total energy of the system is 3 units. Find the number of micro states if the particles obey (i) MB statistics (ii) BE statistics (iii) FD statistics.

Group C: Long Answer Type Questions

Answer any *three* questions.

7. (a) Explain briefly the action of a dichroic polaroid.

(b) In a Young's double-slit interference experiment, distance between the coherent sources are 1.15 mm. Calculate the fringe width that would be observed on a screen placed at a distance of 85 cm from the source. The wavelength of light used is 5893 Å.

(c) Interference phenomenon does not violate the principle of conservation of energy. Justify it.

(d) A film of oil of r.i. 1.70 is placed between a plane glass plate and an equi-convex lens. The focal length of the lens is 1 metre. Determine the radius of the 10th dark ring when light of wavelength is 6000 Å.

(e) A convex lens of focal length 40 cm is employed to focus the Fraunhofer diffraction pattern of a single-slit of 0.3 mm width. Calculate the linear distance of the first order dark band from the central band. The wavelength of the light is 589 mm.

8. (a) Calculate the least width that a grating must have to resolve two components of the sodium D line in the second order, the grating having 800 lines/cm.

(b) Deduce the missing orders for a double-slit Fraunhofer diffraction pattern if the slit width are 0.16 mm and they are 0.8 mm apart.

(c) Two stars situated at a distance of 9.5×10^{12} km from a telescope of diameter 20 cm are sending light of wavelength 600 nm. Find the distance of the separation of the stars for which they are just resolved.

(d) Find out the packing for FCD structure. Find out the expression for the separation between the planes with Miller indices h, k, l and lattice constant a for a simple cubic structure.

9. (a) Derive the expression for Compton shift when a photon interacts with electron. Show that it is impossible to transfer all the energy of the photon to the electron.

(b) Describe an experiment which verifies de Broglie's hypothesis about the wave nature of particles.

(c) Show that the product of group velocity and phase velocity of the de Broglie wave are constant.

10. (a)
   i. A particle of mass m is confined within $x = 0$ to $x = L$.
   ii. Solve the equation to find out the normalized eigenfunctions.
   iii. Show that the eigenfunctions corresponding to two different eigenvalues are orthogonal.
   iv. If p denotes momentum, then find $\langle p \rangle$ as well as $\langle p^2 \rangle$ in the ground state.

(b) Using the principle of operator correspondence write down the operator $\hat{L}_x$, $\hat{L}_y$, $\hat{L}_z$ (components of angular momentum) in terms of position and momentum operators. Then show that $[L_x, L_y] = i\hbar L_z$.

11. (a) Write down the energy eigenfunction of a particle in a three-dimensional box of length L. (You need not derive this). Hence find out the energy as well as degeneracy in the ground and the first excited states.